Digital Europe in the Era of Challenges

Digital Europe in the Era of Challenges

The Role of the European Union Presidency in Shaping Technological Security and Sovereignty

Edited by

Marek Górka and Małgorzata Kamola-Cieślik

BRILL

ISBN 978-90-04-76360-9
ISBN 978-90-04-76362-3 (PDF)
DOI https://doi.org/10.1163/9789004763623

Library of Congress Control Number: 2026948182

Brill and Koninklijke Brill BV are part of De Gruyter Brill.
www.degruyterbrill.com

Questions about General Product Safety Regulation: productsafety@degruyterbrill.com

Cover illustration: Singapore © Robynne O, September 2019. The image was retrieved via Unsplash.

Contents

PART 4

Economic Policy in the Area of Security and New Technologies

Foreword: Digital Security and Innovation: Poland's Priorities in the European Union

The rapid advancement of digital technologies has fundamentally transformed the global economic, political, and social landscape. Digitalization is no longer a marginal trend, but a major driver of innovation, economic growth, and global connectivity. As the European Union (EU) grapples with an increasingly complex geopolitical environment, the combination of digital security and innovation has become a key area of focus. Poland, which takes over the presidency of the Council of the European Union in 2025, has the opportunity to play a key role in shaping the EU's digital agenda. This book examines Polish's priorities for the EU Presidency, with a particular focus on cybersecurity, artificial intelligence (AI) and digital governance, and how these priorities fit into the EU's broader goals of digital sovereignty and global leadership.

The concept of digital sovereignty has grown in importance in recent years as nations and regions seek to gain control of their digital infrastructure, data, and technologies. For the EU, digital sovereignty is not just about technological self-sufficiency; it is also about shaping global norms and standards that reflect European values such as openness, inclusiveness and transparency. The EU's quest to achieve digital sovereignty underlines its commitment to reducing its dependency on external actors while fostering internal innovation. This vision is in line with Polish's goals of strengthening digital resilience, promoting cutting-edge technologies and ensuring equitable access to digital infrastructure.

The Polish Presidency comes at a crucial time when the EU is struggling with evolving threats in cyberspace, the transformative potential of artificial intelligence and the need to harmonize digital regulations in all Member States. The challenges of hybrid threats and the Russia–Ukraine war highlight the vulnerabilities inherent in the digital domain. These conflicts have highlighted the importance of a robust cybersecurity framework that will be able to address both state-sponsored and non-state cyber actors. Polish's experience as a frontline country in the field of regional security provides valuable information on how to deal with these complex challenges and strengthen the EU's collective posture on cybersecurity.

The program of the Polish Presidency of the EU lays emphasis on the integration of cybersecurity, artificial intelligence, harmonization of digital law, digital diplomacy and digital governance. These priorities reflect the EU's broader objectives, while building on Polish's unique position and experience. Each of these priorities prepares the ground for in-depth analyses in this area.

Cybersecurity remains a cornerstone of the EU's digital strategy. Poland is pursuing initiatives such as the EU's Joint Cyberspace Unit and the Blueprint Framework for Coordinated Response to Large-Scale Cyber Incidents. The focus will be on combining the civilian and military dimensions of cybersecurity and on fostering international cooperation. The main focus will be on the challenges of implementing the EU Cyber Resilience Act, in particular, on addressing vulnerabilities in critical infrastructure. Polish's efforts to close these gaps will include not only regulatory measures, but also targeted investments in capacity-building and education. Addressing the shortage of skilled cybersecurity professionals remains a priority, and Polish's proposal for an EU cybersecurity funding mechanism reflects its commitment to sustainable digital development.

The Polish Presidency will present a unified plan for a coordinated response to large-scale cyber incidents, a critical need highlighted by the ongoing Russia–Ukraine conflict. This initiative will build on the Tallinn Mechanism and the existing EU framework, fostering stronger partnerships between the public and private sectors.

Recent cyberattacks on critical infrastructure, such as the attack on the Colonial Pipeline in the United States and the hack of SolarWinds, have demonstrated the global nature of cyber threats. The EU must ensure that its cybersecurity framework is robust enough to withstand such attacks. Polish's focus on cybersecurity education and capacity-building is particularly timely as the demand for skilled cybersecurity professionals continues to outstrip supply. According to a report by Cybersecurity Ventures, there will be 3.5 million unfilled cybersecurity jobs worldwide by 2025. Poland's proposal for an EU cybersecurity funding mechanism could help fill this gap by investing in training programs and educational initiatives.

Artificial intelligence is another priority of the Polish Presidency, which is of key importance in the modern digital world. AI is a transformative force that has the potential to redefine industries, economies and societies. To remain globally competitive, the governance of AI in the EU must balance innovation with regulation. The Polish Presidency prioritizes the harmonization of AI policies in all member states, including the implementation of the Artificial Intelligence Act and the establishment of national supervisory authorities.

Beyond regulations, Poland is trying to position itself as a leader in AI-based solutions for small and medium-sized enterprises (SMEs). The Polish Presidency is prioritizing the implementation of the Artificial Intelligence Act, focusing on harmonizing regulatory sandboxes across member states and creating funding mechanisms to support the development of AI. The Polish approach will also include the use of artificial intelligence to achieve strategic security goals, in particular in the field of countering hybrid threats. The EU AI Act, due for finalization in 2025, will establish a regulatory framework for AI applications, categorizing them based on their risk levels. High-risk AI systems, such as those used in critical infrastructure or law enforcement, will be subject to strict requirements, including transparency, accountability and human oversight. Polish's focus on harmonizing AI policies across Member States will be key in ensuring the effective and consistent implementation of the AI Act across the EU.

Poland's ambitions in the field of artificial intelligence are supported by a growing technological ecosystem. Companies, particularly AI startups, are using AI to develop innovative solutions in education, e-commerce, and data analytics. They promote AI-based solutions for SMEs; Poland can help bridge the digital divide and provide smaller companies with the opportunity to compete in the global market.

In the context of the development of modern technologies and artificial intelligence, particular attention should be paid to the spread of digital legislation in the EU. Digital legislation requires a concerted effort to ensure consistency and effective implementation. The Polish Presidency will advocate for the principle of 'quality over speed' in legislative processes, ensuring the seamless integration of digital rules such as the Digital Services Act (DSA) and the Data Governance Act (DGA). Key priorities include the implementation of the Digital Services Act and the Digital Services Act, as well as the elimination of inconsistencies between existing regulations, such as the General Data Protection Regulation (GDPR), and new sector-specific legislation. These actions will contribute to legal certainty and compliance across the Union, ensuring that businesses and governments can navigate the digital landscape with confidence.

The Digital Services Act, which entered into force in November 2022, aims to create a safer digital space by holding online platforms accountable for illegal content and disinformation. On the other hand, the Digital Markets Act focuses on facilitating the exchange of data between sectors and Member States, promoting innovation and competitiveness. Polish pressure to harmonize these rules will be key to ensuring their effective implementation and not creating additional burdens for businesses.

As part of the Presidency of the Council of the EU, Poland also seeks to strengthen the EU's role in global digital governance. This includes promoting multi-stakeholder cooperation through forums such as the Internet Governance Forum (IGF) and ensuring that European values such as openness and inclusiveness are anchored in global digital policy. Polish's efforts to strengthen cooperation with candidate countries such as Ukraine will play a key role in strengthening Europe's digital and geopolitical stability. These initiatives reflect the broader objectives of ensuring that the digital transformation remains a unifying force for the EU and its partners.

The EU's pursuit of digital sovereignty must take into account both internal cohesion and external partnerships. The Polish Presidency prioritizes a multilateral approach to internet governance, building on the framework for cooperation between the EU and the United Nations and the International Telecommunication Union (ITU). By promoting European values in global digital policy, Poland can contribute to ensuring that the internet remains open, safe and accessible to all.

It should also be noted that Polish's digital transformation over the past decade has been characterized by significant progress in e-government, digitization of the public sector, and the development of a dynamic technological ecosystem. Initiatives such as the mObywatel platform are an example of Polish's ability to provide scalable and innovative digital solutions. By showcasing these achievements, the Polish Presidency can inspire Member States to adopt best practices and strengthen cooperation across the EU. In addition, Polish's focus on education and capacity building in cybersecurity highlights the importance of human capital in achieving digital resilience. Addressing the shortage of skilled cybersecurity professionals remains a priority, and Polish's proposal for an EU cybersecurity funding mechanism reflects its commitment to sustainable digital development.

As the EU tackles the twin challenges of the digital and green transitions in the face of geopolitical uncertainty, the Polish Presidency represents a strategic opportunity to shape a resilient and innovative digital Europe. By prioritizing cybersecurity, promoting artificial intelligence and harmonizing digital governance, Poland can contribute to the EU's long-term goals of technological sovereignty, economic competitiveness and global leadership in digitalization.

The future that awaits us requires not only a strategic vision, but also coordinated action. The Polish Presidency's emphasis on inclusivity, innovation and resilience ensures that the EU remains a beacon of digital progress and a global leader in shaping the future of technology.

In conclusion, the era of digital transformation has brought unprecedented opportunities and challenges to global economies, governance, and social structures. At the heart of this transformation is the urgent need for strategic policymaking to ensure technological innovation, security and resilience. With

Poland taking over the Presidency of the Council of the European Union in 2025, its priorities are closely aligned with the broader EU agenda, reflecting a commitment to cybersecurity, artificial intelligence and harmonized digital governance. Polish leadership in 2025 will not only address the most pressing challenges, but also lay the foundations for long-term resilience and prosperity. With a focus on inclusivity, innovation and resilience, Poland is ready to make a lasting impact on Europe's digital landscape, ensuring that the EU remains a global leader in shaping the future of technology.

Rafał Rosiński, PhD
Undersecretary of State Ministry of Digital Affairs
Republic of Poland

Figures and Tables

Figures

Tables

Abbreviations

AI	Artificial Intelligence
AII	Adriatic-Ionian Initiative
ATO	Anti-Terrorist Operation
AVATAR	Automated Virtual Agent for Truth Assessments in Real Time
BIH	Bosnia and Herzegovina
BSEC	Black Sea Economic Cooperation
CARDS	Community Assistance for Reconstruction, Development and Stabilisation
CBBH	Central Bank of Bosnia and Herzegovina
CBRN-E	Chemical, Biological, Radiological, Nuclear, and Explosives
CCDCE	Cooperative Cyber Defense Center of Excellence
CCDCOE	Cooperative Cyber Defence Centre of Excellence
CDTO	Chief Digital Transformation Officer
CEAP	Circular Economy Action Plan
CEFTA	Central European Free Trade Area
CEI	Central European Initiative
CEN	Comité Europeen de Normalisation
CENELEC	Comité Europeen de Normalisation Electrotechnique
CERT	Computer Emergency Response Team
CHST	Anti-Hybrid Support Team
CMF	(frac. *Compagnie Nationale des Mines de France*) Compagnie Nationale des Mines de France
COMES	(frac. *Comité pour les Métaux Stratégiques*) Committee for Strategic Metals
CRA	Cyber Resilience Act
CREF	the Cyber Resilience Engineering Framework
CRM	Critical Raw Materials
CRMA	Commission's Critical Raw Materials Act
CRR	Cyber Resilience Review
CRRT	Cyber Rapid Response Team
CS	(frac. *Comate Stratégique de Filière*) Strategic Sector Committees
DA	The Digital Acquis
DDoS	Distributed Denial of Service
DHS	Department of Homeland Security
DIAMMS	(frac. *délegation interministérielle aux approvisionnements en minerais et métaux stratégiques*) – Interministerial Delegation for the Supplies of Strategic Minerals and Metals
DMA	Digital Markets Act
DSA	Digital Services Act

DSM	Digital Single Market
DSMS	Digital Single Market Strategy
DT4UA	Digital Transformation for Ukraine
DTC	Digital Transatlantic Trade
EC	European Commission
ECAT	European Center for Algorithmic Transparency
ECDS	European Council of Digital Services
EDES	European Digital Economy Society
EDF	European Defence Fund
EDIDP	European Defense Industrial Development
EDMO	European Digital Media Observatory
EDPS	European Data Protection Supervisor
EEA	European Economic Area
EEAS	European External Action Service
EEC	European Economic Communities
EFTA	European Free Trade Association
EHB	European Hydrogen Backbone
EHDS	European Health Data Space
EIC	European Innovation Council
EMSA	European Maritime Safety Agency
ENISA	European Network and Information Security Agency
ENNOH	European Network of Network Operators for Hydrogen
EPF	European Peace Facility
EPO	European Patent Office
EPP	President of the European Parliament
ERCC	Centrum Koordynacji Reagowania Kryzysowego
ESDC	European Security and Defence College
ESPR	Ecodesign for Sustainable Products Regulation
ETIAS	the European Travel Information and Authorisation System
EU	European Union
EUCO	European Council
EUFOR	European Union Force in Bosnia and Herzegovina
EUGS	European Union Global Strategy
EUSR	European Union Special Representative
FIMI	Foreign Information Manipulation Interference
FoP	Friends of the Presidency Group on Countering Hybrid Threats
GAN	Generative Adversarial Networks
GDPR	General Data Protection Regulation
IBIMS	(pol. *Instytut Badań Internetu i Mediów Społecznościowych*) Institute of Internet and Social Media Research

ICANN	Internet Corporation for Assigned Names and Numbers
IFOR	Implementation Force
IGF	Internet Governance Forum
ILO	International Labour Organisation
ILP	Intelligence-Led Policing
IPA	Intelligent Process Automation
IPAI	International Panel on Artificial Intelligence
IR	International Relations
IRA	Inflation Reduction Act
ITU	International Telecommunication Union
JORA	Joint Operations Reporting Application
LLM	Large Language Models
LNG	Liquefied Natural Gas
MFF	Multiannual Financial Framework
MITRE	Massachusetts Institute of Technology Research and Engineering
ML	Machine Learning
NATO	North Atlantic Treaty Organisation
NIS	Network and Information Systems
NIST	National Institute of Standards and Technology
NRC	National Research Council
NRRP	National Recovery and Resilience Plan
ODS	Civic Democratic Party
OECD	Organisation for Economic Co-operation and Development
PACE	Parliamentary Assembly of the Council of Europe
PADR	Preparatory Action on Defense Research
PESCO	Permanent Structured Cooperation
PHARE	Poland and Hungary: Assistance for Restructuring their Economies
PPP	Public Private Partnership
PRES	Prezydencja Rady Unii Europejskiej
PRIMA	Partnership for Research and Innovation in the Mediterranean Area
PTEZ	(pol. *Polskiego Towarzystwa Elektrociepłowni Zawodowych*) Polish Association of Professional Combined Heat and Power Plants
RES	Renewable Energy Source
RT	Russia Today
RYCO	Regional Youth Cooperation Office
SAA	Stabilization and Association Agreement
SACRED	Strategic Asset Cyber Resilience EU Database
SDN	Software Defined Network
SECI	Southeast European Cooperative Initiative
SEECP	South East European Cooperation Process

SIEM	Security Information and Event Management
SME	small and medium enterprises
SOAR	Security Orchestration Automation and Response
SOC	Security Operations Centers
SSCSU	State Special Communications Service of Ukraine
TFEU	Treaty on the Functioning of the European Union
TIP	Threat Intelligence Platforms
TTC	Trade and Technology Council
TTIP	The Transatlantic Trade and Investment Partnership
UCPM	Union Civil Protection Mechanism
UNESCO	United Nations Educational, Scientific and Cultural Organization
UNIDIR	United Nations Institute for Disarmament Research
US	United States
VLOP	Very Large Online Platforms
VLOPSE	Very Large Online Platform and Search Engines
WFD	Water Framework Directive
WPBiO	Common Security and Defence Policy
WSIS	World Summit on the Information Society
WTO	World Trade Organization

Notes on Editors and Contributors

Marko Aćić
Associate Professor at Slobomir P University, Banja Luka, Republika Srpska, Bosnia and Herzegovina. His research interests focus on public international law, teachers' pedagogical and digital competencies, inclusive values in education, and issues of Serbian national identity in the context of Bosnia and Herzegovina. ORCID: 0000-0002-8020-9482, e-mail: acicdominus@gmail.com.

Nezir Akyesilmen
Professor of International Relations at the Faculty of Economics and Administrative Sciences, Selcuk University, Konya, Türkiye. His research interests include cybersecurity, cyber politics, human rights, theory of international relations, international conflict analysis, and human rights and peace education. He currently serves as the Chair of the Association for Human Rights Education in Turkey and is the Editor-in-Chief of the Cyberpolitik Journal. ORCID: 0000-0001-8184-5280, e-mail: nezmen@yahoo.com.

Michele Barbieri
Ph.D., Research Fellow in Policy Science at the University of Salerno and Adjunct Professor in "The Politics of Privacy and Surveillance" at the University of Tuscia, Italy. His research focuses on public policy analysis, with particular attention to disinformation prevention policies, cybersecurity, and research and innovation policies. He also studies the impact of emerging populism, techno-social systems, and the transformation of multi-level governance structures within the EU. ORCID: 0000-0003-4748-093X, e-mail: mibarbieri@unisa.it.

Alina Bârgăoanu
Ph.D., Professor and Dean of the Faculty of Communication and Public Relations at the National University of Political Studies and Public Administration in Bucharest, Romania. She is a member of the Advisory Board of the European Digital Media Observatory (EDMO) and a member of the "Hybrid Threats" research unit of Folke Bernadotte Academy (Sweden). Her research interests include disinformation, influence operations, resilience, and the impact of technology on democracy. ORCID: 0000-0003-3912-8442, e-mail: alina.bargaoanu@comunicare.ro.

Serban Brebenel
Ph.D., a practitioner in the field of diplomacy having joined the Romanian diplomatic corps in 2024. He has served in various positions in Bucharest and Washington and has been working with EU institutions in Brussels since 2015. He holds a Ph.D. from the University of Amsterdam, with a dissertation on building compromise in the EU and the enlargement process. His research interests include EU enlargement, the withdrawal of the UK from the EU, and EU institutional processes. E-mail: serban.brebenel@gmail.com.

Ladislav Cabada
Professor of Political Science at the Department of Political Sciences and International Relations, University of West Bohemia in Pilsen, Czech Republic. His research interests focus on comparative politics of East-Central Europe, regionalism and development studies, Europeanization, Euroscepticism, and political anthropology. ORCID: 0000-0001-9604-0987, e-mail: cabada@ff.zcu.cz.

Xuechen Chen
Associate Professor in Politics and International Relations at Northeastern University London, United Kingdom. She specializes in research on digital policy, cybersecurity, and the management of digital technologies at the international level. Her work focuses on the regulation of the digital economy, cyberspace norms, and international cooperation in technological security. She also studies the impact of artificial intelligence on international politics, with particular emphasis on China and the European Union. ORCID: 0000-0002-0046-8012, e-mail: xuechen.chen@nulondon.ac.uk.

Viktor Denisenko
Senior Researcher at the Centre for Communication Influences and Propaganda Research, Faculty of Communication, Vilnius University Lithuania. Holds a Ph.D. in Information and Communication from Vilnius University. Research interests include strategic communication, political communication, and theories and practices of propaganda and disinformation. Author of academic, analytical, and popular publications on these topics, including the book *In the Encirclement of Propaganda* (Vilnius University Press, 2021). ORCID: 0000-0001-8646-3740, e-mail: viktor.denisenko@kf.vu.lt.

Iryna Fyshchuk
Associate Professor in the Scholar at Risk program at the Department of Media and Social Sciences, University of Stavanger, Norway. Her research focuses on

digital transformation, cybersecurity, public administration, municipal development, change management, and international relations, particularly in the context of Ukraine and EU integration. ORCID: 0000-0002-7645-3490, e-mail: irafyshchuk@gmail.com.

Xinchuchu Gao
Lecturer in International Relations at the University of Lincoln, United Kingdom. She specializes in the analysis of global political processes, with a particular focus on the dynamics of relations between Asian countries and the West. Her research interests include international security, regional politics of East Asia, and the role of international organizations in conflict management. ORCID: 0000-0001-5664-9122, e-mail: XinGao@lincoln.ac.uk.

Andrew Glencross
Professor of Political Science and Director of ESPOL at Université Catholique de Lille, France. He specializes in European politics, with a particular focus on Brexit, European integration, and EU-UK relations. He is also a Senior Fellow at the Foreign Policy Research Institute in Philadelphia. ORCID: 0000-0001-8320-9181, e-mail: andrew.glencross@univ-catholille.fr.

Miloš Grujić
Ph.D., Associate Professor at University of Business Studies Banja Luka in Bosnia, and Herzegovina.He holds a Ph.D. in business economics and specializes in capital markets, finance, risk management and audit. His research primarily focuses on financial markets, investment strategies and auditing. ORCID: 0000-0001-5566-5921, e-mail: milos.grujic@univerzitetps.com.

Marek Górka
Ph.D., Associate Professor at the Faculty of Humanities, Koszalin University of Technology, Poland. His research interests include security policy, cybersecurity, terrorism, intelligence, and counter-intelligence services, sociology of the Internet, electoral competition, and charismatic leadership. ORCID: 0000-0002-6964-1581, e-mail: marek_gorka@wp.pl.

José Carlos Hernández Gutiérrez
PhD in Social Sciences. He also holds a Master's degree in Latin American Studies and a Bachelor's degree in Political Science. He is an Adjunct Professor in the Department of Political Science and Public Administration at the University of Granada. His research interests include political behavior, violence,

security, and organized crime. ORCID: 0000-0002-2855-1053, e-mail: jchernan dez@ugr.es.

Fuad Jomma

Ph.D., Associate Professor at the University of Szczecin, Poland. He specializes in Middle Eastern affairs, with a research focus on political transformations in Syria and the diversity of its society in terms of religion, ethnicity, and national identity. An important area of his academic work also includes issues related to the Kurds. ORCID: 0000-0002-7981-7114, e-mail: fuad.jomma@usz.edu.pl.

Małgorzata Kamola-Cieślik

Ph.D., Associate Professor at the University of Szczecin, Poland. Her research focuses on political and economic transformations in Poland and the European Union, with particular attention to the shipbuilding industry and energy security, including cybersecurity in the energy sector. Since 2018, she has served as the academic editor of the journal "Acta Politica Polonica." ORCID: 0000-0003-2956-3969, e-mail: malgorzata.kamola-cieslik@usz.edu.pl.

Izabela Kapsa

Ph.D., Associate Professor at the Faculty of Political Science and Administration, Kazimierz Wielki University in Bydgoszcz, Poland. She is a political scientist and psychologist, specializing in the use of ICT to enhance civic participation and democratic engagement. She has been involved in multiple research projects, including electronic voting systems, digital and psychological resilience, and the role of digital technologies in improving civic engagement. ORCID: 0000-0003-2342-3682, e-mail: izabela.kapsa@ukw.edu.pl.

Kiryl Kascian

Head of the Centre for Communication Influences and Propaganda Research at Vilnius University, Lithuania. Holds a Ph.D. in Law from the University of Bremen. Research interests include communication influences, political communication, interethnic relations, electoral behavior, minority rights, and constitutionalism. Has extensively published and provided expert input internationally in these fields. ORCID: 0000-0002-1777-5895, e-mail: kiryl.kascian@kf.vu.lt.

Christian Kaunert

Professor of International Security at Dublin City University and Professor in Policing and Security at the University of South Wales, where he directs the International Centre for Policing and Security. He serves as Jean Monnet Chair and directs the Jean Monnet Centre of Excellence and the Jean Monnet Network on EU Counter-Terrorism. He is Editor of the Journal of Contemporary European Studies. His research interests focus on international security, policing and security, and EU counter-terrorism. ORCID: 0000-0002-4493-2235, e-mail: christian.kaunert@southwales.ac.uk.

Bjørn Olav Knutsen
Ph.D., Chief Researcher at the Norwegian Defence Research Establishment (FFI) and Associate Professor at Nord University in Bodø, Norway. His research interests focus on European and transatlantic security policy, and he has published extensively in books, reports, and professional journals within this area. ORCID: 0000-0002-6820-3679, e-mail: Bjorn-Olav.Knutsen@ffi.no.

Tomáš Kolomazník
Ph.D. Candidate at the Metropolitan University of Prague, co-founder and director of the Center for Security Consulting, and lecturer at Ambis and CEVRO University in Prague, Czech Republic. He is dedicated to security policy, the defense industry, and cyber and information security. He is a member of the Geneva Center for Security Policy Alumni Association. ORCID: 0009-0005-3420-4815, e-mail: T.Kolomaznik@seznam.cz.

Dominika Liszkowska
Ph.D., Assistant Professor at the Department of Regional and European Studies, Faculty of Humanities, Koszalin University of Technology, Poland. Her research interests include political systems of Türkiye and Poland, Türkiye-European Union relations, the role of the European Union and Türkiye in international politics, and securitization of migration. ORCID: 0000-0001-6312-341X, e-mail: dominika.liszkowska@gmail.com.

Guillermo López-Rodríguez
Ph.D., political scientist from the University of Granada, Spain. He works on the cultural dimensions of military-change processes, focusing on the evolution of the Spanish army from the Spanish-Moroccan War to the current Security Force Assistance operations. ORCID: 0000-0001-8704-9007, e-mail: guillermolopez@ugr.es.

Samara López-Ruiz
Ph.D., employed at the Institute of Sustainable Processes at theUniversity of Valladolid (Spain). Her research focuses on water policy, particularly challenges related to water access in Spain, such as water poverty, social accept-

ance of unconventional water use, and the role of water companies in promoting water conservation. ORCID: 0000-0002-9280-9717, e-mail: samaral r@uva.es.

Miroslav Mareš
Professor at the Department of Political Science, Faculty of Social Studies, Masaryk University, Czech Republic. His research focuses on extremism, terrorism, and security policy in Central and Eastern Europe. ORCID: 0000-0002-7102-3205, e-mail: mmares@fss.muni.cz.

Marco Marsili
Ph.D., Professor at the Department of Philosophy and Cultural Heritage at Ca' Foscari University of Venice, Italy, specializing in international relations, security, and defense. His research focuses on international law, humanitarian law, human rights, terrorism, counter-terrorism, hybrid warfare, and emerging military technologies. ORCID: 0000-0003-1848-9775, e-mail: info@marcomar sili.it.

Lucas Miailhes
Ph.D., Researcher at ESPOL, Université Catholique de Lille, France. His research areas include applied computer science, artificial intelligence, and data analysis. He is actively involved in interdisciplinary projects focused on developing innovative AI solutions. ORCID: 0000-0002-1825-0097, e-mail: lucas.miailhes@univ-catholille.fr.

Magdalena Musiał-Karg
Professor at the Faculty of Political Science and Journalism, Adam Mickiewicz University in Poznań, Poland. She serves as President of the Polish Political Science Association, Vice-President of the International Political Science Association (IPSA), a member of the Political Science Committee of the Polish Academy of Sciences, and a member of the OSCE/ODIHR Core Group of Political Party Experts. Her research focuses on political systems, direct democracy, elections, and the use of new technologies in electoral processes (e-voting). ORCID: 0000-0002-6089-1381, e-mail: magdalena.musial-karg@amu. edu.pl.

Mihaela Pană
Ph.D. in Communication Sciences at Faculty of Communication and Public Relations, National School of Political Studies and Public Administration (SNSPA), Bucharest, Romania. Her research interests include strategic com-

munication, cybersecurity, and algorithmic influence. ORCID: 0009-0009-2466-3101, e-mail: mihaelapana@protonmail.com.

Donatas Palavenis
Major in the Lithuanian Armed Forces and Ph.D. Candidate at the General Jonas Žemaitis Military Academy of Lithuania. He is a junior researcher at the Baltic Institute of Advanced Technology and currently serves at the Warfare Institute. His research interests include the defense industry, defense policy, political economy, defense economics, and arms procurement. ORCID: 0000-0003-0908-7512, e-mail: donatas.palavenis@gmail.com.

Elisabeth Pettersen
Associate Professor at the Faculty of Social Sciences, Nord University Bodø, Norway. Her research focuses on security and human security. She teaches International Relations for undergraduate students at Nord University. ORCID: 0000-0002-1108-011X, e-mail: elisabeth.pettersen@nord.no.

Janusz Ruszkowski
Professor at the Institute of Political and Security Sciences, University of Szczecin, Poland. His academic work focuses on topics related to the European Union, European integration, and international relations. He is a member of the European Community Studies Association (ECSA), the European Political Science Network Association (Eps-Net), and the Jean Monnet Community (JMC). ORCID: 0000-0002-4786-7416, e-mail: janusz.ruszkowski@usz.edu.pl.

Zdeněk Rod
Ph.D., Assistant Professor at the Department of Security Studies, CEVRO University in Prague (since 2025). He also serves as Assistant Professor at the Faculty of International Relations, Prague University of Economics and Business. In addition, he is co-director of the Center for Security Consulting. In 2026, he is a Visiting Researcher at the John Lukacs Institute for Strategy and Politics, University of Public Service in Budapest. ORCID: 0000-0002-2958-8288, e-mail: zdenek.rod@icloud.com.

Mauro Santaniello
PhD is an Associate Professor in Political Science at the University of Salerno, where he teaches Internet Governance, Digital Policy, and Local Public Policy. Mauro is also Associate Research Fellow at the United Nations University Institute on Comparative Regional Integration Studies (UNU-CRIS), where

he works within the Digital Governance Cluster. Mauro is the director of the Internet & Communication Policy Center (ICPC), and the scientific coordinator of the Salerno Winter School on Internet Governance (SWING). Mauro is also Principal Investigator of the PRIN project on "Digital Sovereignty in Comparative Perspective: State Authority, Corporate Power and Fundamental Rights in Cyberspace", funded by the Italian Ministry of University and Research. He is also the coordinator of the GIG-ARTS international conference series, and the secretary of the Global Internet Governance Academic Network (Giganet). His main research interests include internet governance, digital sovereignty, digital constitutionalism, cybersecurity policy, and geopolitics of information networks. ORCID iD: 0000-0001-5582-622X e-mail: msantaniel lo@unisa.it.

Antonia-Maria Sarantaki

Ph.D., Researcher at the Institute of European Integration and Policy, National and Kapodistrian University of Athens, Greece. She holds a Ph.D. from Panteion University and an MSc (Econ) from Aberystwyth University. Her primary research interests focus on Frontex, borders, migration and EU security. ORCID: 0000-0002-5472-7243, e-mail: asarantaki@pspa.uoa.gr.

Ali Shoker

Research Associate Professor in the Computer, Electrical and Mathematical Science and Engineering Division (CEMSE) at King Abdullah University of Science and Technology (KAUST), Saudi Arabia. He is a founding member of the Resilient Computing and Cybersecurity Center (RC3) and the Cyber Security and Resilience Community (CriSys). His primary research interests lie in the design and development of secure, resilient, scalable, highly available, efficient, environmentally sustainable, intelligent, and distributed systems. ORCID: 0000-0002-4898-9394, e-mail: ali.shoker@kaust.edu.sa.

Kamila Sierzputowska

Ph.D., Assistant Professor at the Faculty of Political Science and Administration, Kazimierz Wielki University in Bydgoszcz, Poland. Her research focuses on cybersecurity, cyber resilience, cyber stability and international relations, and issues regarding the challenges facing NATO and international security. Her research also applies to Estonia as a pioneer of digitization in public administration. ORCID: 0000-0002-3974-8118, e-mail: kamila.sierzputows ka@ukw.edu.pl.

Chiara Spiniello
PhD is a postdoctoral researcher at ZeMKI, Centre for Media, Communication and Information Research, University of Bremen, where she works on the projects Reignite Multilateralism via Technology (REMIT) and Fostering Digital Pan-Africanism in AI Governance through Evidence and Action (AI PAN-AFRICANISM). She obtained her PhD in Public, Comparative and International Law through a joint Italian -French supervision between Sapienza University of Rome and Lille Droit & Santé. Her research lies at the intersection of political science, comparative public law and digital governance, with a particular focus on digital constitutionalism, digital sovereignty, electoral processes and forms of government. Before joining ZeMKI, she was Research Fellow in Political Science at the University of Salerno and Research Fellow in Comparative Public Law at the University of Florence. She is a member of the Digital Constitutionalism Network (DCN) and of the Internet and Communication Policy Center (ICPC), University of Salerno. ORCID: 0009-0006-4735-2515 e-mail: spiniell@uni-bremen.de.

Kamil Tarhan
Ph.D., Lecturer at Social Sciences University of Ankara, Türkiye. His research focuses on international relations, national security, and cybersecurity, with particular emphasis on the role of cybersecurity in the context of state sovereignty and global security challenges. His interests also include international relations theories and human security. ORCID: 0000-0003-4668-7920, e-mail: kamil.tarhan@asbu.edu.tr.

Johannes Thumfart
Ph.D., Researcher at the Brussels Interdisciplinary Research Centre on Migration & Minorities (BIRMM) and the research group Law, Science, Technology, and Society (LSTS) at Vrije Universiteit Brussels, Belgium. He is the author of the book The Liberal Internet in the Postliberal Era (Palgrave Macmillan, 2024). He holds a Ph.D. from Humboldt Universität Berlin on the intellectual history of international law. ORCID: 0000-0002-5084-1490, e-mail: Johannes.Thumfart@vub.be.

Šárka Waisová
Habilitated Doctor (Dr. habil.) at the Faculty of Humanities, University of West Bohemia in Pilsen, Czech Republic. She specializes in security studies and conflict resolution. Her research focuses on energy security, renewable resources, and environmental peacebuilding. ORCID: 0000-0001-7197-9269, e-mail: waisova@ff.zcu.cz.

Louis Wierenga
Ph.D. Candidate and Lecturer in International Relations at the Baltic Defence College and Political Analyst at the Johan Skytte Institute of Political Studies, University of Tartu, Estonia. His research interests include NATO, European and Baltic Sea Region security issues, US foreign policy, and the global security environment. ORCID: 0000-0002-1269-3153, e-mail: louis.wierenga@baltdef col.org.

Danni Zhang
Ph.D. Candidate in Politics and International Relations at Northeastern University London and the University of Kent, United Kingdom. She is affiliated with the Institute of Cyber Security for Society. Her research focuses on global digital governance, particularly the roles of the European Union and China in data governance, AI governance, and cybersecurity. ORCID: 0009-0006-4697-2691, e-mail: dz3501phd@students.nulondon.ac.uk.

Digital Europe in an Era of Challenges: Introduction

Marek Górka and Małgorzata Kamola-Cieślik

The book *Digital Europe in an Era of Challenges. The Role of the Presidency of the European Union in Shaping Sovereignty in Security and Technology,* poses a fundamental question about the future of Europe and identifies key problem areas that the contemporary European Union (EU) must confront. Understanding current challenges depends on Europe's competitiveness, particularly in the field of new technologies, where EU policy must respond to an increasingly confrontational international environment. This is also linked to a profound shift in US policy. In the context of growing geopolitical, economic, and technological tensions, decision-makers must be aware that the coming years will be marked by strategic rivalry between the United States and China. This rivalry will unfold across multiple domains, with an economy based on new technologies being one of them.

In addressing the contemporary challenges shaping the political landscape, fundamental questions arise about how the EU should function and respond to phenomena that are inherently increasingly complex. For many years, globalization has been the subject of public debate, compelling efforts to build the most competitive economy possible. The challenge, therefore, is to build a strong economy while simultaneously maintaining the values and norms on which the EU is founded. This idea – that the economic environment must be transformed in the digital era – is widely accepted, but it also includes guidelines related, among other things, to the need for a model of state functioning based on cooperation and collaboration, as well as to a global order increasingly driven by competitiveness, especially in the context of the EU's position between the United States and China. In light of this reality, it becomes important not only to adapt economic structures but also to develop political strategies addressing phenomena rooted both in the changing balance of international power and in internal processes within Europe itself.

The events of 2025 and the election of Trump as President of the United States demonstrated that the risk of excessive reliance on the United States has become far more difficult to ignore. Consequently, the idea of European strategic autonomy – not only as a long-term aspiration but also as the capacity of states to concretely build economic and technological potential – has become highly relevant.

 | HTTPS://DOI.ORG/10.1163/9789004763623-002

In such an antagonistic global environment, the challenge lies in finding a balance between strengthening the EU on the international stage and maintaining its competitiveness, while at the same time, minimizing disparities among Member States, which are often the source of many internal tensions and disputes. This is a very serious challenge faced both by individual governments and by the European community as a whole, as global markets are characterized by rapid change and high uncertainty – an important consideration, for example, for investments made by enterprises and international political actors. This issue takes on particular significance in the context of Central European countries, which, in order to keep pace with the economic development of other EU members, must intensively invest in industry, research and development, and many other sectors based on new technologies. Since Russia's attack on Ukraine, it has also underscored how dependence on energy supplies from single entities can affect the economic and energy security of many countries.

The current political and economic reality is a world in which many investments and technological tools supporting the economy will be dominated by artificial intelligence, automation, and the geopolitical interests of the world's largest players. This is not only – as mentioned earlier – a matter of dependence on key raw materials, but also concerns digital technologies, which are currently the subject of competition between the United States and China.

Alongside the geopolitical context, there is also a second, technological dimension that strongly shapes international reality. An example is the heavy reliance on services, including those provided by US cloud providers (such as Amazon AWS, Microsoft Azure, and Google Cloud). In a world so dependent on digitalization, any single technological tool can become a geopolitical weapon or an effective means of exerting political pressure.

The multitude of challenges facing the EU, combined with the lack of clear and detailed priorities in official EU documents, makes it difficult today to precisely define the directions in which the European Union should develop in the coming years.

The monograph poses a question about the future of Europe and the main problems that the contemporary EU must confront. The key to understanding current challenges lies in Europe's competitiveness, whose policies – especially in the area of new technologies – must respond to an increasingly confrontational international environment. The European Union will have to adapt to shifts in the global system of values in the world economy.

It should be noted that despite the lack of legislative initiative, the presidencies of the Council of the EU influence the direction of Union policies in the area of digital security. Poland's Presidency in the first half of 2025 focused on countering cyber incidents, combating disinformation, and the development and use of new technologies and artificial intelligence.

During the six months of the Polish Presidency, numerous key debates took place that shifted the understanding of security towards entirely new issues, demonstrating how complex and multidimensional this phenomenon is. The most important diplomatic events included: the HoSPOC (Heads of Single Points of Contact) Meeting in Warsaw on 2–3 April 2025, devoted to the future of police cooperation in the EU, including the fight against cross-border crime, cybercrime, and hybrid migration; the Meeting of the Joint Parliamentary Scrutiny Group on Europol (JPSG), Warsaw, 23–24 February 2025, which discussed Europol's role in protecting the EU's external borders, combating terrorism, using data to fight crime, and protecting children online; the Meeting of the Europol Management Board in The Hague on 18–19 March 2025, at which the SOCTA 2025 report on serious and organized crime in Europe was presented; the Interparliamentary Conference on the Common Foreign and Security Policy and the Common Security and Defence Policy (CFSP/CSDP), Polish Parliament, Warsaw, 23–25 March 2025, which focused on the security of the EU's eastern flank, the continuation of support for Ukraine, cybersecurity, and the EU's strategic autonomy; the Weimar+ Initiative Meeting, Paris, February 2025, involving Poland, France, Germany, Ukraine, and the United Kingdom, which addressed military and political support for Ukraine, regional security, and EU defence integration; the Informal Meeting of Development Ministers (FAC Development), Warsaw, 10–11 February 2025, which discussed countering forced migration from Africa and the Middle East through development policy, neighbourhood cooperation, and cross-border cooperation; meetings of COREPER I and II (Permanent Representatives to the EU) in Brussels, held regularly in the first half of 2025, during which Poland promoted EU resilience to hybrid threats, combating human smuggling at the Belarusian border, and energy and infrastructure security; the conference "Technological Infrastructures: A Strategic Asset for European Competitiveness," Warsaw, 6–7 May 2025, organized with the participation of the European Commission, industry, and academia, addressing technological security, critical infrastructure, and the reduction of dependence on non-EU technologies; and EUIndTech 2025 – Industrial Technologies for a Sustainable Europe, Kraków, 2–4 June 2025, which discussed raw material security, industrial independence, and innovations enhancing the EU's resilience to external threats.

The monograph is an interdisciplinary work focusing on the role of EU Member State presidencies in shaping policies related to digitalization, security, and geopolitics. The project analyses key challenges facing the EU and the ways in which individual Council presidencies address them, taking into account the context of global technological and geopolitical transformations.

Emphasizing the plurality of perspectives as well as the dominant problems and themes present in public debate – an immanent and recognizable feature of selected European states – makes it possible to better capture the diversity of political and institutional conditions within the European Union. At the same time, it allows for the identification of areas and issues, which, although currently dominant at the regional or national policy level, may in the nearer or more distant future acquire pan-European or even global significance, as is the case with access to energy resources or water resources.

For many of the analyses presented in the book, the starting point is the Presidency of the Council of the European Union, understood not merely as an administrative function but as a key mechanism for aggregating national interests, transferring regional problems to the EU level, and shaping the EU's political agenda. An analysis of the Presidency makes it possible to capture how Member States – through their experiences, priorities, and institutional capacities – influence the direction of EU security and technology policy.

For this reason, it is justified to highlight solutions developed at the national level that may serve both as points of reference and as ready-made models of action capable of being implemented at the level of the European Union. Such examples demonstrate that regional and national policies often shape pan-European policy, and that bottom-up initiatives can assume a pioneering and avant-garde character. Focusing on selected Member States as well as countries aspiring to EU membership thus gives the study a distinctive research character. The cases described in the book function as a kind of testing ground for policy solutions, in which challenges related to cybersecurity, disinformation, infrastructural resilience, and energy and technological security emerge earlier than at the level of the Union as a whole. These experiences – often articulated and institutionalized during Council of the EU presidencies – may subsequently take the form of solutions with pan-European relevance.

The book also presents a broad approach to security policy, which is increasingly interpreted not only in military terms but also in non-military ones, encompassing socio-economic and technological dimensions. Complementing this perspective is an analysis of the evolution of the very nature of cybersecurity, which in recent years has gradually moved away from a narrowly understood technical dimension towards issues of a socio-political character.

These phenomena, increasingly visible in the European space, do not remain without impact on the actions of EU decision-makers, including those representing the European Commission and the Council of the European Union, and consequently become the subject of debate and action undertaken at the international level. An important research challenge therefore remains the analysis of the ways in which the European Union responds to growing digital and technological challenges, which, on the one hand, foster stability and predictability in economic development, and on the other hand, may generate new threats – especially when they are insufficiently taken into account in development processes or their importance is marginalized in the everyday functioning of states.

The added value of the volume thus lies in demonstrating the mutual interpenetration of the national, regional, and EU levels in shaping EU security and technology policy in the 2020s. The individual chapters show not only how the Union responds to new digital and technological threats, but also how the presidencies of the Council of the EU become a key instrument for translating local and sectoral challenges into strategic priorities for the entire Community.

The book is a collective work divided into four parts. The authors of the individual chapters employed a variety of research methods, techniques, and tools used in the social sciences. The first part of the volume is theoretical in nature, analysing the priorities of EU presidencies over the past decade in the context of challenges faced by Member States, such as digitalization and cybersecurity. Through theoretical analysis, it demonstrates the extent to which the EU Digital Single Market contributes to ensuring uninterrupted digital access within Member States and beyond the Union. This part of the book also seeks to show how political tasks and objectives pursued at the pan-European level – particularly within the framework of the Presidency of the Council of the European Union – respond to contemporary security challenges and the ongoing process of digitalization. The studies presented indicate that the dynamic development of digital technologies not only determines the changing nature of security policy but also significantly influences the shaping of political agendas and decision-making processes in the European Union. This makes it possible to capture the directions of transformation of EU policies and to identify key processes, which, although widely recognized, require deeper analysis and better understanding in the context of the role of Member State presidencies. In addition, attention is drawn to their integrative role in the digitalization of EU sectoral policies as well as at the national level. On the basis of the collected research material, the authors conducted a comparative analysis of various EU presidencies in the context of countering disinformation, cyberattacks, the growth of cybercrime, and hybrid warfare. They demonstrate how Poland, upon assuming the Presidency of the Council of the EU in the first half of 2025, drew on the experiences of presidencies in countries such as Estonia, the Czech Republic, and Slovakia to enhance resilience to cyberattacks and to develop EU digital governance. An important issue addressed in this part of the book is the analysis of the extent to which subregional political entities, such as the Visegrad Group and the Three Seas Initiative, have influenced the direction of EU policy in ensuring digital security, fostering technological development, improving economic efficiency, and increasing military potential. The authors also consider whether the role of these organizations is limited merely to implementing decisions adopted at the EU level.

The second part of the book presents an empirical analysis of the challenges facing EU security policy in a changing world. In this case, the analyses focus on two interrelated vectors of influence. The first concerns the impact of the activities of selected Member States on shaping the European agenda and the discourse conducted at the level of the European Union, which forms the basis for decision-making processes. The second vector relates to the impact of common EU mechanisms and security policies on the actions undertaken by Member States in response to contemporary challenges, particularly those associated with new technologies and processes of digitalization. This section begins with a chapter showing how contemporary influence operations use digital platforms (such as Facebook) and clickbait websites to manipulate emotions and limit independent thinking among audiences. The 2024 presidential elections in Romania, in which the first round was annulled by the Constitutional Court, serve as an example illustrating mechanisms of influence on public opinion. The origins of this long-term influence campaign date back to the COVID-19 pandemic, when social dissatisfaction and criticism of the political class were exploited. It was noted that without appropriate procedures and technological tools, detecting activities that influence public discourse and destabilize democratic processes would be impossible. The authors of the remaining chapters in the second part of the monograph analyse 21st-century phenomena that pose challenges to the leadership of the Union and the Council of the EU. The analysis focuses on digital sovereignty and the implementation of defence industry development programs in the context of new technologies and digital systems that are resilient to cyberattacks or make use of artificial intelligence. The monograph also addresses issues related to combating organized crime, expanding the competences of the EU border control agenda, and restricting the right to asylum in situations of security threats at the Union's external borders. In the context of the documents analysed, developed by EU institutions, it is evident how important it is to design EU institutional mechanisms, in response to threats to the stability of the Community. It is worth noting that the authors address not only the EU's current challenges but also those that will gain importance in the coming years in security policy and the development of new technologies, both at the EU level and in the national interests of the Member States. An example confirming this argument are the circumstances and course of the emergence of the concept of European digital sovereignty and the role of the Council of the EU in creating institutions responsible for shaping EU policy in this area. The EU's efforts to establish "digital sovereignty" and the Cyber Resilience Act (CRA) were aimed at guaranteeing digital security by increasing control over digital infrastructure, data governance, and technological innovation. Other challenges for EU policy included migration and the associated rise in organized crime. Consequently, the future of Frontex, in terms of its competences and the legal aspects of restricting the right to asylum, together with the consequences of such actions, constitute important questions. Particular attention was paid to the impact of artificial intelligence on the development of science, technology, and innovation, as well as its use in disinformation campaigns (through recommendation algorithms and deepfakes), manipulation of public opinion, and the undermining of authorities in order to weaken public trust in democratic institutions. The author addressed the limitation of the use of artificial intelligence in ways incompatible with democratic values, pointing to institutional and legal solutions aimed at counteracting this phenomenon. The solutions presented referred to measures taken at the EU level as well as in Italy.

The third part of the study is devoted to the policies of countries applying for EU membership in the context of their cybersecurity, vulnerability to disinformation and hybrid warfare, and the use of modern technologies. This issue is analysed with reference to the policies of countries such as Moldova, Ukraine, Turkey, Bosnia and Herzegovina, as well as Norway as a cooperating non-EU country. In this part of the book, the concept of the Europeanization of states aspiring to EU membership acquires particular analytical significance, examined in the context of changing political and security conditions. The authors indicate that alongside traditional socio-economic criteria, issues related to digitalization, cybersecurity, and resilience to hybrid threats are gaining increasing importance. These aspects are becoming a key element of relations with the European Union and one of the conditions for deepening cooperation and potential accession, while simultaneously aligning with the priorities of EU security policy and the agenda of the Presidency of the Council of the EU. The first chapter shows the extent to which Russian disinformation and manipulation influenced public opinion, interfering in elections in Moldova and in the 2024 referendum campaign on the country's accession to the EU. The actions of the Russian Federation were aimed at blocking Moldova's path towards European integration by exploiting existing internal social divisions and vulnerabilities, seeking to maintain its influence in the country. The next analysed case is Ukraine, which, despite the ongoing war, has made progress in digital development while simultaneously facing numerous cyber threats. The authors present opportunities for the development of cooperation between Ukraine and the EU at the local, national, and international levels to strengthen cybersecurity, including through joint training programs, knowledge exchange, technical support, and the development of human capital in the IT sector. The case of the Subcarpathian (Zakarpattia) region in Ukraine is then discussed in the context of its use by the Russian Federation in hybrid warfare as a bridge to the main theatres of operations in Crimea and Donbas, and later during the invasion of the entire territory of eastern Ukraine. Another interesting example of EU cooperation with a non-Member State is the analysis of relations between Norway and the EU in the field of international security. This partnership includes crisis management, defence industry cooperation, activities in outer space, protection of critical infrastructure, and countering hybrid threats. Two chapters are devoted to Turkey as a strategic partner of the EU in the areas of migration security, economic security, and digital and information systems. One chapter analyses the significance of Turkey in the context of EU security priorities during Council presidencies and at the level of Member State policies, which differed in their assessment of Turkey's importance in this area. The second chapter on Turkey discusses the dynamics of cooperation between Turkey and the EU in cybersecurity, using a comparative method based on documents prepared by both sides. Although the EU and Turkey interpreted Internet-related cyber threats differently in their strategic documents, their cooperation created opportunities to identify shared priorities. The final analysis concerns the role of the EU in the development and stabilization of Bosnia and Herzegovina over the past thirty years, with particular emphasis on key programs and actions aimed at combating irregular migration, ensuring economic security, and developing new technologies, all implemented in the context of the country's integration with the European Union.

The final part of the book broadens the analysis of EU digital security and technological sovereignty by adding an economic dimension and examining the role of the Council of the EU in this process. The guiding premise is that economic and financial issues, as well as their stable and predictable functioning, are increasingly dependent on the effective use and regulation of new technologies. The texts analysed show that, alongside stable economic foundations, the adaptation of digital technologies is gaining key importance as an essential element in building the European Union's economic security and its position as a significant actor in global technological and geopolitical competition. The authors focus on analysing the challenges facing the Union in the energy and climate sectors, hydrogen policy, critical raw materials, and innovation. The first chapter provides an analysis of EU decisions and their effectiveness in countering Russian disinformation in the energy and climate sectors, which intensified after Russia lost its main energy commodity markets – the EU Member States. This problem is also illustrated by the example of Poland, which has undertaken measures to build information resilience in this sector. The next chapter discusses the challenges facing the EU with regard to the infrastructure necessary for hydrogen transport both within Member States and beyond the Union. Using the Czech Republic as an example, it analyses the difficulties that this transit country will have to overcome in order to secure hydrogen supplies from abroad. Another chapter examines the impact of the French Presidency of the Council of the EU in the first half of 2022 on the development of the EU Critical Raw Materials Act. Subsequent contributions discuss the assumptions and implementation of EU water policy, highlighting weak enforcement mechanisms and insufficient cybersecurity safeguards for water infrastructure, as well as challenges related to the cross-border management of water resources. The final chapter presents EU policy on shaping its role in the governance of artificial intelligence in the context of internal political challenges and international competition from the United States and China.

This monograph offers a unique, multidimensional analysis of the role of EU presidencies in shaping responses to contemporary digital, technological, and geopolitical challenges. The editors believe that, thanks to its interdisciplinary approach, the publication contributes to a better understanding of the impact of EU regulation on policy and security domains. We hope that this book will help deepen the debate on the future of the European Union in an era of dynamic digital and geopolitical transformations.

PART 1

Priorities of the European Union Presidency in the Face of Digitalisation and Security Challenges

CHAPTER 1

The European Union's Digital Single Market as a Tool for Europeanisation *ad extra*

Janusz Ruszkowski

Abstract The Digital Single Market (DSM) in the European Union represents a significant initiative aimed at removing barriers to online trade and ensuring seamless digital access across member states and outside of the EU. It is a key component of the broader Europeanisation process, which seeks to integrate diverse national economies and policies into a cohesive European framework. The aim of the study is to examine the impact of the Digital Single Market in the European Union on the process of Europeanisation ad extra (external) i.e outside European Union. In practice Europeanisation ad extra is strictly connected with regulatory extrateritorialism and digital external governance of the EU. Therefore thanks to these concepts, we need to look for an analytical area that will help identify the impact of the DSM on non-EU entities. The Digital Single Market is a critical instrument for Europeanisation ad extra , driving economic cooperation with third parties, fostering innovation, and enhancing consumer trust not just across the EU but outside too. As it evolves, it will likely continue to play a pivotal role in shaping the digital landscape of Europe, promoting unity while addressing the challenges posed by a rapidly changing global digital economy.

Keywords European Union – Digital Single Market – Europeanisation ad extra

1 Introduction

The Digital Single Market in the European Union (DSM) is an initiative to contribute to the removal of barriers to online trade and digital services, as well as to the provision of seamless digital access in the Member States, as well as outside the EU. If the DSM is a noticeable part of the *ad extra* Europeanisation process, it also (and perhaps even more so) serves to integrate the various digital elements of sectoral policies in the EU and national policies the Member States. Only a coherent European digital framework can effectively inter-

HTTPS://DOI.ORG/10.1163/9789004763623-003

act with externally policies. Examining the impact of the Digital Single Market European Union' on its international environment (Europeanisation *ad extra*) must take into account both regulatory extraterritorialism and digital external governance (digital external governance). Regulatory extra-territorialism is the transfer and impact of EU regulations outside the EU. On other hand the digital external *governance* encompasses different areas of digital activity with impacts and new effects. External governance in practice is based on taking coercive action against non-EU digital platforms offering their digital services in the EU, although it can also target third parties that do not operate within the EU. In addition, digital extraterritorial governance involves cooperation (often multi-level) with third countries and international organisations (see Figure 1.1). Therefore, by means of regulatory extraterritoriality and digital external governance, the two paths along which the Europeanisation process takes *ad extra* place, an analytical area should be located to help identify the impact of DSM on non-EU actors.

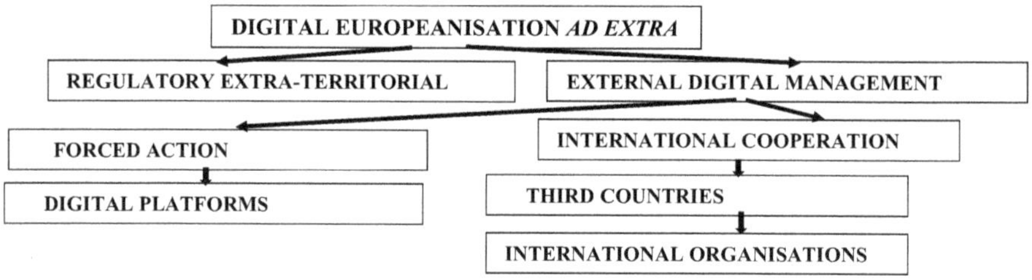

FIGURE 1.1 Digital Europeanisation *ad extra* (Source: Own elaboration).

The external environment of the EU is delimited not only by external Schengen (geopolitical) borders, but also by other types of 'borders' that distinguish the political system of the European Union from its environment, among which are: legal-institutional, transactional, cultural borders (Smith, 1996, p. 5), and axiological. F. Schimmelfennig sees the impact of the Europeanisation processes generated by the EU on non-European countries and other non-European actors, but also on European candidate countries, which are subject to a strong transfer of the EU acquis (*acquis communautaire*) (Schimmelfennig, 2007, p. 3–5).

The aim of the analysis is to identify the impact (also the impact causing new effects) of the EU Digital Single Market externally (*ad extra*), i.e. on the environment located outside the EU through the prism of regulatory extraterritoriality and external digital governance. Thus, it is important to consider whether DSM in the EU is relevant to the process of Europeanisation *ad extra* (external).

The methodological assumptions adopted in the study are based on the important conceptual turn in Europeanisation studies that took place at the turn of the 20th and 21st centuries, which consisted in moving away from a dualistic understanding of Europeanisation (either *top down* or *bottom up*) towards a multi-directional understanding. The object of study defined here makes it necessary to analyse using *an output-oriented approach*, 'oriented' outside EU member states (Ruszkowski, 2010) with an effect *output*. The nature of the EU impact *ad extra* also requires the sequential application of a (*multi-level governance* – MLG) approach, as actors located at national, supranational, and global level are active in this impact.

An important characteristic of Europeanisation *ad extra* is its limited character of legal *coercion*. While such coercion can be seen, for example, in the extra-territorial activities in EU (more on this below), it is much more evident *conditionality*, as from a perspective, rationalist the profit and loss calculations made by national actors within the logic of consequence and logic of benefits, which make them willing to accept EU rules and norms. From a constructivist perspective, on the other hand, adaptive change is the result of the socialisation and internalisation of norms in the process of international and transnational interaction and within the logic *of appropriateness*. This means that EU law cannot be imposed on third countries, but conditions and circumstances may arise that make bottom-up and optional decisions in third countries (or by external actors) produce a Europeanised legal effect. The key for Europeanisation *ad extra* principle of conditionality refers not only to the mechanism of *leverage* (Cianciara et al., 2015), but also the principle of voluntariness.

Europeanisation *ad extra* may thus be a mechanism for projecting the normative power of the European Union onto its external environment (Cianciara et al., 2015, p. 19–38, Ruszkowski, 2019, p. 7). The question here is whether such projection of the EU's normative power onto its external environment occurs in the case of the DSM?

2 Legal and Institutional Framework for the Digital Single Market in the European Union

The DSM was designed in the Digital Single Market Strategy (DSMS), which was published in a European Commission Communication adopted on 6 May 2015 (COM/2015/0192). The idea of the DSM was the European Union's response to the global revolution in the IT sector, which had its first phase between 1995 and 2010, when the digital economy and the digital society were gradually emerging.

The strategy consists of three pillars. The first pillar, covering access to digital goods and services, initiates an agenda of digital policy reforms and legislative actions to foster digital transformation including aiming to intensify cross-border e-commerce between small- and medium-sized enterprises (including the lifting of the geo-blockade). Geo-blocking is a discriminatory practice that may be the result of a unilateral decision by market players preventing customers from accessing and purchasing products and services online through a site located in another Member State (Council Regulation against barriers to e-commerce, 2018, COM/2015/0192, p. 6.). The second pillar is aimed at building a digital environment for online platforms and digital services and networks with trusted infrastructure. In this pillar, the strategy considers the removal of illegal online content and the protection of personal data in digital circulation as extremely important. The third pillar projects the digital market as a growth engine European Digital Economy (COM/2015/0192, p. 3–4), which will imply the creation of a European Digital Economy Society (EDES).

The EU supports digital transformation internationally to achieve complementarity with external partners, inspire them to cooperate and explore new solutions, and enforce EU digital regulations. Digital transformation and building a climate that supports digital networks, including innovative business, are important objectives of the DSM in the EU. Special financial instruments have therefore been established to support it, including the Digital Europe Programme (2021–2027 with a budget of €7.5bn) and Horizon Europe (2021–2027 with a budget of €95.5bn).

The DSMS, as well as the White Paper on How to meet Europe's digital infrastructure needs (White Paper on How to meet, 2024) are acts *sui generis*, as they do not trigger direct legal sanctions for any failure to comply with the obligations contained therein. However, the dynamic development of the DSM and the extension of digital integration to new areas of activity (artificial intelligence, digital services, data management, etc.) have led to the creation and subsequent gradual extension of a hard legal infrastructure (digital *acquis* – DA) governing the functioning of this market in the areas of digital services (Digital Services Act of 2022 – DA) (Regulation (EU) 2022/2065), digital markets (Digital Markets Act – DMA of 2022) (Regulation (EU) 2022/1925), digital security (Cybersecurity Strategy of 2020 and Cyber Resilience Act of 2023) (Joint Communication, JOIN(2020) 18, 2024, Regulation (EU) 2024/2847), digital data governance (Data Governance Act of 2022 and Data Act adopted in 2023) (Regulation (EU) 2022/868, Regulation (EU) 2023/2854), and artificial intelligence (AI Act of 2024) (Regulation (EU) 2024/1689). The

DA defines the normative scope of access to public and private online services and helps create a European digital identity (Regulation (EU) 2024/1183), and thus to protect EU digital sovereignty.

As can be seen from the above, the EU DA is a set of legal regulations (primarily EU derived law) and acts *sui generis* covering the digital activities of the EU and its DSM and dealing with various aspects of its functioning. The DA is the digital part of the EU acquis *communautaire* and, importantly, also contains the rules applicable to the DSM area (digital rulebook).

The DA determined the internal institutionalisation of the DSM. The Digital Single Act of 2022 established two institutions to support its implementation. The first is the European Center for Algorithmic Transparency (ECAT), which, through scientific and technocratic analysis, is to serve the European Commission in its oversight function covering the level of compliance of digital platforms with their obligations under the DA. The second institution acting on the basis of the DSA is the independent European Council of Digital Services (ECDS) (Article 61, DSA, 2022), which has the competence to supervise and enforce the obligations belonging to digital service providers and has the possibility to refer to the EC the supervision of digital service providers and enforce the obligations of digital platforms under this act. ECDS has the ability to refer DSA violations to the EC, and to propose remedies (Legal theme 129 DSA, 2020).

In February 2023, The European Commission established the European Data Innovation Board (EDIB), which is formed by a group of experts facilitating the implementation of the Data Act. In March 2024, the Body of European Regulators for Electronic Communications (BEREC) was established, and on 29 October of the same year, the European Innovation Council (EIC) was established under Horizon Europe by a decision of the European Commission to be a major EU investor in advanced digital technologies and innovation (European Commission Decision C92024)7451).

Two sectoral initiatives Digital Compass and Digital Decade are specific tools to dynamise digital cooperation within the DSM. In its communication of 9 March 2021 entitled 2030 *Digital Compass: The European Way for the Decade* (COM/2021/118), the European Commission set out a vision for the digital development of the EU by 2030, including enabling citizens and businesses in the EU to digitally transform (digital transformation) (COM(2022), 27) while focusing on citizens for a secure digital ecosystem. In 2022, the *Digital Decade* Policy Programme 2030 (Decision (EU) 2022/2482) was adopted, which envisages that at least 75% of business in the EU will use artificial intelligence by 2030. The digital transformation agenda was presented in the European Declaration on Digital Rights and Principles for the Digital Decade, adopted by the

European Parliament, the Council and the European Commission, which closely corresponds to the UNO Global Digital Compact and thus supports the global vision for the digital future at a global level.

2.1 Extraterritorial Impact of Digital Single Market Regulation

The DSM strategy establishes the European Union as a leader in the digital (data-driven) economy, which means that the EU will require external partners (located outside the EU) who wish to operate in the DSM to recognise and comply with its regulations, i.e. the DA communautaire. The DA communautaire is where the potential for the digital impact of the DSM on external parties lies, i.e. regulatory extraterritoriality, which is one of the paths along which the ad extra Europeanisation process takes place. Regulatory extraterritoriality is the normative-based diffusion and enforcement of DA beyond the borders of the EU. On the one hand, this influence is intended to ensure that non-EU third parties take EU regulations into account and establish digital safeguards that will benefit them and the EU (including digital security). On the other hand, it should ensure a level playing field and effective protection for public entities and individuals across the EU.

The process of digital Europeanisation *ad extra* takes place through regulatory extraterritorialism based on the DA, including the extraterritorial impact potential of its key acts. Let us trace this impact through the example of three key acts: on digital services, on the digital market and on artificial intelligence.

2.2 Extraterritorial Impact of the Digital Service Act

The Digital Services Act (DSA) contains obligations that must be complied with by digital service providers offering their services within the European Union (Article 71, DSA.). The main obligation for digital service providers is to protect the EU DSM from disinformation, i.e. the provision of false, incomplete, or misleading content, and to submit to scrutiny in situations of suspected disinformation practices.

The DSA defines very large digital platforms and digital search engines as VLOPSEs (Very Large Online Platform and Search Engines). VLOPSE is a common acronym that was created by combining VLOP (Very Large Online Platform) and VLOSEs (Very Large Online Search Engines). These are subject not only to the DSA, but to all regulations of DA. By 2024, 19 VLOPSEs have applied the DSA regulations. These are companies that are EU or non-EU based and offer their indirect services to more than 45 million EU users per month (i.e. more than 10% of the EU population) and in at least one Member State (benefit logic). VLOPSEs are required to report to the European Commission every 6 months on their progress in combating disinformation and on transparency (conditionality) (Article 2, Article 3, Article 5 and Article 23, DSA, 2022). It is non-EU VLOPSEs that are targeted by the digital extraterritoriality resulting from the DSA. The extraterritorial impact of the DSA on VLOPSE is primarily based on: 1. the fight against disinformation, 2. financial and administrative sanctions, and 3. audits.

The DSA projects the fight against disinformation on two levels: national and EU. Article 9 allows Member States (and in practice their judicial or administrative authorities) to issue an order to take action against specific illegal content issued under EU law or national law compatible with EU law, including an information injunction (Art. 9 and Art. 10, DSA, 2022). In addition, Member States appoint digital services coordinators and have the possibility to sanction VLOPSE for violations of the obligations established in the DSA (logic of consequences).

On the EU level, on the other hand, in the event that violations of the DSA regulations or non-compliance with interim measures possibly refusing to submit to inspections, European Commission may impose sanctions on VLOPSE (Article 73, DSA, 2022). Provisional measures are imposed on the basis of an EC decision in order to prevent further infringements so as not to lead to a risk of serious harm to the recipients of the service (Article 70 and Legal theme 143, DSA, 2022). Sanctions come in two varieties (financial and non-financial). The main financial sanction for breach of the DSA regulations is a monetary fine of up to 6% of the annual worldwide turnover (Article 52, DSA, 2022) or up to 1% of the total annual income or worldwide turnover in the preceding financial year if they have intentionally or negligently provided incorrect, incomplete, or misleading information, failed to respond to a request for information by EC decision within the prescribed period (Article 74, DSA, 2022). In addition, the EC may impose periodic monetary penalties (Article 76, DSA, 2022), the daily amount of which is up to 5% of the average daily income or worldwide annual turnover generated in the preceding financial year and is calculated from the date set in the decision (Legal theme 144, DSA., 2022). It should be emphasised that all decisions taken by the EC are subject to review by the CJEU according to the logic of consequences (Article 261 of the TfUE).

Non-financial sanctions come in the form of court or administrative orders to cease or prevent violations of the DSA and to remove illegal content or possibly prevent access to it (Legal theme 25, DSA, 2022). In addition, the DSA provides for a crisis response mechanism. The EC may require the development of voluntary crisis protocols to coordinate a rapid, collective, and cross-border response in the online environment of large online platforms (Article 36 and Legal theme 108, DSA, 2022).

In turn, the DSA stipulates that the European Commission and the coordinators in the Member States will carry out an independent audit of VLOPSE at VLOPSE's expense and that VLOPSE will be subject to access to all data held (Article 4 and 37, DSA, 2022). In addition, the DSA forces VLOPSE to establish its own cell to ensure compliance with the DSA, which will be independent of its operational cells (Article 41, DSA, 2022).

2.3 Extraterritorial impact of the Digital Market Act

The Digital Market Act (DMA) defines large digital platforms providing digital services (app shops, messaging, social networks, etc.) to the EU market as gatekeepers, who are external (non-EU) actors. There are three main criteria for the decision to designate a company as a gatekeeper. 1. The company has a certain annual turnover in the European Economic Area and provides at least one of the 10 core online platform services listed in the DMA and does so in at least three EU member states. The ten services listed in the DMA include: app shops, networks social, messaging, video sharing, virtual assistant services, web browsers, cloud computing services, operating systems and advertising services. 2. The company provides a core service on an online platform to more than 45 million active end-users per month based in the EU and to more than 10,000 active business users based in the EU. 3. The company meets the second criterion within the last three years (Article 33, DMA, 2022). Gatekeepers have a significant impact on the EU market as they provide the main digital services and are active in many Member States; moreover, they have links with a broad user base and the business world (logic of advantage). The first gatekeepers were admitted to the DSM by a decision of the European Commission taken on 6 September 2023. This group included Alphabet[1] Amazon,[2] Apple,[3] ByteDance,[4] Meta,[5] and Microsoft[6]. They have a

1 It includes eight service platforms, such as the Android operating system, YouTube, Google, the Chrome browser, Google Maps, Google Play, Google Shopping, and Google Search.

2 It offers two service platforms: Amazon and Amazon MarketPlace.

3 It offers three service platforms: the Apple App Store, the Safari browser, and the iOS operating system.

4 It only offers one social network, TikTok.

5 It offers six service platforms: the social networks Facebook and Instagram, the messaging services WhatsApp and Messenger, and Meta and Meta Marketplace.

6 It offers the Linkedin social network and the Windows operating system.

total of 22 service platforms that are supplied and offered in the EU and meet the DMA criteria (Richter, 2024).

The extraterritorial impact of the DMA on Gatekeepers is based on: 1. catalogues of actions that can be performed ('do's'), 2. actions that cannot be performed ('dont's') in the EU DSM, and 3. possible financial or non-financial remedies (sanctions) that may be imposed on them.

The catalogue of example, 'do's' includes indications for gatekeepers who: 1. must allow third parties to interoperate with their services in certain specific situations, 2. must also allow their business users to access the data they generate when using the gatekeeper platform, 3. must provide companies advertising on the gatekeeper platform with the tools and information necessary for advertisers and publishers to carry out a verification of their gatekeeper hosted ads, and 4. must allow their business users to promote their offers and contract with customers outside the gatekeeper platform (The Digital Markets Act: ensuring fair, 2025).

The catalogue of examples of, 'don'ts' includes indications for gatekeepers that: 1. must not give preferential treatment to services and products provided by themselves compared to similar services and products offered by third parties on the gatekeeper's platform. 2. may not prohibit users from uninstalling any pre-installed software or applications if they so wish, 3. may not track end users in their activities outside of the main gatekeepers service platform for the purpose of targeting of advertisements without giving effective consent, 4. may not impede consumers' ability to connect and use tools outside of gatekeeper platforms (The Digital Markets Act: ensuring fair, 2025).

The most important requirement of the DMA arising from both the do's and don'ts is to prevent the gatekeeper from achieving a monopoly position in the EU market (no competition between platforms) (Legal theme 32, DMA, 2022). Gatekeepers often act in a dual role, as providers of operating systems and as sellers of hardware, while reserving access to certain functionalities in their hardware (security elements, software identification mechanisms, etc.) (Legal theme 56, DMA, 2022). Gatekeepers should ensure interoperability for third-party vendor-independent providers of interpersonal communication services, end-users, and business users in the EU (Legal theme 64, DMA, 2022). Importantly, gatekeepers' European competitors can set up alternative app shops, in which case users have more control over their data. Other platforms, SMEs, and innovative start-ups, although not subject to DMA regulation (they are not gatekeepers), will benefit from relevant interoperability opportunities in cooperation with gatekeepers. Gatekeepers, on the other hand, may adopt slightly different business conditions and practices in different EU Member States than those contained in the DMA, which may create discrepancies between the conditions of competition for users of core platform services provided by gatekeepers and thus harm internal market integration (Legal theme 7, DMA, 2022).

If the European Commission takes at least three decisions on regulatory breaches under the DMA within an 8-year period, then it can impose remedies (logic of consequences). The DMA provides for certain remedies (sanctions), which can be financial (the European Commission can impose a fine of up to 10% of the company's total worldwide turnover) or non-financial (behavioural or structural, such as divesting parts of the company or prohibiting the acquisition of additional services related to the systemic breach) (Legal theme 75, DMA, 2022). Gatekeepers do, however, have the possibility to request the lifting of the DMA obligation, but in exceptional circumstances that are beyond their control (proportionality principle). For example, this could be an unforeseen external shock that temporarily eliminates a significant part of end-user demand for the core service offered by the platform (e.g. public health reasons) (Legal theme 66, DMA, 2022). The European Commission (taking into account the mandatory then reports submitted by gatekeepers) may decide that a specific obligation does not apply to a particular digital platform service (Legal theme 67 and 68, DMA, 2022). Decisions taken by the European Commission under the DMA Regulation are subject to review by the EU Court of Justice in accordance with Article, 262 TFEU (Legal theme 94, DMA, 2022).

3 Extraterritorial Impact of the Artificial Intelligence Act

The provisions of this Act (AI Act) with the status of a regulation apply to AI system providers on a non-discriminatory basis, regardless of whether they are based in the Union or in a third country, and to AI system operators based in the Union (Legal theme 21, AI Act, 2024).

The AI Act is often referred to as the Horizontal Act, as it implements a tiered system of regulatory obligations for specifically listed artificial intelligence applications (e.g. Open AI, or Chat GPT from the US, GemAI from Turkey, or DeepSeek from China). In addition, deepfake generation, chatbots, and biometric analysis must be clearly disclosed to those concerned (Meltzer et al., 2022).

The extraterritorial impact of the AI Act on AI system providers is based on compliance with the Code of Practice for General-Purpose AI (GPAI) model providers operating in the EU DSM. Digital platforms, including those defined by the EU as VLOPSE and gatekeepers (V/G) are also dependent on AI as they operate on the basis of network algorithms or recommender systems. The AI Act calls on third parties (here, platforms offering AI) to develop a wide range of standards and to make decisions that safeguard against the misuse of AI. Such regulations can be already seen in the USA, including in particular the efforts of Housing and Urban Development (HUD) and international bodies such as ISO/IEC (International Standardisation Organisation/International Electrotechnical Commission), or CEN (Comite Europeen de Normalisation) and CENELEC (Comite Europeen de Normalisation Electrotechnique) (Legal theme 150, AI Act., 2024).

Due to their digital nature, certain AI systems should fall within the scope of the AI Act, even if they are not marketed, commissioned, or used in the Union. This is the case, for example, of an operator established in the Union that contracts with an operator established in a third country for certain services in relation to an activity to be performed by an AI system that would be classified as high risk. In such circumstances, the AI system used in the third country by the operator could process data that was lawfully collected in the Union and transferred outside the Union.

The AI Act should also apply to suppliers and users of AI systems established in a third country, to the extent that the results produced by these systems are intended for use in the Union. (…). The AI Act should not apply to public authorities of a third country and international organisations acting in a cooperative capacity or pursuant to international agreements on law enforcement and judicial cooperation with the Union or its Member States concluded at Union or national level, provided that the third country or international organisation provides appropriate safeguards with respect to the protection of fundamental rights and freedoms of natural persons (Legal theme 22, AI Act., 2024). Where appropriate, this may include activities of entities entrusted by third countries with specific tasks in support of law enforcement and judicial cooperation. Such cooperation frameworks or agreements have been established bilaterally between Member States and third countries or between the European Union, Europol, and other Union agencies and third countries and international organisations (Legal theme 22, AI Act., 2024).

The scope of application of the AI Act includes, inter alia, AI system providers and AI system users located also in a third country when the results produced by the AI system are used in the Union. Excluding public authorities of a third country or international organisations falling within the scope of this Act if these authorities or organisations use AI systems within the framework of international cooperation or international agreements on law enforcement and judicial cooperation concluded with the Union or with one or more Member States (Art. 2 AI Act., 2024).

3.1 External Governance in Sensitive Areas of the Digital Single Market

Effective external governance in sensitive areas of the EU DSM (effective DSM governance) is, next to regulatory extraterritorialism, the second path followed by the process of digital Europeanisation *ad extra*. Effective external governance is digital in nature, as it is based either on coercive actions (enforcement of digital regulations *acquis* and exerting influence in anticipation of new effects) targeting digital platforms outside the EU (*external actions*) or on multi-level cooperation with third countries or international organisations. Such external actions can be located in a number of particularly sensitive areas, among which are: digital security, use of artificial intelligence, protection of citizens' rights and social minorities, standardisation, and the fight against monopolisation of digital services.

3.2 External Management of Digital Security

The digital security of the European Union may be threatened primarily by attacks or interference by external digital platforms on democracy, including electoral processes in Member States or European Parliament elections, and by disinformation and fake news. The fight against disinformation and fake news is fragmented, as it involves a lot of manipulated or false information imported into the EU and its Member States from outside. The response to such attacks has been coercive towards digital platforms and has taken place at two levels (at EU level and/or at national level).

3.2.1 External Enforcement Action

Let's take a look at the European Commission's coercive action against TikTok in 2024 and its effects on digital security. Back in February, the EC initiated proceedings against TikTok for launching the Lite Rewards programme in the EU market allowing users to earn points for performing various tasks (challenges), which could have had very negative effects on people's health and lives, especially those of teenagers. The proceedings ended in April 2024 with a decision to withdraw Lite Rewards from the EU market and a commitment not to introduce new programmes with similar characteristics. On 24 November 2024, an attack on the presidential election in Romania took place, originating from TikTok, a messenger owned by gatekeeper ByteDance. The attack took place using unlabelled political content, bots, and approximately 26,000 unauthorised accounts created on TikTok by Russia. Here, it should be recalled that VLOPSE, according to the logic of consequences, are obliged to be

TABLE 1.1 European Commission proceedings against TikTok in 2024 and their effects
Source: Author's own elaboration based on European Commission press releases and official communications under the Digital Services Act (DSA) enforcement framework, including European Commission (2024, April 22), European Commission (2024, February 19), and European Commission (2024, December 17).

Date of initiation of the EC procedure	Reason for proceeding	New effect	Level of response
February	Lite Rewards programme	Withdrawal of Lite Rewards from the EU market (August)	EU (completed)
December	Attack on the presidential election in Romania	Annulment of elections (December) data retention order	National (completed) EU (ongoing)

responsible for harmful and misleading content based on the regulations under the Digital Services Act (DSA). These services include, "mere transmission" (e.g. wireless access points, virtual private networks, Internet traffic exchange points) "caching" (e.g. content delivery), and "hosting" (e.g. cloud computing). (Art. 4, 5, 6 DSA, 2022) Hosting services are particularly important for combating and controlling disinformation (limiting fake accounts or the use of bots), as providers of these services not only store content, but may also make it available to the relevant authorities afterwards.

As a result of this attack, there have been reactions at EU level (EC) and national level (Romania). At the EU level, the Commission opened investigations into suspected EU Digital Services Act (DSA) violations. The EC's decision to open proceedings was based on information from Romanian authorities, as well as an analysis of the risk assessment reports submitted by TikTok in 2023 and 2024, including the DSA regulations. The subject of the investigation was TikTok's recommendation systems, which the European Commission checked for vulnerability to coordinated manipulation and disinformation, and the risk of automated use of the service. The EC's vetting activities also covered TikTok's policies on advertising and payment for political content, and the level of compliance with the digital service operator's obligations under the DSA and regarding its assessment of systemic risks related to the integrity of the Romanian elections and actions to mitigate such risks. On 5 December 2024, the EC ordered TikTok to freeze and secure data related to actual or anticipated systemic risk (data retention order). On 27 November 2024 at the national level, Romania's National Audiovisual Media Council asked the European Commission to investigate the role of TikTok in the first round of the presidential election. Two days later, the Commission asked the platform to clarify its management of the risks associated with information manipulation. Therefore the response at the national level was to inform the EC of foreign interference from outside, followed by the annulment of the results of these elections by the Romanian Supreme Court, which took place on 6 December 2024.

Although all VLOPSEs reported in September 2024 on measures to protect European elections, the European Commission was forced to verify allegations of DSA violations against some VLOPSEs, including Portal X and the Meta digital platform. These allegations were made by, among others, former Internal Market and Services Commissioner Thierry Breton, who explicitly accused Portal X of flouting the Code of Practice on Disinformation. The basic version of the code was adopted in 2018 by the European Commission as the first regulation of its kind in the world (EU Code of Practice, 2024). Unfortunately, Twitter (now X) dropped the Code in 2023 after Elon Musk acquired the platform. EP President Roberta Metsola felt that the European elections had become a test of the resilience of EU procedures to disinformation attacks (europarl.europa.eu, 2025).

3.2.2 Multi-Level International Cooperation

In the field of digital security, the EU also cooperates with third countries and their institutions (bilateral level). Here, the relationship with the US should be cited, specifically with the US Trade and Technology Council (TTC). Both parties have agreed to cooperate on information manipulation and interference with the functioning of democracy and fundamental values, thereby supporting the security and resilience of sensitive projects in third countries. In addition, in 2022, the EU entered into bilateral digital partnerships with Japan and South Korea, and in 2023 with Singapore and Canada.

At a global level, the European Commission has been heavily involved in the consultation process with UNESCO on the Guidelines for Governance of Digital Platforms since November 2023.

3.3 External Management for Artificial Intelligence

Artificial intelligence is, in practice, a transnational technology that affects the external environment of the European Union. Here we can speak of AI external *governance*. It involves using AI to transform the economy, sharing good practices, preventing the risks that AI brings, etc. For this reason, AI external governance takes place primarily through cooperation at the global (international) level and in bilateral formats, although these levels can also intersect.

3.3.1 Multi-Level International Cooperation

From 2021, the EU and the US are coordinating a joint contribution to the multilateral initiative AI, which is part of an ongoing UN process to advance and responsibly manage artificial intelligence at a global level. Another global example of the use of AI relates to investment in submarine cables, including fibre optics, which is becoming almost impossible without building a cross-border AI infrastructure. This is why the G7 Partnership for Global Infrastructure and Investment initiative has been proposed. Leveraging the world's largest economies from the G7 (Great 7) is an ideal format for the European Union to achieve the end goal. Investment in the wider digital ecosystem can contribute to economic growth.

In the international management of the risks posed by artificial intelligence, the European Commission also wants to cooperate with the external standardisation agencies CEN and CENELEC. In this regard, it made a request to them in May 2022 for support on security and trust in artificial intelligence (European Commission, Draft standardisation, 2022). In 2024, the European Commission initiated a discussion on the labelling of information crafted by artificial intelligence. If this solution is adopted, it will be almost identical to the existing labelling of genetically modified products. Such a practice will certainly be designed outside the EU as well. The European Union is convinced that digital transformation, including through AI, must contribute to economic growth, in particular when its DSM, in addition to its impetus inwards, will also have an impact outside the EU, on its trading partners, including primarily US actors. In such a bilateral format, we can talk about transatlantic EU–US risk management related to artificial intelligence. An example of such cooperation is the report on the impact of AI prepared by the European Commission and the White House Council of Economic Advisors (European Commission and The White House Council, 2023). EU cooperation with the US Trade and Technology Council (TTC) is also important here. A meeting of this body held in September 2021 in Pittsburgh focused on three joint projects involving advanced AI: 1. measurement and evaluation of trust in AI, 2. cooperation on AI technologies dedicated to privacy protection, 3. joint preparation of economic studies on the impact of AI on the workforce. As a result, on 1 December 2022, the two sides published a roadmap on evaluation, AI trust measurement tools, and AI risk management. The Union and the US also involved the OECD in their efforts, which released its framework for classifying AI systems and a catalogue of AI tools and verification developed in 2023 in collaboration with the US National Institute of Standards and Technology (NIST) (Engler, 2023).

Economic growth through digital transformation is an important objective for the EU in its transatlantic trade relations. The idea of *digital transatlantic trade* (*DTC*) has taken the lead in US–EU relations. 45% of start-ups operating in the EU admit that for their companies, access to artificial intelligence technology is extremely important, but unfortunately, it is achieved through contacts with companies outside Europe (The economic opportunity of generative AI, 2024). American Open AI or ChatGPT applications have made European companies more reactive and less proactive. This problem is recognised by Mario Draghi in his report, in which he highlights the importance of support for strategic research collaboration and the need for technology transfer and international mobility of researchers between the EU and the US (The future of European competitiveness, 2024).

3.4 External Management for the Protection of the Rights of Citizens and Social Minorities

The key EU regulatory solution for the protection of the rights of citizens and social minorities, which has a strong impact on non-EU partners, is the regulation in regard to the (EU) 2016/679 of the European Parliament and of the Council on the protection of natural persons with processing of personal data and on the free movement of such data (*General Data Protection Regulation* – GDPR) of 27 April 2016, Poland is known as RODO) (Regulation (EU) 2016/679).

3.4.1 External Enforcement Action

The RODO is significant because of the massive use of personal data, including sensitive data, by digital platforms from third countries, including those on EU-drawn lists (since VLOPSE and gatekeepers collect users' personal data they are subject to the RODO according to the principle of conditionality) (Legal theme 36 of Regulation (EU) 2022/1925, Legal theme 18 of Regulation (EU) 2016/679), but also because of their activities on the wider internet, and therefore it triggers external EU coercive action aimed precisely at digital platforms. This means that personal data of EU citizens not only flows outside the EU, but is also managed and processed by third parties. The regulations embodied in the RODO must be complied with by all foreign entities operating in the EU DSM and conducting any other business within the EU. Europe's leading cybersecurity corporation NCC Group independently monitors the protection of the personal data of European users of this data, investigating in particular the lack of transparency and oversight of this data. Such measures raise the level of data protection and safeguard against unauthorised access to data and even limit the possibility of foreign interference with such data (Politico, 1.12.2024, p. 27.). Importantly, however, these regulations have also become the model for the protection and processing of personal data outside the EU. The RODO stipulates that algorithmic systems should not be allowed to make significant decisions that affect rights without human oversight and should not be allowed unauthorised access or foreign interference with this data. Based on this clause, in 2021, Uber has been obliged to reinstate six drivers who were dismissed on the basis of the company's algorithmic system (The European Commission, Can I be subject, 2025). Security of data storage and processing, including personal data, is taken care of by the European Data Protection Supervisor (EDPS). His competences also extend to third parties.

Another example of the external impact of the RODO is TikTok's decision in 2023 to invest €12 million to better secure in Europe personal data it collects. Despite this, the European Commission launched an investigation against TikTok in 2024 on the basis of its obligations under the Digital Services Act for breach of minority protection rules, publishing inaccurate content and for breach of privacy. There is no doubt that the RODO has also contributed to modernising data protection in third parties (outside the EU).

The protection of the rights of citizens and social minorities in the digital environment is further regulated by the Code of Conduct for Combating Illegal Hate Speech on the Internet, which was adopted in 2016 (amended 20 January 2025). And then the Code has been incorporated into the DSA. Although the Code is a document *sui generis*, i.e. it has no legal effect, the large digital platforms (V/G) are keen to emphasise their commitment to the Code's recommendations. The document has been adopted by Facebook, Dailymotion, Instagram, Jeuxvideo.co, LinkedIn, and within consumer services hosted by: Microsoft, Snapchat, Rakuten, Viber, TikTok, Twitch, X, and YouTube. Thus, we note the impact of *the ad extra in the Code,* and the implementation of its provisions is monitored and evaluated by the European Commission and the EU Council.

3.4.2 Multi-Level International Cooperation

Data protection is also the subject of UIE international cooperation, in July 2023. The European Commission adopted an implementing decision on the EU–US Data Privacy Framework, in which the US ensures an adequate level of protection for personal data transferred from the EU to the US, which is included in the Data Privacy Framework. This sensitive area has also been the subject of EU cooperation with the US Trade and Technology Council (TTC). This cooperation has allowed the EU to take concrete steps whereby the US side has similarly defined human rights and European values in the conduct of digital business. This is a major achievement for the EU in the changing geopolitical digital environment (digital environment). In particular, there was agreement on the protection of children and young people in the digital environment and on access to data online platforms for research. In addition, common rules were agreed to combat violence on online platforms (Monitoring of the European Declaration, 2024).

Since the 1990s, the European Union has been adopting legal solutions for the protection of personal data, which in practice have become the basis for defining global standards in this matter (see, for example, EP and Council Directive 95/46 of 24 October 1995 on the protection of individuals with regard to the processing of personal data and on the free movement of data). They extend beyond the impact of EU data protection regulations and privacy policies and thus sequentially initiate the process of alignment of third parties with these regulations (including the initiation of legislative work in non-EU countries harmonising national laws with EU rules) and even efforts to establish independent agencies in these countries for the implementation of privacy policies and personal data management (for example, in South Africa or Argentina). For example, the European Commission together with ASEAN is developing a joint guide in the area of international data transfers. The final version of the guide was published in May 2023.

4 External Governance for Standardisation and the Fight against Monopolisation of Digital Services

The European Union's DSM also includes standardisation and antitrust measures, which – in addition to intra-EU applications – should also be required of third parties offering their products in the EU. In this matter, we find the external influence of the EU both through coercive actions towards digital platforms and other third parties and through international cooperation with third countries and international organisations.

4.1 External Enforcement Action

Extortion actions have become the response of the DSM, as part of the EU Internal Market, to monopolistic practices by technology giants. In March 2024, the European Commission initiated proceedings against Alphabet, Apple, and Meta platforms for exposed monopoly practices and failure to comply with obligations under the DA. The EC has also taken legal action against other gatekeepers while continuing to work on desegmentation in parallel. The biggest EU scrutiny involves Google, as many as eight of its service platforms have been admitted to the EU market.

The European Commission's flagship anti-trust push was to force the introduction of a single charging port for all devices. From December 27, 2024, portable electronic devices sold in the EU, such as mobile phones, tablets, digital cameras, readers, headphones, and even computer mice, among others, must be equipped with the same charging port, namely USB-C. And from 2026, these requirements will also apply to laptops sold in the EU (in practice, this means one charger regardless of the type of equipment, faster charging, asmanufacturers will not be able to limit the charging speed of devices). This solution also applies to games consoles, computer keyboards, and portable PDSs. Customers will also be able to buy equipment without a charger attached, reducing the purchase of new chargers and the number of old chargers on the market or unused (as e-waste). Manufacturers will be obliged to clearly and legibly inform consumers about charging characteristics, including the power required by the device and whether it is suitable for fast charging (Kurier Szczeciński, 29.12.2024.).

In 2024, the European Commission initiated formal proceedings against the Chinese platform Temu (listed in VLOPSE on 31 May 2024) for violations of the DSA in areas related to online sales of illegal products (not compliant with EU requirements, potentially addictive service design, systems used to recommend purchases to users, and access to data for researchers (these are violations of Articles 27, 34, 35, 38, and 40 of the DSA). The EC's decision follows preliminary analyses of the risk assessment report submitted by Temu in September 20204, responses to the EC's formal requests for information of 28 June 2024 and 11 October 2024, as well as information provided by third parties. The EC also relied on information made available through a mechanism for collaboration with national organisations within the European Council of

Digital Service Coordinators, in particular the Irish Digital Services Coordinator (Kostrzewska 2024).

4.2 Multi-Level International Cooperation

On a bilateral level, the European Union is cooperating with the US on the standardisation of digital technologies. This issue came to the fore at the EU-US Trade and Technology Council (TTC) in April 2024. At that time, it was decided to deepen the standardisation process of new technologies including digital identity, which is important for both sides.

On the international level, in turn, the European Commission works in particular with international organisations specialised in the area of standardisation, such as CEN and CENELEC. For example, in 2024 the Commission asked the two organisations to harmonise standards to support the implementation of the Digital Product Passport, which was adopted as part of the Ecodesign for Sustainable Products Regulation (ESPR) (Darvas, 2023).

5 Conclusions

The DSM (DSM) is one of the instruments of Europeanisation *ad extra*, driving economic cooperation with third parties (not only those operating within the EU, but also absent from the market) and fostering digital innovation and increasing consumer confidence in digital services not only across the EU, but also beyond. As it has developed, the market has begun to play a key role in shaping Europe's digital landscape and European digital identity, promoting unity while addressing the challenges posed by a rapidly changing global digital economy. The aim of this study was to examine the impact of the DSM in the European Union on the process of Europeanisation *ad extra* (external). The findings of the study suggest that the process of digital Europeanisation *ad extra* is taking place through regulatory extraterritorialism and by means of digital external governance, including algorithms governance, and can contribute to a more perceptible unification of this environment and mitigate the risks of potential international regulatory conflict. Regulatory extraterritorialism is the normative pressure-based spillover and enforcement of digital regulations *acquis* beyond EU borders. On the other hand, digital extraterritorial governance is primarily aimed at influencing digital platforms, which are defined as gatekeepers and as VLOPSE (G/V) in EU legislation, as well as third countries and international organisations. Large G/V digital platforms absorb EU regulations (DA) in order to be able to offer their tools and services within the EU (logic of advantage), but this also means that the requirements for these platforms (conditionality and coercion), and new effects may be different in different markets outside the EU (e.g. USA, China, etc.). In practice, digital platforms operate under several separate regulatory regimes, adding to the complexity of the global digital environment.

Digital Europeanisation *ad extra,* while defining the objectives and scopes of actions targeting the external environment, also affects actors outside the EU, but its sources are intra-EU (endogenous). The impact of *ad extra* the DSM on the G/V is in practice either direct or indirect. The direct impact is through the aforementioned designation of G/Vs and their admission to the EU Internal Market, provided they fulfil their obligations under the DA. The indirect impact is through G/Vs, which tend to adopt global or pan-European business models and algorithmic structures. The DSM exerts pressure on third parties concentrated in G/V groups primarily on issues of digital security (including the fight against disinformation), the use of artificial intelligence, the protection of citizens' rights and social minorities, and the standardisation and monopolisation of digital services. This multi-directional external impact of DSM is to take place while keeping the Internet free and open. One might be tempted to say that the multi-level and multi-directional external impact of the DSM becomes part of the EU's tech diplomacy, or the EU's technical diplomacy, which is based on the EU's positive impact in the broad digital ecosystem (impact in digital ecosystem) favours the cross-border transfer of digital content. As a result, we can speak of digital multi-level external governance (DMLEG).

References

Cianciara, A., Burakowski, A., Wódka, J., & Olszewski, P. (2015). *Europeizacja partii politycznych i grup interesu w wybranych krajach Partnerstwa Wschodniego i kandydujących do Unii Europejskiej*. Warsaw.

Communication from the Commission to the European Parliament, the Council, the European Economic and Social Committee and the Committee of the Regions (2015). *A digital single market strategy for Europe* (COM/2015/192 final). Brussels.

Communication from the Commission to the European Parliament, the Council, the European Economic and Social Committee and the Committee of the Regions (2022). *Establishing a European declaration on digital rights and principles for the digital decade* (COM(2022) 27 final).

Communication from the Commission to the European Parliament, the Council, the European Economic and Social Committee and the Committee of the Regions

(2021). *2030 digital compass: The European way for the digital decade* (COM/2021/118 final).

Council regulation against barriers to e-commerce (2018). COM/2015/0192 final. Brussels.

Darvas, Z. (2023). *The European Union's remarkable growth performance relative to the United States*. Breugel.

Decision (EU) 2022/2482 of the European Parliament and of the Council of 14 December 2022 establishing the Digital Decade Policy Programme 2030. *Official Journal of the European Union, L* 19, 323–324.

Engler, A. (2023, April 25). The EU and U.S. diverge on AI regulation: A transatlantic comparison and steps to alignment. Brookings. https://www.brookings.edu.

EU Code of Practice on Disinformation (2024, January 8). Digital-strategy.ec.europa.eu.

European Commission and The White House Council of Economic Advisors (2023). *The impact of workforces in the European Union and the United States of America*. Washington, D.C.

European Commission (2022). *Draft standardisation request to the European Standardisation Organisations in support of safe and trustworthy artificial intelligence*. Brussels. https://www.europa.eu/docsroom/document/.

European Commission (2024, October 29). *European Innovation Council (EIC) work programme 2025* (Decision C9(2024)7451).

European Commission (2024, February 19). Commission opens formal proceedings against TikTok under the Digital Services Act. Brussels.https://digital-strategy.ec.europa.eu/en/news/commission-opens-formal-proceedings-against-tiktok-under-digital-services-act

European Commission (2024, April 22). Commission opens proceedings against TikTok under the DSA regarding TikTok Lite. Brussels. https://digital-strategy.ec.europa.eu/en/news/commission-opens-proceedings-against-tiktok-under-dsa-regarding-launch-tiktok-lite-france-and-spain

European Commission (2024, December 17). Commission opens formal proceedings against TikTok on election risks under the Digital Services Act. Brussels. https://digital-strategy.ec.europa.eu/en/news/commission-opens-formal-proceedings-against-tiktok-election-risks-under-digital-services-act

European Commission; High Representative of the Union for Foreign Affairs and Security Policy (2020, December 16). *The EU's cybersecurity strategy for the digital decade* (JOIN(2020) 18 final). Brussels.

Kostrzewska, A. (2024, November 22). Komisja wszczyna formalne postępowanie przeciwko Temu. *Applia.co.uk*; *Kurier Szczeciński*, 29.12.2024.

Meltzer, J. P., & Tieleman, A. (2022, June 1). The European Union AI Act: Next steps and issues for building international cooperation in AI. Brookings. https://www.brookings.edu/research.

Monitoring of the European Declaration on Digital Rights and Principles – Support study (2024, December 1). Digital-strategy.ec.europa.eu; *Politico.*

Regulation (EU) 2016/679 of the European Parliament and of the Council of 27 April 2016 on the protection of individuals with regard to the processing of personal data and on the free movement of such data and repealing Directive 95/46/EC (General Data Protection Regulation). *Official Journal of the European Union, L* 119, 4.5.2016. (Applicable from 25 May 2018).

Regulation (EU) 2022/1925 of the European Parliament and of the Council of 14 September 2021 on contestable and fair markets in the digital sector and amending Directives (EU) 2019/1937 and (EU) 2020/1828 (Digital Markets Act). *Official Journal of the European Union, L* 265, 12.10.2022.

Regulation (EU) 2022/2065 of the European Parliament and of the Council of 9 October 2022 on the digital single market for services and amending Directive 2000/31/EC (Digital Services Act). *Official Journal of the European Union, L* 277, 1–55.

Regulation (EU) 2022/868 of the European Parliament and of the Council of 30 May 2022 on European data governance and amending Regulation (EU) 2018/1724 (Data Governance Act). *Official Journal of the European Union, L* 152, 3.6.2022.

Regulation (EU) 2023/2854 of the European Parliament and of the Council of 13 December 2023 on harmonised rules on fair access to and use of data and amending Regulation (EU) 2017/2394 and Directive (EU) 2020/1828 (Data Act). *Official Journal of the European Union, L* 2854, 22.12.2023.

Regulation (EU) 2024/1183 of the European Parliament and of the Council of 11 April 2024 as regards establishing the European Digital Identity Framework. *Official Journal of the European Union, L* 1183, 30.4.2024.

Regulation (EU) 2024/1689 of the European Parliament and of the Council of 13 June 2024 on establishing harmonised rules for artificial intelligence and amendments to Regulations (EC) No 300/2008, (EU) No 167/2013, (EU) No 168/2013, (EU) 2018/858, (EU) 2018/1139, and (EU) 2019/2144 and Directives 2014/90/EU, (EU) 2016/797, and (EU) 2020/1828 (Artificial Intelligence Act). *Official Journal of the European Union, L* 1689, 12.7.2024.

Regulation (EU) 2024/2847 of the European Parliament and of the Council of 23 October 2024 on horizontal cybersecurity requirements for products with digital elements and amending Regulations (EU) No 168/2013 and (EU) 2019/1020 and Directive (EU) 2020/1828 (Cyber Resilience Act). *Official Journal of the European Union, L* 2847, 20.11.2024.

Richter, F. (2024, March 8). The six gatekeepers regulated by the Digital Markets Act. Statista. https://www.statista.com.

Ruszkowski, J. (2010). Europeizacja ad extra w zarządzaniu zewnętrznym (external governance) Unii Europejskiej. *Rocznik Integracji Europejskiej*, 4.

Ruszkowski, J. (2010). *Ponadnarodowość w systemie politycznym Unii Europejskiej*. Warsaw.

Ruszkowski, J. (2019). *Europeizacja: Analiza oddziaływań Unii Europejskiej*. Warsaw.

Schimmelfennig, F. (2007). Europeanization beyond Europe. *Living Reviews in European Governance*, 2(1).

Smith, M. (1996). The European Union and a changing Europe: Establishing the boundaries of order. *Journal of Common Market Studies*, 34(1).

The Digital Markets Act: Ensuring fair and open digital market (n.d.). European Commission. https://commission.europa.eu (accessed January 23, 2025).

The economic opportunity of generative AI in the EU: Beyond the hype – boosting EU competitiveness with generative AI (2024, September 30). Implement Consulting Group (commissioned by Google).

The European Commission (n.d.). Can I be subject to automated individual decision-making including profiling? https://commission.europa.eu/law-topic/data-protection/reform (accessed January 2, 2025).

The future of European competitiveness. (2024, September). European Commission, p. 24. https://commission.europa.eu (accessed January 2, 2025).

Treaty on the Functioning of the European Union (TFEU).

What is the Digital Single Market about? (n.d.). Ec.europa.eu.

White Paper on how to meet Europe's digital infrastructure needs. (2024, February 21). COM(2024) 81 final. Brussels.

CHAPTER 2

European Council (EUCO): Group Dynamics, Behaviour, and Their Impact on Decision-Making

Serban Brebenel and Christian Kaunert

Abstract This chapter analyses and explains group dynamics and behaviour in the European Council (EUCO). This chapter provides an in-depth explanation of decision-making based on theoretical ideas about group dynamics and the organisational culture of this institution. It identifies EUCO as a group where representatives of the Member States, the heads of state and government, respectively, meet with the joint objective of reaching a common decision, realising their political positions and internal member states' considerations. The chapter analyses actual negotiations in two case studies, dealing with negotiations regarding the Multiannual Financial Framework and Ukraine's candidature to join the EU, demonstrating that the European Council represents a group with distinct particularities and dynamics.

Keywords European Council (EUCO) – group dynamics – group behaviour – EU institutions – negotiations

1 Introduction

The European Council is the key decision-making institution of the European Union. The 27 EU heads of state and government meet and provide political guidance to the other institutions in Brussels. This guidance eventually translates into concrete actions and measures, including EU primary and secondary legislation that all member states implement. While separate institutions, the European Council and the Council of the EU, which each Member State chairs for 6 months, with Poland exercising this function from January to June 2025, work closely together. The Council works on, makes decisions and implements, putting into practice, the political guidance of the European Council.

To exemplify, the European Council meeting of 26–27 June 2025 included on its agenda the topic of EU competitiveness, a broader term to refers also to the twin digital and climate transitions. The discussion on this topic was even-

HTTPS://DOI.ORG/10.1163/9789004763623-004

tually deferred to the European Council meeting of 23 October. Section IV of the European Council Conclusions specifically refer to the "competitiveness and twin transition". A separate section within section IV refers to "a sovereign digital transition", paragraphs 48 to 52 of the overall Conclusions. These substantial paragraphs present the overarching political guidance for increasing EU's competitiveness through digitalisation (European Council Conclusions, 2025). This means reinforcing EU's sovereignty in the field, "addressing risks arising from hybrid threats, cyber security challenges, and strategic dependencies" or increasing innovation (European Council Conclusions, 2025).

Building on this guidance, in the second half of 2025, the Council of the EU adopted several key packages of conclusions to take the work forward. Of relevance were the conclusions to strengthen Europe's digital competitiveness (December 2025), covering many of these topics, with an increased focus on cyber-attacks. The conclusions on a strategy for the EU's engagement in global digital affairs, adopted in November 2025, and putting forward the principles for including the digital sector in EU's external actions are also an example of how the Council of the EU translates into action the political guidance of the European Council, also in the digital sector. To reach decisions that give the overall direction of the EU in internal and external affairs, the 27 members of the European Council meet as a group to discuss every three months (in theory) or more (in practice) the topics that are part of the agenda of meetings. This chapter, thus, argues that the most critical decision-making institution in the EU, the European Council, is a group of individuals bearing particular characteristics, dynamics and behaviour. To substantiate this, the chapter first defines the characteristics of the group and analyses its structure, the types of intragroup interaction, and the types of interdependence. Interdependence is a common denominator of what a group represents within an organisation (Jex & Britt, 2014). Smaller intragroups formed by Member States within the EUCO follow the same objectives. As Jex and Britt (2014) showed, the outcomes of the group depend on each member of the collective. Jex and Britt (2014) also list other critical defining characteristics of a group, such as social interaction and the commonality of purpose. These are all relevant for the case in point of EUCO, where individuals meet, socially interact over the length of the reunions, and have a commonality of purpose: to reach a common decision, a joint common position for the EU. The latter is what they are trying to accomplish collectively, namely, to reach conclusions and adopt the way forward.

The chapter also examines the existing literature on group structure by replicating this to the specific groups represented by EUCO. Several elements are relevant for analysing the group structure: roles, norms, values, communication patterns, and status differentials. There are specific roles in EUCO, including that of the dean of the group (and how this links with the role of the leader) and that of the president, as well as those linked with the power representation of each Member State. Status differentials are also linked to roles and power relations within these two groups. Norms can be assimilated into procedures, while some overall norms and values govern the decisions made by these institutions. Through case studies, the chapter examines situations when norms are broken and what happens to the group members (Dentler & Erickson, 1959; Levine et al., 2001). The chapter also looks at communication patterns and discourse elements to describe how EUCO communicates internally and externally. This analysis of group structure leads to a study of group interaction, as reflected in EUCO. This starts from Thomas's (1976) proposed matrix regarding intergroup interaction that, according to him, depends on the degree of compatibility between the goals of the different groups and the degree to which the groups must interact, respectively. The resulting situations are competition, collaboration, avoidance, accommodation, and compromise. The chapter aims to reflect on these negotiation situations within EUCO and explain how various forms of intragroup behaviour occur.

The current literature proposes several studies that deal with the functioning of EUCO. Some are constructed as an introductory handbook to the functioning of the EU and, in this context, provide a presentation of the interaction between institutions in the decision-making process. In this category, one can include books such as "the Oxford Handbook of the European Union" (Jones, Menon & Weatherill, 2012), "The European Union Decides" (Thomson et al., 2009), "How the EU Institutions Work" (Akse, 2020), and "Understanding the European Union" (McCormick, 2020). Some books regarding EU institutions and the decision-making process go into more detail. They are more focused on the factors that impact the policy process (Versluis, van Keulen & Stephenson, 2010), as well as on aspects related to governance (Sabel & Zeitlin, 2012). At the same time, others propose a comprehensive perspective as regards the EU. For example, Lelieveldt and Princen (2023) use the tools of comparative politics to explore the history, theories, institutions, key actors, politics, and policymaking of the EU. Some books regarding EU institutions were explicitly developed to cover certain institutions, such as EUCO. The categorisation is like the one presented previously. Some, such as Westlake et al. (2004), favour a more general, descriptive approach, covering elements such as the legal basis, composition and preparation, and process (de Schoutheete & Wallace, 2002), while others (Wessels, 2015) go into more detail and the books encourage an in-depth understanding of each institution. Some works provide an analysis of a particular aspect related to one of the institutions, such as Tallberg's analysis of the bargaining power in EUCO (2007) or the role of agenda-shaping in decision-making (Tallberg, 2003).

The primary purpose of this chapter is to analyse decision-making in EUCO as it relates to group dynamics and group behaviour. The premise is that the EUCO is a group of individuals representing the heads of state and government of the 27 EU Member States. As such, EUCO, as a group, has many of the characteristics one recognises universally in other groups, such as the elements of the group structure. Furthermore, the group comprises various smaller groups, which is best reflected during negotiations on multiple topics. Depending on the issues discussed, member states, represented by their elected presidents or prime ministers, group themselves in discussions, following similar group objectives and aiming to obtain a compromise solution as close to their position as possible. Previous literature has focussed on EUCO and its functionality as a European institution. This chapter, however, aims to take a different investigative approach and looks at EUCO as a group with its characteristics, structure, dynamics, and behaviour.

This analysis is divided into three main sections. First, we look at the characteristics of EUCO as a group, focusing on its interdependence, perception of the group, and commonality of purpose. Subsequently, we discuss the structure of EUCO as a group and its aspects such as roles, norms, and communication patterns. Finally, we also discuss types of intergroup interaction, focusing on negotiation as the main instrument of communication to reach a compromise solution. These theoretical considerations are then exemplified in two case studies dealing with negotiations regarding the Multiannual Financial Framework and Ukraine's candidature to join the EU.

2 Theoretical Considerations: EUCO as a Group

Definitions of what a group is vary. Interdependence is an essential element that transforms a collection of people into a group (Jex & Britt, 2008). Interdependence means that the potential outcomes of the group are based on the different participations and interactions of the members and that, in various ways, individuals' actions will affect the group's performance. Forsyth (2009) also points out the intrinsic connections between the group members, which are essential for outcomes and determine specific actions, feelings, or experiences between members. At the same time, he also shows that interdependence is not necessarily equal in the group. In a unilateral interdependence, one member of the group influences others. In sequential interdependence, one member influences another, who, in turn, influences the next (for example, when it comes to an assembly line activity, where each worker needs to complete his part of the work before the other can do his). Reciprocal but unequal interdependence is also a particular situation. These are all relevant for the empirical analysis because EUCO's complex and dynamic nature allows all these dependencies to manifest themselves at different moments. These variations will depend on the discussed topic, the formation of the institution and its members, which changes regularly, and external dynamics such as foreign policy developments.

Social interaction is another characteristic of a group, and it refers to "verbal and nonverbal communication" (Jex & Britt, 2008). However, relevant to EUCO meetings, various determinants can affect social interaction, including intensity, seating arrangements, communication patterns, or controllability (Forsyth, 2010). A third defining characteristic is the internal and external perception of being a group, also closely linked to entitativity. Forsyth (2009) defined this concept, referring to Campbell (1958), as "the extent to which an assemblage of individuals is perceived to be a group rather than an aggregation of independent, unrelated individuals". Finally, the last characteristic of groups is the commonality of purpose. This means the group aims to collectively accomplish an objective (Jex & Britt, 2008). The group has some common objectives on which they are working to achieve. In the case of EUCO, meetings can be formal or informal. Still, the result is the same: the meetings take place to discuss topics of interest for the EU and to reach decisions on these particular subjects. The purpose is to translate these group characteristics to EUCO and determine how these operate in this particular case. The first essential characteristic that defines EUCO as a group is its *interdependence*. Several elements reflect this. The members of EUCO have a common objective: to provide political guidance for the European Union. Art.15(1) of the Treaty on the EU describes this broadly. Still, the main objective of EUCO is to provide strategic and political direction for the EU's future development and actions.

The interdependence is also reflected in Art. 15(4). Decisions in EUCO are taken by consensus. There is no effective voting, but the European Leaders comment, objecting when they disagree, until a compromise is reached. If this is not the case or not possible, the decision is deferred to another reunion. In critical cases, such as adopting the Multiannual Financial Framework (MFF) and the EU's budget, the Heads of State or Government will remain in meetings as long as necessary to reach the needed consensus. In line with the general acceptance in the EU, even in cases where unanimity is not the voting rule, consensus, namely the understanding that all member states agree with a solution, remains the preferred way forward. This is even more the case for EUCO, where its members represent the executive leadership of their member states. Given this, the interdependence between the members of EUCO becomes even more evident. The 27 members of EUCO, its President and the President of the Commission need to work together to achieve a consensus. When this is not possible, solutions need to be found among the members, and this can only be done through interaction, both in the official, formal sessions and in informal meetings on the margins. Derogation is sometimes a solution that can be identified. In December 2019, EU leaders endorsed the objective of making the EU climate-neutral by 2050, in line with the Paris Agreement. One member state, Poland, could not commit to this objective, so leaders had to return to the issue in June 2020.

Regarding the *internal and external perception as a group*, EUCO is seen by all outside actors and individuals as a group of 27 heads of state and government representing one of the 27 member states of the European Union. They perceive the group as such. This results less from the personal individual in that particular member state position for some time and more from the institutional framework defined by the EU Treaties. According to these, each member state is allocated a seat in EUCO. The head of state or government sits in that seat when he has this national capacity. The projection is thus less on the individual as a member of the group and more on a transposition that the individual changes, but the position as such remains inside the group. Some elements that come with that particular member state's elements as part of group characteristics will remain unchanged. In contrast, those personal ones will be reflected in some respects, such as social influence.

Regarding the *commonality of purpose*, this concept reflects that the group aims to accomplish an objective collectively. Previously, we have shown that EUCO gives political guidance regarding the way forward for the EU. What this means in practice is that EUCO invites other institutions (this is generally the polite term used in EUCO Conclusions as opposed to actual tasking) to go forward with particular legislation, to adopt specific policies, to look into ways that relations with a third country can be further developed, etc. This commonality of purpose means that EUCO guides the EU in the short-, medium-, and long-term, and fundamentally, it reflects that EUCO is the main decision-making body of the EU.

3 Group Structures in EUCO

The group structure is a "set of dimensions along which any group can be described" (Jex & Britt, 2008). This section will focus on roles, norms, values, communication patterns, and status differentials. *Roles* are "coherent sets of behaviours expected of people in specific positions within a group or social setting" (Forsyth, 2009). As such, roles are more than tasks that individuals assume or assign within the group; they are also about how individuals act about these tasks and how they interact with other participants in the group. At the same time, while the set of behaviours is generally well-defined, it is not strict. Forsyth (2009) presents an excellent example of the actor's role in a play. While the role is defined by some general coordinates, such as what defines Juliet in a play, it is up to the actor to adapt to tone, recitation, and overall performance. This similarly happens in the group.

Roles generally come about in the process of *role differentiation*. This starts with the conclusion that the group can be more efficient in achieving its objectives if people do different things than the entire group performing the same activity. All groups have a leader figure, which provides the overall coordination of the group's activity and ensures that this goes towards a final positive goal. This can often be either formal (the President of EUCO is the formal leader of the group) or informal (the Dean of EUCO, for example, is the head of state or government that has been longest present in EUCO and benefits from an informal leadership of the process). The leader figure generally joins other persons in the group that revolves around a task role: organising the group, providing expertise in a specific direction, giving opinions, and providing evaluations. The task role is one of the typologies Benne and Sheats (1948) proposed. The other two roles are the socio-emotional role or the relationship role, which aims to "satisfy the emotional needs of the members" (Forsyth, 2009), and the individual roles, generally those who stand out from the group in terms of their objectives, its involvement, and its socialisation.

In EUCO, roles are linked to tasks the members of EUCO need to fulfil. In principle, each of the members of this group has two tasks: (1) defend the interests of the member state in the group, as endorsed by voters in national elections and by various other national stakeholders, and (2) defend the interests of the EU and ensure that the negotiations and discussions, as related to point 1, lead to decisions and conclusions that are in that note. As such, the members of EUCO already attend all the meetings, fulfilling a role that stays the same for the duration during which they still have this national role before the next election. Their role is the head of state or government of one of the member states. The other roles that are filled are that of the President of EUCO and that of the President of the European Commission, as per Art. 15(2), that has been discussed previously. The High Representative of the Union for Foreign Affairs and Security Policy takes part in the meetings of EUCO and, as such, he has an assigned role as well (other than attending), which is, upon request of the members, to brief on particular situations or developments regarding his area of action and competence. In EUCO, some roles are fixed, and some depend on the tasks at hand, namely on the topics under discussion and on which some of the members can provide more elements, given the interests of the member state in a specific topic. Some fixed roles include the president, the dean, and the agenda setter. The President of EUCO has the role of convening and chairing EUCO meetings. It is not a leadership role but rather a primus inter pares. His informal role is, however, much more complex: by chairing the meetings, his role is to guide the discussion towards consensus, to obtain solutions, sometimes also through informal sessions on the margin of the official reunion, and to aim for a favourable resolution of the discussion.

Sometimes, more than the President of EUCO is needed for this and the informal role of guiding the discussions towards a successful completion is taken by someone with influence in the group. This could be the dean of EUCO, but this is not always enough. The former German Chancellor Merkel had served without interruption for the past 15 years as the German Chancellor and has represented Germany in this capacity at EUCO reunions. This type of experience in EUCO had a distinct impact on her expertise as a negotiator and compromise seeker in the group. Her expertise on files, such as the Multiannual Financial Framework (MFF), where she had been previously engaged in negotiations for the 2021–2027 cycle and the previous MFF cycles starting with 2005, and her representing Germany only increased her clout. Without taking over the President's role, she could help forge a compromise on complex negotiations. Some members of EUCO, such as Prime Minister Orban, are veto players. For Tsebelis, veto players with greater control over the possibilities to replace the status quo are labelled agenda setters (Tsebelis 2002, p. 13). In this role, Prime Minister Orban blocked a consensus on opening accession negotiations with Ukraine. Depending on the task/topic discussed during the meetings, there are also examples of temporary roles, such as the role of the influencer. The interest of a member state in a topic can give that member of EUCO a temporary influencing role in shaping discussions. When EUCO discusses Turkey, for example, the expertise of the Cypriot members gives it particular importance, as these members will be able to influence discussions. In terms of the task or topic being discussed, there are also thematic roles in EUCO, such as, for example, the human rights defenders (heads of state and government that take a keener interest in this subject and, as such, their interventions include elements regarding the protection of human rights) or climate promoters.

Another dimension relevant to group structure is group norms. *Norms* are "emergent, consensual standards that regulate group members' behaviours" (Forysth, 2009). The key word is "regulate", while the complementary ones are "emergent" and "consensual". Norms regulate the group's activity, the interaction between the members, and how the individuals participate. They define what is appropriate, as well as what is prohibited. They are also emergent because they are not imposed outside the group but constructed through the members' interaction and eventual compromise-building. They are also consensual because, in principle, they are accepted by all group members. As for norms as part of the group structure of EUCO, two categories can be identified: formalised and informal. Both categories of norms have the same role: to regulate the group's functioning. The formalised norms are laid down in EUCO Rules of Procedure, which EUCO adopted through a Decision of 1 December 2009 (European Commission, 2009). These formalised norms regulate the place of meeting (in Brussels; only exceptionally, the President of EUCO, with the agreement of the General Affairs Council or the Committee of Permanent Representatives, acting unanimously, may decide that a meeting of EUCO will be held elsewhere), the frequency of meetings (twice every six months, convened by its President; when the situation requires, a special meeting can be convened) and the way decisions are taken (by consensus, except where the Treaties provide otherwise).

The Rules of Procedure also reflect the norms regarding the preparation of EUCO, including how the agenda is defined and adopted, and even norms regarding the total size of the delegations authorised to have access to the building where the meeting of EUCO is held (20 persons for each member state and the Commission, and to five for the High Representative of the Union for Foreign Affairs and Security Policy.). An important norm is also that the meetings of EUCO are not public (although Politico, the famed information instrument of Brussels, is always able to present what the leaders are eating for dinner (Barigazzi, 2024) but also how discussions evolve in their live blog during the reunions) (Barigazzi & Martuscelli, 2023).

The informal norms are not laid down or adopted but are known to the members, who act upon them in their meetings to reach decisions. These complement the formalised norms and permit the proper works of the reunion, facilitating consensus. One such norm is that, for example, when blockages occur regarding the reunion's outcome, the group members may have informal meetings in smaller groups to try and work out differences. Another informal norm is that meetings regarding foreign policy are generally held over dinner in a classified environment. When one of the members of EUCO leaves the group, it is usually the dean of the group that has a farewell speech.

These different roles and how they develop within the group, depending on the impact on achieving the group's objectives and other factors, lead to status differentials, another dimension of group structure that will be briefly described. As Jex and Britt (2008) point out, there are two large categories of status differentials: diffuse status characteristics and task-specific status characteristics. The former is not strictly or directly related to the group's task. The examples given refer, for instance, to societies in which professional occupations such as medicine are awarded a higher status than manual labour. At the same time, logically, status in the group is also granted, given the individual's impact on the task's achievement. Citing also Strodtbeck and Lipinski (1985), Jex and Britt highlight that an individual is likely to have a higher status in the group if his contribution is more significant.

Status differentials lead to various consequences. One example is a different reward level for the individual perceived as contributing more to the group's activity. Another is linked to respect for the norms. Higher status in the group potentially allows for greater derogation from the group's standards. Citing Hollander (1971) Jex and Britt (2008) refer to idiosyncrasy credits. Like financial credits, these are banked and can be used by group members whenever they need to break a norm. In EUCO, status differentials are linked to influence in the group. It is both given by status characteristics, such as the aggregate structural power of the member state that the head of state or government represents, but also by task-specific elements, such as the knowledge of the topic, the social clout of the member of EUCO, and the diffuse influence he or she can exercise within the group. Evidence shows that while all EUCO members are equal, interdependence is not equal in this group. Tallberg (2007) notes that each leader carries the same weight in EUCO, given the principle of consensus in the decision-making process. However, the difference is in the influence a leader can exercise during negotiations to produce an outcome that reflects his or her starting negotiating position most closely. This influence and member state bargaining power in EUCO is captured, according to Tallberg (2007), by three dimensions: state sources of energy, institutional sources of power, and individual sources of power.

Tallberg's description of aggregate structural power derives from realist theory, which proposes that aggregate structural power results from the respective state's resources and capabilities. Population, economy, governance, and political stability are components of aggregate structural power. The general assumption proposed by realist theory and Tallberg reflects positioning and the relationship with other member states in defending the respective positions. Member states with greater aggregate structural power will have greater bargaining power and can manoeuvre negotiations towards a compromise closer to their starting negotiating position. Translating this to EUCO, member states with more considerable aggregate weight can better support their negotiating mandate, push for a successful solution, and use various tactics, such as tabling their proposal in the lack of an institutional one. Tallberg is also keen to reflect that state sources of power are not solely formed of aggregate structural power but also of issue-specific power, which tends to balance the bargaining framework more. He argues that if a member state's resources are sufficiently concentrated on one particular issue, this could cover the existing gap in terms of aggregate structural power, permitting smaller and medium member states to gain in negotiations on these specific issues. Issue-specific power and influence are reflected in the higher commitment and higher stakes of a member state on a particular problem that allows that state to try to obtain the most out of the negotiations.

Tallberg divides the institutional sources of power into two categories: the power of the veto and the power of the chair. Each reflects roles in EUCO, so these sources will be discussed when referring to these roles. Individual sources of power strongly affect the influence of EUCO members. Tallberg considers personal authority and expertise to be the two elements that could impact member states' bargaining power. Personal authority is a synergy of various components, such as seniority, personal trust, and personal relations with the other leaders in EUCO. As a complementary comment to Tallberg's studies, seniority also reflects political success in the respective Member states, showing that the respective leader's repeated presence in EUCO results from his or her success in national elections. In Brussels, this is an important matter, as well as aspects such as the political families that form the European Parliament and of which the leaders and their respective parties are part. Aspects of seniority and additional qualities that form personal authority are also critical individual power sources.

Tallberg also highlights expertise as one element that shapes individual power. Often, negotiations at EUCO are also related to technical subjects and require technical expertise to process these. Perhaps no example is so clear as the EU budget or the Multiannual Financial Framework. A higher technical understanding of some of these aspects increases leaders' negotiating and bargaining power in EUCO meetings since they can discuss in more detail and in a more informed manner. Information is an instrument of power that balances the weight of smaller and medium Member states in EUCO negotiations. Tallberg highlighted two primary sources of power: personal authority and expertise. A past member of EUCO, Angela Merkel, embodied both these qualities and, complementary to other factors, wielded the most significant influence in

EUCO. Others benefit from being former ministers in various fields with increased sectorial expertise.

What Forsyth (2009) labels as a social influence – "interpersonal processes that produce, sometimes directly, but often very subtly and indirectly, changes in other people" – is less critical in the decision-making process in EUCO. Social influence is generally based on peer pressure and the capacity to change the decisions of others in the group in this manner. The members of EUCO are less prone to this type of influence. This is, in principle, because of their political influence, having been elected in nationwide elections and, as such, representing the interests of their member state. The influence and the way this is propagated in EUCO are much more diffuse and complex, with the overall intention of reaching a consensus within the group. Concessions are sometimes minimal, and negotiating positions are sometimes immovable precisely because of the numerous national stakeholders and the responsibility the leaders have towards them.

However, even leaders will tend to look at those in the group they perceive as having more influence. Compromise may sometimes be acceptable around a general direction towards which such a leader with influence (or a group of leaders) could aim. This compromise would generally not be the same as any of the positions, but represent a way forward. Sometimes, social influence complements the other forms of influence discussed, allowing an otherwise smaller member state to increase its influence within the group, notwithstanding its aggregate power. Former Prime Minister of Luxemburg, J.P. Juncker, was well known for punching higher than the aggregate power of the member state he represented. His social influence combined institutional knowledge, sectorial and economic knowledge (he has served as the Minister of Finance in Luxemburg and President of the Eurogroup), and social presence in the group. By 2013, he was also the longest-serving head of any national government in the EU. Forsyth (2009) also mentions the concept of conformity in the context of majority influence. This means that people tend to agree to something if many others in the group follow the same idea to avoid standing out. This seldom happens in EUCO. The political background of the leaders and their role as elected representatives of member states limit this possibility. The example given previously, with Prime Minister Orban, is eloquent. There was no intention of conformity on his side when the decision to open accession negotiations with Ukraine was discussed. He preferred to leave the room, allow the decision to be made and return for the rest of the points on the agenda of discussions. In Nail, MacDonald, & Levy's model (2000), the Hungarian Prime Minister can be labelled as an independent: he disagreed with the group before the discussion (as he publicly expressed on many occasions) and disagreed openly and privately with the group after each debate. His opinion cannot be changed; his standards and positions are clear.

Communication patterns or networks are another way a group can be described. These regular patterns determine how the information is exchanged between the group members. The type of communication pattern varies, depending on the number of persons in the network and the centrality of one of the positions. As such, a communication network with a high degree of centrality is the hub-and-spokes model (Shaw, 1964), which plays one position in the middle of the communication channels. Decentralised channels include the "Comcon", where information moves back and forth between the members of a penta-shaped network. As Forsyth (2009) points out, citing Leavitt (1951), the type of communication pattern also provides information about role attribution in the respective group. For example, a hub-and-spokes model, with a high degree of centrality, would tend to reflect that the leader is the core communicator and is placed in the middle of the model. Similarly, those away from the central communicator are likely to play marginal roles in the model and, simultaneously, have a higher chance of leaving the group because of more limited access to information.

In EUCO, communication patterns are not easy to reflect on because of their specific characteristics. It is a centralised structure, like the wheel communication pattern. However, the wheel pattern is described as everybody talking to the central person of the group but not talking among themselves. This differs from the communication model that EUCO proposes. It could thus be assimilated also to a Comcon (a decentralised structure), but with a central person who plays the hub position role, which is to "use the hub position as the data processing centre, and its occupant typically collects information, synthesises it, and then sends it back to others" (Forsyth, 2009). This hub position belongs to the President of EUCO because his role in managing the reunions also involves guiding discussions towards consensus, obtaining and disseminating information from other participants, observing where the compromise could be achieved and, eventually, concluding the meeting on these lines. The other members of EUCO attend the reunions and intervene on the various topics of discussion, thus interacting during the formal parts of the reunions. They also have complementary and informal talks between themselves in various bigger or smaller groups.

EUCO communicates both formally and informally. It communicates to the public and other institutions through its EUCO Conclusions. These provide general political guidance, meaning that through its Conclusions, EUCO tasks the other institutions to undertake various measures to advance with different objectives. Each EUCO member communicates separately from the group through press statements, interviews (upon arrival and at the end of the reunions), etc. In general, this communication is for the internal audience and explains the positioning during the meetings and how this defended national preferences. Informally, members of EUCO meet regularly on the margins of meetings, thematically related to the political family or when they belong to geographical groups (for example, Hungary, Poland, Slovakia, and the Czech Republic occasionally meet as the V4 on the margins of important EUCO, such as those discussing climate or the multiannual budget).

These different roles and how they develop within the group, depending on the impact on achieving the group's objectives and other factors, lead to status differences, another dimension of group structure that will be briefly described. As Jex and Britt (2008) point out, there are two large categories of status differentials: diffuse status characteristics and task-specific status characteristics. The former is not strictly or directly related to the group's task. The examples given refer, for instance, to societies in which professional occupations such as medicine are awarded a higher status than manual labour. At the same time, logically, status in the group is also granted, given the individual's impact on achieving the task. Citing also Strodtbeck and Lipinski (1985), Jex and Britt highlight that an individual is likely to have a higher status in the group if his contribution is more significant.

4 Types of Intergroup Interaction

To exemplify some aspects of intergroup interaction, we now look at EUCO not as a group but as an organisation or institution formed of several smaller groups. This will allow us to analyse intergroup behaviour and interactions. The groups within the negotiations in EUCO are formed considering political affiliation and thematic affiliation. Belonging to a political family is essential in configuring a group in EUCO. They represent the main groups of parties such as the EPP, the S&D, Renew, etc. Leaders of the same political family meet before EUCO reunions to discuss general preparations for EUCO. For example, EPP leaders met in Brussels on 21 March 2024 before the EUCO reunion. EPP heads of state and government joined the President of the European Commission (also EPP) and the President of the European Parliament (EPP) in a morning meeting before the afternoon EUCO. EPP President Manfred Weber hosted the conference.

At the same time, negotiations in EUCO particularly emphasise the thematic criterion for configuring groups within the institution. The thematic criterion groups the heads of state and government according to their position vis-à-vis a specific topic. For example, when climate targets negotiations were held, a group of Central European members argued for a more flexible transitional approach, considering the different starting points and ensuring subsidies to support the economic impact. On the other hand, other member states that had started the green transition earlier advocated for a quicker pace and greater ambition from the EU. Generally, one tends to see different groups forming depending on the topics discussed: it is impossible that two heads of state or government share the same position across all topics. Configurations change. A corollary to this is the geographical criterion, which refers to groups that form depending on the geographical location of the member state. However, this criterion is trumped by the thematic criterion, mainly because the positions that member states have on various topics also reflect, in broad lines, their geographical location in the EU.

Using an interdependence -compatibility matrix, the conceptual framework identifies four principal modes of interaction: accommodation, avoidance, competition, and collaboration, with compromise occupying an intermediate position. Intergroup behaviour is generally located between collaboration and compromise, although occasional instances of competition can also be observed. The level of interdependence in EUCO is high: the members sit together in reunions where they must reach decisions that will guide the EU over the next period. Furthermore, they need to make these decisions by consensus. The overarching goal of the various groups in EUCO is to achieve this shared understanding in the interest of the EU. This is why collaboration, through constructive negotiations, is the general approach, where even if positions may seem different, the overarching objective pulls them together.

In most situations, however, there is moderate compatibility between the groups in EUCO, which usually leads to situations that can generally be mapped in the compromise area on the matrix. There are also situations of higher levels of competition, particularly when resources are discussed, such as the Multiannual Financial Framework negotiations. This does not necessarily mean that the competition is taken out of the collaboration–compromise context but that this situation of a zero-sum game will increase competition. Situations with higher competition between the groups in EUCO also occur on topics for which there is a potentially more considerable political gain.

Regarding the causes of intergroup conflict, we will refer to two of them, namely, goal incompatibility and time incompatibility. This has been covered previously on several occasions in terms of goal incompatibility. During negotiations in EUCO, various groups of heads of state or government find themselves in opposing positions. Suppose one refers to climate negotiations (groups of member states favour a faster approach to reaching the climate goals, while others prefer a slower transition). In that case, MFF negotiations (cohesion vs. frugal approach), Ukraine or negotiations regarding nominations for the leadership positions in the EU (all-out negotiations) always find various groups arguing for different, often opposing positions. There is a goal incompatibility. However, it is also more of a perceived incompatibility since the heads of state and government can find venues towards compromise and identifying a common position. Time compatibility, in the case of negotiations within EUCO, is often related to time constraints. EUCO occurs most often between Thursday at 3 pm and Friday at noon/afternoon. During this period of 24 hours, the heads of state and government need, in principle, to agree on the topics on the agenda. Time pressure often carries negotiations forward, as the leaders are more likely to agree to a compromise solution in the late night of the talks, during the early hours before adjourning, or before calling an end to EUCO, knowing that they will either go home or go to another official state visit. As such, time incompatibility creates the premises for conflict between the groups in EUCO; however, at the same time, it also makes a potential impulse to finalise negotiations through time pressure.

There is a general acceptance that group members tend to become more cohesive when the groups conflict. This is valid for negotiations in EUCO but needs to be appropriately nuanced, and it usually depends on the talks at hand, the degree of salience, and the importance attributed to that particular topic. If the topic is deemed of strategic and long-term significance, such as those regarding the EU's multiannual budget, you would tend to see the groups becoming more cohesive. This means the groups will coagulate around their common position and promote this in the negotiations. On other occasions, such as in discussions regarding leadership nominations, one may also see cohesion built around the political family lines as a fallback.

5 Case Study 1: A Case of Norm Violation – Viktor Orban's Blocking of Consensus

EUCO of 14–15 December 2023 decided to open accession negotiations with Ukraine and the Republic of Moldova. However, this was done against the will of Hungarian Prime Minister Viktor Orban, who, despite the unanimity, refused to agree to the opening of accession negotiations with Ukraine. The problematic situation was solved in a manner not previously used during EUCO, where, by norm, the decision is taken by consensus and unanimity, and, by law, this is mainly the case when it comes to foreign policy, security, enlargement, and finances. The Hungarian Prime Minister was asked to leave the room (reportedly by the German chancellor, something that will be further discussed below when referring to influence), allowing the 26 remaining leaders to make a unanimous decision. Abstention, as could thus be considered the case of Viktor Orban leaving the room, allows for the decision to be taken. The solution worked at that point, although this must be balanced, particularly since, as discussed, consensus is generally the rule in EU decision-making processes, particularly in EUCO.

Forsyth (2009) also mentions the concept of conformity in the context of majority influence. This means that people would agree if others in the group followed the same idea, to avoid standing out. Again, this seldom happens in EUCO for the reasons previously presented. The political background of the leaders and their role as elected representatives of member states limit this possibility. The example with Prime Minister Orban is eloquent in that sense. There was no intention of conformity on his side when the decision to open accession negotiations with Ukraine was discussed. He preferred to leave the room, allow the decision to be made and return for the rest of the points on the agenda of discussions. In Nail, MacDonald, & Levy's model (2000), the Hungarian Prime Minister can be labelled as an independent: he disagreed with the group before the discussion (as he expressed on many occasions) and disagreed publicly and privately with the group after each debate. His opinion cannot be changed; his standards and positions are clear.

Forsyth (2009) rightly asks what happens when norm violation is more intentional and occurs repeatedly. In other words, this seemed to be a one-only situation, and it is less conceivable that EUCO could continue to make decisions by asking one of its members to step out of the room whenever there is a controversy. The German chancellor made this point when she said that constructive abstentions should be limited to "exceptional cases" and that EUCO" should decide whenever possible and come together" (Von der Burchard, 2023). There are several ways forward, including changing the norm violator's behaviour, eliminating the violator, or allowing him to remain a marginal group member. Today, one can evaluate that the Hungarian Prime Minister, Viktor Orban, is a regular violator of the norms. Primarily, he opposes consensus and working towards achieving consensus. At the same time, he tends to find himself in antagonistic positions to the EU27 consensus on most issues, from imposing sanctions on Russia to opening accession negotiations with Ukraine. With the Hungarian Prime Minister, there are usually no possible negotiations to attempt to reach a compromise. His stance is almost always one that refuses a negotiation and a compromise. What to do in this case?

It is impossible to exclude a member of EUCO: the right of each member state to send its head of state or government to this institution is given through the Treaties. However, the others have increasingly taken the marginalising approach. It seems that marginalisation through temporary exclusion could be the way forward when there is no way to convince a group member to agree to an apparent compromise. However, given all the elements previously presented, this must be balanced.

6 Case Study 2: Aggregate vs. Issue-Specific Power – MFF Negotiations in 2020

By July 2020, the UK had already left the European Union and the group of member states with the highest aggregate structural power was reduced to Germany and France. They sought to steer member states through the post-Covid recovery process and, as the most significant economies, put together a joint proposal on May 18 (Bayer, von der Burchard & Smith-Meyer, 2020; also Germany Chancellery's press release no. 173/20 of 18 May 2020). This included a 500 billion euro Recovery Fund in the Next Generation EU. It implied that the Recovery Fund would provide this sum only in grants to Member states, while the remaining 250 billion of the Next Generation EU would be loans to Member states.

The approach that Germany and France took to the negotiations validates some of Tallberg's reflections, mainly that one tactic used by member states with higher aggregate structural power is to table their proposal. In this case, this did not occur instead of an institutional one (the prerogative of the President of EUCO), but ahead of his, to influence the final draft of the former.

In terms of issue-specific power, there were several groups of member states. First was the group of so-called frugal Member states: Denmark, Sweden, the Netherlands, and Austria. Finland joined informally later in the negotiations. These favoured a smaller budget, and a higher proportion of loans to grants strongly reflected the rule of law conditionality in the final agreement. This group of member states has a low aggregate structural power. However, they benefited from high issue-specific power and from pooling these together as a group. The issue-specific power also resulted from the strength of their mandate and the formation of national positions, as mentioned in the literature review of Moravcsik's work. For example, in the case of the Netherlands, the mandate of the Prime Minister was directly validated by the Dutch Parliament, where he would return to report on the results obtained in the negotiations. The fact that his governmental coalition was formed of several parties made the Prime Minister particularly keen to get into the talks as close as possible to the mandate given to him. He and the other frugal Member state leaders would also have to report to national parliaments as to the way the rule of law conditionality was reflected in the final package; these parliaments supported a strong conditionality and link between the respect for the rule of law and the funds received.

A second group of Member states included those hit the hardest by the Covid-19 crisis, particularly Italy and Spain. These member states have a higher-than-average aggregate structural power. Still, their argumentation in the negotiations was supported by the particular interest in having the financial package approved as quickly as possible and with the amounts proposed by the President of EUCO. Additionally, their interest in the negotiations is to have a streamlined procedure when approving the national recovery plans and fund disbursements. The issue-specific power of their negotiating position was further strengthened because all Member states agreed that funds had to be approved as quickly as possible to help all Member states, particularly those that had suffered the most. In a way, the issue-specific power was further complemented by the morality of the issue.

An interesting third group of Member states, further reflected upon when discussing the institutional power sources, were Poland and Hungary. This group did not agree with a rule of law conditionality that would allow the European Commission to propose the suspension of funds in situations where it judged that there were breeches against the rule of law in Member states that would affect the financial interests of the Union. Poland and Hungary believed this would be a subjective instrument in the hands of the Commission and would not agree to language in EUCO Conclusions referring to the rule of law. As can be seen, this group has a small aggregate structural power. Its strength in the negotiations on this topic relies on the issue-specific power (exceptionally high interest in the rule of law component), also deriving from the formation of the national position. In the case of the Netherlands, previously mentioned, the Hungarian Parliament had adopted, before the EUCO meeting, a resolution that strongly argued against any such conditionality. The strength of the group's position also resulted from institutional power sources, as presented below.

Sometimes, creativity leads the decision-making process towards areas that some could call a dangerous precedent, although others refer to it as a creative solution: constructive abstention. In December 2023, Hungarian Prime Minister Viktor Orban refused to agree to the opening of accession negotiations with Ukraine. He was asked to leave the room (reportedly by the German chancellor, something that will be further discussed below when referring to influence), allowing the 26 remaining leaders to take a unanimous decision. The solution worked at that point, although this must be balanced, particularly since, as discussed, consensus is generally the rule in EU decision-making processes, particularly in EUCO.

To exemplify, we will use again the negotiations in 2020 regarding the MFF, which took place over five days. The President of EUCO proposed sessions of discussions (usually daily) on different topics: the governance of the package, the volume of funds, the split between grants and loans, etc. After each negotiating session within the entire group, the President would hold separate negotiations with individual member states or groups of member states supporting similar ideas (like the "frugal"). Other groups of members would meet informally during these breaks. The role of the President in these informal meetings was to understand where the compromise could lay in the negotiations on given topics. The group would revert to a formal setting for further discussions, with new proposals coming from the President. In his role, he obtained more information on where concessions could be made and could propose a way forward in the compromise-building process.

7 Conclusion

This chapter has aimed to translate some of the main theoretical elements of group theory, particularly dynamics and behaviour, to the European Council. The chapter introduced the theoretical hypothesis that the European Council is a group and that, as a group, it has some of its essential characteristics. Among these characteristics, the chapter looks at interdependence, commonality of purpose, a specific communication pattern, social influence within the group, and intragroup conflict to reach a compromise. At the same time, EUCO has certain particularities that make it a distinct group. It comprises heads of state and government and politically elected leaders in the member states. Their representativeness gives them political influence, making them all equal. At the same time, power structures, social and individual impact, and thematic salience differentiate among them and the roles they are likely to play in the group.

The chapter has proposed to exemplify these theoretical findings through two practical case studies, one dealing with negotiations regarding the Multiannual Financial Framework and the other with opening EU accession negotiations for Ukraine. Both of these case studies validated the theoretical conclusions, among others, in what concerns norms and their application, power relations in the group, including the importance of diffused power and influence, the use of informal channels of communication to reach a compromise, the roles played by the main actors in these and other EUCO negotiations.

References

Akse, E. (2020). *How the EU Institutions Work: Your handbook and guide to EU decision-making*. London: John Harper Publishing Ltd.

Barigazzi, J. (2024, June 27). *Eat, pray, summit: What's on EU leaders' plates*. POLITICO. https://www.politico.eu/article/eat-pray-summit-whats-on-eu-leaders-plates/.

Barigazzi, J., & Martuscelli, C. (2023, March 23). *Live blog: European Council summit – 23 March 2023*. POLITICO. https://www.politico.eu/article/live-blog-european-council-summit-23-march-2023/.

Benne, K. & Sheats, P. (1948). *Functional roles of group members*. Journal of Social Issues: Volume 4, Issue 2.

Dentler, R. A. & Erikson K. T. (1959). *The Functions of Deviance in Groups*, Social Problems. 1959. Vol. 7. No. 2. pp. 98–107.

deSchoutheete, P. & Wallace, H. (2002). *EUCO*. Reserch and European Issues No. 19.

European Commission. (2009). *Commission Decision 2009/882/EC of 30 November 2009 concerning the technical specification of interoperability relating to the 'Infrastructure' subsystem of the trans-European conventional rail system* (Document No. C(2009) 9917). *Official Journal of the European Union, L* 331, 1–162. https://eur-lex.europa.eu/legal-content/EN/TXT/?uri=celex%3A32009D0882.

Forsyth, D. (2010). *Group dynamics*. 5th edition Belmont: Wadsworth, Cengage Learning.

Hollander, E. P. (1971). *Principles and methods of social psychology* (2nd ed.). Oxford U. Press.

Jex, S. & Britt, T. (2008, 2014). *Organizational psychology: a scientist-practitioner approach*. Hoboken, New Jersey: John Wiley & Sons.

Jones, E., Anand, M. & Weatherill, S. (eds), (2012). *The Oxford Handbook of the European Union*, online edn, Oxford Academic.

Leavitt, H. J. (1951). Some effects of certain communication patterns on group performance. Journal of Abnormal and Social Psychology, 46(1), pp. 38–50.

Lelieveldt, H. & Princen, S. (2023). The Politics of the European Union. 3rd edition. Cambridge: Cambridge University Press.

Levine, R., Norenzayan, A. & Philbrick, K. (2001). *Cross-cultural differences in helping strangers*. Journal of cross-cultural psychology. Vol. 32. No. 5. pp. 543–560.

McCormick, J. (2020). *Understanding the European Union*. London: Bloomsbury Publishing.

Nail, P. R., MacDonald, G. & Levy, D. A. (2000). Proposal of a Four-Dimensional Model of Social Response. Psychological Bulletin, 126(3), pp. 454–470.

Shaw, M. E. (1964). *Communication Networks*. Advances in Experimental Social Psychology, 1, 111–147.

Sabel, C. & Zeitlin, J. (2012). *Experimentalism in the EU: Common ground and persistent differences.* Regulation & Governance, vol.6, issue 3, pp. 410–426.

Strodtbeck, F. L. & Lipinski, R. M. (1985). *Becoming first among equals: Moral considerations in jury foreman selection.* Journal of Personality and Social Psychology, 49(4), 927–936.

Tallberg, J. (2003). *The Agenda-Shaping Power of the EU Council Presidency. Journal of European Public Policy,* 10/1, 1–19.

Tallberg, J. (2007). Leadership and Negotiation in the European Union. Cambridge: Cambridge University Press.

Thomas, K. W. (1976). Conflict and conflict management. In Dunnette, M. D. (ed.) Handbook of Industrial and Organizational Psychology. Chicago: Rand McNally College Publishing Company. pp. 889–935.

Thomson, R., Stokman, F. N., Achen, C. H. & König, T. (eds.) (2006). The European Union Decides. Cambridge: Cambridge University Press.

Thomson, R., Stokman, F. N., Achen, C. H. & König, T. (2009). The European Union Decides. Cambridge: Cambridge University Press. Published online: 22 September 2009. DOI: 10.1017/CBO9780511492082.

Tsebelis, G. (2002). *Veto Players: How Political Institutions Work.* Princeton: Princeton University Press.

Versluis, E., van Keulen, M. & Stephenson, P. (2010). *Analysing the European Union policy process.* London: Red Globe Press.

Wessels, W. (2015). *EUCO.* London: Bloomsbury Publishing.

Von der Burchard, H. (2023, December 19). *Will EU leaders continue to sidestep Orbán by asking him to leave the room?* Euronews. https://www.euronews.com/my-europe/2023/12/19/will-eu-leaders-continue-to-sidestep-orban-by-asking-him-to-leave-the-room.

Westlake, M. & Galloway, D. (2004). The Council of the European Union. 3rd edition. London: John Harper Publishing Ltd.

CHAPTER 3

Guarding Europe's Digital Future: The 2025 Poland's EU Presidency and Countering Disinformation: Lessons from Estonia's Experiences

Magdalena Musiał-Karg and Louis Wierenga

Abstract As Poland assumed the Presidency of the Council of the European Union in 2025, addressing disinformation and cybersecurity threats became a priority. This paper examined how Poland drew on the experience of Estonia's 2017 EU Presidency to strengthen the EU's response to digital challenges. Estonia's counter-disinformation measures and resilience against cyber and influence operations provided a valuable model for enhancing EU-wide digital security. By analysing Estonia's approach, the study identified strategies relevant to Poland's Presidency, including advancing cybersecurity cooperation, promoting media literacy, and strengthening regulatory standards. The paper highlighted the importance of collaboration, innovation, robust regulatory frameworks, and cognitive resilience in addressing digital threats. It first outlined the EU's digital threat landscape, then examined Estonia's 2017 Presidency as a case study, and finally assessed how its lessons could be applied during Poland's 2025 Presidency. The study concluded with policy recommendations supporting stronger EU cybersecurity and media literacy, contributing to a more resilient digital Europe.

Keywords Poland – EU Council Presidency – Estonia cybersecurity – disinformation – digital security

1 Introduction

Cybersecurity has rapidly evolved into a critical issue at the forefront of political agendas worldwide. As governments, international organizations, and policymakers recognize the growing threats posed by cyberattacks, data breaches, and digital warfare, cybersecurity has transitioned from a relatively obscure

 HTTPS://DOI.ORG/10.1163/9789004763623-005

topic to a central pillar of national security, economic stability, and democratic resilience (Munkøe & Mölder, 2022). This shift is particularly evident in the European Union, where rapid digital transformation has created an increasingly complex cybersecurity landscape, with disinformation emerging as one of the most pressing threats to democratic stability.

Recognizing these challenges, the European Union has underscored the urgent need to strengthen its cybersecurity framework in response to evolving and multifaceted cyber threats (Munkøe & Mölder, 2022; Renda, 2022). These threats encompass a wide spectrum, ranging from disinformation and fake news to direct cyberattacks on government infrastructure and electoral interference in third countries (Meulen et al., 2015; Tasheva & Kunkel, 2022). To address these risks, the EU has emphasized the necessity of a coordinated and robust response (Nenadić, 2019), reinforcing its commitment to safeguarding digital security, democratic processes, and critical infrastructure across member states. As foreign and domestic actors increasingly exploit digital platforms to manipulate public opinion, undermine trust in institutions, and destabilize political processes (Tenove et. al., 2018), the urgency of a unified and resilient European strategy has never been greater.

Amid escalating digital security threats, Poland's EU Presidency in the first half of 2025 presents a critical opportunity to shape the Union's approach to countering AI-driven disinformation, cyberattacks on democratic institutions, and the weaponization of social media. Positioned at the intersection of the EU's digital and geopolitical security landscape, Poland has the opportunity to take a leading role in reinforcing Europe's resilience against these threats. While the challenge is substantial, it is not without precedent. Estonia's 2017 EU Presidency demonstrated how a digitally advanced yet smaller member state could successfully place cybersecurity and disinformation countermeasures at the core of the EU agenda (Eichhorn at al., 2020). Drawing on Estonia's experience (Petkova, 2018) – particularly its initiatives in promoting cybersecurity cooperation, digital sovereignty, and public-private partnerships- provides a strategic foundation for Poland to navigate its own presidency. By leveraging its regional expertise and advocating for stronger EU-wide mechanisms, Poland can contribute to a more robust and coordinated digital security framework, reinforcing the Union's long-term resilience in an era of evolving cyber threats. This chapter explores how Poland can build on Estonia's 2017 EU Presidency to strengthen its response to disinformation and digital threats in 2025. It analyses Estonia's key policies and initiatives, assesses how Poland can apply these lessons to address current challenges such as AI-driven manipulation and cyber conflicts, and offers practical recommendations to reinforce EU resilience through coordinated counter-disinformation strategies. This chapter advances the understanding of how member states can exercise leadership in strengthening the European Union's collective capacity to safeguard digital democracy. Poland's 2025 EU Presidency constitutes not only a moment of heightened political responsibility but also a critical opportunity to shape the Union's strategic response to cyber threats in the context of accelerating technological change.

2 Disinformation Campaigns in Europe's Digital Threat Landscape

2.1 Russia's Strategic Use of Information Warfare

Russia has long employed information warfare as a tool for destabilizing EU member states, particularly through disinformation campaigns designed to exploit societal divisions, influence elections, and undermine trust in democratic institutions. These operations often involve coordinated efforts between state-sponsored media outlets, cyber operatives, and networks of automated social media accounts, amplifying false or misleading narratives. Russia has established itself as a central actor in the deployment of information warfare tactics aimed at destabilizing the European Union and its member states. This strategy is not new, but has intensified significantly in recent years, particularly since the annexation of Crimea in 2014 and throughout the ongoing war of aggression against Ukraine (Musiał-Karg & Łukasik-Turecka, 2023). By leveraging a coordinated ecosystem of state-sponsored media, cyber operatives, and social media manipulation – including bot networks and troll farms – Russia seeks to undermine NATO and the EU's political cohesion, erode public trust in institutions, and influence electoral outcomes.

At the core of Russia's approach is the dissemination of disinformation designed to amplify existing social tensions. These narratives are often tailored to resonate with specific political, cultural, or historical grievances with individual member states. Whether targeting immigration policies, vaccine scepticism, or national security debates, Russian information operations aim to polarize public opinion and delegitimize democratic processes (Musiał-Karg & Łukasik-Turecka, 2023).

Moscow's use of hybrid tactics blends overt propaganda – disseminated via outlets such as Russia Today (RT) and Sputnik – with covert influence campaigns, including hacking and leaking of sensitive data. The disinformation ecosystem also frequently employs "astroturfing" strategies where fake grassroots movements are manufactured to give credibility to fringe opinions (Kalniete & Pildegovičs, 2021).

Despite increasing recognition of these dangers, EU institutions and member states still struggle to effectively counter Russian information warfare. Efforts are hindered by the decentralized structure of digital platforms, the rapid dissemination of false information, and the challenges of reliably attributing responsibility for such operations. Europe has faced several high-profile examples in recent years, including disinformation surrounding the 2016 Brexit referendum, false narratives about the COVID-19 pandemic and vaccination campaigns, and influence operations following Russia's renewed military aggression against Ukraine.

2.2 The Role of AI and Social Media Algorithms in Spreading Disinformation

The rise of artificial intelligence (AI) has transformed disinformation campaigns by enabling the rapid creation, personalization, and dissemination of false content (Gilbert & Gilbert, 2024). Tools such as deepfakes – realistic but fabricated audio-visual materials – blur the line between truth and deception, posing new challenges for verification and accountability. AI-powered language models and chatbots further amplify this threat by generating convincing, targeted narratives that mimic genuine public engagement. AI also enhances microtargeting, exploiting user data to deliver highly personalized messages that reinforce biases and deepen echo chambers. This fragmentation of public discourse makes it harder to establish shared facts, increasing societal polarization (Gaborit, 2024). At the same time, social media algorithms, designed to maximize user engagement, often prioritize sensational or emotionally charged content – regardless of its accuracy. This amplifies the reach of disinformation, giving falsehoods a significant advantage over fact-based reporting. The low cost and high speed of algorithmic amplification allow disinformation to spread rapidly (Warin, 2024), often outpacing fact-checking efforts and creating an asymmetric information environment.

Together, AI technologies and social media platforms have become powerful enablers of disinformation, threatening the integrity of democratic debate and undermining informed public decision-making (Kreps & Kriner, 2023).

3 Addressing Digital Threats: Towards a Coherent EU Response

Recognizing the scale and complexity of the digital threat landscape, the European Union has increasingly sought to develop a coordinated policy framework to counter disinformation, mitigate cyber risks, and safeguard democratic resilience. Yet, as technological advancements continue to outpace regulatory efforts, the EU faces the urgent task of strengthening its collective capacity to respond effectively to these evolving challenges.

3.1 Strengthening Regulatory Frameworks

One of the EU's primary instruments for addressing digital threats is the Digital Services Act (DSA) (Regulation (EU) 2022/2065), which introduces binding obligations on large online platforms to curb the spread of illegal content, enhance algorithmic transparency, and improve user protections. While the DSA marks a significant regulatory milestone, its success depends on the robust enforcement of its provisions and the political will of member states to ensure meaningful compliance. Strengthening the capacity of national regulators, such as Digital Services Coordinators, and fostering cross-border cooperation will be essential to translating the DSA's principles into tangible outcomes across the Union.

Complementing the DSA, the Digital Markets Act (DMA) (Regulation (EU) 2022/1925) targets the structural power of the largest digital platforms identified as "gatekeepers" – those providing core platform services such as online search engines, app stores, and messaging services. The DMA establishes a set of objective criteria to designate these platforms and imposes a series of obligations and prohibitions designed to prevent anti-competitive practices and ensure fair access to digital markets. While the DMA does not replace existing EU competition rules, it provides a proactive regulatory framework aimed at curbing the market dominance of major tech companies, thereby reducing their capacity to exploit informational or economic asymmetries that may facilitate disinformation or undermine democratic processes. The Artificial Intelligence Act (Regulation (EU) 2024/1689) likewise introduced a risk-based governance framework for AI applications (Romero Moreno, 2024). Ensuring this legislation effectively tackles the misuse of AI for disinformation – such as deepfakes, automated propaganda, and algorithmically generated false content – will be essential to preventing the exploitation of emerging technologies to undermine democratic institutions and erode public trust (Łabuz, 2023).

3.2 Enhancing Cybersecurity and Information Resilience

In addition to regulatory measures, the EU must recognize cybersecurity as one of the critical priorities, necessitating a multifaceted approach to address evolving threats (Munkøe & Mölder, 2022), and prioritize investments in enhancing cybersecurity capabilities at both institutional and societal levels. This includes bolstering the protection of critical infrastructure, electoral processes, and government communications from cyberattacks, hacking, and data breaches. While existing frameworks, such as the EU Cybersecurity Strategy and the work of the European Union Agency for Cybersecurity (ENISA), provide a solid foundation, they require sustained political backing, sufficient funding, and the active involvement of all member states to be truly effective.

Equally important is the need to build democratic (Cupać & Sienknecht, 2024) and societal resilience to disinformation and influence operations (Dragomir et al., 2024). Achieving this requires comprehensive efforts to promote media literacy, support independent journalism, and implement public awareness campaigns that empower citizens to critically evaluate the information they encounter. Such efforts must be pursued through broad-based collaboration involving public authorities, civil society, academic institutions, and technology providers, working together to strengthen a more informed and resilient European public sphere capable of resisting manipulative tactics. These initiatives should form part of a wider commitment to ensuring that technological progress is aligned with ethical and legal standards – safeguarding the integrity of information environments and reinforcing public trust in digital ecosystems (Schmitt et al., 2024).

4 Fostering Strategic Leadership through the EU Presidency

The rotating Presidency of the Council of the European Union offers member states a unique opportunity to influence the Union's policy direction and legislative agenda. While the Presidency's formal powers are largely procedural – managing Council meetings, mediating interinstitutional negotiations, and overseeing the legislative calendar – it also provides space for agenda-setting, coalition-building, and policy entrepreneurship. By choosing which issues to elevate, shaping how debates are framed, and facilitating compromises, the Presidency can exert considerable informal influence on the EU's collective priorities (Cross & Vaznonytė, 2020). Research indicates that the Presidency can shape the Council's agenda to a significant extent (Warntjen, 2007) and steer legislative activity according to national priorities (Vaznonytė, 2020; Tallberg, 2006).

Effective Presidencies go beyond procedural management, by anticipating strategic challenges, consensus building, and political mobilization across the Union (Coussens & Crum, 2003). This balancing of national leadership with European responsibility becomes even more crucial in times of geopolitical uncertainty and technological disruption, when coordinated EU action is both more necessary and more difficult to achieve.

4.1 Estonia's 2017 EU Presidency

Estonia held the presidency of the Council of the European Union for six months from 1 July to 31 December 2017 (Toneva-Metodieva, 2020). As the first Baltic state to hold the presidency, it sought to demonstrate its commitment to a strong and united Europe while promoting its own expertise in digital governance and technological innovation (Papp-Váry, 2018). The Presidency offered a timely and high-profile platform to showcase this digital identity to a broader European and international audience. Taking place in a period marked by Brexit-related uncertainties (Bukovskis & Kasekamp, 2022), geopolitical tensions, and debates about the EU's future direction, Estonia used its presidency to advance key priorities that reflected its digital strengths. These included pushing forward the EU's Digital Single Market agenda, promoting cybersecurity cooperation, enhancing cross-border digital services, and encouraging the adoption of e-governance models across member states. Events such as the Tallinn Digital Summit reinforced Estonia's image as a digital frontrunner, positioning digital transformation as a strategic priority for the EU as a whole (Haugevik & Rieker, 2017).

Beyond digital matters, Estonia's presidency also focused on strengthening security cooperation, supporting the development of the EU's defence policy, and maintaining momentum on the EU enlargement process, particularly with respect to the Western Balkans and the Eastern Partnership (Made, 2011). Estonia has developed one of the most secure and digital-forward societies in the world (Gat, 2018). Known for its e-Governance and e-Residency programs (Kattel, & Mergel, 2019; Ernsdorff, & Berbec, 2007). Estonia also recognized early the security challenges posed by digitalization and built a comprehensive national cybersecurity strategy to protect infrastructure, economy, and citizens from both cyber threats and disinformation campaigns. This awareness was sharpened by the large-scale cyberattacks it suffered in 2007 (Guadagno et al., 2010) following the government's decision to relocate a Soviet-era war memorial – the Bronze Soldier of Tallinn (Musiał-Karg et al., 2025). The attacks, which disrupted government services, financial institutions, and media outlets, highlighted the vulnerabilities of Estonia's digital infrastructure (Czosseck et al., 2011; Herzog, 2011; Ottis, 2008; Robinson & Hardy, 2021). The incident also underscored the geopolitical tensions surrounding historical memory and national identity, as it drew protests from the Russian-speaking minority and condemnation from Russia. This event represented a turning point in Estonia's national narrative and security policy (Ehala, 2009), driving greater attention to cyber defence and broader societal security challenges across the Baltic region (Šešelgytė & Bladaitė, 2021).

As a result, Estonia emerged as a pioneer in cybersecurity policy and resilience, taking a leading role in shaping cyber defence protocols within NATO (Crandall & Allan, 2015) and the European Union. The country's comprehensive response included the adoption of national legislation on data protection, securing digital infrastructure, and establishing strategic partnerships with other EU member states for information sharing (Warchoł, 2023). Estonia's leadership was further cemented by hosting the NATO Cooperative Cyber Defence Centre of Excellence (CCDCOE) in Tallinn, a key institution for cybersecurity research, training, and incident response within the alliance (Lorents & Matsak, 2011).

Estonia's efforts did not stop there. Recognizing the ongoing threat of disinformation campaigns, Estonia actively contributed to broader EU initiatives aimed at countering information manipulation. It was among the leading advocates for the creation of the East StratCom Task Force, established by the European External Action Service in 2015, to improve the EU's capacity to detect and respond to disinformation, particularly from external actors (Hedling, 2021).

Estonia, which joined the European Union in 2004, assumed the Presidency of the Council of the EU for the first time in the second half of 2017. Under the slogan "Unity through Balance," Estonia focused on four overarching priorities: (1) fostering an open and innovative European economy, (2) ensuring a safe and secure Europe, (3) advancing a digital Europe and the free movement of data, and (4) promoting an inclusive and sustainable Europe (European Council, 2017). Estonia's leadership particularly emphasized digital transformation, positioning itself as a pioneer in e-governance and cybersecurity. It organized the Tallinn Digital Summit to set the digital agenda for the EU and highlight the importance of technological leadership.

In the area of security and defence, Estonia aligned its priorities with the EU Global Strategy, advocating for enhanced military capabilities and deeper European defence cooperation (European External Action Service, 2016). As one of the few Member States meeting the NATO guideline of spending over 2% of GDP on defence, Estonia sought to encourage other Member States to reach this benchmark (NATO, 2017). Cybersecurity was a cornerstone of Estonia's security agenda, building on its own experiences with large-scale cyberattacks in 2007. The Presidency prioritized strengthening EU-NATO cooperation on cyber defence, supporting cyber exercises such as EU CYBRID 2017, and advancing the EU Cyber Diplomacy Toolbox (European Union) to deter and respond to malicious cyber activities (Eichhorn et al., 2020).

Estonia prioritized digital resilience and cybersecurity innovation, recognizing the increasing threats posed by cyberattacks and disinformation campaigns (Gajewski, 2020). By embedding cybersecurity at the core of its EU agenda, Estonia sought to establish long-term strategic policies for digital security across member states. One of its key objectives was to integrate counter-disinformation measures into the EU's broader security strategy, ensuring that digital threats were addressed as part of European governance frameworks (Fuster & Jasmontaite, 2020; Kasper, 2020).

During its 2017 Presidency of the Council of the European Union, Estonia introduced a series of concrete measures aimed at enhancing the EU's cybersecurity posture and democratic resilience. Recognizing the complexity of digital threats, Estonia facilitated strong public–private partnerships, bringing together governments, technology companies, and civil society organizations to jointly address cybersecurity risks (Górka, 2023). This collaborative approach promoted information sharing, capacity building, and joint response mechanisms across sectors.

In parallel, Estonia launched educational initiatives and media literacy programs designed to empower EU citizens with the critical thinking skills needed to identify and resist disinformation. These efforts were targeted particularly at youth and vulnerable groups, aiming to build societal resilience against manipulation in the information space (Harro-Loit et al., 2023).

Estonia also championed the development of harmonized EU-wide regulations on digital security, advocating for greater consistency in the implementation of the NIS Directive and calling for common cybersecurity standards across all Member States. These efforts helped to align national cybersecurity policies with EU-level strategic objectives, reinforcing Europe's collective digital security architecture (Carrapico & Farrand, 2024).

Estonia's proactive leadership in digital security produced tangible achievements. Most notably, its measures contributed to mitigating Russian disinformation campaigns, demonstrating that coordinated EU action can contain foreign influence operations when supported by strategic communication and citizen engagement. The Estonian Presidency also strengthened the EU's cybersecurity frameworks, facilitating improved cooperation among Member States and advancing institutional reforms such as the enhancement of the

European Union Agency for Cybersecurity (ENISA). These achievements laid the groundwork for future cybersecurity policy developments at the EU level.

Perhaps most importantly, Estonia succeeded in elevating digital security, cybersecurity, and disinformation countermeasures to the top of the EU's political agenda (Vériter, 2024). By doing so, it set a precedent for future EU Presidencies to continue prioritizing these challenges, recognizing that digital resilience is essential for the security, stability, and democratic integrity of the European Union (Hedling, 2021).

4.2 Poland's 2025 Presidency: A Strategic Opportunity for Digital Democracy and Security

Poland assumed the rotating Presidency of the Council of the European Union in the first half of 2025 under the guiding motto "Security for Europe!". This agenda was shaped by the broader geopolitical context of Russia's ongoing aggression against Ukraine, rising global instability, and intensifying hybrid threats aimed at undermining European democracy and security. Poland's Presidency came at a critical moment when the European Union faced an increasingly hostile digital environment, marked by coordinated disinformation campaigns, cyberattacks on critical infrastructure, and technological manipulation targeting electoral processes and democratic institutions. The misuse of advanced technologies, particularly artificial intelligence, to spread false narratives and manipulate public debate further exacerbated these risks.

While addressing the EU's broader priorities – such as defence, energy security, and economic resilience – Poland placed particular emphasis on strengthening digital and technological security. Key areas of action included cyber security, artificial intelligence, emerging technologies, the development of the telecommunications market, digital diplomacy, and transatlantic cooperation. Safeguarding the integrity of information spaces and enhancing the Union's collective resilience to digital threats became one of the Presidency's central objectives, requiring coordinated action at both European and national levels. These efforts aimed not only to counter hybrid warfare, AI-driven disinformation, and cyber threats, but also to reinforce Europe's position as a secure, democratic, and competitive player in the digital age (Bochyńska, 2025).

Drawing on its direct experience as a frontline state exposed to Russian information warfare and hybrid tactics (Gasztold & Gasztold, 2020), Poland was well-positioned to lead the EU's response. It sought to elevate digital security from a technical challenge to a strategic priority for the entire Union, promoting the need for a comprehensive and coordinated approach.

Throughout its Presidency, Poland worked to advance the implementation of key legislative frameworks, including the Digital Services Act and the Artificial Intelligence Act, ensuring they addressed AI-generated disinformation, algorithmic manipulation, and other emerging threats. However, despite these declared priorities, Poland faced criticism for delays in fully implementing the Digital Services Act at the national level (Euronews, 2025), leading the European Commission to initiate infringement proceedings for failing to designate a Digital Services Coordinator and establish enforcement mechanisms (REUTERS, 2025; Polish Press Agency, 2025). Nevertheless, Poland remained active in promoting stronger cybersecurity cooperation by advancing investments in cyber defence, critical infrastructure protection, and cross-border information sharing, while reinforcing the role of ENISA and other relevant EU bodies, including through initiatives such as hosting the 2025 European Cybersecurity Certification Conference (ENISA, 2025).

Recognizing the importance of societal resilience, Poland emphasized the need to complement regulatory and technological measures with public engagement. While official documents referred to priorities such as media literacy, civic education, and support for independent journalism, concrete initiatives in these areas were less visible or lacked coordinated implementation at the EU level. Nevertheless, Poland sought to foster dialogue among governments, civil society, and technology companies, positioning itself as a consensus-builder within the EU by promoting awareness that defending democracy in the digital age requires not only technological safeguards but also an informed and engaged public (doItinPOLAND, 2025).

The 2025 Presidency also provided a platform for strengthening strategic partnerships beyond the Union, including with NATO, the United States, and Eastern Partnership countries. Acknowledging that digital threats are part of wider geopolitical confrontations, Poland worked to align the EU's digital security strategy with its external security and foreign policy goals. For instance, during the Eastern Partnership Business Forum held in Warsaw, the European Commission reaffirmed its commitment to deeper economic cooperation and strategic connectivity with Eastern Partnership countries (European Commission, 2025).

To succeed in these ambitions, Poland had to build broad-based coalitions among member states, balancing diverse interests while keeping digital resilience high on the EU's strategic agenda. Framing digital security as a shared European responsibility, Poland aimed to leave a lasting impact by strengthening the Union's ability to defend its democracy, institutions, and societies in an increasingly contested digital and geopolitical environment.

In the area of cybersecurity, it was recognized that securing the EU's digital infrastructure required not only regulatory action but also practical cooperation with industry, which plays a key role in promoting cybersecurity and ensuring secure infrastructure. To this end, the Polish Presidency emphasized the need to involve representatives of companies developing security solutions in shaping and implementing EU regulations and strategies. By fostering collaboration between governments, industry, and civil society, Poland supported efforts to strengthen cybersecurity through information sharing, joint development of best practices, and coordinated actions aimed at eliminating vulnerabilities across critical systems.

Moreover, Poland's Presidency provided momentum for strengthening international partnerships, particularly in the spirit of transatlantic cooperation and NATO–EU coordination. Recognizing that cyber and hybrid threats have become a core element of modern geopolitical confrontations – especially in light of Russia's aggression against Ukraine – Poland highlighted the importance of building the EU's capabilities for joint defensive and offensive cyber operations alongside its allies. NATO's growing focus on cybersecurity as a domain of operational relevance underscored the need for deeper EU–NATO collaboration. Poland's initiatives aimed to expand this cooperation, including in the area of countering disinformation and hybrid threats, thus contributing to the security of the Union and its allies in an increasingly volatile international environment (Union of Entrepreneurs and Employers, 2024).

4.3 Poland's 2025 EU Presidency in the Footsteps of Estonia: Continuity and Divergence in Responding to Hybrid Threats

Poland assumed the Presidency of the Council of the European Union in the first half of 2025 under circumstances in many ways similar to those of Estonia's presidency in 2017, but marked by a significantly intensified threat environment. The ongoing war of Russian aggression against Ukraine, entering its fourth year, fundamentally reshaped Europe's security perceptions, shifting the focus of EU institutions toward defending against hybrid warfare, including cyberattacks, disinformation, and sabotage operations. As Ukraine's direct neighbour and a frontline EU Member State, Poland fully understood the gravity of these threats and made cybersecurity one of the top priorities of its presidency, alongside traditional security, defence, and energy concerns (see Table 3.1 for a comparative overview of the Estonian and Polish EU Presidencies).

Poland not only promoted initiatives aimed at protecting critical infrastructure and enhancing rapid response systems to cyber incidents but also placed special emphasis on citizen education, particularly for young people, in countering disinformation. It highlighted the importance of media literacy, the

TABLE 3.1 Estonia 2017 vs. Poland 2025 EU Presidency (Source: Author's own compilation based on Government Office of Estonia (2017), Estonian Presidency of the Council of the European Union 2017, and Republic of Poland (2025), Polish Presidency of the Council of the European Union 2025.)

Dimension	Estonia 2017	Poland 2025
Geopolitical Context	Brexit, rising hybrid threats, early AI disruption	Ongoing war in Ukraine, intensified Russian hybrid operations, return of President Trump in the US
Presidency Motto/ Framing	"Unity through Balance"	"Security for Europe!"
Digital Focus Areas	Digital Single Market, cybersecurity, e-governance	Cybersecurity, AI regulation, disinformation, digital resilience
Key Threats Addressed	Russian disinformation, cyberattacks (post-2007 attack experience)	Russian hybrid warfare, AI-driven disinformation, cyberattacks on critical infrastructure
Policy Leadership	EU Cyber Diplomacy Toolbox, strengthening ENISA, promoting media literacy	Advancing DSA and AI Act, promoting cyber defence investments, supporting media literacy
Partnerships Emphasized	EU-NATO cooperation, Baltic and Eastern Partnership engagement	EU-NATO cooperation, Transatlantic relations (challenged by Trump's Ukraine policy), Baltic solidarity
Institutional Delivery/ Weaknesses	Strong agenda-setting, clear follow-up through ENISA and cyber exercises	Failure to appoint Digital Services Coordinator, slow DSA implementation
Public Engagement Initiatives	Media literacy programs, public-private partnerships	Promoted but limited visibility or coordination at EU level
Legacy/ Impact	Elevated cybersecurity and disinformation as strategic EU priorities	Reinforced digital security on agenda, but weakened credibility due to regulatory gaps

development of educational programs, and support for independent media as key elements in building societal resilience against information manipulation. These efforts also served as an expression of solidarity with the societies of Ukraine and the Baltic states, which – like Poland – were directly exposed to ongoing Russian destabilization campaigns. Poland positioned itself within the EU as a spokesperson for these countries, advocating for stronger protection of the Union's informational and cyber borders. Poland's presidency also coincided with the return of President Donald Trump to the White House, whose re-election dominated transatlantic political discourse. Trump's stance on the war in Ukraine, particularly his pressure for rapid peace negotiations with Russia – often disregarding Ukraine's position – caused concern among many EU Member States and shaped the diplomatic climate throughout Poland's term. As one of Washington's closest allies in Europe, Poland faced the delicate task of balancing loyalty to the United States with the defence of the European approach to supporting Ukraine, which was rooted in respect for Ukraine's sovereignty and its right to self-determination.

Despite ambitious declarations and an active international agenda, Poland's presidency was not without institutional shortcomings. One of the most criticized aspects was its failure to fully implement the EU's Digital Services Act, particularly the failure to appoint a national Digital Services and Cybersecurity Coordinator. This delay prompted the European Commission to launch infringement proceedings against Poland. This failure weakened Poland's credibility in the area of digital regulation and demonstrated that political declarations must be matched by consistent domestic implementation. Nevertheless, Poland's presidency played an important role in keeping digital security and informational resilience high on the EU's political agenda, seeking to develop a comprehensive strategy to address hybrid threats, in close cooperation with NATO, the United States, and the Eastern Partnership countries.

5 Conclusion

The comparative analysis of Estonia's 2017 and Poland's 2025 EU Presidencies highlights both the continuity of digital security challenges facing the European Union and the evolving complexity of hybrid threats in today's geopolitical environment. While Estonia's presidency succeeded in placing cybersecurity and disinformation resilience at the centre of the EU's strategic agenda, Poland's presidency built on this legacy in a period of far greater geopolitical instability, marked by Russia's ongoing war of aggression against Ukraine and intensified technological manipulation, including AI-driven disinformation.

Poland's presidency demonstrated that digital security is no longer a niche or technical issue, but a core element of European security, democracy, and resilience. By emphasizing cybersecurity, AI regulation, media literacy, and transatlantic cooperation, Poland sought to elevate these priorities to the level of Union-wide strategic action. Its efforts to strengthen cooperation with NATO, defend the interests of Ukraine and the Baltic States, and engage industry and civil society reflect a recognition that defending Europe's digital democracy requires broad-based partnerships across sectors and borders.

However, Poland's presidency also revealed critical institutional weaknesses, notably its failure to fully implement the Digital Services Act and appoint a national Digital Services Coordinator, which undermined its credibility as a digital policy leader. This gap between political declarations and domestic implementation serves as an important lesson for future presidencies: strategic ambition must be matched by concrete institutional delivery.

Looking ahead, the European Union must continue to build on the foundations laid by Estonia and Poland, ensuring that digital resilience is institutionalized as a permanent strategic priority. This requires closing regulatory gaps, investing in public engagement, fostering cross-sectoral collaboration, and strengthening international partnerships. Only through coordinated and sustained action can the EU effectively defend its democratic institutions, protect its citizens, and secure its digital future in an era of accelerating technological and geopolitical change. Poland's experience underscores that the role of the EU Presidency is not only to manage legislative processes but to shape long-term strategic visions. By learning from both the successes and shortcomings of recent presidencies, future EU leaders can continue to advance a more secure, resilient, and democratic Europe.

References

Bochyńska, N. (2025, January 1). Rusza polska prezydencja w UE. Oto priorytety. *Cyber Defence24.pl*. https://cyberdefence24.pl/polityka-i-prawo/rusza-polska-prezydencja-w-ue-oto-priorytety.

Bukovskis, K., & Kasekamp, A. (2022). Brexit coping strategies of the Baltic states. *Journal of Contemporary European Studies*, 31(3), 657–669. https://doi.org/10.1080/14782804.2022.2110044.

Carrapico, H., & Farrand, B. (2024). Cybersecurity trends in the European Union: Regulatory mercantilism and the digitalisation of geopolitics. *JCMS: Journal of Common Market Studies*, 62(1), 147–158. https://doi.org/10.1111/jcms.13654.

Coussens, W., & Crum, B. (2003). *Towards effective and accountable leadership of the Union: Options and guidelines for reform* (EPIN Working Paper No. 3).

Crandall, M., & Allan, C. (2015). Small states and big ideas: Estonia's battle for cybersecurity norms. *Contemporary Security Policy*, 36(2), 346–368. https://doi.org/10.1080/13523260.2015.1061765.

Cross, J. P., & Vaznonytė, A. (2020). Can we do what we say we will do? Issue salience, government effectiveness, and the legislative efficiency of Council Presidencies. *European Union Politics*, 21(4), 657–679. https://doi.org/10.1177/1465116520950829.

Cupać, J., & Sienknecht, M. (2024). Regulate against the machine: How the EU mitigates AI harm to democracy. *Democratization*, 31(5), 1067–1090. https://doi.org/10.1080/13510347.2024.2353706.

Czosseck, C., Ottis, R., & Talihärm, A.-M. (2011). Estonia after the 2007 cyber attacks: Legal, strategic and organisational changes in cyber security. *International Journal of Cyber Warfare and Terrorism*, 1(1), 24–34. https://doi.org/10.4018/ijcwt.2011010103.

doItinPOLAND (2025, January 8). Poland's EU Presidency: Focus on cybersecurity and AI. https://doitinpoland.com/polands-eu-presidency-focus-on-cybersecurity-and-ai.

Dragomir, M., Rúas-Araújo, J., & Horowitz, M. (2024). Beyond online disinformation: Assessing national information resilience in four European countries. *Humanities and Social Sciences Communications*, 11, 101. https://doi.org/10.1057/s41599-024-02605-5.

Ehala, M. (2009). The Bronze Soldier: Identity threat and maintenance in Estonia. *Journal of Baltic Studies*, 40(1), 139–158. https://doi.org/10.1080/01629770902722294.

Eichhorn, N., Nedea, A., & Smed, U. T. (2020). When Technology Becomes Geopolitics: The Eu's Response to Cyber Threats. In M. Westlake (Ed.), *The European Union's new foreign policy* (pp. 329–348). Palgrave Macmillan. https://doi.org/10.1007/978-3-030-48317-3_14.

Ernsdorff, M., & Berbec, A. (2007). The Short Road to E-Government and E-Democracy. In P. G. Nixon & V. N. Koutrakou (Eds.), *E-government in Europe: Re-booting the state* (1st ed.). Routledge. https://doi.org/10.4324/9780203962381.

Euronews (2025, February 17). Poland risks legal action for failing to appoint digital regulator. *Euronews*. https://www.euronews.com/next/2025/02/17/poland-risks-legal-action-for-failing-to-appoint-digital-regulator.

European Commission (2025, April 16). Eastern Partnership Business Forum reaffirms EU's commitment to economic ties and connectivity in uncertain times. https://enlargement.ec.europa.eu/news/eastern-partnership-business-forum-reaffirms-eus-commitment-economic-ties-and-connectivity-uncertain-2025-04-16_en.

European Council (2017). Priority dossiers under the Estonian EU Council Presidency. https://www.consilium.europa.eu/media/56238/2017-jul-dec-ee-priorities-eparliament.pdf.

European External Action Service (2016, June). Shared vision, common action: A stronger Europe – A global strategy for the European Union's foreign and security policy. https://www.eeas.europa.eu/sites/default/files/eugs_review_web_0.pdf.

European Parliament & Council of the EU (2022, October 19). Regulation (EU) 2022/2065 (Digital Services Act). https://www.consilium.europa.eu/en/policies/digital-services-act/.

European Parliament & Council of the EU (2022, September 14). Regulation (EU) 2022/1925 (Digital Markets Act). https://eur-lex.europa.eu/legal-content/EN/TXT/?uri=uriserv%3AOJ.L_.2022.265.01.0001.01.ENG.

European Parliament & Council of the EU (2024, June 13). Regulation (EU) 2024/1689 (Artificial Intelligence Act). https://eur-lex.europa.eu/legal-content/EN/TXT/?uri=CELEX%3A32024R1689.

European Union Agency for Cybersecurity (ENISA) (2025). 2025 European Cybersecurity Certification Conference. https://www.enisa.europa.eu/events/2025-european-cybersecurity-certification-conference.

European Union (n.d.). The Cyber Diplomacy Toolbox. https://www.cyber-diplomacy-toolbox.com/.

Fuster, G. G., & Jasmontaite, L. (2020). Cybersecurity Regulation in the European Union: The Digital, the Critical and Fundamental Rights. In M. Christen, B. Gordijn, & M. Loi (Eds.), *The ethics of cybersecurity* (Vol. 21, pp. 109–132). Springer. https://doi.org/10.1007/978-3-030-29053-5_5.

Gaborit, P. (2024). A sociopolitical approach to disinformation and AI: Concerns, responses and challenges. *Journal of Political Science and International Relations*, 7(4), 75–88. https://doi.org/10.11648/j.jpsir.20240704.11.

Gajewski, T. (2020). Towards resilience: European cybersecurity strategic framework. *Ante Portas – Security Studies*, 1(14). https://doi.org/10.33674/3201911.

Gasztold, A., & Gasztold, P. (2020). The Polish counterterrorism system and hybrid warfare threats. *Terrorism and Political Violence*, 34(6), 1259–1276. https://doi.org/10.1080/09546553.2020.1777110.

Gat, O. (2018). Estonia goes digital: Residents of the tiny Baltic nation are going all in on techno-governance. *World Policy Journal*, 35(1), 108–113. https://doi.org/10.1215/07402775-6894885.

Gilbert, C., & Gilbert, M. A. (2024). Leveraging artificial intelligence (AI) by a strategic defense against deepfakes and digital misinformation. *International Journal of Scientific Research and Modern Technology*, 3(11), 62–78. https://doi.org/10.38124/ijsrmt.v3i11.76.

Government Office of Estonia (2017). Estonian Presidency of the Council of the European Union 2017. Tallinn, Estonia: Government Office of Estonia.

Górka, M. (2023). Baltic States cyber security policy: Development of digital capabilities in 2017–2022. *Stosunki Międzynarodowe – International Relations*, 3, 15. https://doi.org/10.12688/stomiedintrelat.17684.1.

Guadagno, R. E., Cialdini, R. B., & Evron, G. (2010). Storming the servers: A social psychological analysis of the first internet war. *Cyberpsychology, Behavior, and Social Networking*, 13(4), 447–453. https://doi.org/10.1089/cyber.2009.0134.

Harro-Loit, H., Lõhmus, E., & Loit, U. (2023). Capability of monitoring media-related competences of the public in Estonia. *Media Literacy and Academic Research*, 6(2).

Haugevik, K., & Rieker, P. (2017). Autonomy or integration? Small-state responses to a changing European security landscape. *Global Affairs*, 3(3), 211–221. https://doi.org/10.1080/23340460.2017.1377625.

Hedling, E. (2021). Transforming practices of diplomacy: The European External Action Service and digital disinformation. *International Affairs*, 97(3), 841–859. https://doi.org/10.1093/ia/iiab035.

Herzog, S. (2011). Revisiting the Estonian cyber attacks: Digital threats and multinational responses. *Journal of Strategic Security*, 4(2), 49–60. https://doi.org/10.5038/1944-0472.4.2.3.

Kalniete, S., & Pildegovičs, T. (2021). Strengthening the EU's resilience to hybrid threats. *European View*, 20(1), 23–33. https://doi.org/10.1177/17816858211004648.

Kasper, A. (2020). *EU cybersecurity governance – Stakeholders and normative intentions towards integration* [PDF]. https://www.um.edu.mt/library/oar/bitstream/123456789/52308/1/EU_cybersecurity_governance%E2%80%93stakeholders_and_normative_intentions_towards_integration_2020.pdf.

Kattel, R., & Mergel, I. (2019). Estonia's Digital Transformation: Mission Mystique and the Hiding Hand. In P. 't Hart & M. Compton (Eds.), *Great policy successes* (online ed., chap. 8). Oxford University Press. https://doi.org/10.1093/oso/9780198843719.003.0008.

Kreps, S., & Kriner, D. (2023). How AI threatens democracy. *Journal of Democracy*, 34(4), 122–131. https://doi.org/10.1353/jod.2023.a907693.

Łabuz, M. (2023). Regulating deep fakes in the Artificial Intelligence Act. *Applied Cybersecurity & Internet Governance*, 2(1), 1–42. https://doi.org/10.60097/ACIG/162856.

Lorents, P., & Matsak, E. (2011). Situations, Deduction, Plausibility. *Proceedings of the 2011 International Conference on Artificial Intelligence* (*ICA4909*). https://worldcomp-proceedings.com/proc/p2011/ICA4909.pdf.

Made, V. (2011). Shining in Brussels? The Eastern Partnership in Estonia's foreign policy. *Perspectives*, 19(2), 67–79. http://www.jstor.org/stable/23616145.

Meulen, N. V., Jo, E., & Soesanto, S. (2015). Cybersecurity in the European Union and beyond: Exploring the threats and policy responses.

Munkøe, M., & Mölder, H. (2022). Cybersecurity in the era of hypercompetitiveness: Can the EU meet the new challenges? *Revista CIDOB d'Afers Internacionals*, 131, 69–94. https://doi.org/10.24241/rcai.2022.131.2.69/en.

Musiał-Karg, M., & Łukasik-Turecka, A. (2023). Disinformation in the Media Space during the War in Ukraine: How did Kremlin's Fake News Blame Ukraine, the USA and NATO for the Invasion? In *The war in Ukraine. (Dis)information – Perception – Attitudes* (pp. 13–38). Peter Lang.

Musiał-Karg, M., Wierenga, L., & Maluzinas, M. (2025). Why do Electoral Meddlers Fail to Succeed in Estonia? Explaining Resilience to Interference in Democracy. In J. Rak & R. Bäcker (Eds.), *Democracy and its enemies in Europe: Successes and failures in combating autocratic threats* (1st ed.). Routledge. https://doi.org/10.4324/9781003585091.

NATO (2017, June 29). Defence expenditure of NATO countries (2010–2017) [Press release]. https://www.nato.int/nato_static_fl2014/assets/pdf/pdf_2017_06/20170629_170629-pr2017-111-en.pdf.

Nenadić, I. (2019). Unpacking the "European approach" to tackling challenges of disinformation and political manipulation. *Internet Policy Review*, 8(4). https://doi.org/10.14763/2019.4.1436.

Ottis, R. (2008). Analysis of the 2007 cyber attacks against Estonia from the information warfare perspective. Cooperative Cyber Defence Centre of Excellence. https://ccdcoe.org/uploads/2018/10/Ottis2008_AnalysisOf2007FromTheInformationWarfarePerspective.pdf.

Papp-Váry, Á. F. (2018). A successful example of complex country branding: The 'E-Estonia' positioning concept and its relation to the Presidency of the Council of the EU. *Acta Universitatis Sapientiae, European and Regional Studies*, 14, 87–115.

Petkova, T. V. (2018). Estonia of E-Estonia: Digitalization as the Highest Priority for its European Presidency. In T. V. Petkova & A. Debeljuh (Eds.), *1st International e-Conference on Studies in Humanities and Social Sciences* (pp. 79–87). Center for Promoting Ideas. https://centerprode.com/conferences/1IeCSHSS/1IeCSHSS.pdf.

Polish Press Agency. (2025, May 7). EC takes Poland to CJEU over digital act and road charge law. *PAP*. https://www.pap.pl/en/news/ec-takes-poland-cjeu-over-digital-act-and-road-charge-law.

Renda, K. K. (2022). The development of EU cybersecurity policy: From a coordinating actor to a cyber power? *Ankara Avrupa Çalışmaları Dergisi*, 21(2), 467–495. https://doi.org/10.32450/aacd.1226890.

Republic of Poland (2025). Polish Presidency of the Council of the European Union 2025. Warsaw, Poland: Chancellery of the Prime Minister of the Republic of Poland. https://poland25.eu.

Reuters. (2025, May 7). EU sues several countries for not properly implementing Digital Services Act. *Reuters*. https://www.reuters.com/technology/eu-sues-several-countries-not-properly-implementing-digital-services-act-2025-05-07/.
Robinson, N., & Hardy, A. (2021). Estonia: from the "Bronze Night" to Cybersecurity Pioneers. In S. N. Romaniuk & M. Manjikian (Eds.), *Routledge companion to global cyber-security strategy* (1st ed.). Routledge. https://doi.org/10.4324/9780429399718.
Romero Moreno, F. (2024). Generative AI and deepfakes: A human rights approach to tackling harmful content. *International Review of Law, Computers & Technology*, 38(3), 297–326. https://doi.org/10.1080/13600869.2024.2324540.
Schmitt, V., Tesch, J., Lopez, E., Polzehl, T., Burchardt, A., Neumann, K., Mohtaj, S., & Möller, S. (2024). Implications of regulations on large generative AI models in the super-election year and the impact on disinformation. *LEGAL*.
Šešelgytė, M., & Bladaitė, N. (2021). How to Defend Society? Baltic Responses to Hybrid Threats. In A. M. Brady & B. Thorhallsson (Eds.), *Small states and the new security environment* (Vol. 7, pp. 89–108). Springer. https://doi.org/10.1007/978-3-030-51529-4_6.
Tallberg, J. (2006). *Leadership and negotiation in the European Union*. Cambridge University Press.
Tasheva, I., & Kunkel, I. (2022). In a hyperconnected world, is the EU cybersecurity framework connected? *European View*, 21(2), 186–195. https://doi.org/10.1177/17816858221136106.
Tenove, C., Buffie, J., McKay, S., & Moscrop, D. (2018). Digital threats to democratic elections: How foreign actors use digital techniques to undermine democracy. Centre for the Study of Democratic Institutions, University of British Columbia. https://papers.ssrn.com/sol3/papers.cfm?abstract_id=3235819.
Toneva-Metodieva, L. (2020). Evaluating the Presidency of the Council of the EU: Towards a comprehensive evidence-based framework for performance assessment. *European Politics and Society*, 21(5), 650–667. https://doi.org/10.1080/23745118.2020.1712539.
Union of Entrepreneurs and Employers (ZPP). (2024). *Poland at the forefront of digital transformation in the EU: Priorities for the Polish presidency of the Council of the European Union* [Memorandum]. https://zpp.net.pl/wp-content/uploads/2024/10/Poland-at-the-forefront-of-digital-transformation-in-the-EU.-Priorities-for-the-Polish-presidency-of-the-Council-of-the-European-Union.pdf.
Vaznonytė, A. (2020). The rotating Presidency of the Council of the EU – Still an agenda-setter? *European Union Politics*, 21(3), 497–518. https://doi.org/10.1177/1465116520916557.
Vériter, S. L. (2024). Small-state influence in EU security governance: Unveiling Latvian lobbying against disinformation. *JCMS: Journal of Common Market Studies*, 62(6), 1538–1559. https://doi.org/10.1111/jcms.13601.

Warchoł, A. (2023). Od ofiary do światowego lidera. Estonia po cyberatakach z 2007 roku. *Politeja*, 5(86), 307–327. https://doi.org/10.12797/Politeja.20.2023.86.14.

Warin, T. (2024). Disinformation in the digital age: Impacts on democracy and strategies for mitigation (2024PR-03, For reflection). CIRANO. https://doi.org/10.54932/GQWB1497.

Warntjen, A. (2007). Steering the Union: The impact of the EU Presidency on legislative activity in the Council. *JCMS: Journal of Common Market Studies*, *45*(5), 1135–1157. https://doi.org/10.1111/j.1468-5965.2007.00763.x.

CHAPTER 4

Polish Presidency of Council of the European Union: Cybersecurity Priorities in the Face of Contemporary Digital Challenges

Kamila Sierzputowska and Izabela Kapsa

Abstract The European Union is currently confronting unprecedented cybersecurity challenges, including state-sponsored cyberattacks, vulnerabilities in emerging technologies, and a surge in cybercrime. These evolving threats are shaping the strategic agenda of the Polish Presidency of the Council of the European Union in 2025. Set against the backdrop of hybrid warfare and geopolitical instability -particularly in the Baltic Sea region – this chapter examines how Poland can contribute to strengthening the EU's cybersecurity and digital governance. Using institutional and legal analysis of EU legislation, along with a systemic review of governance mechanisms, the study identifies key priorities for enhancing regulatory coherence, coordination, and technological sovereignty across member states. The chapter highlights several strategic objectives of the Polish Presidency, including the operationalization of the EU Joint Cyber Unit, harmonized implementation of the NIS2 Directive, balanced regulation of artificial intelligence, and increased investment in cybersecurity education and research. Preliminary findings suggest progress in EU-level coordination, improved response to cyber incidents, and deepened cooperation with NATO. These developments are seen as crucial steps toward building Europe's digital immunity, addressing hybrid threats such as disinformation and cyberattacks, and safeguarding democratic stability. Ultimately, the chapter argues that Poland's leadership presents a timely opportunity to shape a future-oriented EU cybersecurity framework.

Keywords Polish Presidency – EU Council – Cybersecurity – Cyberdefence – Digital Challenges – EU Digital Policy

 HTTPS://DOI.ORG/10.1163/9789004763623-006

1 Introduction

The demanding global context and the third year of full-scale war in Ukraine, coupled with the persistence of hybrid warfare, are shaping the political priorities of the Polish Presidency of the Council of the European Union in 2025. The current geopolitical landscape underscores growing threats in Europe's digital sphere. Recent incidents in the Baltic Sea mirror a broader pattern of attacks on critical infrastructure and ICT systems. The escalation of such activities – often driven by states hostile to the EU – signals a continued increase in pressure on information technology and industrial systems. The European Union's cybersecurity landscape now faces unprecedented challenges, including sophisticated state-sponsored cyberattacks, vulnerabilities in emerging technologies, and the growing threat of cybercrime. In parallel, the rapid digitalization of public institutions and businesses has exposed critical gaps in regulatory compliance, cross-border coordination, and infrastructure resilience. As Poland assumes the EU Council Presidency, it is presented with a strategic opportunity to address these challenges and help shape the future direction of European cybersecurity policy amid mounting geopolitical instability, technological competition, and regulatory complexity.

Cybersecurity and digital governance have become cornerstones of EU policy, not only in the context of external relations and cyber defence but also in building resilient digital economies and secure public institutions. This chapter explores the priorities of the Polish Presidency in cybersecurity and digital technologies, analysing how Poland can contribute to the EU's strategic goals while addressing current political and technological challenges. Using institutional and legal analysis of EU legislation and policy declarations, as well as a systemic examination of cybersecurity governance mechanisms, the chapter seeks to answer the following questions: What are the cybersecurity and digital technology priorities of the Polish Presidency? What challenges does the EU face, given the current geopolitical situation and technological advancements in cybersecurity and digital governance?

From a historical and geopolitical perspective, Poland's role in countering hybrid threats, strengthening cyber resilience, and promoting cross-border cooperation was examined. The key cybersecurity priorities of the Polish Presidency included: operationalizing the EU Joint Cyber Unit to ensure a coordinated response to cyber threats; strengthening the implementation of the NIS2 Directive, fostering harmonized cybersecurity standards across member states; enhancing investments in cybersecurity education and innovation, reinforcing Europe's technological sovereignty. Additionally, Poland played a role in shaping EU regulations on artificial intelligence (AI), digital markets, and emerging technologies. The implementation of the AI Act was a critical challenge, requiring a balance between ensuring security and fostering innovation. As Deputy Minister for Digital Affairs, Rafał Rosiński, emphasised: "Our presidency's priorities focus on cybersecurity, artificial intelligence, emerging technologies, the development of the telecommunications market, digital diplomacy, and transatlantic cooperation. These are our main areas of focus" (Bochyńska, 2024).

By fostering regulatory efficiency, promoting innovation-friendly policies, and ensuring that cybersecurity measures strengthen rather than hinder technological progress, Poland's Presidency has the potential to contribute significantly to the EU's digital resilience and global competitiveness. The chapter examines how these strategic objectives can be translated into concrete policy actions and assess the broader implications of Poland's leadership in shaping the future of Europe's digital transformation.

2 Political Context of the Polish Presidency in 2025

At the beginning of 2025, for the second time, after 13 years, Poland took the presidency of the Council of the European Union. And, like in 2011, it was creating a presidential trio, together with Denmark and Cyprus, chairing council meetings at all levels and taking care of the continuity of EU works. Each country, covering the presidency of the EU Council for half a year, setting its own priorities (in accordance with the long-term EU strategy), sets the political direction of activities, determines the areas particularly promoted, and setting its goals. In shaping the priorities of the Presidency, the international situation plays a key role, forcing the adaption of the objectives of the conductivity to current challenges. When in the second half of 2011, Poland exercised its first presidency, its priorities were a response to the global challenges at the time, such as: euro area crisis, Arab revolutions, energy security, strengthening Eastern partnership, and work on the new EU budget (Programme, 2011).

Similarly, in 2025, this global situation determined the determination of its priorities. Difficult international conditions have shaped the political areas of Polish presidency. The key element was the third year of the full-scale war in Ukraine, which was a huge challenge, changing the paradigm in the field of international security. The impact of events in Ukraine on the safety of Europe, and especially the countries of the eastern NATO`s flank or the EU, (Kuczyńska-Zonik & Sierzputowska, 2023, pp. 42–44) was largely determined by geographical conditions, and also their consequences: military, political, social, or economic were translated into security and stability of all of Europe.

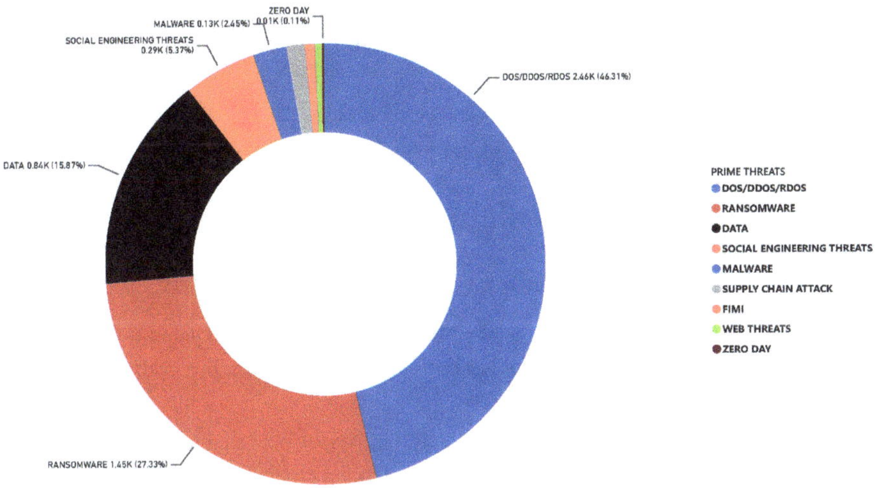

FIGURE 4.1 EU breakdown of number of threats by threat group (Source: European Union Agency for Cybersecurity (ENISA), ENISA Threat Landscape 2024, September 2024, Figure 5: "EU breakdown of number of threats by threat group", p. 12.).

Hybrid activities had a significant impact on the modern landscape of political challenges, in the region scale, as well as globally, next to Russian military activities. Hybrid war and disinformation, i.e. the key elements of the Russian strategy, in particular to weaken Ukraine, were the main tools of the destabilization and polarization of Western countries, undermining the purposefulness of international support for Ukraine (Reginia-Zacharski, 2015, p. 32). The severity of hybrid activities after 2022, also directed against the EU and NATO, constituted one of the serious consequences of the war of Ukraine, shaping its scope and nature of actions.

The cyber threat landscape in 2024 remains dynamic and concerning. Due to their effectiveness, DDoS (Distributed Denial of Service) attacks are among the most prevalent cyber threats. DDoS incidents represent one of the most critical risks to IT systems, with their scope and complexity continuously increasing. These attacks target both the public and private sectors and have become the most widely used tool in the ongoing Russia–Ukraine cyberwarfare (see Figure 4.1 for a breakdown of threats by threat group).

Ransomware operations, involving data hijacking and ransom demands for access restoration, continue to be a significant cyber threat. In 2024, the number of such attacks remained high but did not increase significantly compared to previous years (ENISA, 2024b, pp. 10–11).

Meanwhile, artificial intelligence (AI) plays an increasingly crucial role in cyberattacks and disinformation campaigns, which Russia has been actively

TABLE 4.1 Priorities of the Polish Presidency in the Council of the European Union 2025 (Source: *Programme of The Polish Presidency of The Council of The European Union,* January 1–June 30, 2025, pp. 4–7.)

Pillar	Pillar name	Challenges under the priority
I.	Defence and security	– Common and bold activities for European defence – Stronger defence industry – Supplementing the deficiencies in defence skills – Large debate on the financing of defence in the EU
II.	Protection of people and borders	– Internal security of EU citizens – Safety of EU's external boundaries – Limiting illegal migration – Strengthening the effectiveness of returns – Fight with hybrid threats – Ensuring the proper functioning of the Schengen zone – Increasing the potential of the EU and Member States in the field of population protection, resistance to disaster, rescue and humanitarian aid
III.	Resistance to foreign interference and disinformation	– Strengthening the immunity of democracy – Inhibition of polarization and radicalization – The ability to recognize and eliminate disinformation and foreign manipulation – Long-term operated in the field of civic education – Strengthening civil society – Combating disinformation, information manipulation and improving the EU's – Ability to prevent and reduce the effects of hostile activities in cyberspace
IV.	Ensuring security and freedom of business	– Improving support mechanisms for industry in areas important for the safety and construction of economic advantages, while ensuring maintaining equal opportunities within the EU

TABLE 4.1 Priorities of the Polish Presidency in the Council of the European (*cont.*)

Pillar	Pillar name	Challenges under the priority
V.	Energy transition	– Actions aimed at full departure from the import of Russian energy sources – EU energy security revision
VI.	Competitive and resilient agriculture	– Strong common agricultural policy
VII.	Health security	– Digital healthcare transformation – Improving EU drug safety – Diversification of drug supply chain – Support for their production in the EU

leveraging and adapting to both local and global events. Cybercriminals are using AI to generate phishing emails and malicious scripts, such as those created with FraudGPT. According to the ENISA report, EU member states, particularly Poland, are at heightened risk of cyber incidents due to geopolitical factors. These activities are closely linked to the geopolitical situation (ENISA, 2024b, p.12).There is no doubt that the winning direction of EU policy also had a win in the US presidential election – Donald Trump in November 2024. As a consequence, the existing international order was finally destroyed. Reconfiguration of the US place and role in the world has mobilized the EU to determine subsequent directions of activities, especially in matters related to international security. Although 2025 was expected to represent a breakthrough in EU -US relations, it ultimately reversed the positive trend that had characterized transatlantic relations over the previous four years. The victory of the Republican Donald Trump and his actions demonstrated the US departure from the formula of multilateralism, which, in consequence, caused political turbulence, a kind of international order and seriously influenced the shaping of the Union's political directions.

These progressive changes in safety conditions: the prolonged war of Russia with Ukraine, the escalation of global tensions, an increase in hybrid threats and disinformation actions, and dynamically progressing development of new technologies, supporting the military strategies of states increased the period of European instability in the uncertain war times.

In the face of the current international situation, the Polish Presidency in the EU Council in 2025 was focused on the broadly understood security, both in military, energy, economic and digital dimensions. The motto of today's

Presidency: *Security, Europe!* Is a reflection of the priorities of activities, focusing on ensuring stability in key areas. In the seven main areas that most of them were designated, focusing around the security issue (Programme, 2025).

3 Cybersecurity Priorities Under the Polish Presidency

Therefore, the priority of the Polish presidency was to guarantee security in various fields. The context of Poland's international conditions and therefore the location on the eastern flank of NATO, in the face of a difficult geopolitical situation, in the immediate vicinity of hostilities, played a fundamental role in the assumptions. The war in Ukraine has realized that the Russian cyberspace could be used as a tool for supporting military operations – the full-scale invasion was preceded by attempts to disturb the functioning of the Ukrainian energy and banking sector. After the failure of the cybernetic war against Ukraine, Russia intensified the attacks on its allies (Thales, 2024). Even before the annexation of Crimea in 2014, as well as after it, most incidents concerned primarily in Ukraine. Meanwhile, in the years 2016–2023 an increasing trend was observed, and the number of incidents increased by 80% (ENISA, 2024b). Since the outbreak of the war in Ukraine, European Union countries have become a target for cyber criminals from Russia.

Together with the invasion in 2022, NATO states, as well as countries supporting Ukraine, are the main purpose of attacks in cyberspace. Last year, a rapid increase in incidents was recorded in the European Union countries. These activities are part of the strategy of the hybrid war, which aims to destabilize NATO and EU member countries, and above all to weaken their political unity. The geopolitical situation indicates the growing threats in the digital space of Europe. The incidents recently observed in the Baltic Sea are reflected in attacks on ICT systems. The escalation of activities aimed at critical infrastructure, inspired by states hostile to Europe, is a signal that pressure on IT and industrial systems will continue to increase. Public administration, financial sector, transport sector, telecommunications, and energy sector are most often attacked. Although earlier the conflict was relatively not very severe, the European healthcare sector, public administration, industry, IT services, and the aviation sector are increasingly the target of attacks. The most threats come from Russia, which uses activities in cyberspace as part of the activities of the implemented strategy as part of the hybrid war.

In 2024, the activities of entities sponsored by the state were strongly focused around places of active armed conflicts or regional tensions. Hence, about 70% of Russian goals were in Ukraine or in NATO Member States

TABLE 4.2 Nation – state threat actor activity (Source: Own work based on: *Microsoft Threat Intelligence, 2024.*)

Sector	Region	Percentage
1	Europe & Central Asia	**68%**
2	North America	20%
3	Middle East & North America	5%
4	East Asia & Pacific	3%
5	Latin America & Caribbean	3%
6	South Asia	1%
7	Sub – Saharan Arica	1%

TABLE 4.3 Regional sample of activity levels observed (Source: Own work based on: *Microsoft Threat Intelligence, 2024.*)

Position	The most attacked country	Region
1	USA	North America, Latin America & Caribbean
2	Israel	Middle East & North Africa
3	Ukraine	Europe & Central Asia
4	United Kingdom	Europe & Central Asia
5	Poland	Europe & Central Asia
6	Germany	Europe & Central Asia
7	France	Europe & Central Asia

(Table 4.2). This is due to the fact that the Kremlin tries to collect intelligence data on western policy towards this war. The latest Microsoft report indicates Poland as one of the countries most exposed to cyber attacks (Microsoft, 2024). It takes the third place in Europe, right after Ukraine and Great Britain and the fifth on a global scale (Table 4.3). As a member of the European Union and NATO, Poland has become the target of criminal and hactic groups and those supported by the states. The attackers focused primarily on key state institutions, including ministries, government administration, services, and critical infrastructure, especially in energy and transport sectors. Among the most frequently recorded threats, attacks were dominated by interfering with the availability of websites and public services, attempts to infiltrate ICT systems, and violations of network security and communication platforms, such as e-mail (CSIRT RP, 2023). In this context, cyber security becomes one of the key priorities of the Polish Presidency in the European Union. In addition to political and energy issues, digital security has gained special importance at the EU level.

Strategic goals in the area of digitization focus around the four main aspects: cybersecurity, Artificial Intelligence (AI) and new digital technologies, effective implementation of digital regulations, and digital diplomacy (Programme, 2025). When treating cybersecurity as the foundation of Europe's stability in the 21st century, the basic cyber priority is to develop a common strategy in the protection of digital space and resistance to threats in the era of growing geopolitical challenges. Formal and informal meetings (e.g. Brussels, Warsaw, and Bydgoszcz,) are intended to establish joint activities in EU security. Investments in cybersecurity and improving cooperation with the private sector and industrial sector are key to strengthen European cybersecurity by exchanging knowledge, experience and best practices. The key priority of the Polish presidency is also the development of a uniform reacting system to large-scale incidents and a standardized model of reporting violations. Poland announced work on the update of the EU coordination framework in crisis situations in cyberspace (Cyber BluePrint) and improving cooperation between the civil and military sectors.

To enhance preparedness, detection, and response to cyber incidents across the EU, the Cyber Solidarity Act was adopted on February 4, 2025. Its primary objective was to strengthen and improve EU's capabilities in preparedness, response (mitigation), and identification of major cybersecurity threats, particularly large-scale attacks. The act introduced a European early warning system, co-developed by interconnected security monitoring centres in individual EU countries, which forms a key component of the cyber incident management mechanism.

Furthermore, the need to strengthen cybersecurity in the face of browning threats and geopolitical tensions also encompassed efforts to implement and enforce the NIS2 Directive and the Cyber Resilience Act. Poland's EU Council Presidency coincided with a crucial phase in the implementation of both regulations, which aimed to enhance the cyber resilience of EU member states and the private sector. Therefore, their effective implementation and enforcement posed a significant challenge for Poland as a leader in EU policy. The supervision of their implementation primarily focused on monitoring the adoption of NIS2 in member states and coordinating actions to enforce new cybersecurity standards, including incident reporting systems and sanctions against entities failing to comply with regulations.

The need to strengthen cybersecurity in all sectors was defined as the most urgent need, while the war in Ukraine was considered the main determinant of further strengthening by the EU resistance to cyber threats and intensify activities for cyber security and cyber defence (Warsaw Call, 2025). A clear signal of the determination of the Member States is a message on the challenges related to cybersecurity, developed in March 2025 at the informal meeting of ministers for digitization of EU countries, constituting guidelines for further actions as part of the implementation of the priorities of the Polish Presidency in the EU Council.

4 Digital Challenges Facing the EU and the Polish Presidency's Response

Cybersecurity is a cornerstone of the EU's current policy, not only in the context of external threats and geopolitical tensions but also in fostering a secure and digitally resilient European ecosystem. As the EU strives to develop digitally advanced member states and organizations, the increasing frequency of cyberattacks, data breaches, and AI-driven security threats presents a major challenge. Since the beginning of the 21st century, the European Commission has adopted numerous strategic documents and legal regulations, setting digitalization as a key objective. These include initiatives such as a cheaper, faster, and more secure Internet (2002); modern online public services (2005); e-government development (2016); investment in digital skills and the protection of civic rights in cyberspace (2020); and the most recent measures -DESI-based digital goal monitoring, biennial action plans, and multi-country projects (2021).

Current EU Digital Strategy (2021), the Digital Decade Policy Programme 2030 (2022), and the NIS2 Directive (2022) highlight the critical role of digitalization in public administration, businesses, and essential services while ensuring robust cyber resilience. Regulations such as the General Data Protection Regulation (GDPR, 2016) and the Cybersecurity Act (2019) establish legal frameworks for data protection, secure digital infrastructure, and risk management across the EU. As digital transformation accelerates, initiatives like the European Health Data Space (EHDS, 2022) and the Digital Services Act (DSA, 2022) further demonstrate the EU's commitment to modernizing institutions, enabling cross-border digital services, and safeguarding the integrity of digital markets against growing cyber threats. The transition towards fully digital governance, e-commerce, and AI-driven services requires stronger cyber resilience mechanisms, regulatory harmonization, and cross-border cooperation.

In this context, the Polish Presidency has prioritized strengthening cybersecurity legislation and policy implementation, aiming to advance harmonized cybersecurity regulations across member states and promote the effective enforcement of cybersecurity laws at both the national and EU levels. The key challenge here is finding a balance between introducing more regulations and avoiding excessive legal constraints that hinder digital development. The EU's complex regulatory landscape, among the most extensive globally in the technology sector, imposes strict compliance requirements, often perceived as limiting innovation rather than fostering it. For example, while the GDPR ensures strong data protection, it also restricts AI research due to limited access to training datasets. Similarly, the AI Act introduces new bureaucratic hurdles, slowing down technological deployment. As indicated by the Union of Employers and Entrepreneurs (ZPP, 2024), rather than introducing more regulations, Poland should focus on modernizing regulatory approaches, such as regulatory sandboxes that allow controlled experimentation with emerging technologies. Taking these aspects into account, Poland is focusing on improving the efficiency and coherence of digital regulations while striving to reduce administrative burdens for businesses. A key priority will be systematically reviewing the implementation of regulations and promoting the exchange of best practices.

The global competition in technology highlights the need to balance data privacy and AI development with innovation and competitiveness. While Europe prioritizes ethical AI and data protection, China and the USA. benefit from more flexible regulations, greater access to big data, and faster AI advancements. The Draghi Report (2024) underscores Europe's need to reduce reliance on foreign technology and strengthen its digital economy through an annual €750–800 billion investment agenda to keep pace with global leaders. Likewise, the CEE Digital Coalition (2024) emphasizes openness to new technologies, particularly AI, which will shape market trends for years to come and offer significant economic opportunities for Europe. In response, the Polish Presidency has made the digital economy, innovation, and AI development as key priorities. Its initiatives include supporting AI research and competence centres across the EU, adapting regulations to emerging technologies, and promoting digital transformation, especially for SMEs. Additionally, AI development and regulation will be central to Poland's agenda, ensuring both innovation and security.

To foster seamless technological collaboration between public administration, academia, and the business sector, the Polish Presidency recognizes the necessity of building international partnerships while simultaneously emphasizing the uniqueness of the European market and its strategic autonomy. As outlined in its priorities, key focus areas include digital trade and market competitiveness, which entail developing regulations for international digital trade, advancing work on regulatory impact assessments and digital tools to support the Single Market, strengthening the Capital Markets Union to finance digital transformation, and facilitating the introduction of the digital euro. Another priority, The Future of Work in a Digital Europe, addresses the role of AI in the workplace, automation, and its impact on employment, as well as regulatory discussions on teleworking and the right to disconnect. Moreover, the Presidency aims to support digital upskilling and reskilling initiatives, in alignment with the EU Strategic Framework on Health and Safety at Work 2021–2027, a policy that is also relevant to Digital Transformation in Healthcare. In this sector, priorities include expanding e-health services and cross-border digital health solutions, reinforcing the European Health Data Space (EHDS) to enhance healthcare data integration, addressing mental health challenges for children and adolescents in the digital era, and combatting climate disinformation to improve public awareness and health security.

Beyond these areas, digital technologies remain a core focus in digital governance, justice, and fundamental rights, with efforts to support the digitalization of justice under the European e-Justice Strategy 2024–2028, improve access to digital information for law enforcement while safeguarding fundamental rights, and promote gender equality in the digital society, particularly through the Digital Services Act. Additionally, digital transformation in key sectors remains a priority, covering transport (enhancing digital infrastructure within the Trans-European Transport Network (TEN-T)), maritime transport (supporting the digital transformation of the European Maritime Safety Agency (EMSA)), agriculture (promoting research, innovation, and digitalization to enhance food security), postal and telecom services (evaluating the regulatory framework for postal and courier services in the digital era), and telecommunications (continuing the reform of the EU telecoms market to ensure secure and resilient communication). As evident from these priorities, digital technologies are becoming an increasingly integral part of economic development and the functioning of state institutions, shaping the EU's path towards greater digital resilience and competitiveness.

5 Conclusion

The Polish Presidency of 2025 focuses primarily on broadly understood security. The war in Ukraine and the growing hybrid threats, including cyberattacks, forced the intensification of activities to increase Europe's resistance to the destabilizing activities of third countries, especially the Russian Federation. Hence, cybersecurity is the starting point and the foundation of contemporary EU policy, both in the face of external threats and geopolitical tensions, as well as in the context of building safe and cyber-resistant digital ecosystem in Europe.

Given the growing threats, cyber security has become the highest priority for the presidency of the EU Council in Poland, along with political and energy issues. The Polish program emphasizes cyber resistance, AI and digital technologies, digital regulations, and digital diplomats, aimed at increasing the defence of Europe against cybernetic threats. To achieve this, Poland aims to strengthen civilian and military (NATO) cooperation, improve the mechanisms of crisis response, and update EU Cyber BluePrint. Strengthening of cyber security in the EU through Cyber Solidarity Act, the NIS2, and Cyber Resilience Act directive are of key importance to increasing the resistance of the EU Member States to cyber threats, especially in the context of growing large-scale attacks. Polish 2025` Presidency played an important role in monitoring the implementation of these regulations and coordination of activities aimed at their effective enforcement.

Equally important, the Polish Presidency's cybersecurity and digital technology agenda reflects the EU's broader ambition to balance security, innovation, and economic growth in an increasingly digitalized and geopolitically complex world. As cybersecurity threats escalate and AI-driven transformations reshape economies, Poland's leadership is tasked with ensuring that regulatory frameworks enhance rather than hinder Europe's technological progress. The challenge lies in maintaining the EU's commitment to data protection and cybersecurity while fostering an environment conducive to innovation and investment in emerging technologies.

One of the key takeaways from the Polish Presidency's priorities is the recognition that Europe must adapt its regulatory approach to remain competitive on the global stage. The EU's current regulatory landscape, while ensuring high security and ethical standards, risks limiting the ability of European businesses and research institutions to scale up and compete internationally. Poland's focus on streamlining regulations, promoting regulatory sandboxes, and reducing administrative burdens aligns with the need for a more dynamic digital policy framework.

Ultimately, Poland's role in shaping EU digital policies will be judged by its ability to balance security with innovation, regulatory oversight with economic freedom, and European autonomy with global cooperation. The success of its initiatives will depend on how effectively it can build consensus among member states, encourage investment in strategic digital sectors, co-shape digital resilience, optimise reaction systems under EU cyber defence, and position Europe as a leader in digital governance without stifling its technological ambitions. Moreover, the Polish Presidency underlines the importance of digital governance as a cross-sectoral issue, integrating digital transformation into areas such as healthcare, labour markets, and critical infrastructure. The inclusion of cybersecurity within broader discussions on economic and geopolitical resilience indicates a shift towards a more holistic approach to digital security, recognizing its role not only in preventing cyber threats but also in ensuring the EU's long-term stability and competitiveness.

References

Bochyńska, N. (2024). *Prezydencja Polski w UE. Wiceminister wymienia priorytety.* https://cyberdefence24.pl/polityka-i-prawo/prezydencja-polski-w-ue-wiceminister-wymienia-priorytety. Accessed January 16, 2025.

CEE Digital Coalition (2024). *CEE Digital Coalition recommendations on improving the EU's competitiveness.* https://cyfrowapolska.org/wp-content/uploads/2024/08/August_2024_CEE-Digital-Coalition-recommendations-on-Improving-the-EUs-competitiveness.pdf. Accessed January 16, 2025.

CSIRT (2023). Raport o stanie bezpieczeństwa cyberprzestrzeni RP. https://csirt.gov.pl/cer/publikacje/raporty-o-stanie-bezpi/980,Raport-o-stanie-bezpieczenstwa-cyberprzestrzeni-RP-w-2023-roku.html.

Draghi, M. (2024). *The future of European competitiveness.* https://commission.europa.eu/topics/eu-competitiveness/draghi-report_en#paragraph_47059. Accessed January 16, 2025.

ENISA (2024a). ENISA Threat Landscape 2024. July 2023 to June 2024. September 2024. https://www.enisa.europa.eu/sites/default/files/2024-11/ENISA%20Threat%20Landscape%202024_0.pdf.

ENISA (2024b). Annual Report Trust Services Security Incidents 2023. https://www.enisa.europa.eu/sites/default/files/2024-12/2023_Trust_Services_Incident_Reporting_week50-2024_en_10.12.2024.pdf.

European Commission (2001). *eEurope 2002: Impact and Priorities.* Brussels, Belgium: European Union.

European Commission (2002). *eEurope 2005 action plan: an information society for everyone*. Brussels, Belgium: European Union.
European Commission (2016). *EU eGovernment Action Plan 2016–2020*. Brussels, Belgium: European Union.
European Parliament (2016). *General Data Protection Regulation*. Brussels, Belgium: European Union.
European Parliament (2019). *Cybersecurity Act*. Brussels, Belgium: European Union.
European Commission (2020). *Shaping Europe's digital future*. Brussels, Belgium: European Union.
European Commission (2021). *2030 Digital Compass: the European way for the Digital Decade*. Brussels, Belgium: European Union.
European Parliament (2022). *Digital Decade Policy Programme 2030*. Brussels, Belgium: European Union.
European Parliament (2022). *Measures for a high common level of cybersecurity across the Union, amending Regulation (EU) No 910/2014 and Directive (EU) 2018/1972, and repealing Directive (EU) 2016/1148 (NIS 2 Directive)*. Brussels, Belgium: European Union.
European Commission (2022). *Proposal for a Regulation of the European Parliament and of the Council on the European Health Data Space*. Brussels, Belgium: European Union.
European Parliament (2022). *Digital Services Act*. Brussels, Belgium: European Union.
European Union Agency for Cybersecurity (ENISA) (2024, September). ENISA Threat Landscape 2024. Athens, Greece: ENISA. https://www.enisa.europa.eu/publications/enisa-threat-landscape-2024.
Kuczyńska-Zonik, A., & Sierzputowska, K. (2023). The Baltic States in the Face of Russian. Aggression in Ukraine. In A. Kasińska-Metryka, K. Pałka-Suchojad (Eds.), *The Russia-Ukraine War of 2022. Faces of Modern Conflict*, Routledge.
Microsoft Digital Defence Report (2024). *The foundations and new frontiers of cybersecurity*. A Microsoft Threat Intelligence report. https://cdn-dynmedia-1.microsoft.com/is/content/microsoftcorp/microsoft/final/en-us/microsoft-brand/documents/Microsoft%20Digital%20Defense%20Report%202024%20%281%29.pdf.
Ministerstwo Cyfryzacji (2024). *Cyfrowe priorytety polskiej prezydencji w Radzie Unii Europejskiej*. https://www.gov.pl/web/cyfryzacja/cyfrowe-priorytety-polskiejprezydencji-w-radzie-unii-europejskiej.
Polish Presidency official website. https://polish-presidency.consilium.europa.eu/pl/wiadomosci/119-posiedzenie-hma-w-ramach-polskiej-prezydencji-w-radzie-ue/.
Programme of the Polish Presidency of the Council of the European Union (2011). 1 July -31 December. https://oide.sejm.gov.pl/oide/en/index.php?option=com_content&view=article&id=14741:tasks-of-the-parliament-during-the-polish-presidency&catid=45.

Programme of the Polish Presidency of the Council of the European Union (2025). 1 January 1- June 30. https://polish-presidency.consilium.europa.eu/media/zkcn0325/programme-of-the-polish-presidency-of-the-council-of-the-european-union.pdf.

Reginia-Zacharski, J. (2015). Ukraina 2014–2015: wojna (nie)hybrydowa. *Zeszyty Naukowe Akademii Obrony Narodowej*, *100*, Article 3.

Thales (2024). Data Threat Report 2024 Global Edition. https://cpl.thalesgroup.com/sites/default/files/content/DTR_pages/2024/2024-thales-data-threat-report-global-edition.pdf.

Warsaw Call on cybersecurity challenges (2025). https://polish-presidency.consilium.europa.eu/media/bkmbggso/warsaw-call_en.pdf.

Związek Pracodawców i Przedsiębiorców (2024). *Polska na czele cyfrowych przemian w UE. Priorytety dla Polskiej Prezydencji w Radzie Unii Europejskiej*. https://zpp.net.pl/polska-na-czele-cyfrowych-przemian-w-ue-priorytety-dla-polskiej-prezydencji-w-radzie-unii-europejskiej/.

CHAPTER 5

Countering Hybrid Threats in the European Union: Policy Evolution, Challenges, and Strategic Responses among the European Union Presidencies

Zdeněk Rod and Tomáš Kolomazník[1]

Abstract This chapter examines the evolution of EU policies aimed at countering hybrid threats across successive Council presidencies. The importance of this issue increased significantly following Russia's annexation of Crimea in 2014, prompting the EU to strengthen and institutionalise its response. Key developments included the creation of dedicated structures such as the Horizontal Working Party on Enhancing Resilience and Countering Hybrid Threats and the development of a "hybrid toolbox" under the leadership of France, Czechia, and Sweden. The chapter highlights how differing threat perceptions among Member States have shaped policy development, with Central and Eastern European countries generally advocating stronger measures than many Mediterranean states. It also examines evolving conceptual approaches to hybrid threats and the EU's efforts to enhance resilience through new institutional mechanisms and crisis-management practices. As the understanding of hybrid threats expanded to include cyberattacks, disinformation, and the instrumentalisation of migration, the EU's policy framework became increasingly comprehensive. The chapter concludes by assessing the strengths and limitations of the EU's approach and its implications for future Council presidencies.

Keywords hybrid threats – European Union – EU Council Presidency – resilience – disinformation – cybersecurity – Russia

1 Tomas Kolomaznik while writing this chapter was supported by Metropolitan University Prague research project no. E105-115 &2024 "Strategic communication as a tool for eliminating disinformation campaigns. Possibilities of applying foreign experience to the Czech environment" (2024 and 2025) funded by The Ministry of Education, Youth and Sports from the Funds for Support of Specific Research.

 | HTTPS://DOI.ORG/10.1163/9789004763623-007

1 Introduction

In the years following Russia's annexation of Crimea in 2014 and its aggression in the Donbas, the European Union found itself navigating a rapidly shifting security environment (Mälksoo, 2019; Dolya, 2016; Kalniete & Pildegovičs, 2021; European Council, 2024a; Hindrén, 2021). The annexation of Crimea – and perhaps more significantly, the collective lessons drawn from the conflict by US and European policymakers – fundamentally altered how European security, particularly along NATO's eastern flank, was conceptualised and pursued (Tallis & Šimečka, 2016, p. 2). What were once perceived as isolated incidents, from cyberattacks and disinformation campaigns to covert operations, were increasingly understood as components of a broader and more systematic challenge: hybrid threats. These activities, which deliberately operate below the threshold of open armed conflict, have presented a profound challenge to European security and political cohesion, compelling both EU institutions and Member States to reassess their approaches to resilience, deterrence, and response (Hindrén, 2021, p. 9, 12; Tallis & Šimečka, 2016). This shift occurred within a broader transatlantic context, where NATO and the United States had, after 2014, formally recognised hybrid warfare as a core security concern (Szymański, 2020) – a concept that, while gaining prominence after Crimea, had already been discussed within US military circles since the early 2000, particularly following the Iraq invasion in 2003, with the term itself popularised by US military thinker Frank Hoffman (2007).

Over EU presidencies, the imperative to address hybrid threats gained increasing prominence. The establishment of dedicated working groups, most notably the Horizontal Working Party on Enhancing Resilience and Countering Hybrid Threats (Lasoen, 2022, p. 7; European Council, 2024b), signalled a growing institutional commitment to tackling these challenges. Further momentum was provided by a tripartite initiative spearheaded by France, Czechia, and Sweden, culminating in the development of the EU Hybrid Toolbox (Bryjka, 2022; European Council, 2024c; Lasoen, 2022; Atlamazoglou & Moyer, 2022). Widely regarded as an effective instrument, its implementation has significantly strengthened the EU's capacity to counter hybrid threats (Jungwirth & Smith eds, 2023, pp. 5, 38). However, its relative novelty poses a challenge in terms of demonstrating its reliability in practice, as its effectiveness remains contingent on evolving security dynamics and the EU's ability to coordinate responses across its MS.

However, the development of policy in this domain has been far from straightforward. Perceptions of hybrid threats have varied considerably across EU member states, shaped by historical legacies, geopolitical realities, and differing security priorities. Central and Eastern European countries, due to their geographical proximity to Russia and direct exposure to Russian hybrid tactics since 2014, have been among the most vocal advocates for stronger countermeasures. This stance is further reinforced by their historical experiences of Soviet domination during the Cold War, which continues to shape a deep-seated strategic distrust towards Moscow. In contrast, Mediterranean states such as Spain and Italy initially exhibited greater caution, perceiving hybrid threats as less immediate and secondary to more pressing challenges, particularly irregular migration, which has dominated their security agendas since the onset of the European migration crisis in 2015. For these states, Russia has traditionally played a less prominent role in their geopolitical thinking and strategic planning (Authors' note).

Disagreements also emerged over terminology: whereas Germany and France preferred the terms "hybrid warfare" and "hybrid threat", Czechia and others pushed for "hybrid interference". Why? According to Bergaust and Sellevåg (2024), for instance, the distinction between hybrid warfare and hybrid interference centres on the level of visibility, intent, and the threshold of armed conflict. Hybrid warfare blends conventional military force with non-military means, often in the context of open confrontation or imminent conflict, while hybrid threats encompass a broad range of coercive actions that exploit societal vulnerabilities and may escalate into conflict. Hybrid interference, by contrast, refers specifically to covert activities aimed at manipulating a state's strategic interests and undermining social cohesion, all while remaining below the threshold of armed conflict. It focuses on non-military instruments, such as disinformation, economic coercion, and cyber operations, and operates in the grey zone, where plausible deniability and long-term strategic influence take precedence over immediate, visible confrontation. This distinction is critical for understanding how states exert influence without triggering conventional defensive responses.

Amid these debates, practical steps were taken to integrate hybrid threat considerations into EU decision-making. Several MSs introduced "hybrid threat attachés" within the Council, embedding expertise at the diplomatic level. Hybrid threats also became a key focus of EU crisis management exercises, such as EU Integrated Resolve 2024 (Insight EU Monitoring, 2022; European Council, 2024d), underscoring their growing importance within the Union's security framework. Moreover, the definition of hybrid threats itself has expanded. Initially centred on cyber operations and disinformation, it now encompasses broader tactics such as the instrumentalisation of migration, reflecting an evolving threat landscape.

As discussions on hybrid threats continue within the Council, it is crucial to analyse the trajectory of these policies and their impact on EU security strategy. This chapter seeks to explore how different EU presidencies have navigated the complexities of hybrid threat management, shedding light on the interplay between national perspectives, institutional responses, and emerging challenges. By examining these dynamics, we can extract valuable insights to inform future policymaking in an era where hybrid threats are likely to remain a persistent and evolving challenge.

This chapter examines how the evolution of countering hybrid threats policies and developments were shaped in the context of the European Union presidencies between 2016 and 2024. Particular attention is paid to how these presidencies influenced policy evolution, navigated emerging challenges, and shaped strategic responses at the EU level. The following research question guides the Chapter: *How did the countering hybrid threats policies evolve among the European Union presidencies from 2016 to 2024, and what challenges and strategic responses emerged during this period*? In addressing this question, the chapter seeks to illuminate both the formal processes of policy development and the political, institutional, and strategic factors that have shaped the EU's evolving response to hybrid threats over the past decade. Moreover, 2016 was chosen as a starting date since the evolution of hybrid threat policies in the EU was kicked off by adopting the *Joint Framework on Countering Hybrid Threats: A European Union Response* in 2016.

In terms of methodology, this chapter adopts a qualitative and process-oriented methodology, combining document analysis, process tracing, and comparative case studies to examine how successive EU presidencies shaped the Union's evolving response to hybrid threats between 2014 and 2024. The analysis draws on primary sources, including presidency programmes, Council conclusions, EU policy documents, and communications from key institutions such as the EEAS and Hybrid Fusion Cell, complemented by academic literature, think-tank reports, and institutional evaluations from bodies such as the EU Hybrid Centre of Excellence. Process tracing is used to identify key moments of policy change, focusing on how hybrid threat agendas were shaped by the national preferences of successive presidencies, the disruptive impact of external crises – particularly Russian aggression and disinformation campaigns – and evolving institutional dynamics within the EU. The chapter focuses on a targeted selection of presidencies that played key roles in advancing the EU's hybrid threat agenda, including the Netherlands (2016), and Estonia (2017), as well as later presidencies – Bulgaria and Austria (2018), Finland (2019), France and Czechia (2022), Sweden (2023), and Belgium and Hungary (2024). Their contributions range from the recognition of hybrid threats in EU security policy to the creation of coordination mechanisms, strategic frameworks, and operational tools such as the Hybrid Rapid Response Teams. The comparative analysis highlights both continuities and divergences across these presidencies, offering insight into the broader institutional and strategic evolution of the EU's hybrid threat policy over the past decade.

The chapter finds that EU presidencies have played a crucial role in shaping the Union's evolving response to hybrid threats between 2016 and 2024, with certain presidencies acting as key agenda-setters, particularly those from member states with direct experience of hybrid operations, such as Latvia, Estonia, Finland, Poland, and the Czech Republic. The evolution of EU hybrid threat policy has been driven by a combination of external crises – most notably Russian aggression – and the political agency of individual presidencies that sought to institutionalise resilience, improve situational awareness, and enhance the EU's strategic communication capabilities. The 2014 annexation of Crimea and the 2022 full-scale invasion of Ukraine were particularly pivotal moments, accelerating both policy innovation and institutional consolidation, including the creation of the EU Hybrid Fusion Cell, the Hybrid Toolbox, and the Hybrid Rapid Response Teams. While these developments highlight growing institutional capacity, the chapter also notes that the long-term prominence of hybrid threats within the EU's security agenda remains contingent on the political priorities of future presidencies, the evolving technological landscape – particularly the disruptive potential of artificial intelligence – and broader debates around the regulation of disinformation and the protection of fundamental rights.

Lastly, the chapter is structured into three main parts. The first part examines the underlying assumptions of the European Union's policy on countering hybrid threats, tracing its evolution from 2014 to 2024. The second part focuses on the development of EU hybrid threat policies, with particular attention to the Dutch Presidency in 2016, which served as a pivotal moment and foundation for subsequent progress leading up to 2024. The final part offers a summary of the key findings, alongside a critical reflection on the strengths, limitations, and broader implications of the EU's hybrid threat policies over the past decade.

2 The Assumptions of the European Union's Policy on Countering Hybrid Threats until 2015

The basis of the European Union's approach to counter hybrid threats can be traced to 2014, when the cornerstone for the upcoming policy initiatives to counter hybrid threats was established. The EU has addressed hybrid threats in various Council meetings since 2014. Since 2014, the EU has adopted a range of legislation in this field, including in policy areas such as energy security, safeguarding of critical infrastructure, data protection, screening of foreign investments, and transparency of political funding, among others (Kalniete & Pildegovičs, 2021, p. 26; Hindrén, 2021, p. 12; Szymański, 2020).

If we zoom in on the particular initiatives in 2014, it can be found that in 2014, the EU adopted several important initiatives aimed at strengthening its overall security framework across multiple domains. These included the introduction of a new approach to the detection and mitigation of CBRN-E risks, enhancing the EU's capacity to address chemical, biological, radiological, nuclear, and explosive threats. In the same year, the EU adopted a comprehensive European Energy Security Strategy, which reflected growing concerns over the resilience of the Union's energy supplies, particularly in light of geopolitical instability in its eastern neighbourhood. Additionally, the EU adopted the Maritime Security Strategy and its accompanying Action Plan, which provided a coordinated framework to safeguard the Union's maritime interests and address a range of emerging threats, including hybrid challenges. Together, these initiatives marked an important step towards a more integrated and cross-sectoral approach to European security policy (Fiott & Parkes, 2019, p. 14, 29–30).

Following 2014, 2015 recorded further developments to deepen initiatives to counter hybrid threats. In 2015, the EU took several important steps to strengthen its response to hybrid threats, emphasising disinformation, border security, and resilience-building. One of the most notable developments was the establishment of the East StratCom Task Force within the EEAS, specifically tasked with countering Russian disinformation campaigns targeting the EU and its eastern neighbourhood (ibid, p. 5–7, 14) – notably Baltic states and Poland were instrumental in advocating for a coordinated EU response to counter disinformation and enhance strategic communications in the Eastern Partnership region (authors' note). This marked a significant institutional recognition of the link between strategic communication and hybrid threats. Alongside this, the EU adopted the European Agenda on Security, which set out priorities for strengthening internal security, including cyber resilience and countering hybrid tactics. The year also saw a growing awareness of the vulnerabilities in the Schengen system, particularly in light of the Paris terrorist attacks and the instrumentalisation of migration flows in Russian disinformation narratives. These developments reflected a growing understanding within the EU that hybrid threats encompass a broad spectrum of tools and tactics, requiring a more coordinated and adaptive response across multiple policy areas (Fiott & Parkes, 2019, p. 5–7, 14).

In the context of 2015 events, the EU took other significant steps to institutionalise its response to hybrid threats, recognising them as a key horizontal security challenge requiring enhanced coordination between internal and external security policies. The Council Conclusions 8971/15 on the Common Security and Defence Policy (CSDP) called for the development of a joint framework to counter hybrid threats, to be prepared by the High Representative in cooperation with the Commission, the European Defence Agency, and MSs. This framework was intended to enhance resilience through concrete proposals, incorporating cyber defence, early warning, strategic communications, and cooperation with external actors such as Europol, FRONTEX, and the European Gendarmerie Force.

The Council Conclusions 8971/15 marked an important step in recognising hybrid threats, but several key limitations hindered its effectiveness. It lacked a clear, commonly accepted definition of hybrid threats, allowing for divergent interpretations among MSs and complicating efforts to develop a coherent response. While it called for better coordination between internal and external security, it offered few concrete mechanisms to achieve this, leading to fragmented implementation across policy areas. The document also downplayed the role of strategic communication, leaving the EU poorly equipped to counter disinformation campaigns. Its reliance on MS-led implementation, rather than embedding strong EU-level operational processes, further weakened coherence. Finally, the decision to mandate the High Representative to draft a joint framework only after adoption reflected a reactive approach, leaving the EU without the immediate tools needed to address rapidly evolving hybrid threats (authors note).

Although considerable efforts were undertaken between 2014 and 2015, as outlined above, the EU's capacity to effectively counter hybrid threats remained limited (Fiott & Parkes, 2019, p. 42). One the one hand, a relative success was the establishment of East StratCom Task Force which became responsible for analysing disinformation trends and exposing disinformation narratives originating from Russia (Szymański, 2020). On the other hand, what was largely missing was a set of concrete instruments to translate strategies and legislative frameworks into practical operational measures. The absence of clear operationalisation left the EU poorly equipped to respond to the increasingly sophisticated and multidimensional nature of hybrid threats. Moreover, it can also be concluded that at that time, the EU was not able to deter Russian hybrid attack. These shortcomings were progressively addressed in the years that followed, as successive EU presidencies, in close cooperation with MSs, played a central role in refining the institutional architecture and developing practical tools to enhance resilience, preparedness, and deterrence. The following part maps this policy evolution from 2016 to 2024, demonstrating how the EU's response to hybrid threats gradually matured through a combination of institutional learning, political prioritisation, and external pressures.

3 Perspectives on the Evolution of Hybrid Threats Based among European Presidencies in 2016–2024

3.1 Key Developments under the Dutch Presidency and the 2016 Joint Framework

The year 2016 witnessed further developments in the EU's approach to hybrid threats, with the Council adopting new conclusions on the matter at the close of the Dutch Presidency in June. These conclusions underscored the increasing use of hybrid strategies by both state and non-state actors in the EU's immediate and wider neighbourhood, highlighting the need for a swift and coordinated response to protect the EU, its MSs, and its partners. While reaffirming the primary responsibility of MSs for security and defence, the Council stressed the necessity of mobilising EU instruments in accordance with the European Council's conclusions of June 2015 and the CSDP conclusions of May 2015 (Council of the EU, 2016). Moreover, by 2016, awareness of hybrid threats and foreign interference in democratic processes had increased significantly within Western political and academic circles. This heightened attention also gave rise to a rapidly expanding body of research on hybrid warfare, though its widespread use has led some scholars to criticise the term as a buzzword and question its value as a meaningful analytical concept (Kalniete & Pildegovičs, 2021, p. 24).

Furthermore, as a result of the 2014–2015 Council meeting, the *Joint Framework on Countering Hybrid Threats: A European Union Response* (2016) was adopted. The framework provided an initial shared understanding of these threats and served as the first blueprint for the EU's response (Hindrén, 2021, p. 12). The framework outlines the EU's strategic approach to hybrid threats, recognising them as a combination of coercive and subversive tactics employed by both state and non-state actors, often operating below the threshold of formal warfare (European Commission & High Representative, 2016, p. 2; Szymański, 2020). The document underscores the necessity of a coordinated response, structured around improving awareness, building resilience, and enhancing cooperation. To enhance situational awareness, it proposes the establishment of the EU Hybrid Fusion Cell within the EU Intelligence and Situation Centre (EU INTCEN), tasked with analysing and sharing intelligence on hybrid activities. Strategic communication is identified as a key pillar in countering disinformation, advocating for a proactive approach to media monitoring and information campaigns (ibid, p. 4–5). Those initiatives laid the first steps towards a system of coordination at the EU and national levels in case of a hybrid attack. This was significant not only because EU MSs and institutions themselves could become targets of a hybrid campaign, but also because, if an MS were to invoke the Solidarity Clause (Article 222 TFEU) in response to such an attack, the EU institutions would be expected to coordinate the response – a task they would struggle to fulfil if incapacitated (Fiott & Parkes, 2019, p. 7, 22).

In terms of resilience-building, the framework highlighted the need to protect critical infrastructure, including energy networks, transport, and space assets, alongside strengthening cybersecurity through closer industry cooperation and EU-wide cyber strategies. Addressing hybrid threat financing and enhancing defence capabilities are also central to the strategy, with the European Defence Agency identified as a key actor in capability development (European Commission & High Representative, 2016, p. 6–9). Additionally, the document emphasises the role of crisis response mechanisms, proposing the application of the EU Solidarity Clause (Article 222 TFEU) and the Mutual Defence Clause (Article 42(7) TEU) in the event of significant hybrid attacks (ibid, p. 16–17). However, zooming in on the solidarity and mutual defence clause, at that time, both clauses have faced several limitations. The EU's mutual defence (Article 42.7 TEU) and solidarity (Article 222 TFEU) clauses are hamstrung by deliberate vagueness, lack of clear triggers, and the need for unanimity, which renders them slow and impractical in crises. Their flexibility, designed to respect national sovereignty and accommodate diverse security postures, ultimately weakens their credibility and operational value. Subordination to NATO further undermines Article 42.7, while Article 222's procedural hurdles and lack of defined military commitments make it little more than a symbolic gesture. Both clauses reflect the deeper ambiguity over the EU's role in defence, leaving member states unwilling to trust the EU as a reliable security guarantor (Autio, 2024).

Recognising the transnational nature of hybrid threats, the framework also called for closer EU–NATO cooperation in intelligence-sharing, cybersecurity, and joint crisis management exercises (European Commission & High Representative, 2016, p. 2). Regarding the EU–NATO cooperation, there were promising advances in strengthening resilience against hybrid threats. As natural partners with aligned strategic outlooks and shared concerns over Russian hybrid activities, both organisations formalised their cooperation through a 2016 joint declaration, which identified countering hybrid threats as a key priority (Szymański, 2020; Kalniete & Pildegovičs, 2021, p. 28). However, the most tangible manifestation of EU–NATO cooperation in the realm of hybrid threats materialised in 2017 with the establishment of the European Centre of Excellence for Countering Hybrid Threats (Hybrid CoE), Finland being instrumental beyond the whole initiative. This Helsinki-based institution, bringing together participating MSs from both the EU and NATO, serves as a platform for enhancing strategic dialogue, fostering information exchange, and strengthening joint analytical and training capabilities to improve resilience against hybrid challenges. Its creation symbolised a practical step forward in aligning transatlantic and European approaches to these increasingly complex threat landscapes (Kalniete & Pildegovičs, 2021, p. 28).

The framework ultimately laid the groundwork for a more structured EU response to hybrid threats, reflecting an evolving security environment that demands greater strategic coordination and institutional resilience. To ensure ongoing progress, the Council tasked the Commission and the High Representative with delivering a report by July 2017 to evaluate the implementation of these measures (Council of the EU, 2016).

3.2 The Estonian Presidency and the Formation of FoP in 2017

Things in the domain of hybrid threats within the EU further developed gradually under the Estonian presidency in 2017, reflecting also discussions conducted in 2016 as mentioned above. The most significant effort during this time was setting the Friends of the Presidency Group for Countering Hybrid Threats (FoP) up. FoP was meant to play a significant role in advancing the EU's efforts to address hybrid threats, which encompass a broad spectrum of tactics ranging from disinformation campaigns to the use of military force. Initially conceived as a coordination mechanism, the FoP had since been granted an extended and expanded mandate, allowing it to oversee ongoing EU initiatives, identify gaps, prevent duplication of efforts, and provide support for political decision-making within Council structures (Council of the EU, 2020, p. 2; Hagelstam, 2018).

However, although the terms of reference were agreed and work commenced, the group's overall effectiveness rested almost entirely on the extent of contributions from individual MSs – a structural weakness given the uneven levels of engagement and commitment across the Union. There remained a pronounced divide between Mediterranean MSs, which tended to place less emphasis on hybrid threats, and the Baltic states, for whom such threats represented a far more immediate and existential concern – however, due to the classified nature of FoP meetings, it is not possible to name exact MS which supported or opposed FoP work. Ultimately, the success of the initiative hinged upon the speed and thoroughness with which Member States undertook their respective national surveys on hybrid threats, as these would have provided crucial insights into the wider state of Europe's vulnerabilities and preparedness (EU Monitor, 2017).

Since April 2017, the group had seen a substantial increase in participation, with the number of MSs involved having doubled, reflecting growing recognition of the importance of countering hybrid threats at the EU level, although some areas such as disinformation lacked to be considered as important security challenges (Fiott & Parkes, 2019, p. 5). Complementing these efforts, in July 2019, the establishment of the Horizontal Working Party on Enhancing Resilience and Countering Hybrid Threats (HWP ERCHT) further institutionalised coordination, information-sharing, and resilience-building measures within the EU framework, underscoring the Union's commitment to addressing hybrid threats in a systematic and cooperative manner (ibid).

3.3 Policy Developments under the Bulgarian and Austrian Presidencies in 2018

Following a brief hiatus, discussions on hybrid threats resumed in 2018 under the Bulgarian and Austrian Presidencies, marking another policy development phase within the Council. However, before we delve deeper into the events of 2018, it is vital to highlight the context of this period. In 2018, the European Parliament expressed concern regarding Russia's disinformation operations and cyberattacks targeting European nations. In June 2018, the Parliament adopted a resolution 2018/2004(INI) on cyber defence, highlighting the serious repercussions of cyberattacks against EU infrastructure and the role of such activities in aiding disinformation campaigns. The resolution underscored the necessity for a joint EU response to these threats. Besides, looking overseas to the US, in 2018, Russia's interference operations continued, with ongoing disinformation campaigns, cyberattacks by GRU officers, and social media manipulation aimed at deepening divisions in American society. A DOJ indictment revealed that Russia's state-sponsored Internet Research Agency (IRA) significantly increased its budget after 2016, showing that Russian interference was not limited to elections but part of a sustained strategy to weaken U.S. democracy (GMF, 2019).

Coming back to the EU, during this period, several key documents were adopted. Notably, in March 2018, the European Council meeting (EUCO 1/18), chaired by Bulgaria, reaffirmed the urgency of strengthening the EU's and MSs capabilities to address hybrid threats. Particular emphasis was placed on enhancing resilience in the areas of cybersecurity, strategic communication, and counter-intelligence. Recognising the evolving nature of hybrid threats, the European Council called upon the European Commission and the High Representative to advance efforts in these domains and to report on progress by the June European Council. This signalled a renewed commitment to reinforcing the EU's strategic response, ensuring that hybrid threats remained a central concern within the broader framework of European security policy (European Council, 2018a, p. 4).

The end of Bulgarian presidency confirmed that hybrid threats are an important matter and that the EU should continue working on operationalisation of concrete measures to counter them (European Council, 2018b, p. 5). Concretely, the EU took several key steps to address hybrid threats and disinformation, beginning with Exercise Chimera in February, a DG SANTE tabletop exercise focused on hybrid threats, followed in March by the launch of a cyber platform for education, training, and exercises under the ESDC. In April, the European Council issued conclusions in response to the Salisbury attack, highlighting the threat of hybrid activities, and in June, the Commission published a communication on tackling online disinformation, setting out a European approach to countering information manipulation. Later in June, the EU released a report on the implementation of the 2016 Joint Framework on Countering Hybrid Threats, followed by a communication on increasing resilience and strengthening capabilities, further embedding hybrid threat awareness into the EU's security agenda (Fiott & Parkes, 2019, p. 15).

On top of that, the European Council underscored the necessity of enhancing resilience and strengthening capabilities to counter hybrid threats, welcoming the Joint Communication of the High Representative and the Commission, along with the accompanying reports adopted on 13 June 2018. The Council also stressed the importance of further developing the EU's strategic communication strategy in cooperation with MSs, recognising that countering hybrid threats requires a comprehensive and long-term approach, including sustained resource allocation. Furthermore, the EU's efforts should be complemented by close coordination with international organisations, particularly NATO, while ensuring the Union's own analytical autonomy. The Council called for continued progress in these areas, with regular reviews to assess advancements (Council of the EU, 2018, p. 9).

The second half of 2018 experienced several initiatives to strengthen the hybrid threats resilience. In December 2018, the EU intensified its efforts to combat hybrid threats and disinformation with the EU Hybrid Exercise 2018, which tested institutional responses to hybrid scenarios (Fiott & Parkes, 2019, p. 7, 15). Although the nature of the exercise is classified, it can be concluded that the EU Hybrid Exercise was a comprehensive, biennial double exercise designed to enhance the EU's capacity to respond to complex hybrid crises involving internal and external dimensions. The exercise combined a CSDP planning phase with an event-driven crisis response scenario, simulating hybrid threats from both state and non-state actors. It tested the coordination between political, politico-strategic, strategic, and operational levels and improved liaison procedures between the EU and NATO under the Parallel and Coordinated Exercise (PACE) concept. The scenario spanned Europe, maritime regions (Mediterranean, Atlantic, and Baltic), and Africa, requiring a coordinated civilian and military CSDP deployment. Greece hosted the EU Operational Headquarters (OHQ) in Larissa and the Force Headquarters (FHQ) in Nea Santa, marking the first time that an EU OHQ led a multilayer hybrid exercise under the PACE framework (EL EU OHQ, 2018).

Shortly after, the Action Plan against Disinformation was unveiled, outlining coordinated measures to detect, prevent, and respond to disinformation campaigns (Fiott & Parkes, 2019, p. 15). The Action Plan highlighted that the component of hybrid warfare constitutes a strategic challenge to democratic governance, necessitating a sustained and coordinated response that remains firmly rooted in respect for fundamental rights. Acknowledging the multidimensional nature of this threat, the Council emphasised the need for a comprehensive strategy that integrates both internal and external dimensions, ensuring effective coordination and adequate resource allocation. Central to this approach is the expedited and systematic implementation of the Joint Action Plan on Disinformation, designed to enhance EU capacities, reinforce cooperation between the EU and MSs, engage private sector stakeholders, and build societal resilience against malign influence operations. Furthermore, in light of the significant risks posed to democratic processes, the European Council stressed the urgency of decisive action at both European and national levels to safeguard electoral integrity, thereby fortifying democratic institutions against external interference and manipulation (Bryjka, 2022; European Council, 2018c, p. 4).

In 2018, the EU's hybrid threat policy became more operational, building on the 2016 Joint Framework, with new tools such as EU Hybrid Exercise Multilayer 18 to test EU–NATO coordination, a cyber training platform under the ESDC, and the formal establishment of ENISA to strengthen cybersecurity. The

Action Plan against Disinformation also marked a shift toward practical responses to foreign information manipulation. However, while the Bulgarian and Austrian presidencies successfully highlighted the need for better institutional coordination and closer EU–NATO cooperation, they fell short in addressing critical enforcement gaps, particularly in mandatory intelligence sharing, consistent disinformation countermeasures, and binding operational mechanisms. Their main contribution lay inlaying the groundwork for later advancements – such as the EU Hybrid Toolbox and Hybrid Rapid Response Teams – with limited immediate impact on the EU's operational security capabilities, leaving much of the practical follow-through to subsequent presidencies.

3.4 The Establishment of HWP ERCHT in 2019

The establishment of the Horizontal Working Party on Enhancing Resilience and Countering Hybrid Threats (HWP ERCHT) in July 2019 marked a significant development in the EU's institutional approach to hybrid threats. Emerging as a culmination of efforts under the Romanian Presidency, with contributions from preceding presidencies, its formalisation coincided with the commencement of the Finnish Presidency (Council of the EU, 2020, pp. 2, 7).

The HWP ERCHT was designed to enhance strategic and horizontal coordination among MSs, focusing on key areas such as state and societal resilience, strategic communication, and countering disinformation. Its role extends to facilitating information exchange, promoting best practices, and strengthening the EU's collective capacity to address hybrid threats. Notably, the report highlights that the Friends of the Presidency Group, which previously oversaw hybrid threat-related discussions, was dissolved, with its mandate rescinded. This transition reflects a shift towards a more structured and institutionalised framework for coordination, reinforcing the EU's commitment to a comprehensive and systematic response to hybrid threats (ibid).

The benefit of HWP ERCHT is undoubtedly the improved coordination among EU member states and the sharing of information and best practices. On the other hand, there are still several challenges to the effective functioning of HWP ERCHT. First, it is the consistent involvement of all countries in the group's activities. Second, it is the quality and flexibility of shared information and cooperation with other partners like the European Centre of Excellence for Countering Hybrid Threats. Last, it is the implementation of individual EU policies within the member states (Hybrid threats, 2023). The effectiveness of the functioning of the HWP ERCHT is also influenced mainly by national priorities. Eastern European states are more concerned with Russian hybrid threats than Western European states. The decisions taken have no legal impact, so it is necessary to ensure their transformation into legislation. However, this process requires broad acceptance among member states (Bajarūnas, 2024).

3.5 Strategic Compass 2022: Beginnings of Taking Hybrid Threats Seriously

In the context of geopolitical shifts creating more significant pressure on the EU in the economic, energy, and security areas, and especially in the context of the war in Ukraine and the growing risks of hybrid threats, the European Union has been forced to intensify steps to ensure the security and resilience of society. In January 2022, France took over the EU presidency. Security and hybrid threats were among the key topics of the presidency. As part of it, the European Council adopted the "Strategic Compass for Security and Defence" on March 25, 2022. The document represents a roadmap of steps and measures to strengthen the EU's security and defence (IISS, 2022).

The approval of this document under the French presidency is no coincidence. The Strategic Compass expresses, among other things, Emmanuel Macron's vision for the future of Europe, particularly the pursuit of "European sovereignty". The Strategic Compass is thus a tool to defend European values and interests on the global stage. The adopted compass was also an expression of the strengthening of Emmanuel Macron's ambitions as a successful international leader who managed to promote French interests in Europe, especially as a key player in building the European security architecture (Ruppert, 2022). In this context, however, we must note that this document was initiated during the German presidency in the second half of 2020. Germany's approach to the Strategic Compass reflects its long-term position, where, on the one hand, it calls for strengthening European security and, on the other hand, it advocates a strong Euro-Atlantic bond (Mölling, Schütz2020). It was Germany's initial role in this process that was reflected in this document, especially when it comes to strengthening EU–NATO cooperation.

The Compass is a comprehensive document that defines EU measures in response to crises, developing a rapid deployment capacity of up to 5,000 troops, strengthening command and control structures, increasing investments and defence capacities, relations with other partners outside the EU, etc. A significant part of the compass is dedicated to hybrid threats, cyber diplomacy and foreign influence. The aim is to strengthen the EU's resilience and its ability to counter hybrid threats and cyber attacks. This part of the document responds to the growing activities of state and non-state actors who use various tools to weaken the EU.

In this context, the EU decided to create a toolbox based on the experiences of individual MSs, which should be a combination of civilian and military tools that can be used against hybrid threats, the so-called EU Hybrid Toolbox. The next step was strengthening the EU's cyber diplomacy toolbox, adopted by the EU in 2017. The aim was to improve preventive measures and sanctions against external actors carrying out non-military activities in cyberspace against MSs. The document also mentions the EU's Joint Cyber Unit, which has been under development since 2021 (Council of the EU, 2022a).

Suppose we compare the Strategic Compass with the Global Strategy on Foreign and Security Policy (EUGS), 2016. The Strategic Compass is more detailed in assessing the full spectrum of threats. The Compass outlines over 80 concrete actions with clear deadlines, providing a roadmap for enhancing EU security and defence capabilities over the next decade and further focusing on concrete steps to enhance EU security and defence capabilities (Vázquez,-quez, 2022).

3.6 The EU Hybrid Rapid Response Teams

The EU Strategic Compass also highlighted the importance of strengthening the Union's ability to respond to a hybrid crisis. In this context, it should be noted that attitudes towards hybrid graves have varied across the EU. On the one hand, there are a group of countries for which this topic is very relevant, such as the Baltic States, Poland, Finland, and Sweden. On the other hand, there are countries for which this topic is not a major priority, such as Hungary, Austria, or Italy (Gressel, 2019).

In this context, the EU announced the creation of Hybrid Rapid Response Teams (EURHRT). These teams aim to support EU MSs and partner countries when they face hybrid attacks. The aim is to create teams of experts with various specialisations, such as experts in strategic communication, cybersecurity, counterintelligence, energy security, and critical infrastructure protection, which military experts can also supplement. The idea was that the aforementioned experts could then be sent to individual countries and act there as an advisory team (Bryjka, 2022). The teams were to be modelled after the NATO Counter-Hybrid Support Teams (CHST) established in 2018.

However, it cannot be said that these teams are entirely similar to the NATO CHST. The differences between the teams are mainly in the deployment scope – the NATO CHST is limited to NATO MSs, while the European team can also be deployed to partner states. The EU teams focus mainly on "non-military" threats (Lasoen, 2022). The EU Hybrid Rapid Response Team was first activated in late February 2022 to support Ukraine, but it was not deployed due to the Russian invasion. The team was only deployed in Moldova in November 2022. In March 2023, a Cyber Rapid Response Team was deployed in Mozambique, addressing issues primarily in cybersecurity (Grossman, 2023). If we were to evaluate the effectiveness of the deployment in Moldova, it would be difficult to assess the results from a long-term perspective due to the situation in the region. However, recommendations for improving the functioning and coordination of state administration offices resulted. Lack of coordination and a lack of quality personnel led to reactive measures rather than a long-term strategy (Sweijs, 2023).

In May 2024, the EU's General Affairs Council (GAC) approved the guiding framework for the practical establishment of the EU Hybrid Rapid Response Teams. At the same time, the creation of the Emergency Response Coordination Centre (ERCC), which is a coordination hub for deployment to partner countries, was approved (Council of the EU, 2024e).

3.7 France–Czechia–Sweden Tripartite Initiative and the Development of the "Hybrid Toolbox" 2022–2023

The EU Presidencies of France, the Czech Republic, and Sweden represent a significant stage in the shift in the approach to hybrid threats. This shift was driven by the dynamics within the security environment, particularly the war in Ukraine. Another factor was that security and hybrid threats were important issues for all three countries. France has long been trying to strengthen the European security identity. The Czech Republic was and still is a very active supporter of Ukraine, having experience with influence operations and disinformation campaigns by Russia. Sweden is also in a position where it faces influence operations and has long-term experience in building a resilient society. All these facts were reflected in the mentioned countries' activities during their presidency.

The French Presidency brought a basic outline and direction of activities in this area. Above all, in the form of the Strategic Compass, which defined specific steps in the form of the Hybrid toolbox and the building of EU Hybrid Rapid Response Teams.

The Czech Presidency has taken another significant step on the issue of hybrid threats. The Foreign Information Manipulation and Interference (FIMI) Toolbox was approved as part of the Czech Presidency. This is a set of measures aimed at identifying, analysing, and combating foreign manipulation of information and disinformation. This concept refers to the activities of foreign actors who seek to influence public opinion, democratic processes, or the stability of EU MSs through disinformation campaigns, false narratives, or cyber operations (Council of the EU, 2022a). In December 2022, the Implementation Guidelines for the EU Coordinated Response Framework to Hybrid Campaigns were approved. These guidelines are a key part of the EU Hybrid Toolbox. The approval itself meant introducing specific procedures and measures to counter hybrid threats. It enabled better coordination between MSs and EU institutions and supported using sanctions and other defensive tools against state and non-state actors involved in hybrid campaigns (Council of the EU, 2022c).

Sweden, during its presidency in the first half of 2023, has significantly contributed to the fight against hybrid threats with its experience in building a resilient society. The Swedish Presidency strengthened EU–NATO cooperation in the fight against hybrid threats, leading to more intensive information sharing and coordination of activities. Furthermore, emphasis was placed on developing strategic communication to counter disinformation campaigns and strengthen society's resilience to manipulative information. Last but not least, it was about strengthening cybersecurity (Baranowski& Kintsurashvili & Bronert & Kaminski, 2023).

The Information Sharing and Analysis Centre was created in 2023. It is a platform for sharing information related to the spread of disinformation and influence operations. The centre systematically detects and evaluates disinformation narratives, and its contribution to this activity is undeniable. However, it faces challenges such as limited funding, international coordination, or balancing member participation (ISAC, 2023).

The FIMI Toolbox evolved through multiple follow-up projects. In November 2023, the ATHENA project was launched to analyse at least 30 cases of FIMI, examining the tactics, techniques, and procedures used by perpetrators. ATHENA focuses on developing strong countermeasures and assessing their societal impact (EU Disinfo Lab, 2023).

Other projects include, for example, the De-Conspirator project, a Horizon Europe-funded project, focused on analysing FIMI tactics and developing specific tools (Tackling FIMI in Europe, 2023). In general, the efforts within the FIMI Toolbox focused mainly on monitoring and analytical activities. Annual reports are published within the FIMI framework, which map the situation in this area and give and continue to give recommendations in the form of "Reactive Response". However, these recommendations are more of a guide for individual states on how to react in the event of information influences. In other words, they are only a list of them (2nd EEAS, 2024).

3.8 Hungarian Presidency: Finishing Operationalisation of Hybrid Support Teams, 2024

The Hungarian presidency was the last of the trio of EU Council presidency, composed of Spain, Belgium, and Hungary. For 18 months (since July 1, 2023), these countries focused on strengthening the EU's resilience and strategic autonomy as part of their agenda, with one of the main goals being to strengthen security in all its dimensions (Council of the EU, 2023). During the Belgian Presidency, on 21 May 2024, the EU General Affairs Council (GAC) approved the main framework for the practical establishment of the EU Hybrid Rapid Response Teams (Council of the EU, 2024e).

Given that Hungary has long had a reserved stance on disinformation and influence operations, it has focused its attention on cybersecurity during its presidency. It included the continuation of the implementation of the Strategic Compass, strengthening defence capacities and technologies, and the defence industry (HU2024, 2024). The position of the EU Agency for Cybersecurity (ENISA) has been strengthened under the Hungarian presidency (European Council, 2024f). In this context, Hungary is among the countries that were the first to implement NIS2. The Horizontal Working Party on Enhancing Resilience and Countering Hybrid Threats met during the Hungarian Presidency. The aim was to strengthen cooperation between MSs and improve the sharing of information and best practices in this area (HU2024, 2024b).

Although the Hungarian presidency focused less on disinformation and the elimination of foreign influence, it must be said that it further advanced the agenda in cybersecurity, especially the further strengthening of EU organisations such as the aforementioned ENISA agency.

4 Conclusion

Since 2014, the EU has undoubtedly adopted a number of key documents and created various teams and expert groups under its various presidencies. The key question remains: To what extent have these steps contributed to eliminating disinformation and influencing operations? To summarise, it should be emphasised that between 2014 and 2016, and in the following years, the countries holding the Presidency of the Council of the European Union undertook a series of initiatives aimed at countering hybrid threats. Notable developments include: the Dutch Presidency (2016), which adopted a joint framework for countering hybrid threats focused on awareness, resilience, and cooperation, and established the EU Hybrid Fusion Cell; the Estonian Presidency (2017), which created the Friends of the Presidency group on hybrid threats and introduced a coordination mechanism for EU initiatives; the Bulgarian and Austrian Presidencies (2018), which emphasised cybersecurity, strategic communication, and counterintelligence, and launched a joint action plan against disinformation; the Finnish Presidency (2019), which formally established the Horizontal Working Party on Enhancing Resilience and Countering Hybrid Threats (HWP ERCHT); the French Presidency (2022), which adopted the EU Strategic Compass with a clear focus on hybrid threats, cyber diplomacy, and an EU toolbox, as well as launching hybrid rapid response teams; the Czech Presidency (2022), which approved a toolbox for countering Foreign Information Manipulation and Interference (FIMI) and issued guidelines for implementing the EU's coordinated response framework; the Swedish Presidency (2023), which strengthened EU–NATO cooperation on hybrid threats, with particular emphasis on strategic communication and cybersecurity; the Belgian Presidency (2024), which adopted guideline frameworks for the EU's hybrid rapid response teams; and the Hungarian Presidency (2024), which operationalised these teams with a focus on cyber defence and enhancing the role of ENISA.

The intensity of the individual activities corresponded to national priorities and the relevance of these threats, specifically the Baltic and Nordic states. On the other hand, in the case of the Hungarian presidency, the topic of foreign interference was not given such attention. Nevertheless, the EU currently has the tools and capacities to monitor and analyse various campaigns, their progress, implementation, and recommendations for countering them. The aforementioned different levels of prioritisation within Member States may present certain limitations.

On the other hand, there is criticism of the hyperpoliticisation of countering hybrid threats. In this context, it is pointed out that, for example, limiting Russia and China in the economic sphere under the arguments of hybrid threats is more a manifestation of reducing economic rivalry than eliminating security risks (Bazarkina, 2022).

In conclusion, it is necessary to ask a question. Will individual instruments be operationalised within the framework of individual presidencies, or will there be any qualitative change? The intensity of these topics has depended and will depend on two factors. The first factor is the security situation. In this context, 2014 was a breakthrough year, after which the EU began addressing this issue, including institutionally and allocating capacities. The second moment was 2022, the year of Russia's invasion of Ukraine, after which the EU intensified its steps in the form of the creation of the EU Hybrid Toolbox, the Foreign Information Manipulation and Interference (FIMI) Toolbox and EU Hybrid Rapid Response Teams. The second factor was the initiative of states that significantly pushed this agenda forward and, at the same time, had experience with this type of threat, as was the case with the Nordic countries, Poland, and the Czech Republic.

The EU is currently facing several challenges related to hybrid threats, especially the spread of disinformation and foreign information manipulation and interference. The first challenge is undoubtedly technology, especially artificial intelligence and quantum technologies, which will bring a completely different dimension to the spread of disinformation and will need to be responded to somehow. The aforementioned technologies will also enable the development of more advanced deepfake videos and spread disinformation narratives more quickly. The response to this situation is undoubtedly the implementation of the Digital Service Act (DSA, 2025) or the Artificial Intelligence Act (AIA, 2025). Adopting these regulatory measures represents a concrete step to use the findings and recommendations within the framework of European and national legislation.

The second level is political. To what extent will this topic be politically "attractive" for the presidency in the future, especially regarding the spread of disinformation and foreign information manipulation and interference? At the Munich conference, US Vice President JD Vance criticised the restriction of freedom of speech by EU regulatory measures. At the same time, he mentioned the annulment of the results of the presidential elections in Romania due to election manipulation by a foreign state actor, which he considers unprecedented (Vance, 2025). To what extent will this criticism strengthen European politicians who agree with him? If this were to happen, the hypothesis is that the topic of hybrid threats will undergo some corrections in the future.

References

Atlamazoglou, C., & Moyer, J. C. (2022). *A strategic compass: The European Union expands its toolbox*. Wilson Center. https://www.wilsoncenter.org/article/strategic-compass-european-union-expands-its-toolbox.

Autio, L. (2024). Why do EU solidarity and mutual defence clauses struggle to be implemented? *The Grimshaw Review of International Affairs*, *1*(2), 2–37.

Bajarūnas, E. (2024). How to counter Russia's hybrid threats? First, make them a priority. https://cepa.org/article/how-to-counter-russias-hybrid-threats-first-make-them-a-priority/?utm_source=chatgpt.com.

Baranowski, M., Kintsurashvili, E., Bronert, M., & Kaminski, M. (2023). *How Sweden can use its EU Presidency to build the civilian security dimension of the Eastern Partnership*. GMF. https://www.gmfus.org/news/how-sweden-can-use-its-eu-presidency-build-civilian-security-dimension-eastern-partnership?

Bazarkina, D. Y. (2022). Countermeasures for hybrid threats: The experience of the European Union and its member states. https://pmc.ncbi.nlm.nih.gov/articles/PMC9492455/.

Bergaust, J. C., & Sellevåg, S. R. (2024). Improved conceptualising of hybrid interference below the threshold of armed conflict. *European Security*, *33*(2), 169–195. https://doi.org/10.1080/09662839.2023.2267478.

Bryjka, F. (2022). *Tracing the development of EU capabilities to counter hybrid threats*. Polish Institute of International Affairs. https://pism.pl/publications/tracing-the-development-of-eu-capabilities-to-counter-hybrid-threats.

Council of the EU (2015). *Council conclusions on CSDP no. 8971/15*. Brussels.

Council of the EU (2018). *Council conclusions on security and defence in the context of the EU Global Strategy – Council conclusions (25 June 2018) (10246/18)*. Brussels.

Council of the EU (2020). *Joint staff working document: Mapping of measures related to enhancing resilience and countering hybrid threats*. Brussels.

Council of the EU (2023). *Taking forward the Strategic Agenda 18-month Programme of the Council (1 July 2023–31 December 2024) (20 June 2023) (10597/23)*. Brussels.

Defence innovation and the European Union's Strategic Compass (2022). *The International Institute for Strategic Studies (IISS)*. https://www.iiss.org/publications/strategic-comments/2022/defence-innovation-and-the-european-unions-strategic-compass/.

Dolya, A. (2016). *The annexation of Crimea: Lessons for European security*. Fondation Robert Schuman.

EEAS (2024). *2nd EEAS report on foreign information manipulation and interference threats*. https://www.eeas.europa.eu/node/437511_fr.

EL EU OHQ (2018). *EU HEX-ML 18 (PACE)*. https://eleuohq.mil.gr/eu-hex-ml-18-pace/.

EU Disinfo Lab (2023). ATHENA kicks off a fight against foreign information manipulation and interference. https://www.disinfo.eu/athena-kicks-off-a-fight-against-foreign-information-manipulation-and-interference/.

EU Monitor (2017). Legal provisions of JOIN(2017)30 – Implementation of the Joint Framework on countering hybrid threats – a EU response. https://www.eumonitor.eu/9353000/1/j4nvhdfcs8bljza_j9vvik7m1c3gyxp/vkg2gamhmolz.

European Commission & High Representative (2016). *Joint communication to the European Parliament and the Council: Joint framework on countering hybrid threats – A European Union response*. Brussels.

European Council (2015). *European Council meeting (25 and 26 June 2015) – Conclusions (EUCO 22/15)*. Brussels.

European Council (2016). *Council conclusions on countering hybrid threats (19 April 2016)*. https://www.consilium.europa.eu/en/press/press-releases/2016/04/19/fac-conclusions-hybrid-threats/.

European Council (2018a). *European Council meeting (22 March 2018) – Conclusions (EUCO 1/18)*. Brussels.

European Council (2018b). *European Council meeting (28 June 2018) – Conclusions (EUCO 9/18)*. Brussels.

European Council (2018c). *European Council meeting (13 and 14 December 2018) – Conclusions (EUCO 17/18)*. Brussels.

European Council (2022a). *A Strategic Compass for security and defence*. https://www.consilium.europa.eu/en/policies/strategic-compass/.

European Council (2022b). *Council conclusions on Foreign Information Manipulation and Interference (FIMI)*. https://data.consilium.europa.eu/doc/document/ST-11429-2022-INIT/en/pdf.

European Council (2022c). *Implementing guidelines for the Framework for a coordinated EU response to hybrid campaigns*. https://data.consilium.europa.eu/doc/document/ST-15880-2022-INIT/en/pdf.

European Council (2024a). *Sanctions against Russia*. https://www.consilium.europa.eu/en/policies/sanctions-against-russia/.

European Council (2024b). *Horizontal Working Party on Enhancing Resilience and Countering Hybrid Threats (HWP ERCHT)*. https://www.consilium.europa.eu/en/council-eu/preparatory-bodies/horizontal-working-party-on-enhancing-resilience-and-countering-hybrid-threats/.

European Council (2024c). *Council conclusions on a framework for a coordinated EU response to hybrid campaigns*. https://www.consilium.europa.eu/cs/press/press-releases/2022/06/21/council-conclusions-on-a-framework-for-a-coordinated-eu-response-to-hybrid-campaigns/.

European Council (2024d). *EU Integrated Resolve 2024: EU concludes wide crisis response exercise*. https://www.consilium.europa.eu/en/press/press-releases/2024/10/25/eu-integrated-resolve-2024-eu-concludes-wide-crisis-response-exercise/.

European Council (2024e). *Hybrid threats: Council paves the way for deploying Hybrid Rapid Response Teams*. https://www.consilium.europa.eu/en/press/press-releases/2024/05/21/hybrid-threats-council-paves-the-way-for-deploying-hybrid-rapid-response-teams/.

European Council (2024f). *ENISA: Council approves conclusions for a stronger EU agency for cybersecurity*. https://www.consilium.europa.eu/en/press/press-releases/2024/12/06/enisa-council-approves-conclusions-for-a-stronger-eu-agency-for-cybersecurity/?utm_source=chatgpt.com.

Fiott, D. & Parkes, R. (2019). Protecting Europe: The EU's Response to Hybrid Threats. Chaillot Paper, No. 151. Paris: European Union Institute for Security Studies.

Foreign Information Manipulation and Interference (FIMI) – Information Sharing and Analysis Centre (ISAC). (2023). https://fimi-isac.org/index.html.

Global Strategy (EUGS) (2016). https://www.coe-civ.eu/kh/a-global-strategy-for-the-european-unions-foreign-and-security-policy.

GMF (2019). *Fact sheet: What we know about Russia's interference operations.* https://www.gmfus.org/news/fact-sheet-what-we-know-about-russias-interreference-operations.

Gressel, G. (2019). *Protecting Europe against hybrid threats.* https://ecfr.eu/publication/protecting_europe_against_hybrid_threats/.

Grossman, T. (2023). *Cyber Rapid Response Teams: Structure, organization, and use cases.* Center for Security Studies (CSS), ETH Zürich. https://css.ethz.ch/content/dam/ethz/special-interest/gess/cis/center-for-securities-studies/pdfs/Cyber-Reports-2023-11-Cyber-Rapid-Response-Teams.pdf.

Hagelstam, A. (2018). Cooperating to counter hybrid threats. *NATO Review.* https://www.nato.int/docu/review/articles/2018/11/23/cooperating-to-counter-hybrid-threats/index.html.

Hindrén, R. (2021). *Calibrating the compass: Hybrid threats and the EU's Strategic Compass* (Hybrid CoE Working Paper No. 12). Hybrid CoE. https://www.hybridcoe.fi/wp-content/uploads/2021/10/Hybrid_CoE_Working_Paper_12_Calibrating_the_compass_WEB.pdf.

Hoffman, F. G. (2007). *Conflict in the 21st century: The rise of hybrid wars.* Potomac Institute for Policy Studies.

Horizontal Working Party on Enhancing Resilience and Countering Hybrid Threats. HU2024 (2024). https://hungarian-presidency.consilium.europa.eu/en/events/horizontal-working-party-on-enhancing-resilience-and-countering-hybrid-threats/.

Hybrid threats: A comprehensive resilience ecosystem (2023). https://www.hybridcoe.fi/wp-content/uploads/2023/04/CORE_comprehensive_resilience_ecosystem.pdf.

Insight EU Monitoring (2022). Hybrid threats: EU concludes EU Integrated Resolve 2022 exercise. https://ieu-monitoring.com/editorial/hybrid-threats-eu-concludes-eu-integrated-resolve-2022-exercise/394001?utm_source=ieu-portal.

Jungwirth, R., Smith, H., Willkomm, E., Savolainen, J., Alonso, V. M., Lebrun, M., Aho, A., & Giannopoulos, G. (2023). *Hybrid threats: A comprehensive resilience ecosystem.* Publications Office of the European Union. https://doi.org/10.2760/37899JRC129019.

Kalniete, S., & Pildegovičs, T. (2021). Strengthening the EU's resilience to hybrid threats. *European View, 20*(1), 23–33. https://doi.org/10.1177/17816858211004648.

Lasoen, K. (2022). *Realising the EU hybrid toolbox: Opportunities and pitfalls.* Netherlands Institute of International Relations. https://www.clingendael.org/sites/default/files/2022-12/Policy_brief_EU_Hybrid_Toolbox.pdf.

Mälksoo, L. (2019). The annexation of Crimea and balance of power in international law. *The European Journal of International Law*, *30*(1), 303–319. https://doi.org/10.1093/ejil/chz019.

Mölling, C., & Schütz, T. (2020). The EU's Strategic Compass and its four baskets. https://dgap.org/en/research/publications/eus-strategic-compass-and-its-four-baskets.

Polish Presidency Council of the European Union (2025). https://polish-presidency.consilium.europa.eu/en/programme/priorities/.

Programme of the Hungarian Presidency of the Council of the European Union in the second half of 2024, HU2024 (2024a). https://hungarian-presidency.consilium.europa.eu/media/32nhoeop/programme-and-priorities-of-the-hungarian-presi dency.pdf.

Ruppert, A. (2022). A strategic compass leading where? RLS Brussels. https://rosalux.eu/en/2022/import-2099/.

Sweijs, T. (2023). Moldova's response to hybrid attacks: A learning-by-doing strategy. https://hcss.nl/report/moldovas-response-to-hybrid-attacks/.

Szymański, P. (2020). Towards greater resilience: NATO and the EU on hybrid threats. OSW. https://www.osw.waw.pl/en/publikacje/osw-commentary/2020-04-24/towards-greater-resilience-nato-and-eu-hybrid-threats.

Tackling FIMI in Europe (2023). https://www.disinfo.eu/wp-content/uploads/2024/06/Tackling-FIMI-in-Europe-brochure.pdf.

Tallis, B., & Šimečka, M. (2016). *Collective defence in the age of hybrid warfare*. Institute of International Relations Prague.

The Digital Services Act package (DSA) (2025). https://digital-strategy.ec.europa.eu/en/policies/digital-services-act-package.

The EU Artificial Intelligence Act (AIA) (2025). https://artificialintelligenceact.eu/.

Vance turns on European allies in blistering speech that downplayed threats from Russia and China (2025, February 14). *CNN World*. https://edition.cnn.com/2025/02/14/europe/jd-vance-munich-speech-europe-voters-intl/index.html.

Vázquez, G. (2022). EU Strategic Compass: Has the EU advanced in its threat-analysis capacity? https://www.unav.edu/web/global-affairs/eu-strategic-compass-has-the-eu-advanced-in-its-threat-analysis-capacity.

CHAPTER 6

Framing Security-Related Policy Agendas in EU Council Presidencies: Trends and National Political Dynamics from 2018 to Mid-2025

Michele Barbieri

Abstract This chapter offers an analysis of the security-related policy agendas addressed by EU Council presidencies from 2018 to 2025. The study focuses on how security issues are framed during these presidencies and investigates the existence of convergent trends based on the national political affiliations of selected EU countries' presidencies. This exploration employs a quantitative-and-qualitative analysis through a constructivist policy science approach, i.e., an inductive analysis on secondary empirical materials. To do so, the analysis is based on EU Council Presidencies' programs, to evaluate the framing of security by the EU presidencies. Through this perspective, this policy documents are understood as empirical artifacts resulting from a process of production and reproduction of the issue as a social construct, from which it is possible to identify the problems, solutions, and dynamics of the actors involved. Particular attention is given to key policy areas such as cyberspace & digital technologies, defense & geopolitics, and health & biopolitical measures. By examining these issues across presidencies, the study tries to reveal whether certain political affiliations correlate with specific approaches to security governance and whether overarching trends emerge over time. Hence, findings discover a consistent trajectory in security policy agenda, due to national political configurations' ideology. Such insights offer new perspectives for exploring how political dynamics at the national level influence the supranational policy-making process within the EU. This chapter intends contributing to the broader literature on EU security-related agenda and governance, offering a preliminary framework for future research on the interplay between national politics and European-level security-related policy strategies.

Keywords EU Council Presidency – agenda setting – framing – security issue – Europeanization – bottom-up-down strategies

1 Introduction

The growing complexity of global threats, crises, geopolitical conflicts, and mutating cyberthreats are part of the factors that convert the concept of security as a priority issue in the EU institutions and its member states' agenda. While the security assumes a crucial role in their decision-making process, it concept varies on its definition and perception, where also cultural, institutional, political, and socio-economic factors affect its frame. In this mutating context, this chapter intends to analyse how the concept of security is framed by the EU Council Presidencies between 2018 and the first semester of 2025. The aim is to examine the securitization dynamics in the EU by identifying the recurring related topics and the frame-setting by each presidency, as well as the impact of national governments on the EU Council Presidency's agenda-setting process. The chapter is articulated as following. Besides this introduction, the first chapter provides a conceptual literature review of public issues and its configuring process in the decisional agenda, as well as the different theoretical approach on security studies. Subsequently, a methodological framework grounded on the literature reviewed is provided to explain the hypotheses and the research questions, as well as materials and the analysis methodology underlying this investigation. Its core is represented by the main findings and their discussion to present the empirical evidences detected. Finally, the conclusive section tries to answer research questions posed and shed light on the relevance of this exploratory analysis, its positioning in the related policy literature, and future research trajectories. Hopefully, the results could provide useful foundation for academic debate and future policy guidelines in European security-related policies.

2 Framing Security Issues in Policy Agendas: A Conceptual Review

The meaning of public issues is controversial, as it represents a social circumstance that requires the intervention through a public policy. While until the sixtieth of prior century public issues were identified as a dysfunction element of the social system by the lens of the objectivist approach (Neveu, 2015), from the seventieth of the 20th century, the constructivist approach emerged (Blumer, 1971; Hilgartner & Bosk, 1988), arguing that issues are socially constructed through claims and contexts-in-motion. By this perspective, its definition is relative, mutating over time, and affected by actors involved and by contexts in which they occur. When a condition is recognized as a public issue, it has to enter in the political agenda in order to be solved. This represents the first stage of a circular public policy cycle, the so called "identification of the issue" through the agenda-setting process: "a concentric dynamic that allow a progressive selection of issues to deal with by policy-makers" (Cerruto, 2022, p. 36). According to Cobb & Elder (1972), from this dynamic emerge three types of agenda: the systemic agenda, the institutional agenda, and the decisional agenda. Due to public resources scarcity, as well as institutional and contextual constraints, merely when an issue reaches the decisional agenda, policy-makers start the decision-making process, to formulate and, therefore, implement the related public policy. Three main models can thrust issues in the decisional agenda (Cobb, Ross & Ross, 1976; Knoepfel et al., 2011): outside initiation, inside initiation, and silent corporatist action.

Notwithstanding, during the agenda-setting process, the framing (or cognitive framing) mechanism plays a key role in the categorization of public issues, as it constructs a defining narrative on who benefits or suffers from a specific social condition, as well as who is responsible for resolving it (Schoen & Rein, 1994). As a matter of fact, the focus of decisional agenda depends on several features (i.e., focusing events mechanisms, clashes, or symbolic rhetoric), which boost or weaken the salience of a public issue (Cobb & Elder, 1972; Kingdon, 1984). Furthermore, Kingdon's Multiple Streams Theory (1984) describes how a public issue enters the decisional agenda on the basis of Coehn et al.'s (1972) Garbage Can model. Accordingly, a public issue enters in the agenda-setting process thanks to the convergence of three main streams (Kingdon, 1984): the problem stream; the policy stream; and the political stream. Their convergence is almost always casual and requires the ability of a policy entrepreneur to exploit the occurrence of a "window of opportunity" and, therefore, promoting policy solutions. However, empirical investigations (see for instance: Guaschino, 2019; Di Mascio et al., 2021) show how policy makers can also adopt evidence-based policy strategies to address the complexity of public issues, as well as deliberative approaches to foster dialogue between different actors' stances. In a nutshell, during this first phase of the circular policy cycle, it is crucial for policy entrepreneurs to combine resources, timing, formal legitimacy, personal relationships, and specialized knowledges.

The securitization concept is usually referred to Copenhagen School and its studies on the security issue (Wæver, 1995; Buzan et al., 1998). This theoretical approach seeks to transcend the debates between who objectively approach to threats (meant as a "real" danger to national and international security) and those who subjectively approach to threats (meant as a "perceived" danger). By these perspectives, the security issues should be acknowledged as a "speech act", which socially construct certain issues as threats, thus requiring an action against them. According to Wæver (1995) and Buzan et al. (1998), to qualify a speech act as a security issues, three rhetorical criteria have to occur: (i) an actor claims that something is substantially threatened; (ii) the actor demands the legitimacy to adopt extraordinary measures to address the threat; (iii) the actor convinces the people that extraordinary measures are justified. The security frame on a public issue could also legitimize the bypass of public and political debate, as well as democratic procedures (Van Munster, 2005).

At a first period, the securitization process was analysed across five key sectors, such as military, political, societal, economic, and environmental. Afterwards, new studies have expanded these sectors, as well as have introduced the concept of desecuritization, managing issues without framing them as security threats (Williams, 2003; Emmers, 2007; McDonald, 2008; Balzacq, 2010). In the EU context, the mutation of security issues represents a relevant example of securitization. The first joint initiative was launched in the 1998 and culminated in the 2001 Treaty of Nice, recognizing a first autonomous European role in security and defence policy (Mauer, 2009). From this moment, this highly discussed policy represents a dynamic example of configuration of new strategies, institutions, and international operations (Howorth, 2007). On the one hand, some scholars have highlighted a process of "militarization" in EU security policies (Manners, 2006). On the other hand, other scholars have emphasized the role of soft balancing on transatlantic relations (Posen, 2006). Moreover, constructivist scholars focus on the emergence of a common strategy based on European norms, values, and beliefs (Meyer, 2005), shaping the construction of threats and political responses.

Despite this, contemporary processes and events have affected the framing of security. Indeed, the radicalization and polarization of public debate (Norris & Inglehart, 2019), triggered by the rise of populist rhetoric (Ruzza, 2018), as well as by the escalation of social media, and the increasing pervasiveness of complex techno-social systems (Chadwick, 2017; Fuchs, 2021), are some of factors that lead towards a progressive increase of insecurity perceptions (Przeworski, 2019). According to Boin et al. (2020), a "creeping crisis" is triggering several heterogeneous global trends, potentially affecting values, processes, and institutions of EU democracies. Indeed, investigations proceeding from Science and Technology Studies (STS) mean the concept of security as a "surplus of ambiguity" (Gusterson, 2008), which can be exploited by several actors. Hence, the security can be framed and reframed on multiple phenomena, leading to an open and co-produced securitization process (Vogel et al., 2016; Martins & Jumbert, 2020).

Eventually, the concept of securitization provides an insightful analytical tool for understanding how issues can become security priorities, and find a window of opportunity in the decisional agenda, to legitimize the implementation of public policies and, consequently, the adoption of policy instruments. The analysis of EU Council Presidencies' agenda could shed light in understanding how securitization processes shape values, interests, and power dynamics in the EU context, to construct the perception of threats and related political responses. The next section outlines the methodological framework underpinning the proposed investigation.

3 Theoretical and Methodological Framework

Against this background, this chapter focuses on the mutation of security frame in EU Council Presidencies' agenda. Although this type of analysis is already well conducted in recent literature (Princen, 2024), a lack emerge in highlighting the national political power dynamics developed by the Presidencies, as a bottom-up strategies in the multi-directional Europeanization process (Radaelli, 2003). Moreover, the attempt is to explore the existence of convergent or divergent strategies between national political-and-ideological stances in framing security priorities in the European agenda-setting process. This explorative research moves from the historical neo-institutionalism theoretical approach (Pierson, 2004), to emphasize the role of actors' behaviours, belief, and knowledge in the construction of security frames, and to consider the effect of critical junctures in explaining how the context-in-motion affects the securitization process. Likewise, the analysis also integrates the critical realism perspective (Cruickshank, 2016), to emphasize the imperfection of epistemological approaches in interpreting the reality. Thus, the aim is to refine its interpretation rather than seek absolute truths.

As a matter of fact, if (H1) the convergence of problem, policy and political streams could generate a window of opportunity to configurate the decisional agenda (Kingdon, 1984), and if (H2) the securitization is an ambiguous and involved actors' co-produced process (Gusterson, 2008; Vogel et al., 2016) exploited by actors (Van Munster, 2005), the research questions are threefold: In the EU Council Presidencies' Agenda, (RQ1) what is the salience of security? (RQ2) Which issues are framed through the securitization process? (RQ3) Are there trends of convergence or divergence between national political party ideologies and specific security frames? To do so, a quantitative-and-qualitative content analysis is conducted on policy documents collected on the official website of EU Council Presidency, from the 2018 to the first semester of

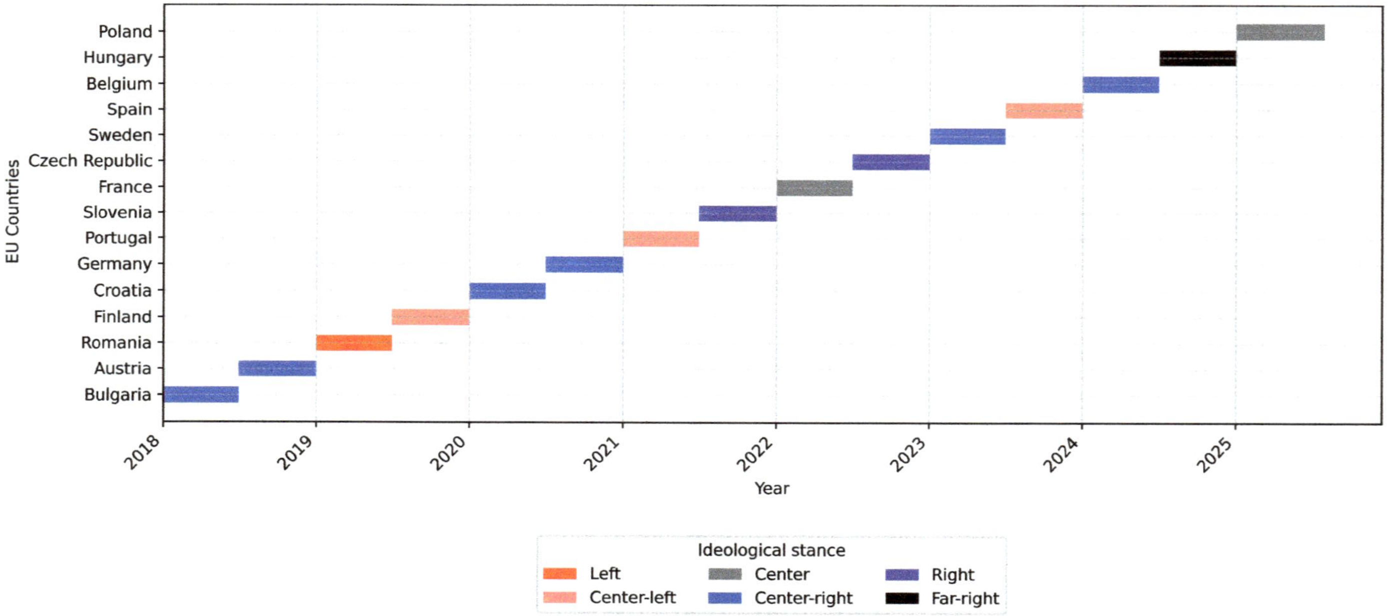

FIGURE 6.1 Gantt chart of EU Council Presidencies from 2018 to mid-2025 (Source: Author's graphical elaboration on Council of the EU (2025) official website data.)

2025. The chosen timeframe covers several critical junctures that have significantly influenced the salience of security-related issues, such as: the Facebook-Cambridge Analytica scandal; the COVID-19 pandemic and its infodemic; the ongoing Russo-Ukrainian conflict. The Figure 6.1 shows the presidencies during this timeframe and their political-ideological stances. Policy documents chosen correspond to "Presidency Programmes", which is consistent with a policy analysis of agenda setting.

To address the RQs posed the analysis proceeds in three steps. Firstly, a quantitative analysis of the semantic root term "secur" is conducted, to understand the scope and the mutation of the security occurrence over time. The application of the stemming approach to the semantic root is useful to capture the full range of terms related to securitization process, such as "security", "secure", "securitization", and other related terms. It is a fundamental technique in Natural Language Processing (NLP), allowing for the detection of word derivatives, enhancing the consistency of text content analysis, and effectively tracking thematic patterns in the securitization discourse. Secondly, a qualitative analysis is performed, to identify and codify thematic issues within the securitization process. The selection of issues follows an inductive approach based on open content analysis (Van Gorp, 2010), which is consistent with the theoretical approach outlined, to allow the identification of emerging issues triggered by context-in-motion and focusing events. The coding activity is performed by the Python software, which extracts text segments containing the semantic root "secur" from each document, to unpack the textual context and detect thematic associations framed as securitized. The coding was tested through a preliminary manual coding of 3 out of the 15 policy documents collected (20% of the total). The intercoder reliability rate achieved an agreement percentage ranging between 0.98 and 1 (approx.), according to Holsti's formula. Finally, a network analysis is developed, to examine the relationships between ideological stances and recurring topics, aiming to detect possible elements of convergence or divergence between national political dynamics and agenda-setting processes at the European level. The next sections show the relevant findings and their discussion on the basis of the methodological framework outlined.

4 Findings: The Security Salience, the Emerging Issues, and the National Politics Patterns

The analysis of word occurrences with "secur" semantic root in the agenda of EU Council Presidencies shows interesting dynamics related to the securitization of the EU institutional and political discourse between 2018 and 2025. Findings highlight a mutation of the security centrality, both in terms of its scope and thematic association. This section shows how the security perception turns in a multifaceted concept, which salience is affected by global events and crises, as well as underling elements of convergence between national ideological stances and specific topics.

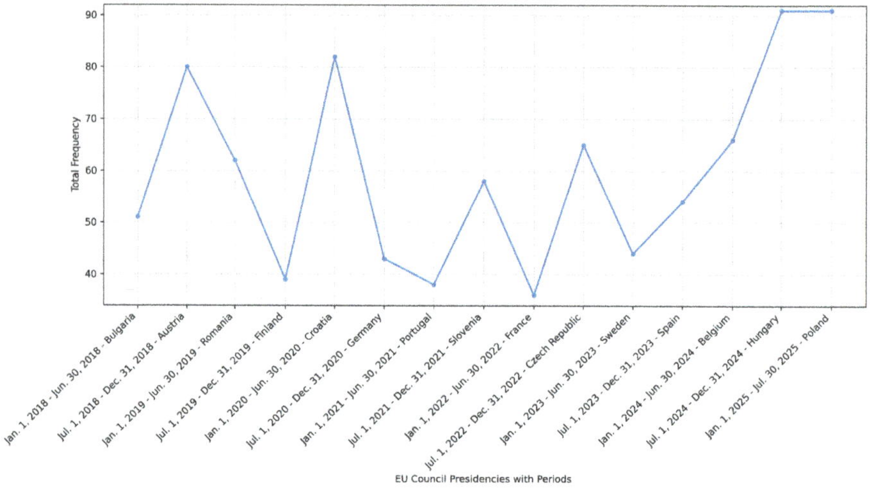

FIGURE 6.2 Diagram of frequencies of security-related words over EU Council Presidencies (2018–mid-2025) (Source: Author's data elaboration on policy document collected.)

The analysis of security salience reveals significant fluctuation in terms of occurrences in the EU Council Presidencies' policy document collected (Figure 6.2). Although it represents a permanent issue in the agenda-setting process, there are focusing events that trigger more emphasis to the securitization process in the EU institutional debate.

From 2018 to 2020, the security emerges as a concept related to cybersecurity issues and protection of techno-social systems' infrastructure. This incidence mainly appears during the Croatian presidency, where the "cybersecurity" word significantly increases rather than prior presidencies. The salience of digital-related threats arises from the Facebook-Cambridge Analytica scandal of 2018, which foster concerns related to data privacy protection and security of online information system. From 2020 to 2022, the security discourse evidently shifts towards public health issues, due to the COVID-19 pandemic and its counteracting policy. This broad policy entails the adoption of several pol-

TABLE 6.1 Frequencies of securitized thematic groups (relative percentage values) (Source: Author's data elaboration based on policy document collected).

EU Council Presidencies	Thematic association						
	Democracy, cohesion and EU institutions	Sustainability, energy and climate change	Migration and boarder control	Health and biopolitical measures	Economic, financial and market	Cyberspace and digital technologies	Defence and geopolitics
Bulgaria	17.65%	3.92%	21.57%	0%	21.57%	11.76%	23.53%
Austria	8.75%	3.75%	32.5%	0%	8.75%	27.5%	18.75%
Romania	16.13%	11.29%	16.13%	4.84%	11.29%	17.64%	22.58%
Finland	5.13%	15.38%	12.82%	0%	7.69%	25.64%	33.33%
Croatia	9.76%	1.22%	23.17%	1.22%	4.87%	32.92%	25.61%
Germany	6.97%	6.97%	9.30%	6.97%	13.95%	27.91%	27.91%
Portugal	18.42%	15.79%	7.89%	0%	5.26%	21.05%	31.58%
Slovenia	18.96%	10.34%	6.89%	1.72%	10.34%	36.21%	15.51%
France	27.78%	2.77%	8.33%	8.33%	5.5%	22.22%	25%
Czech Republic	9.23%	18.46%	3.08%	1.53%	9.23%	27.69%	30.77%
Sweden	18.18%	9.09%	6.81%	0%	9.09%	13.63%	43.18%
Spain	20.37%	3.70%	12.96%	0%	22.22%	16.66%	24.07%
Belgium	7.57%	7.57%	9.09%	10.60%	15.15%	25.75%	24.24%
Hungary	6.59%	16.48%	14.29%	0%	24.17%	9.89%	28.57%
Poland	12.09%	14.29%	3.29%	6.59%	19.78%	23.08%	20.88%

icy instruments over EU countries, such as biopolitical measures to tackle the spread of the virus, regulatory measures to control EU boarders, and procedural measures to better manage health-related data. Actually, and during the German and Portuguese presidencies (2020–2021), the frequencies of "biosecurity" and "health security" terms notably increase, emerging the need of an integral approach to counteract the sanitary crisis and emergency. From 2022, the escalation of the Russian–Ukraine war triggers new emphasis to the security concept, framing it in defence and geopolitical related issues. During the French and Czech presidencies (2022), the analysis detects a peak with the use of security word linked with the protection of international geopolitical context and to counteract hybrid military threats. This trend continues during the following presidencies, particularly with Belgic, Hungarian and Polish' presidencies (2024–mid-2025), as European defence policies are focused on the debate on "energy security", as the reduction of Russian gas supplies.

The analysis of association between the security frame and topics confirms the dynamic nature of securitization process, with well fit by events and contexts occurring (Table 6.1). Particularly, three stages of securitization process can be detected, depending by the context-in-motion and related to specific policy arena.

One of the most recurring topic groups is the "Cyberspace & Digital Technologies". After the global data scandal of 2018, and the main digital policy reforms converted in European acts (for example, the Digital Service Act, the Digital Market Act, the Cybersecurity Act, and the Artificial Intelligence Act), EU Council Presidencies have focused their attention on cybersecurity, data protection, regulation of digital platforms, and of AI. Likewise, "Defence and Geopolitics" topic groups represents another recurring one, due to the geopolitical tensions and the role of EU in defending European citizens, values and democracy principles. Frequently, both topic groups merge, due to the warfare developed in the cyberspace. Recently, the analysis detects a renewed interest for "Sustainability, Energy and Climate Change" topic group, especially in the Belgic, Hungarian, and Polish presidencies, which dealing with the Russia's threats to cut off gas supplies as a result of sanctions. Worth of attention is the emerging attention on "Health & Biopolitical Measures" topic group, during and after the pandemic crisis. As a matter of fact, the COVID-19 outbreak has reconfigured security priorities, shifting the focus towards the sanitary protection and counteracting strategies to decrease the contagious rate. Thus, health security-related topics represent a central stake in European policies from 2020, and during the German presidency the "biosecurity" emerge as a new security-related topic in EU resilience strategy.

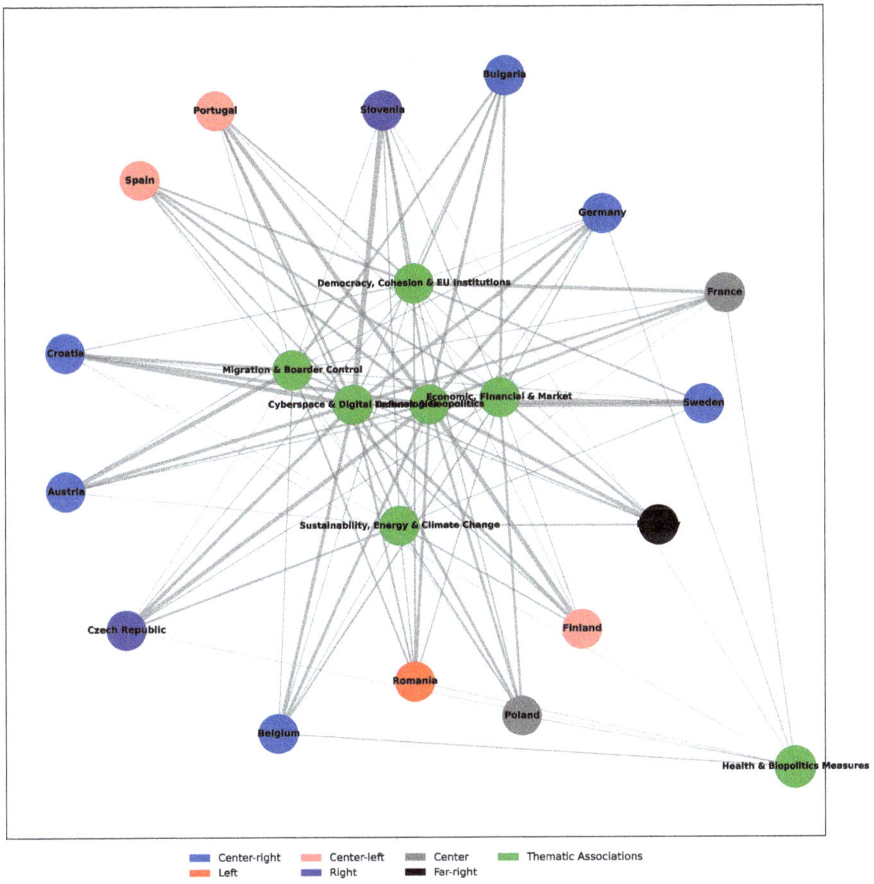

FIGURE 6.3 Network analysis scatter chart of political ideologies and security-related topic groups (Source: Author's data elaboration on policy document collected.)

The network analysis between ideological stances and thematic associations highlights points of convergence between national political positions and security priorities as Figure 6.3 shows.

Findings reveal three main convergences between ideological stances and security priorities. Centre-right EU Council presidencies (Austria, Germany, Croatia, Belgium, Sweden, and Bulgaria) focus on cybersecurity and financial security, promoting restrictive data protection approaches and digital platforms regulations, to promote competitiveness and ensure economic stability. Centre-left EU Council Presidencies (Portugal, Spain, Finland, and Romania) emphasise democratic cohesion and sustainability, highlighting the importance of safeguarding the EU institutional framework as well as the resilient response as tools for socio-economic stabilization. Centre (France, and Poland), Right-wing (Slovenia, and Czech Republic), and far-right (Hungarian) EU Council presidencies underline the importance of defence and boarder security. Indeed, right-wing-oriented national governments raise to the EU level the role of military security and migration control, face with the recent geopolitical crises. These findings suggest that although the concept of security is a transversal political topic, its framing addresses different priorities, depending on the ideological orientation of national governments, thus potentially affecting the bottom-up-down Europeanization process.

5 Conclusions

The analysis of EU Council Presidency' agenda from 2015 to the first semester of 2025 highlights the development of the security concept, as well as how its framing could orient the EU decision-making process. Through a quantitative-and-qualitative content analysis of presidency programs collected, now the effort is to answer to the three exploratory RQs posed. Firstly, findings related to the salience of security in EU Council Presidencies' agenda (RQ1) show that, although this concept is always included in the programming documents, its variations depend from historical context and global events. Three key moments act as focusing events on the securitization processes, such as: (i) the Facebook-Cambridge Analytica Scandal, which trigger new concerns related to the platformization process; (ii) the COVID-19 pandemic and the concerns related to public health; (iii) the Russian–Ukraine war, and its related geopolitical tensions. Actually, frequencies of security-related terms increase during these critical junctures, confirming the role of the focusing event mechanism (Kindon, 1984) in the definition of EU Council Presidencies' Agenda. Secondly, findings related to the securitization of thematic issues (RQ2) reveal a partial reconfiguration of security frames during the timeframe choose. Although the security concept is well absorbed on cybersecurity and defence topic groups, new thematic associations emerge, focusing from 2020 to 2022 on health and biosecurity, and from 2022 to ongoing on sustainability and energy security. These shifts underline a progressive framing process on international concerns related to EU stability threats. Thirdly, findings also reveal convergent streams between national ideological stances and security frames (RQ3). Cybersecurity and financial concerns seem to be the main security topics of centre-right presidencies, while EU democratic principles and sustainability are the focus of left-centre presidencies, whilst defence and border control represents the main worries of right and far-right presidencies. The thematic association with the security approach appears in line with the traditional ideologies involved investigated (Princen, 2024), thus confirming the bi-directional Europeanization process (Radaelli, 2003) in EU policy-making.

These findings are consistent with the social construction theory on public issues (Blumer, 1971; Hilgartner & Bosk, 1988), and with the key role of cognitive framing in agenda-setting process (Schoen & Rein, 1994). The securitization formulated by the Copenhagen School (Wæver, 1995; Buzan et al., 1998) emerges as a dynamic and co-produced process, where policy actors shape the security-related discourse to legitimize their political action (Van Munster, 2005). According to Gusterson (2008), the "surplus of ambiguity" helps the incessant redefinition of security, explaining the role of context factors and strategic interests of policy-makers. Moreover, the Kingdon's Multiple Stream Approach (1984) is useful to reveal how the convergence of problem, political, and policy flow determine the security frame in the decisional priorities. Eventually, this explorative investigation could lead to future analysis, which dives into specific insights detected. Some suggestions are therefore proposed. An initial development could enlarge the dataset, including other policy documents and public talks of EU Council Presidencies, to understand the link between the presidencies' agenda and the actual implementation of security-related policies. A temporal extension of the analysis could observe the evolution of the securitization process to highlight the role of context-in-motion on a long-term period. Last but not least, it would be interesting to explore the role of techno-social systems in the spread of securitization frames, by analysing the impact of public debate on polarization and radicalization, as well as the effect of emerging information technologies on security perceptions.

References

Blumer, H. (1971). Social problems as collective behavior. *Social Problems*, 18(3), 298–306. https://doi.org/10.2307/799797.

Boin, A., Ekengren, M., & Rhinard, M. (2020). Hiding in plain sight: Conceptualizing the creeping crisis. *Risk, Hazards & Crisis in Public Policy*, 11(2), 116–138. https://doi.org/10.1002/rhc3.12193.

Buzan, B., Wæver, O., & de Wilde, J. (1998). *Security: A new framework for analysis*. Lynne Rienner.

Cerruto, M. (2022). *Politiche pubbliche. Vincoli, attori e idee*. Mondadori Education.

Chadwick, A. (2017). *The hybrid media system: Politics and power*. Oxford University Press.

Cobb, R. W., & Elder, C. D. (1972). *Participation in American politics: The dynamics of agenda building*. Allyn & Bacon.

Cobb, R. W., Ross, J. K., & Ross, M. H. (1976). Agenda building as a comparative political process. *American Political Science Review*, 70(1), 126–138. https://doi.org/10.2307/1960328.

Cohen, M. D., March, J. G., & Olsen, J. P. (1972). A garbage can model of organizational choice. *Administrative Science Quarterly*, 17(1), 1–25. https://doi.org/10.2307/2392088.

Cruickshank, J. (2016). Critical Realism. In L. McIntyre & A. Rosenberg (Eds.), *The Routledge companion to philosophy of social science* (pp. 290–300). Routledge.

Di Mascio, F., Natalini, A., Barbieri, M., & Selva, D. (2021). The role of regulatory agencies in agenda-setting processes: Insights from the Italian response to the COVID-19 infodemic. *Swiss Political Science Review*, 27(2), 271–282. https://doi.org/10.1111/spsr.12465.

Emmers, R. (2007). Securitization. In A. Collins (Ed.), *Contemporary security studies* (pp. 109–125). Oxford University Press.

Fuchs, C. (2021). *Social media: A critical introduction*. Sage.

Guaschino, E. (2019). The role of regulatory agencies in framing public problems: An empirical analysis. *European Policy Analysis*, 5(2), 285–316. https://doi.org/10.1002/epa2.1053.

Gusterson, H. (2008). Nuclear futures: Anticipatory knowledge, expert judgement, and the lack that cannot be filled. *Science and Public Policy*, 35(8), 551–560. https://doi.org/10.3152/030234208X370639.

Hilgartner, S., & Bosk, C. L. (1988). The rise and fall of social problems: A public arenas model. *American Journal of Sociology*, 94(1), 53–78. https://doi.org/10.1086/228951.

Howorth, J. (2007). *Security and defence policy in the European Union*. Palgrave Macmillan.

Kingdon, J. (1984). *Agendas, alternatives, and public policies*. HarperCollins.

Knoepfel, P., Larrue, C., Varone, F., & Hill, M. (2011). *Public policy analysis*. Policy Press.

Manners, I. (2006). Normative power Europe reconsidered: Beyond the crossroads. *Journal of European Public Policy*, 13(2), 182–199. https://doi.org/10.1080/13501760500451600.

Martins, B. O., & Jumbert, M. G. (2020). EU border technologies and the co-production of security 'problems' and 'solutions'. *Journal of Ethnic and Migration Studies*, 48(6), 1430–1447. https://doi.org/10.1080/1369183X.2020.1851470.

Mauer, V. (2009). The European Union and the Security Dilemma. Palgrave Macmillan.

Mauer, V. (2010). The European Union: From Security Community Towards Security Actor. In M. Dunn Cavelty & V. Mauer (Eds.), *The Routledge handbook of security studies* (pp. 371–381). Routledge.

McDonald, M. (2008). Securitization and the construction of security. *European Journal of International Relations*, 14(4), 563–587. https://doi.org/10.1177/1354066108097553.

Meyer, C. O. (2005). Convergence towards a European strategic culture? A constructivist framework for explaining changing norms. *European Journal of International Relations*, 11(4), 523–549. https://doi.org/10.1177/1354066105057899.

Neveu, E. (2015). *Sociologie politique des problems publics*. Armand Colin.

Norris, P., & Inglehart, R. (2019). *Cultural backlash: Trump, Brexit, and authoritarian populism*. Cambridge University Press.

Pierson, P. (2004). *Politics in time*. Princeton University Press.

Posen, B. (2006). European Union security and defense policy: Response to unipolarity? *Security Studies*, 15(2), 149–186. https://doi.org/10.1080/09636410600829356.

Princen, S. (2024). Agenda-Setting in EU Policy-Making. In G. J. Brandsma (Ed.), *Handbook on European Union public administration* (pp. 216–228). Edward Elgar Publishing.

Przeworski, A. (2019). *Crises of democracy*. Cambridge University Press.

Radaelli, C. M. (2003). The Europeanization of Public Policy. In K. Featherstone & C. Radaelli (Eds.), *The politics of Europeanization* (pp. 30–41). Oxford University Press.

Ruzza, C. (2018). Populism, Migration, and Xenophobia in Europe. In C. de la Torre (Ed.), *Routledge handbook of global populism* (pp. 201–216). Routledge.

Schoen, D. A., & Rein, M. (1994). *Frame reflection: Toward the resolution of intractable policy controversies*. Basic Books.

Van Gorp, B. (2010). Strategies to Take Subjectivity Out of Framing Analysis. In P. D'Angelo & J. A. Kuypers (Eds.), *Doing news framing analysis: Empirical and theoretical perspectives* (pp. 84–109). Routledge.

Van Munster, R. (2005). *Logics of security: The Copenhagen School, risk management and the war on terror* (Political Science Publications No. 10). University of Southern Denmark.

Vogel, K. M., Balmer, B., Weiss, S. W., Kroener, I., Matsumoto, M., & Rappert, B. (2016). Knowledge and Security. In U. Felt, R. Fouché, C. A. Miller, & L. Smith-Doerr (Eds.), *The handbook of science and technology studies* (pp. 973–1001). MIT Press.

Wæver, O. (1995). Securitization and Desecuritization. In R. D. Lipschutz (Ed.), On Security (pp. 46–86). Columbia University Press.

Wæver, O. (2005). Securitization and Desecuritization. In R. D. Lipschutz (Ed.), *On security* (pp. 46–86). Columbia University Press.

CHAPTER 7

The Role of the Visegrád Group Countries in the Presidency of the Council of the EU: Between Symbolism and Real Agency

Marek Górka

Abstract The subject of this study concerns the analysis of factors that currently, or in the near future, may become catalysts for changes in the digital policy of the European Union (EU). Accordingly, the aim of this chapter is to examine whether there is a correlation between the political preferences of the Visegrad Group (V4) states, as expressed in their presidency documents, and the themes included in the programmes of their Presidencies of the Council of the European Union. Such a comparison makes it possible to determine whether there is convergence between the political preferences of the V4 countries and the priorities contained in their Council Presidency programmes. The characterization of the digital policy of Central European states is based on an analysis of the similarities and differences between digital policy priorities formulated at the regional level and those expressed within the framework of EU policy. The findings demonstrate that the V4 can act as a catalyst for change in the fields of digital policy and cybersecurity, particularly in situations that may indirectly affect the stability of the region and the European Union as a whole, as exemplified by the war in Ukraine. An examination of digital policy reveals that the Presidency of the Council of the European Union reflects regional political priorities. This also leads to the conclusion that the EU Presidency provides an opportunity to use the six-month term as an extension of regional political agendas.

Keywords Visegrad Group (V4) – European Union digital policy – Presidency of the Council of the European Union – cybersecurity – European integration

1 Introduction

This paper explores the factors that currently, or in the near future, may become catalysts for change in the European Union's (EU) digital policy. The research is based on issues addressed during the rotating presidency of

 HTTPS://DOI.ORG/10.1163/9789004763623-009

the Council of the EU and the Visegrád Group (V4) presidency in the years 2019–2025. The article emphasizes the growing importance of new technologies and their application in the public sphere, as well as their role in the EU's strategic plans – an influence that significantly shapes the European integration process. Since the outbreak of the Russian–Ukrainian war, the role of defense and the Eastern Neighborhood has increased in the EU's decision-making process. Consequently, countries that form the eastern border of the EU and NATO – as well as subregional organizations such as the Visegrád Group and the Bucharest Nine – have become key actors in the EU's security policy. The challenges facing European security have revived the academic debate on regional cooperation and prompted renewed reflection on the concept of subregionalism. Most of the relevant literature consists of case studies focusing on a single group, while others are comparative studies involving two or three countries (Poznar & Havlík, 2025, pp. 1–18; Grgić, 2023, pp. 216–238; Abels, 2017, pp. 542–543; Pop, 2021).

Researchers have noted that well-established regional groupings can enhance the political weight of member states by facilitating coalition-building (Wivel, 2021, pp. 489–505) or strengthening their political significance in the international arena (Szalai & Garai, 2024, pp. 30–51). This is particularly important in the case of so-called "small states," whose resources – such as population size or territorial reach – are limited. Nevertheless, by participating in broader political projects, such states can reinforce the legitimacy of their political interests, especially when those interests are presented as the collective position of the organization (Lang & von Ondarza, 2018, pp. 1–8). Can the V4, then, serve as a catalyst for change in EU security policy, or can it act as a representative of Central European interests in the broader European debate?

In this context, understanding the interplay between regional needs or interests and EU-wide policy allows for a deeper analysis of the processes shaping the future of the Union. Most current security policy analyses examine the V4 in relation to migration and foreign policy issues (Holányi, 2025, pp. 1–22; Strnad, 2021, pp. 72–100; Vaagland & Zaun, 2024, pp. 559–579), whereas studies on cooperation dynamics in other policy areas and among other active regional groupings remain marginal. The aim of this analysis is to determine whether a correlation exists between the political preferences of V4 countries – expressed in their presidency documents – and the topics featured in the Council of the EU presidency programs led by these states. Within this framework, the article reviews the past five years with a focus on the Central European region in the field of digital policy, in order to answer the following questions: In which specific areas of digital policy are V4 members active? How do they define the use of new technologies, and which aspects do they emphasize? What role does digitalization play in the individual programs formulated by Central European countries?

Although previous research has identified certain key features of the Visegrád Group's approach to digital policy (Kovacs & Domonkos, 2024, pp. 432–463; Perarnaud, 2022, pp. 55–71; Urbanovská, Chovančík, & Brajerčíková, 2022, pp. 402–425), it has not yet been examined whether these positions are consistently pursued and reproduced at the European level – in other words, how the regional perspective aligns with the international dimension. Addressing this question will help determine whether V4 countries possess the capacity to cooperate and shape relationships in the field of digital policy both among themselves and within the EU framework. Therefore, this study examines the interaction between EU policy and regional priorities by posing the following research question: Is there a convergence between the political preferences of V4 countries and the themes included in their Council of the EU presidency programs? To explore the connection between the policy themes articulated by the V4 and their presence in the broader European context, the first section outlines the theoretical foundations for understanding the Council of the EU presidency and the presidency within the Visegrád Group. The second section analyzes the Council presidency programs of V4 countries and examines the content of their Visegrád Group presidencies. The conclusion of the study will offer a characterization of the digital policy priorities of Central European states, based on an analysis of similarities and differences between the digital policy priorities formulated at the regional level (V4) and those expressed within EU-wide policymaking.

2 The Role and Significance of the Presidency of the Council of the European Union: Priorities, Compromises, Realities

This article aims to answer the question of whether the priorities and goals set out during a Council of the EU presidency – often framed as ambitious declarations – can be effectively translated into practical actions and implemented within the political reality. This question is particularly relevant, as public debate sometimes raises doubts about the actual significance of the presidency. Therefore, the analysis of the Council presidency seeks to verify this hypothesis and provide an informed answer.

Holding the presidency of the Council of the EU offers the presiding country an opportunity to indicate the direction in which the European community should develop in political, economic, social, and geopolitical terms. For this reason, the presidency also has a symbolic dimension. Observers often assume that the presidency can lead to an enhanced international position for the presiding country. It is also worth noting that a common and natural reaction is to question whether the presidency has been used effectively and whether the presiding country has been perceived as a credible political partner (Eisele et al., 2022).

Some analysts argue that the presidency is largely a bureaucratic event. This view is partly based on the observation that the presidency tends to focus on routine, procedural matters, and thus rarely sets out ambitious objectives (Vaznonytė, 2020, pp. 497–518). However, given the complexity of today's international situation, even routine matters pose significant political, economic, military, and technological challenges. In current times, the presidency requires considerable flexibility and creativity – not only to foster compromise in a highly polarized environment but also to keep pace with the rapidly evolving global landscape, where digital technology acts as a primary catalyst for change. The presidency of the Council of the European Union does not directly include the conduct of the EU's common foreign policy. According to the Treaty of Lisbon, which entered into force on December 1, 2009, responsibility in this area lies with the High Representative of the Union for Foreign Affairs and Security Policy. This office leads the Union's common foreign and security policy and represents the EU internationally. Nevertheless, global events – such as President Donald Trump's policy toward Europe and the ongoing war in Ukraine – have posed major challenges for the entire EU and have impacted the role and expectations placed on the Council presidency.

The presidency of the Council of the European Union also has a technical dimension – it is where ministers from individual policy areas meet to negotiate legislative proposals put forward by the European Commission. These negotiations often last for months, due to the fact that this is the arena where national interests most directly clash. It is also where national capitals form alliances and work out compromises. Taking this into account, it can be said that the EU is both an entity aimed at integration and a venue for negotiating national interests – within a framework that also considers the interest of the Union as a whole. This process is closely monitored by Members of the European Parliament (as co-legislators in most areas) as well as by all other EU institutions. The presence of these often opposing positions highlights the underlying tension between national interests and the broader interest of the Union's future as a collective entity (Eihmanis, 2024, pp. 3299–3323; Truchlewski & Schelkle, 2024, pp. 958–971).

However, the Council of the EU is not limited to ministerial meetings; it also comprises a wide array of working groups operating at lower levels. These are meetings of civil servants from the member states, who prepare decisions for the ministers. Ministerial meetings, therefore, represent the final stage of the decision-making process, with prior work conducted by national officials. In order for this process to run smoothly, one member state coordinates the entire effort. This coordination occurs horizontally as well – ensuring that the activities of various working groups across different policy domains remain coherent. For example, environmental and economic issues must progress in a parallel and coordinated fashion (Leuffen, Malang, & Wörle, 2014, pp. 616–631; Wratil & Hobolt, 2019, pp. 511–531).

It is commonly accepted that the strength of a Council presidency lies primarily in a member state's ability to pursue its own priorities under the banner of the European Union (Huhe, Naurin, & Thomson, 2017, pp. 25–51). There is certainly merit to this view. One example is Poland's presidency program, which emphasized priorities such as the Eastern Partnership, the Energy Union – as a strategic goal of energy solidarity through cross-border interconnections – and broadly defined security. This last area refers not only to the militarization of the Union and related expenditures, but also to combating cybercrime and disinformation, and supporting the energy transition.

The EU presidency is tasked with ensuring that all actors involved in shaping EU policy – while simultaneously representing national interests – are able to reach a common position. As such, the primary role of the presiding country is to act as a mediator that facilitates agreement among member states on complex and contentious issues (Bailer, Mattila, & Schneider, 2015, pp. 437–456; Carammia, Princen, & Timmermans, 2016, pp. 809–825). However, it is important to note that not all presidencies carry the same weight. For instance, after Denmark, the presidency may pass to Cyprus – but it is difficult to compare Cyprus's political influence with that of more powerful states like Poland, France, Germany, the Netherlands, or Spain. In general, the stronger a country's political position, the greater its influence over the agenda, the pace of negotiations, and the ability to build compromises. In this context, a country's ability to mobilize and unite other member states around common objectives plays a crucial role, as do its levels of economic, technological, and expert development – which directly affect its staffing capacity and real influence within the Council of the EU.

Thus, the nature of the presidency should not be reduced solely to its formal, rotational character. An equally important – though often underemphasized – task of the presiding country is the effective coordination of work within the EU's complex institutional mechanisms. This is particularly significant given that the presidency lasts only six months – a relatively short period. However, with clearly defined priorities and effective diplomacy, a country can, in cooperation with the European Commission, influence the prioritization of issues and gain the support of a majority of member states to swiftly address them – or at least to move them forward within the institutional process (Dinan, 2013, pp. 1256–1273; Häge, 2017, pp. 695–713). This is precisely the mechanism that captures the attention and mobilizes national governments. EU policymaking is a complex system in which many issues require time, dialogue, and a willingness from all participants to work toward compromise.

The process of setting priorities requires a specific level of engagement from political decision-makers. A clear illustration of this is Poland's presidency, which has maintained fairly consistent priorities within the work of the European Union – focusing, among others, on neighborhood policy, the future accession of Eastern neighbors, and the EU's energy independence. It is therefore noticeable that both in 2011, during Poland's first presidency, and in 2025, during its second, the main thematic orientation shaping the presidency program has remained largely consistent. However, when a country begins designing its presidency program, it must also ensure that its priorities are understandable and broadly supported by the domestic political environment. This makes the national dimension a particularly important aspect to consider. Nonetheless, it should be noted that presidency priorities are not static; they are shaped by the evolving political context. Under pressure from certain events, presidency goals must be revised to ensure the program remains current and aligned with changing circumstances. This is, of course, a natural process of evolution, where political tools – such as legislative priorities, topics for debate in the Council of the EU, foreign policy initiatives, and mechanisms of intergovernmental cooperation – must be adapted to the shifting landscape. In this context, it becomes clear that international realities are the primary driver of EU policy and therefore influence the choice of specific objectives by the presiding country. The presidency is not merely a technical exercise in organizing legislative processes – it is also a test of the ability to respond to rapidly unfolding changes on the international stage. It serves as a platform through which a country is assessed based on how well it handles any crises that may emerge during its presidency of the Council of the EU (Schout, 2017, pp. 54–63; Toneva-Metodieva, 2020, pp. 650–667). Another important factor influencing the presidency program is the EU's institutional work calendar, which includes a predetermined set of tasks for each period. One such cyclical challenge is the negotiation of the Multiannual Financial Framework, which coincided with Poland's presidency. This was a period of intense political debate on future EU priorities, during which member states aimed to clearly assert their positions and secure favorable financial outcomes.

It is also worth noting that member states' preferences are not fixed – they evolve as European actors influence the positions adopted by national governments. The differences among countries often lead to the search for strategic alliances, in which political decision-makers form temporary coalitions based on shared national interests. Sometimes, these agreements may contradict long-standing geopolitical alignments or traditional axes of cooperation. As a result, the EU decision-making process may become prolonged, and its outcomes may prove controversial when member states try to reconcile their domestic concerns with the broader goals of the Union (De Ruiter & Vliegenthart, 2018, pp. 649–668). The need to align national expectations with EU-wide objectives often leads to a slower, more contentious process – one that may fuel or reinforce existing populist and anti-EU sentiments.

3 Regional Cooperation Formats in the European Union: The Case of the Visegrad Group

Individual countries, as well as subregional organizations, can play a key role in shaping policy proposals by engaging in lobbying, debates, public consultations, or campaigns promoting specific issues (Dür & Mateo, 2012, pp. 969–987). A clear illustration of this can be found in the activities undertaken within frameworks such as the Central European Defence Cooperation, the Slavkov/Austerlitz Triangle, the Salzburg Forum, the Three Seas Initiative, the Visegrad Group, and the Bucharest Nine. These platforms serve to coordinate interests and amplify the voice of member states in both regional and EU-level policymaking. These entities may represent a broad spectrum of interests, which may result in member states expressing conflicting views within the same organization. As a consequence, such divergence can weaken the ability to present a unified position, thereby reducing the influence of the group or organization in advancing its priorities on the international stage.

The establishment and subsequent cooperation within the Visegrad Group were particularly significant in the context of the 1990s – a period marked across Central and Eastern Europe by ongoing political, economic, and social transformation, along with a strongly articulated ambition of joining the EU and NATO. The group also represented an attempt to strengthen the voice of the region during negotiations with Western institutions. The strategic rationale behind this approach was to demonstrate that a broader coalition of like-minded states existed and could act together, thereby enhancing the region's bargaining power in discussions with the EU and NATO (Szalai, 2022, pp. 210–232). Through participation in the Visegrad Group, the four member states gained an improved geopolitical position and an advantage over other post-socialist countries in the region. Since joining the European Union, they have recorded impressive economic growth – much stronger in nearly all cases than initially predicted or feared at the time of accession. After successful EU integration, the V4 countries began to promote themselves as a successful model of post-socialist transformation and regional cooperation – one that should serve as an example for Eastern European and Balkan countries. This was reflected in strong support for the EU's Eastern Partnership programme and the European Neighbourhood Policy (Cabada and Waisová, 2018, pp. 9–20). However, it should be noted that the gap between the so-called "old" Western European states and those that joined the EU in 2004 and 2007 remains evident – particularly in the area of economies based on new technologies (Lakócai, 2025, pp. 425–440).

Since the Visegrad Group countries achieved their primary objective – accession to the EU and NATO between 1999 and 2004 – the V4 has sought to redefine its purpose. Part of this has involved regional cooperation in areas such as high-tech-based economic development and the shaping of a common security policy.

The importance of Eastern regions within the European Union has grown not only in economic terms. Russia's war against Ukraine has shifted the entire focus of European politics eastward. In 2022, shortly after the outbreak of the war and during the Czech Presidency of the EU Council, Poland and the Czech Republic gained international respect and an influential position regarding support for Ukraine. This was not self-evident, considering their economic status and Poland's frequent opposition to EU decisions and values – particularly in areas such as the rule of law, migration policy, and European integration (Grgič, 2023, pp. 216–238). Meanwhile, Hungarian Prime Minister Viktor Orbán confirmed his country's role as a key adversary of EU policy and also as an opponent to other V4 members. This was particularly visible in threats to veto decisions concerning support for Ukraine and sanctions against Russia. Following the parliamentary elections in Slovakia on 30 September 2023, the government of Robert Fico joined the political balancing act between Russia and the European Union, which, among other things, led to a cooling of relations between the Czech Republic and Slovakia. These differences provoke the question of whether the Visegrad Group can build a constructive and progressive regional identity in the coming years. The existing divisions among V4 members are particularly evident in how they define their security policies. Although all four countries share a strong interest in energy security and support both technological neutrality and nuclear energy, they have not been able to find specific issues on which to formulate joint policy proposals (Osička et al., 2021). Secondly, the V4 countries are divided on how to approach Russia, traditionally the region's main energy supplier. While Hungary has developed strong ties with Russia, the other Visegrad Group members – mainly Poland – have taken a more reserved and critical stance.

It should be emphasized that cooperation within the Visegrad Group is not institutionalized in any formal way. In practice, it mainly consists of regular meetings between prime ministers, ministers, and representatives of national administrations. The absence of bureaucracy and a formal institutional framework means that the functioning of the alliance depends primarily on the current constellation of governments in Prague, Warsaw, Bratislava, and Budapest. As a result, the level of engagement and interest in cooperation from individual governments is highly variable (Kaniok & Hloušek, 2025, pp. 1–19). The V4 operates on three main levels of cooperation among its members: the first concerns parliamentary cooperation, the second involves meetings of heads of state, and the third mechanism is multi-level governmental cooperation. This level includes rotating presidencies, meetings of prime ministers and ministers, regular consultations between national V4 coordinators, and permanent representations in international organizations (*Declaration of the Prime Ministers*…, 2004). Although the V4 has effectively defended its economic interests in EU budget negotiations and demonstrated cohesion in migration policy – by advocating for stronger control of the EU's external borders and warning about the religious and cultural issues that could arise from accepting large numbers of refugees – it has failed to present a unified position on other key issues such as relations with Russia, European integration, or adherence to the principles of democracy, freedom of speech, and media independence. The lack of an institution responsible for an effective internal cooperation mechanism within the V4, combined with significant differences among its members, hinders – and will continue to hinder – the V4's ability to play a significant role in the EU.

Central European governments – including those of Poland, Romania, and the Baltic states – issued urgent warnings about the Russian threat well before Putin's annexation of Crimea in 2014. An analysis of the current situation suggests that their regional policies were largely accurate. In this context, a question arises as to the extent to which the voices, opinions, and positions of countries considered small or peripheral – based, among other things, on historical experience – carry weight in overall EU policy. Do the signals coming from these countries hold any real value, and do they influence how European decision-makers assess international situations? An analysis of the key points in each V4 country's presidency program, as well as their agendas during their respective Presidencies of the Council of the EU, may help answer these questions. It will also help identify the interactions occurring in the fields of security policy and digital technologies between Central European states and the priorities developed at the EU level.

4 Cybersecurity in the Presidency Programs of the Visegrád Group Countries (2019–2025)

To analyze how the countries formulate digital policy, the next objective of this study is to present the main programmatic assumptions of the Visegrád Group presidencies over the last five years. The research perspective will be based on the content outlined in successive V4 presidencies, which can be seen as the official positions of these countries regarding developments in the digital, political, and economic environment. The analysis uses the presidency programs of the Visegrád Group countries, focusing on excerpts related to digital policy, specifically: Poland's documents from 2020–2021 (2020–2021 Polish Presidency) and 2024–2025 (2024–2025 Polish Presidency); the Czech Republic's from 2019–2020 (2019–2020 Czech Presidency) and 2023–2024 (2023–2024 Czech Presidency); Slovakia's from 2022–2023 (2022–2023 Slovak Presidency); and Hungary's from 2021–2022 (2021–2022 Hungarian Presidency).

Challenges arising from digitalization – with varying intensity and levels of engagement – were repeatedly addressed during the V4 presidencies by all members. The Czech presidency in 2019–2020, similarly to previous declarations by the Visegrád Group states, also emphasized the importance of sharing experiences in building cyber capabilities in the region. A strong point was the support for Ukraine and Georgia as NATO and EU partners, aligning with one of the goals of the Eastern Partnership program. The Czech side stressed the integrative character of the V4 in the security dimension, which was to be based especially on regional cohesion with EU policies and on NATO membership in the context of securing the eastern border of the North Atlantic Alliance (2019–2020 Czech Presidency, p. 2). The main goal of the Czech presidency regarding security policy was the coordination of different positions to develop a common regional stance. The Czech Republic promoted the idea of creating a V4 strategy for digital transformation, grounded in research on artificial intelligence (AI) development, joint actions in the digital market, and the implementation of digital infrastructure through initiatives such as Industry 4.0, smart cities, and start-ups (2019–2020 Czech Presidency, p. 4).

The next Visegrád Group presidency in 2020–2021 was held by Poland, which continued efforts to further develop security policy based on cooperation with Central and Eastern European countries. The Polish presidency during this period was eclectic, as on the one hand it tried to combine the ideas of European community in facing challenges posed by the Covid-19 pandemic, digitalization, and EU border protection, and on the other hand, strongly emphasized opposition to the forced relocation of migrants within the EU. The Polish V4 presidency indicated significant differences between the countries in the region and EU policy (2020–2021 Polish Presidency, p. 2). The program pointed to the promotion of digitalization and AI solutions aimed at facilitating the introduction of technological considerations into the public sphere, but within autonomous national actions without strategic coordination with the rest of the EU (2020–2021 Polish Presidency, p. 2). This demonstrates how doctrinal attitudes of individual governments influence the hierarchy and priorities set during the V4 presidency.

In 2021–2022, Hungary held the presidency of the Visegrád Group and largely distanced itself from the actions and declarations made in previous presidencies regarding the building of a common policy both at the regional and European level. The priority was mainly issues referring to national sovereignty (2021–2022 Hungarian Presidency, p. 5). This stance clearly challenged Hungary's integrative approach within the common EU policy in sensitive areas such as migration and the rule of law. During Hungary's presidency, a situation similar to Poland's tenure occurred, but with greater intensity and clarity, demonstrating how perceptions of the development of information and communication technologies can depend on political perspectives. The government in Budapest limited itself to its own actions, emphasizing the need to preserve national technological sovereignty at the cost of abandoning initiatives aimed at building common digital policy frameworks with the other V4 countries (2021–2022 Hungarian Presidency, p. 6).

The next V4 presidency, held by Slovakia in 2022–2023, focused – contrary to Hungary's previous leadership – on an integrative approach, promoting V4 unity in the face of external threats, which was particularly pronounced in the context of the Russian Federation's aggression against Ukraine. The need to build a coherent policy with the EU and NATO in the field of cybersecurity was also emphasized (2022–2023 Slovak Presidency, p. 2). The program highlighted innovation as a crucial element for further cooperation in cybersecurity. According to this direction, the digitalization of the economy became an important source of economic growth. However, to stimulate further development in this area, it was emphasized that future success would largely depend on the level of research, development, and innovation carried out in the Visegrád countries (2022–2023 Slovak Presidency, p. 5).

Further development of conceptual frameworks for building bonds at both the regional and European levels was articulated during the Czech presidency in 2023–2024. During this period, attention was drawn to the need to continue integrative efforts, especially in the context of the response to the Russia–Ukraine conflict. During this presidency, issues related to climate change and digital challenges as dimensions of contemporary security were strongly highlighted (2023–2024 Czech Presidency, p. 14). The Czech position indicates actions aimed at deepening cooperation with the EU in security and defense and ensuring complementary relations with NATO based on the principle of partnership. It is also characteristic that aspects of security policy are closely linked with building technological and digital potential. Therefore, the Czech presidency aimed to initiate the idea of developing a common digital infrastructure, which would simultaneously promote Central Europe as a center of innovation (2023–2024 Czech Presidency, p. 5).

The Polish presidency in 2024–2025 shows that governments' attention to cybersecurity is not only increasing, but the area of cybersecurity policy being addressed is becoming increasingly complex. This may also indicate that the Visegrád countries are trying to broadly define the phenomena occurring in cyberspace, as well as that cybersecurity policy requires continuous updating of knowledge and adjustment of actions in response to new forms of cyber threats. The promotion and emphasis on the need to build a common policy corresponded with reactions to hybrid threats and information warfare (2024–2025 Polish Presidency, p. 28). It is worth noting that in 2025 Poland's presidency of the Visegrád Group coincided with its presidency of the Council of the European Union, which provided an opportunity to strengthen the regional demands articulated regarding cybersecurity on the European agenda. References were also made to demands previously articulated during the Czech and Slovak presidencies concerning scientific cooperation in the field of new technologies. The need for digital transformation in both infrastructure and business operations was also strongly emphasized, which can be seen as a continuation of earlier demands – especially from the Czech side – related to the ideas of Industry 4.0, smart cities, and the implementation of artificial intelligence solutions in the public sphere (2024–2025 Polish Presidency, p. 14).

The Visegrád Group presidency is largely declarative in nature, and due to its focus on cooperation within the V4, considerably more attention is devoted to this issue. However, the declarations themselves are not binding and do not carry the pressure of actual implementation; they are rather proposals that provide guidance on possible specific actions. Despite all four countries declaring a willingness for regional cooperation, their attitudes largely depend on the political environments in power. Examples of this include Hungary and Poland, which during its V4 presidency in 2020–2021 adopted a strategy of selective integration – promoting the idea of common policy while simultaneously avoiding sensitive areas such as migration and legislative sovereignty. Summarizing the positions of all V4 members, it should be noted that the Czech Republic plays the role of the actor that most strongly promotes the idea of integration and community both at the regional and European levels. In Poland's case, a noticeable change, especially after the 2023 elections, points to a significantly more pro-EU stance. Nevertheless, it should be noted that the Polish side oscillates between openness and integration and national pragmatism, for example regarding migration. The third country is Slovakia, which emphasizes the desire for regional cooperation; however, the stance of some political circles – particularly associated with Prime Minister Robert Fico – contributes to disintegration both within the V4 relations and the EU itself. His critical stance, especially on migration, member state sovereignty, and sanctions against Russia, brings this politician closer to the position held by Hungarian Prime Minister Viktor Orbán. These controversial positions have become the main cause of tensions among EU and NATO members and are also one of the reasons for the lack of trust between political elites, which translates into disintegration – especially in digital, legislative, and geopolitical cooperation within the V4 framework.

5 Digital Policy in the Presidency Programs of the Visegrád Group Countries

In this part of the article, the analysis will use the EU presidency programs of four Central European countries, namely: Poland – "Security, Europe" (Poland 2025); Hungary – *Programme of the Hungarian Presidency of the Council of the European Union in the Second Half of 2024* (Hungary 2024); the Czech Republic – *Europe as a Task: Rethink, Rebuild, Repower* (Czech Republic 2022); and Slovakia – *Programme of the Slovak Presidency of the Council of the European Union* (Slovakia 2016).

In all Council of the EU presidency programs, the importance of new technologies (AI, 5G, digital services) is strongly emphasized. The V4 countries point to their role in building a competitive economy (Poland 2025, p. 41; Czech Republic 2022, p. 42; Hungary 2024, p. 24). This also includes the development of research centers focused on digital tools while ensuring stable development of the public sphere based on strengthening the resilience of digital infrastructure (Slovakia 2016, p. 8; Czech Republic 2022, p. 50). This is the main trend in all programmatic and strategic content, as well as for the majority of European countries. In this respect, the V4 countries do not diverge from the main current of digital policy development indicated by the EU.

All four entities state that the concept of security encompasses not only military aspects but also political, economic, and social ones. Consequently, the perception of cyber threats has also been broadened. Thus, countries leaning toward a non-military approach – due to their high dependence on information infrastructure – view the stable functioning of this area as a key condition of national security (Hungary 2024, p. 22; Czech Republic 2022, p. 33). In this context, combating disinformation as well as hybrid and terrorist threats is emphasized, which is a common denominator in the articulated content of the EU presidency programs. V4 members highlight their belonging to the EU and NATO. In all four presidency programs, cyberspace is perceived in both military and civilian contexts, with an emphasis on a competitive economy (Poland 2025, p. 6; Hungary 2024, p. 31). They also indicate that their actions in digital policy correspond to the transformational actions of the EU, fitting within the framework of national security by strengthening digital skills (Poland 2025, p. 42; Slovakia 2016, p. 34).

Reading the EU presidency programs reveals that despite many similarities in defining digital policy, each country sought to emphasize and distinguish a different area of this field. Poland highlights the importance of competitiveness in developing and implementing new technologies (Poland 2025, p. 6), similarly to the Czech Republic, which emphasizes innovation in public administration aimed at improving the functioning of the public sphere (Czech Republic 2022, p. 43). Slovakia perceives new technologies as a challenge for public education, with a strong emphasis on openness and inclusion of all social groups (Slovakia 2016, p. 34). Such an approach to digitization also contributes to increasing regional cohesion while marginalizing geopolitical significance. Hungary presents a different perspective, viewing technologies as tools to build competitive advantages in the international arena (Hungary 2024, p. 6).

An important concept for characterizing digital policy in the presidency programs is the definition of cybersecurity, where each author places different emphasis. Poland and the Czech Republic perceive the term in close coordination with EU and NATO policy, with the difference that the Polish presidency places stronger emphasis on the resilience of critical infrastructure (Poland 2025, p. 7), whereas the Czech document highlights international cooperation in combating disinformation (Czech Republic 2022, p. 9). The Slovak presidency more often associates cybersecurity with the perspective of stable and predictable development of the private sector (Slovakia 2016, p. 6). Hungary, on the other hand, defines the concept through the prism of national values, emphasizing the development of its own cyber defense capabilities, including primarily the stable functioning of state structures (Hungary 2024, p. 15).

All V4 countries' documents also address international cooperation in digital policy. Building an alliance in this area is strongly emphasized in the Polish document, though the goal of this call is to build the EU's resilience and independence from China and Russia (Poland 2025, p. 16). The Czech document, in turn, points more toward reducing barriers between EU countries (Czech Republic 2022, p. 42). Slovakia also supports these initiatives; however, for Slovakia, strengthening digital solidarity is more an opportunity to reduce the innovation gap in the economy (Slovakia 2016, p. 27). The Hungarian program shows a distance toward technological cooperation, where political decision-makers emphasize protecting national interests in EU digital policy, thereby omitting geopolitical issues (Hungary 2024, p. 16).

The content of the presidency programs articulated by the V4 countries allows characterization of their stance toward digital policy in terms of both security and the functioning of the economy and society. In Poland's case, there are efforts to maintain a balance in actions focused on both internal and foreign policy, as well as in the application of cyber technologies at military and non-military levels (Poland 2025, p. 4). The Czech presidency also strongly emphasized belonging to the EU, with digital policy development serving as an area integrating European countries. The dominant security perspective in the Czech presidency was the non-military aspect related to innovative solutions in the private sector (Czech Republic 2022, p. 22). Conversely, the Hungarian presidency documents strongly highlight the controlling and monitoring aspect of the existing EU digital policy. Cross-border digital services are meant to support EU security, build the European Digital Identity (eIDAS 2), and implement new solutions, including artificial intelligence, aligning with the general current of EU digital policy. At the same time, their cross-border nature and the need for control give these documents a perspective emphasizing a greater role for member states in shaping and supervising the EU's digital policy (Hungary 2024, p. 21). Slovakia's Council presidency promotes the implementation and acceptance of technological solutions that foster greater integration of EU countries, but with an emphasis on supporting digitization and innovation as conditions for economic development and success. Slovakia sees this as an opportunity for regional and economic development, especially in implementing technological solutions aimed at improving public services' functioning, particularly in less developed regions of Europe (Slovakia 2016, p. 11).

6 Conclusion

An analysis of the Council of the EU presidencies' programs in the context of digital policy indicates not only differences in how individual goals and tasks related to the application of new technologies in the public sphere are defined but also reveals how particular countries position themselves in this area on the international stage. It turns out that the objectives set out in the Polish program – such as developing European AI research and competence centers, promoting EU digital diplomacy, and ensuring energy security – suggest that Poland has the ambition to be a significant player in cybertechnology in the region. By contrast, the Czech Republic represents moderate ambitions in cybertechnology, mainly within the EU context. Slovakia does not show ambitions to be a dominant regional actor in digital policy but rather focuses on implementations carried out within the EU framework. This stems from the described asymmetries compared to other EU countries in terms of digital skills development and historical experiences. Therefore, it is understandable that smaller Central European countries more often refer to international organizations like the EU and NATO and the norms and standards they set to strengthen their own economic, technological, and military development. Hungary presents a completely different stance in its Council presidency. Although it does not reject the general goal of developing digital skills and implementing innovative solutions in the economy and cybersecurity standards, it distances itself from unequivocally identifying with the EU's vision of a comprehensive and harmonized digital policy. Instead, Hungary approaches it selectively, using only those areas beneficial from the perspective of national interest.

The analysis carried out in this work allows the conclusion that the V4 could act as a catalyst for changes in digital policy and cybersecurity, especially in situations that may indirectly affect stability in the region and the entire EU, as exemplified by the war in Ukraine. However, differing perceptions of threats and varying interpretations of Russia's policies – primarily by Hungary – limit the V4's ability to express a unified position. In a similar vein, internal differences weaken the organization's role on the international stage as a signatory of Central European regional interests.

Characterizing digital policy, it can be seen that the Council presidencies reflect national political priorities. Therefore, it can be concluded that there is a correlation between the political preferences of the V4 countries and the topics included in their Council presidency programs. This also allows the inference that the EU presidency is an opportunity to use the six-month term as an extension of regional political agendas. Examples of this are the contents of the documents: Poland emphasizes the resilience of critical infrastructure; Hungary stresses digital sovereignty, while the Czech and Slovak programs promote integration and technological cooperation.

Assessing the convergence between the political preferences of the V4 and the themes of their Council of the EU presidencies, it can be stated that these countries use the EU presidency to promote their regional or national interests, as seen, for example, in Poland's program ("Security, Europe") and Hungary's focus on digital sovereignty. The analysis of documents and the international situation clearly shows that it is impossible to separate technological development from political issues. The Czech Republic and Poland, by combining these areas at the European and regional levels – especially during the war in Ukraine – gained international recognition. Meanwhile, the government in Budapest has deepened its distance in relations with neighbors and the EU, weakening its perception as a reliable partner. Thus, holding the Council of the EU presidency can strengthen a country's position, provided it aligns with the dominant current of EU policy.

References

Abad, L. A., & Maurer, N. (2020). Do pandemics shape elections? Retrospective voting in the 1918 Spanish Flu pandemic in the United States. *SSRN*. http://dx.doi.org/10.2139/ssrn.3680286.

Abels, G. (2017). A 'macro-regional' Europe in the making: Theoretical approaches and empirical evidence. *Europe-Asia Studies*, 69(3), 542–543. https://doi.org/10.1080/09668136.2017.1299918.

Agamben, G. (2005). *A Brief History of the State of Exception*. https://press.uchicago.edu/Misc/Chicago/009254.html.

Amat, F., Arenas, A., Falcó-Gimeno, A., & Muñoz, J. (2020). Pandemics meet democracy. Experimental evidence from the COVID-19 crisis in Spain. *SocArXiv*. https://osf.io/preprints/socarxiv/dkusw/.

Aviña, M. M., & Sevi, S. (2021). Did exposure to COVID-19 affect vote choice in the 2020 presidential election? *Research and Politics*, 8(1). https://doi.org/10.1177/20531680211041505.

Bailer, S., Mattila, M., & Schneider, G. (2015). Money makes the EU go round: The objective foundations of conflict in the Council of Ministers. *Journal of Common Market Studies*, 53(3), 437–456.

Bol, D., Giani, M., Blais, A., & Loewen, P. J. (2021). The effect of COVID-19 lockdowns on political support: Some good news for democracy? *European Journal of Political Research*, 60(2), 497–505.

Çabada, L., & Waisová, S. (2018). The Visegrád Group as an Ambitious Actor of (Central-)European Foreign and Security Policy. *Politics in Central Europe*, 14(2), 9–20. https://doi.org/10.2478/pce-2018-0006.

Campante, F. R., Depetris-Chauvin, E., & Durante, R. (2020). The virus of fear: the political impact of Ebola in the U.S. *National Bureau of Economic Research*, Working Paper 26897. http://www.nber.org/papers/w26897.

Carammia, M., Princen, S., & Timmermans, A. (2016). From summitry to EU government: An agenda formation perspective on the European council. *Journal of Common Market Studies*, 54(4), 809–825.

Czech Presidency 2019–2020. (n.d.). *Visegrád Group*. https://www.visegradgroup.eu/home/documents/presidency-programs/presidency-programs.

Czech Presidency 2023–2024. (n.d.). *Visegrád Group*. https://www.visegradgroup.eu/home/documents/presidency-programs/presidency-programs.

De Ruiter, R., & Vliegenthart, R. (2018). Understanding media attention paid to negotiations on EU legislative acts: A cross-national study of the United Kingdom and the Netherlands. *Comparative European Politics*, 16, 649–668. https://dx.doi.org/10.1057/s41295-017-0090-1.

Debbage, N., Gonsalves, N., Shepherd, J. M., & Knox, J. A. (2014). Superstorm Sandy and voter vulnerability in the 2012 US Presidential Election: A case study of New Jersey and Connecticut. *Environmental Hazards*, 13(3).

Declaration of the Prime Ministers of the Czech Republic, the Republic of Hungary, the Republic of Poland and the Slovak Republic on Cooperation of the Visegrad Group Countries after their Accession to the European Union. (2004, May 12). *Visegrád Group*. http://www.visegradgroup.eu/2004/declaration-of-prime.

Dinan, D. (2013). The post-Lisbon European Council Presidency: An interim assessment. *West European Politics*, 36(6), 1256–1273.

Dür, A., & Mateo, G. (2012). Who lobbies the European Union? National interest groups in a multilevel polity. *Journal of European Public Policy*, 19, 969–987.

Eihmanis, E. (2024). EU transnational coalitions in polycrisis: the Visegrád-4 before and after the Russo-Ukrainian war. *Journal of European Public Policy*, 31(10), 3299–3323. https://doi.org/10.1080/13501763.2024.2357268.

Eisele, O., Heidenreich, T., Kriegler, N., Syed Ali, P., & Boomgaarden, H. G. (2022). A window of opportunity? The relevance of the rotating European Union presidency in the public eye. *European Union Politics*, 24(2), 327–347. https://doi.org/10.1177/14651165221142504.

Grgić, G. (2023). The changing dynamics of regionalism in Central and Eastern Europe: The case of the three seas initiative. *Geopolitics*, 28(1), 216–238. https://doi.org/10.1080/14650045.2021.1881489.

Häge, F. M. (2017). The scheduling power of the EU Council Presidency. *Journal of European Public Policy*, 24(5), 695–713.

Holányi, Á. (2025). Unpolitics, domestic pressures, or else? The opposition of Visegrád Group countries to the EU's Pact on Migration and Asylum. *Journal of European Integration*, 1–22. https://doi.org/10.1080/07036337.2025.2519043.

Huhe, N., Naurin, D., & Thomson, R. (2017). The evolution of political networks: Evidence from the Council of the European Union. *European Union Politics*, 19(1), 25–51. https://doi.org/10.1177/1465116517740228.

Hungarian Presidency 2021–2022. (n.d.). *Visegrád Group*. https://www.visegradgroup.eu/home/documents/presidency-programs/presidency-programs.

Kaniok, P., & Hloušek, V. (2025). Visegrad four as an institution in times of EU crises. *European Politics and Society*, 1–19. https://doi.org/10.1080/23745118.2025.2488815.

Kovacs, O., & Domonkos, E. (2024). Deindustrialisation and reindustrialisation patterns in V4 countries – industry 4.0 as a way forward? *Post-Communist Economies*, 36(4), 432–463. https://doi.org/10.1080/14631377.2024.2323834.

Lakócai, C. (2025). Wellbeing and environmental performance of the Visegrad countries at the beginning of the 21st century. *Journal of Contemporary Central and Eastern Europe*, 33(2), 425–440. https://doi.org/10.1080/25739638.2025.2507607.

Lang, K.-O., & von Ondarza, N. (2018). Minilateralismen in der EU – Chancen und Risiken der innereuropäischen Diplomatie. *SWP-Aktuell*, 7, 1–8.

Leuffen, D., Malang, T., & Wörle, S. (2014). Structure, capacity or power? Explaining salience in EU decision-making. *Journal of Common Market Studies*, 52(3), 616–631.

Osička, J., et al. (2021). Ready to React: Contextualizing the Visegrád Group's Energy Cooperation. *Politologický časopis – Czech Journal of Political Science*, 28(1), 59–84. https://doi.org/10.5817/PC2021-1-59.

Perarnaud, C. (2022). Power to the connected? Determinants of member states' bargaining success in the making of the EU Digital Single Market. *Journal of Cyber Policy*, 7(1), 55–71. https://doi.org/10.1080/23738871.2022.2030382.

Polish Presidency 2020–2021. (n.d.). *Visegrád Group*. https://www.visegradgroup.eu/home/v4-presidency.

Polish Presidency 2024–2025. (n.d.). *Visegrád Group*. https://www.visegradgroup.eu/home/documents/presidency-programs/presidency-programs.

Pop, A. (2021). Subregional Security Arrangements in Central and Southeastern Europe. *Wilson Center*. https://www.wilsoncenter.org/publication/204-subregional-security-arrangements-central-and-southeastern-europe.

Poznar, A., & Havlík, V. (2025). From complementary and opportunistic sub-regionalism to anti-systemic attitudes? Exploring the post-2022 attitudes of Visegrad member states to cooperation. *European Politics and Society*, 1–18. https://doi.org/10.1080/23745118.2024.2446813.

Slovak Presidency of the Council of the European Union. (2016). Programme of the Slovak Presidency of the Council of the European Union. https://eu2016.sk/wp-content/uploads/2016/06/programme-sk-presidency.pdf

Slovak Presidency 2022–2023. (n.d.). *Visegrád Group*. https://www.visegradgroup.eu/home/documents/presidency-programs/presidency-programs.

Strnad, V. (2021). Les enfants terribles de l'Europe? The 'Sovereigntist' Role of the Visegrád Group in the Context of the Migration Crisis. *Europe-Asia Studies*, 74(1), 72–100. https://doi.org/10.1080/09668136.2021.1976730.

Szalai, M. (2022). Norm localisation in the process of crafting national security strategies – the case of the Visegrád countries. *European Security*, 32(2), 210–232. https://doi.org/10.1080/09662839.2022.2124370.

Szalai, M., & Garai, N. (2024). Status-seeking with a negative image – The changing position of Visegrád countries in the international community between 2004 and 2020. *New Perspectives*, 32(1), 30–51. https://doi.org/10.1177/2336825X231221999.

Toneva-Metodieva, L. (2020). Evaluating the presidency of the council of the EU: Towards a comprehensive evidence-based framework for performance assessment. *European Politics and Society*, 21(5), 650–667.

Truchlewski, Z., & Schelkle, W. (2024). Beyond the North–South divide: Transnational coalitions in EU reforms. *New Political Economy*, 29(6), 958–971. https://doi.org/10.1080/13563467.2024.2356552.

Urbanovská, J., Chovančík, M., & Brajerčíková, S. (2022). Minilateral Cooperation in the EU's Post-Brexit Common Security and Defence Policy: Germany and the Visegrád Countries. *Europe-Asia Studies*, 74(3), 402–425. https://doi.org/10.1080/09668136.2022.2037520.

Vaagland, K., & Zaun, N. (2024). Strategising solidarity: An examination of the Visegrád group's role and motivations in EU migration policies. *Journal of European Integration*, 47(4), 559–579. https://doi.org/10.1080/07036337.2024.2383954.

Vaznonytė, A. (2020). The rotating Presidency of the Council of the EU – Still an agenda-setter? *European Union Politics*, 21(3), 497–518. https://doi.org/10.1177/1465116520916557.

Wivel, A. (2021). The Grand Strategies of Small States. In T. Balzacq & R. R. Krebs (Eds.), *The Oxford Handbook of Grand Strategy* (pp. 489–505). Oxford University Press. https://doi.org/10.1093/oxfordhb/9780198840299.013.32.

Wratil, C., & Hobolt, S. B. (2019). Public deliberations in the Council of the European Union: Introducing and validating DICEU. *European Union Politics*, 20(3), 511–531. https://doi.org/10.1177/1465116519839152.

CHAPTER 8

EU Presidency in the Times of Crises: Slovenian and Czech Presidencies

Ladislav Cabada

Abstract There exist a number of prejudices against small Member States, including their ability to effectively manage the Council of the EU, given their limited capacity. This prejudice is even more pronounced in the case of countries which, for various reasons, are considered to be far from the EU mainstream, or which in the past have shown shortcomings or outright problems in the conduct of the Presidency (PRES). This was the case in both dimensions for the Czech PRES in 2009, which was led from a Eurosceptic position and by a government that was censured midway through its term. The comparison between Slovenian and Czech PRES in 2021 and 2022 therefore makes sense, not least because a country that has long been regarded as one of the most pronounced Eurosceptic actors is being compared with a country that, on the contrary, represents the most pro-European case among the new EU member states in a number of respects. The third dimension of the comparison then places the PRES in a situation of responding to a major crisis – in the first case a post-pandemic socio-economic crisis, in the second a crisis related to Russian aggression against Ukraine. Our analysis shows that small member states have all the potential for a successful PRES, including through the trio platform. At the same time, we show that security has become a crucial issue for both PRES, albeit much more selectively for the Slovenian PRES than in the case of a major crisis after February 2022.

Keywords EU Presidency – crisis – resilience – priorities – security – Slovenia – Czech Republic

1 Introduction

The Presidency of the Council of the European Union (hereinafter referred to as PRES) is one of the most important instruments through which a Member State can introduce and address, in a relatively short period of time, the issues

HTTPS://DOI.ORG/10.1163/9789004763623-010

that it considers to be priorities. "The Council presidency is particularly important for smaller EU member states. It ensures a place for them in the spotlight that allows them to increase their profile domestically, at the EU level and beyond. And it places them right in the middle of the EU decision-making with a much better position to influence the final policy" (Kovář & Weiss, 2023, p. 85). Furthermore, "when holding the presidency, member states can materialise national interests by carrying out the following roles: administrative and coordinative, agenda-setting (standing out as the most important instrument), mediating, leadership and representative" (Svetličič & Cerjak, 2015, p. 11).

In this regard, it is worth mentioning the extensive theoretical literature devoted to the role of the presiding state. For instance, Tallberg (2006) identifies three fundamental roles of the PRES: the agenda-setting, acting as a mediator/broker, and representing the Council externally. Although the PRES is formally defined as neutral, the scholars stress that states try to promote their national preferences directly and, more often, indirectly, especially on long-term sensitive issues (Elgström, 2003; Schout & Vanhoonacker, 2006; Tallberg, 2003, 2006). In this context, the tension between national and European interests presents a permanent and dynamic structural feature. Presidencies must strike a balance between the need for neutrality and pressure from domestic political actors who expect concrete political outcomes in return for their support. Many authors also emphasise that PRES is a tool to bring European affairs closer to the domestic public and thus strengthen EU cohesion (Lange & Svetličič, 2024, p. 369).

The different roles that the member states providing PRES can take on are related to the general assumptions of the theory of Europeanisation, which focuses on the two-way interaction between the Union and its member states. While top-down Europeanisation describes the adaptation of national policies and actors to European rules, the bottom-up perspective emphasises member states' efforts to project their interests and norms onto the EU level. The PRES naturally represents a key channel for bottom-up Europeanisation, accelerating a member state's potential (Börzel & Risse, 2003; Kaniok, 2008).

At the same time, too much emphasis on one's own priorities poses a risk in at least two ways. The first risk is that by changing priorities and agendas too often, the EU could be weakened or even undermined. The second risk is that if a member state pursues priorities or positions that are too far out of the mainstream, it could find itself isolated (as happened quite significantly in the case of the Hungary in 2024) and the PRES would be a wasted period. Both risks are mitigated by the fact that, in addition to the national PRES, a permanent position of President of the European Council was created in 2009 to ensure continuity of processes and agendas. At the same time, the Trio

Presidency system, which involves cooperation among three successive Presidencies, was established to facilitate continuity and foster consensus on the medium-term agenda. (Kaniok, 2008; Kaniok, 2010).

Notwithstanding these mechanisms for ensuring continuity, it remains true that each member country brings its own approaches to the PRES, including by specifically defining its actions in relation to long-term trends and general issues in the development of the European integration process. As Svetličič and Udovič (2024) show in the case of the Slovenian PRES in 2021, "the Trio was more important during the preparation stages, while less important during the operational stage where the presiding state is 'basically left on its own'." In this respect, the first Czech PRES in the first half of 2009 was illustrative, where the main ruling party, the Civic Democratic Party (ODS), dominated by Eurosceptic politicians, tried to push its scepticism towards the EU mainstream onto the agenda and even more strongly into symbolic politics and rhetoric – main motto *We will make it sweet for Europe*[1]. The scepticism towards the Czech PRES was so strong that at least in some agendas (e.g. the escalation of the Israeli–Palestinian conflict), the previous presiding country, i.e. France, showed a tendency to push out or outright ignore Czech politicians or diplomats. The problems faced by the Czech PRES in the first half of 2009 were compounded by a vote of no confidence in the government around the middle of the year and an interregnum of almost a month (Kaniok, 2010).

Thus, not only in the case of the Czech PRES 2009, there exists scepticism towards (some) small member states and their ability to staff and capacitate PRES, but also scepticism towards (some) new member states from the CEE region and their sufficient socialisation into the EU environment. Svetličič and Cerjak (2015, p. 13) describe this double dilemma or double scepticism quite succinctly when they state: "In addition to their small size, new member states face the challenge of adapting to and integrating into the EU's system of decision-making." In this respect, the second PRES of the countries of the so-called New Europe represented an even greater challenge in terms of confirming or, on the contrary, disproving the prejudices and expectations associated with the situation of democratic backsliding, the undermining of the rule of law and de-Europeanisation (Cabada, 2024).

This chapter compares the main programmes and achievements of the two Central European PRES that took office in 2021 and 2022 respectively: the Slovenian PRES in the second half of 2021 and the Czech PRES, which followed

1 The slogan or hidden message "defies translation as it centres on an idiom, meaning of which is to give someone (Europe, in this case) a hard time" (Czech EU Presidency campaign uses bitter(sweet) pun, 2009).

exactly one year after the Slovenian PRES. Both countries belong to the group of so-called new member states, for which analysts have long emphasised specific legacies and path dependencies. At the same time, both PRES had to respond to dramatic challenges – the Slovenian one to the COVID-19 pandemic and the sudden change of priorities, and the Czech one to the Russian aggression against Ukraine and the new challenges related to the changing geopolitical situation and increased security uncertainty.

With regard to the digital agenda, it should be emphasised that both countries and their governments attach great importance to digitisation, cybersecurity, and the regulation of AI. The Czech government has emphasised the key role of cybersecurity and, more generally, digital resilience in a number of strategic documents, primarily in the strategic plan Digital Czechia (The Path to Europe's Digital Decade, 2023) and also in the new security strategy (Bezpečnostní strategie České republiky, 2023). Similarly, Slovenian doctrinal documents also reflect issues of strategic autonomy and digital sovereignty, primarily in the Digital Slovenia 2030 Strategy (2023). However, this document places much greater emphasis on topics such as a free and open internet and a democratic digital society. Although the latest versions of these documents were produced after the PRES had ended, both PRES clearly reflected a focus on digitalisation as a tool for modernisation, as well as on resilience linked to crisis framing (e.g. pandemic, Russian aggression against Ukraine, cyber threats, and digital and technological sovereignty). It is also evident that the digital agenda is firmly embedded in both countries' governments and linked to European policies.

The two countries are different cases in many respects. While Slovenian political parties and the Slovenian public have long shown pro-integration attitudes and, on the contrary, only very moderate scepticism towards the EU, the Czech public has consistently been among the most (euro)sceptical societies in the EU, and the same is true of Czech political parties. While in the case of Slovenia, both left-liberal and national-conservative oriented cabinets have never questioned the country's belonging to the EU core (Cabada, 2005; Cabada & Hlaváčková, 2016; Cabada & Waisová, 2010), the opposite is the case of Czech Republic. The centre-left governments that brought the country into the EU did not present themselves proactively at the European level, while the right-wing parties have had a Eurosceptic profile since the beginning of the new millennium. In the last decade, the traditional left has been replaced by Eurosceptic national populists, which has further marginalised pro-European forces (Drulák & Šabič, 2010; Cabada, 2024).

However, this picture is not complete and may take on different modalities. Slovenia has thus repeatedly found itself in the group of countries criticised for weakening the rule of law. The Prime Minister and President of the Slovenian Democratic Party (SDS) Janez Janša was placed (and placed himself) alongside Viktor Orbán and his style of governance was perceived as authoritarian (Fink-Hafner, 2022; Fink-Hafner, 2024). In contrast, the strongly Eurosceptic Czech ODS took over the government in late 2021 and early 2022, together with four other parties, to seek to protect democracy and the country's pro-Western direction in the face of a populist and anti-liberal opposition (Cabada, 2024).

Nevertheless, from a purely de-ideologized point of view and focusing on the functional and content-related framework, both PRES show important similarities. This is undoubtedly mainly due to the (post)crisis situation in which both PRES took place – the Slovenian one in the wake of the COVID-19 disease pandemic in 2020/2021 and the search for ways to a new economic and social stability, while the Czech one, in addition to this long-term challenge, had to reflect a quite fundamental impulse – the Russian military aggression against Ukraine. In this respect, it will undoubtedly be interesting to see how the practical steps of the Czech PRES have dealt with the change that has occurred in the course of its implementation, and to what extent the intentions and priorities presented before this game changer have been maintained or modified. Last but not least, we find it useful to compare two PRES that took place within a short period of time, as the Czech PRES was launched only one year after the Slovenian one. In this sense, the two countries coordinated their activities in a trio, with France acting as liaison.

Methodologically, we rely primarily on the analysis of policy documents and public policies, using both diachronic and synchronic comparisons. We set the priorities of both PRES, as well as their operationalisation and implementation, as the main objects of comparison. Specifically, we focus on goals and priorities clearly related to digitisation, cybersecurity, resilience, and strategic autonomy. In addition to the comparative method, we apply the process-tracing method, particularly given that the French presidency was inserted between the two others. At the same time, we examine the programmes of the trio in both cases. Applying the process-tracing method will allow us to identify synergies, continuities, and discontinuities relating to the analysed priorities and objectives.

In addition to the policy documents of both PRES, we make use of existing analyses of both PRES. A particularly useful source of information for us was the special issue of the Czech Journal of International Relations (CJIR, issue 1/2023) in the case of the Czech PRES and, in the case of Slovenia, especially the articles authored or co-authored by Marjan Svetličič and Boštjan Udovič.

2 Slovenian EU Presidency 2021

The Slovenian PRES in the second half of 2021 was based on the programme of the Germany–Portugal–Slovenia trio, whose priority theme was overcoming the crisis moments associated with the impact of the pandemic. The programme focused on building a roadmap for recovery. The document on this general objective stressed the need for solidarity, cohesion and convergence, flexibility, agility, and development. The document placed particular emphasis on restoring and further deepening the internal market, driving forward the green transition and digital transformation, the EU's digital sovereignty and strategic autonomy, building resilient infrastructure, especially in the health sector, and producing critical goods in Europe to reduce over-dependence on third countries (Council of the EU, 2020, p. 3). The trio focused intensively on artificial intelligence in all its dimensions, "including research and innovation, applications in education, ethical and human-centred aspects, its global governance, risk-based regulatory frameworks and the aspect of liability for artificial intelligence" (Council of the EU, 2020, p. 9). These activities should reflect efforts to better protect European societies from malicious cyber activities, hybrid threats, and disinformation. The trio presented the intention to strengthen the EU as "a responsible, innovative, sustainable and competitive digital ecosystem", recognising that "artificial intelligence is a key technology with enormous potential for value creation". It therefore declared its intention to significantly strengthen the legislative regulation of this sector (Council of the EU, 2020, p. 16).

The Slovenian PRES in the second half of 2021 took place in the immediate aftermath of more than a year of global crisis associated with the COVID-19 pandemic. While previous PRES focused primarily on ensuring sufficient new vaccines, the Slovenian PRES has already been able to develop more systematically the intentions of the Joint European Roadmap for lifting COVID-19 containment measures and the Roadmap for Recovery. The three key principles of the latter document – a resilient, sustainable, and fair Europe – have become the key themes of the Slovenian PRES programme.

The programme, quoted by the motto Together. Resilient. Europe, underlined the agreement with the main priorities of the New Strategic Agenda 2019–2024, highlighting four key points and/or priorities: 1) The resilience and recovery of the EU. A strategically autonomous EU; 2) The Conference on the Future of Europe; 3) A union of the European way of life, the rule of law, and the same criteria for all; 4) A credible and secure EU. Our analysis will not address the second point, which has no clear policy definition.

With regard to the first group of priorities, the PRES wanted to focus on "strengthening capacities to successfully deal with pandemics and various forms of modern and complex security risks and threats, such as large-scale cyber-attacks" (Programme of the Slovenian Presidency, 2021, p. 3). This cumulative focus shows the importance of pandemics and the resulting crisis for the conceptualisation of security. The link between the two reflected risks was the effort to reduce dependence on external actors and, on the contrary, to strengthen the capacity, sustainability, and resilience of European health systems and critical infrastructure, as well as to respond effectively to major cyber attacks and crises, and to strengthen the cyber resilience of critical infrastructure and the digital single market (Programme of the Slovenian Presidency, 2021, p. 4). A specific contribution of the Slovenian PRES has been the targeting of activities to build and support cybersecurity capacities in the Western Balkans. A visible element of the intentions of the Trio, including the Slovenian PRES, is the emphasis on strategic autonomy and sectoral autonomy of the EU, such as the supply of medicines, vaccines and medical equipment, and in the area of industrial policy and energy, as well with regard to food supply, as well as strengthening the sustainability of European value chains (Programme of the Slovenian Presidency, 2021, p. 5). As Lange and Svetličič note (2024, p. 379), "within the first priority, the majority of pressing EU dossiers were in line with the Commission's work programme (on green and digital transitions, and on recovery and resilience following the COVID-19 crisis) and international agenda (COP26Glasgow)." The programme thus promises not only to maintain the European Green Deal, but also to strengthen it, including through new sources: "We will devote particular attention to the proposal for digital taxation and a new own resource based on a carbon border adjustment mechanism." At the same time, we see in the programme a special emphasis on the freedom of each Member State to choose an energy mix according to its capabilities and preferences, "which also includes the possibility of exploiting the potential of safe nuclear energy" (Programme of the Slovenian Presidency, 2021, p. 6). Repeatedly, the programme emphasises the promotion of e-mobility and alternative fuels, and we specifically see an emphasis on public transport – in addition to the development of TEN-T networks, the programme mentions 2021 as the European Year of Rail, which is somewhat paradoxical, given the very poor level of public transport in the country, including train transport.

As regards the second cross-cutting policy – digitisation – the document emphasises in particular the legislative framework, in addition to the technological aspects: focus on making progress in negotiations on the Artificial Intelligence Act, which will provide the general framework for further regulation in this field, formally define artificial intelligence, and limit the risks posed by its wider use. In the areas of re-use, processing and exchange of data, and the data economy, the PRES will continue consideration of the Data Governance Act and launch deliberations on the Data Act (Programme of the Slovenian Presidency, 2021, p. 8).

According to Svetličič and Lange (2024, pp. 379–380), the third priority "strongly demonstrates an ideological vision of the government in office, which was struggling at home and facing continued protests due to the deteriorating rule of law situation". However, at just over two pages, the programme remains very general and superficial. It mentions the need to raise awareness of the European way of life, high-quality inclusive education, demographic challenges, respect for the fundamental values and principles of the EU, rule of law situation, the freedom and plurality of the media and the fight against fake news, and proper functioning of the judicial system. At the end of this chapter, the Slovenian PRES reflects on the establishment of a European foundation for constitutional democracy. This idea is illustrated by a passage from the Slovenian Presidency programme: "The foundation would include an institute to bring together experts from all EU member states, who would provide the European Commission and the member states with autonomous and independent analyses of specific issues concerning the rule of law, which EU institutions and member states could, if necessary, rely on in various procedures. The Slovenian government proposes that the headquarters of the new institution could be located in Slovenia" (Programme of the Slovenian Presidency, 2021, pp. 12–14). The declarative objectives of the Slovenian PRES in the area of rule of law were perceived by analysts as formal and without any real interest in promoting clear measures and procedures against actors or activities undermining democratic resilience within the EU (Bojinović Fenko & Svetličič, 2022; Fink-Hafner, 2022). As Lange and Svetličič (2024, p. 375) stress, the second Slovenian Presidency was marked by political volatility and civil society protests against the government that took over at the start of the pandemic in March 2020. Though lead by the same Prime Minister (Janez Janša) as during the first Presidency, the opposition refused to enter a similar "non-attack" arrangement, viewing foreign and European policies as being instrumentalised for domestic consolidation of power, and turning away from the more liberal centre of the EU. With regard to the fourth priority, the PRES plan, in addition to the repeated emphasis on the Western Balkans, generally focuses on the EU's role as a soft power in its neighbourhood (Mediterranean, Eastern Partnership), as well as on the positive aspects of development cooperation and, last but not least, on the need to strengthen transatlantic relations (Programme of the Slovenian Presidency, 2021, pp. 15–17; Lange & Svetličič, 2024, p. 379–380).

As the most significant outcome of the Slovenian PRES we can undoubtedly mention the 22 national recovery plans that have been adopted. In retrospect, there has been quite significant progress in the negotiations between the Western Balkan countries and the EU on the region's integration processes. Work has started on the Artificial Intelligence Act and other cyber security-related regulations. Last but not least, improvements achieved in minimum wages in the EU and in terms of equal pay for women and men can be considered a success. However, the European Commission's (2021) Rule of Law Report and the European Parliament resolution on the fundamental rights and the rule of law in Slovenia overshadowed the Presidency. The reports stated grave violations by the government of the rule of law, media independence, and anti-corruption measures (Lange & Svetličič, 2024, pp. 381–382).

3 Czech EU Presidency 2022

The programme of the Czech PRES was set in the middle of the trio, which also included Sweden and France. The trio's programme was divided into four pillars: 1) Protecting citizens and freedoms; 2) Building economic foundations: a European model for the future; 3) Building a climate-neutral, green, fair, and social Europe; 4) Promoting European interests and values in the world. The thematic priorities of the trio thus reflect a wide range of political, social, and economic challenges. Post-pandemic recovery, mitigating the economic and social impact of the crisis, and building resilience across a range of interlinked policy areas continue to be seen as key challenges (Bureš, 2023, p. 122). The document does not weaken the emphasis on two cross-cutting themes – ensuring an effective recovery for green and sustainable growth, and digital transitions (Council of the EU, 2021, pp. 7–8). If we focus specifically on the issue of digitalisation, then Trio emphasises the commitment "to ensuring the Union's digital sovereignty in a self-determined and open manner, promoting cultural content and a strong media sector, reliable industrial solutions and strengthening the Union's cyber resilience and responsiveness, support for a robust and innovative digital ecosystem for businesses, and excellence in research in the field of new technologies. Digital development must safeguard our values, fundamental rights and security, and be socially balanced. Fostering the development of technologies, including supercomputers and quantum computing, blockchain, Artificial Intelligence and semi-conductor chips will contribute to that objective" (Council of the EU, 2021, p. 10).

The analysis of the Czech PRES is fundamentally influenced by the fact that in the context of the Russian military aggression against Ukraine, which began on 24 February 2023, key priorities or intentions had to be fundamentally changed. In addition to the geopolitical consequences, the fundamental impact of the war, or of the EU's moral stance based on support for Ukraine and the establishment of a sanction's regime against Russia, on the availability and prices of oil and gas, and more generally on energy sustainability and security, should be mentioned in particular. Increased cyber-attacks and other hybrid threats initiated and supported by Russia (and its partners) against EU countries and other Western democracies have become another major challenge. Last but not least, the large-scale migration wave, which has brought approximately four million Ukrainian war refugees to the EU within a very short period of time, primarily to Poland, Slovakia, and the Czech Republic, has added a further layer of complexity to the EU policy agenda and placed considerable pressure on the capacities of national administrations (Kovář & Weiss, 2023, pp. 86–87).

In addition to overcoming these challenges, it is worth mentioning the need to maintain consensus among the member states and, at the same time, the symbolic policy that began with the joint visit of the Prime Ministers of Poland (Mateusz Morawiecki), the Czech Republic (Petr Fiala) and Slovenia (Janez Janša) to the Ukrainian capital of Kiev on 15 March 2022. At that moment, the Czech and Slovenian Prime Ministers represented, among others, the country that ended its PRES less than three months ago and the country that will take it over in less than four months. In their short speeches, all three Prime Ministers stressed that their visit was part of the European and EU policy framework[2]. The symbolic significance of the visit of the three Central European prime ministers (and Lithuanian Foreign Minister Gabrielius Landsbergis), less than a month after the start of the Russian invasion, can be seen as a symbolic impulse to "a shift in the balance of power in the EU, where more attention and informal political power began to shift to the countries of Central and Eastern Europe … The Czech credibility with regard to the EU response to the Russian war also facilitated the role of the Presidency in the sanctions agenda" (Sus, 2023, p. 104).

Observers and analysts point out some similarities between the first and second Czech PRES, precisely in connection with the security problems arising from developments within the Eastern Partnership. As Tungul states, in 2009,

2 Among other things, Fiala said: "You are not alone, our country is standing by your side. Europe is standing by your side". Morawiecki and Janša stressed the need to grant Ukraine candidate status as soon as possible (ČT24a, 2022).

the centre-right government of Mirek Topolánek had to deal with the gas dispute between Russia and Ukraine, but also oversaw the first ever Eastern Partnership Summit.The Petr Fiala government presides over the EU when the EU ´s relationship with the eastern neighbours is being reshaped, and it has not been a surprise to anyone that the EU Eastern Partnership and enlargement have become a priority of the Czech Presidency. Czechia has supported further EU enlargement; it has adopted a strong stance on Russia's aggression and has identified China as a threat in its broader concept of security (Tungul, 2022, p. 10). Jirušek also highlights the similarities between the two Czech PRES in the energy crisis involving Russia and Ukraine. "However, in almost all other aspects, the situation in 2022 was markedly different" (Jirušek, 2023, p. 148).

The Czech PRES in the second half of 2022 can be seen primarily through the lens of the critical challenges posed by Russia's aggression against Ukraine, but also within the broader framework of the so-called polycrisis that the EU and the group of Western democracies have been facing since 2008. Under the slogan, Europe as a task: rethink, rebuild, repower, the Czech government has declared its desire to use the negative effects of war and other crises to strengthen the resilience of the EU and its member states, using new technological possibilities and ideological approaches. Let's recall that the new cabinet of Prime Minister Petr Fiala took power at the turn of 2021 and 2022, and two months after its arrival, it had to respond to the fundamental challenges of the mass migration of the Ukrainian population, the energy crisis caused by sanctions against Russia and efforts to break free from dependence on Russian energy resources, a new wave of hybrid activities associated with the use of disinformation campaigns, and a strong affective polarisation of society.

The five programmatic priorities presented by the Czech government and coordinated by the Trio on 15 June 2022 reflected both the new challenge of Russian aggression against sovereign Ukraine and the longer-term goals of fighting populism and strengthening democratic institutions, the need to weaken Russian and Chinese influence (not only) in Central European countries, and the clear anchoring of the Czech Republic and the so-called New Europe in Western structures. These priorities were represented by the following items: 1) Managing the refugee crisis and Ukraine´s post-war recovery; 2) Energy security; 3) Strengthening Europe's defence capabilities and cyberspace security; 4) Strategic resilience of the European economy; and 5) Resilience of democratic institutions. Kovář and Weiss (2023, p. 90) point out that the Czech PRES has deviated somewhat from the intentions of the trio by suppressing climate and environmental issues, including at the expense of energy security. Therefore, it was likely that the long-term EDG agenda, such as energy transition, decarbonisation of the economy, reduction of greenhouse gases, and the carbon border adjustment mechanism, would give way to more short-term energy security agenda, such as the reduction of the EU's dependence on Russian fossil fuels and security of supply. The European Policy Centre think-tank even renamed the Czech PRES, the "Energy Presidency". With regard to the further development of the Czech position in the EU, it is certainly worth noting that Josef Síkela, from his position as Minister of Industry and Trade of the Czech Government to his new role as a member of the European Commission, has provided decisive solutions related to fundamental changes in the area of importing and securing energy resources for the EU (Havelka, 2023, p. 7). Soon after the Russian attack on Ukraine, the Czech Republic became a refuge for about half a million refugees, while the Czech government's principled stance against further extensive use of Russian oil and gas led to a significant rise in energy prices and populist attacks by the opposition, threatening an energy crisis and uncontrolled rise in fuel prices; "a topic intensely politicized in domestic political battles. For this reason, energy market functioning and energy prices became the most closely observed issue of the Czech presidency. On the EU level, the energy-related agenda was also among the most followed topics, as energy prices and supply shortages were the main real-life war impacts in the EU countries. In fact, outside of the countries dealing with the influx of refugees, the energy price hikes were the only noticeable impact on everyday life" (Jirušek, 2023, p. 149).

Diversification of resources, energy savings, and accelerating the energy transition were key elements of the broader REPower EU plan. Replenishing natural gas storage and promoting joint purchases became the Czech Republic's second energy security priority, while implementing measures to tackle high energy prices became the third. Under the first priority, the Czech PRES focused on a voluntary reduction of gas consumption by 15% in the 2022/2023 heating season, assuming that resource availability would be the instrument to reduce prices. It was possible to meet the gas storage needs, but the increased demand here could be related to the sharp price increase in the summer of 2022. Coordinated activities under the Czech PRES became an important tool to mitigate the negative impacts. "Joint gas, LNG, and hydrogen purchases via the EU Energy Platform were agreed on towards the end of the presidency in November 2022" (Jirušek, 2023, pp. 150–151). Notwithstanding the emphasis on the energy dimension of security, in general, most of the activities of the Czech PRES can be related to security in its various forms. Bureš (2023, p. 122) lists the key (external) security issues as follows: "the implementation of the Strategic Compass; strengthening the EU-NATO cooperation, the development of capabilities in this respect and the strengthening of European defence capacities; the development of rapid response capabilities and streamlining CSDP missions and operations; resilience against disinformation and strategic communication at the EU level; and the external aspects of the fight against terrorism." At the same time, he notes that the external impulse in the form of Russian aggression against Ukraine forced the Czech PRES to deviate from the intentions of the trio, which had dealt only marginally with external security issues. The completion of negotiations and the conclusion of the agreement on the UK's participation in EU military mobility projects are considered some of the very positive results of the Czech PRES (Sus, 2023, pp. 105–106). The prioritisation of the resilience of democratic institutions can undoubtedly be included in the complex of security issues, which is why Czech PRES, in addition to the general issues of the protection of the rule of law, media freedom, and dialogue with citizens, has also included several related issues of digital political parties and functioning in the social media environment. Most importantly, "the focus of the Czech presidency in this area was likely to be hybrid threats, particularly fighting disinformation in both online and offline environments, given the public administration's experience with the working of the Centre against Hybrid Threats within the Czech Ministry of Interior" (Kovář & Weiss, 2023, pp. 91–92). Cybersecurity was also addressed by the Czech PRES in the framework of the Strategic Compass and the newly established EU Hybrid Toolbox (EUHT). Among the concrete actions taken was the organisation of a conference on strengthening resilience and countering hybrid interference and, most importantly, the preparation of the Directive on measures to achieve a high common level of cyber security across the Union (the NIS 2 Directive), which was adopted by the European Parliament by the end of December 2022 and entered into force in January 2023 (Sus, 2023, p. 106).

The Artificial Intelligence Act, which was handed over to the French PRES by the Slovenian one, has become an important contribution in the field of digital policy, with an overlap with both security and the protection of fundamental human rights and freedoms. Although implementation has been slowed by the emergence of generative artificial intelligence chatbots and the search for consensus among Member States on how to respond, the AI Act represents an important contribution to regulating the use of AI in ways that would threaten national and European security or restrict the rights of EU citizens. The same also applies to a reworked Chips Act. Finally, Czech negotiators also agreed on a Digital Decade Policy Programme, and introduced a new, negotiated Cyber Resilience Act (Havelka, 2023, p. 7).

In order not to cover only security issues in our analysis, let us mention as one of the achievements of the Czech PRES the organisation of the first summit of the European Political Community on 6–7 October 2022 in Prague. This activity was initiated by the previous French PRES as a platform for coordinating the policies of European states, regardless of their membership or other relations with the EU (ČT24a, 2022).

Prior to the start of the PRES, many observers, including the new government that inherited the problem, pointed to the underfunding and insufficient human resources that the outgoing Prime Minister Andrej Babiš had devoted to the organisation of the Czech PRES (Kovář & Weiss, 2023, p. 82; Havelka, 2023, p. 3). In this sense, the Czech PRES itself became an unexpected success for many analysts and politicians, strengthening the Czech position within the EU and at the same time, disrupting the one-sided prejudiced view of some actors towards the new member states of Central Eastern Europe, primarily the Visegrad Group. As Havelka (2023, pp. 3–4) states, some could even argue that Czech term outperformed the average presidency. The Czech Republic managed to finalize 41 pieces of new legislation and keep the EU united through 2022. The former is well above what the Czech public administration had estimated in terms of new legislation; with an initial target of 15 finalised proposals. Sus (2023, p. 109) makes a similar assessment of the Czech PRES, stating the following: By skilfully setting and scheduling the agenda and staying on top of the key policy dossiers, Prague succeeded in leveraging the window of opportunity triggered by the Russian invasion of Ukraine to push for tangible progress within security and defence issues.

4 Conclusion

As our analysis has shown, even the PRES of a small country provides a proxy for setting its own priorities and agendas, which can be seen as part of the national interest of the presiding Member State. At the same time, it does not automatically follow that small member states do not have sufficient potential to push through their agendas. Thus, the difference between the two presidencies lies primarily in the fact that the Slovenian PRES was based on the experience of a crisis that had already been sufficiently profiled, while the Czech PRES had to improvise more, although its start more than four months after the beginning of the Russian aggression against Ukraine can certainly not be considered as the impact of a black swan effect.

While the Slovenian presidency effectively built on the two previous presidencies of the trio in responding to the pandemic and its effects, primarily in socio-economic terms (albeit with a clear overlap into the security sphere in terms of building robust resilience mechanisms and strategic autonomy), the Czech PRES was forced to build its activities around a key security pillar. The

Slovenian Presidency made the focus on the Western Balkans a key element of its agenda, while the Czech Presidency had to deal primarily with the strategic agenda of EU-Russia and EU-Ukraine relations. In both key relationships, the Czech government succeeded in building a strong European consensus. It has been an important actor in building sanctions regimes against Russia and limiting the dependence of EU member states – including the Czech Republic itself – on Russian resources. At the same time, together with key partners across the EU, it has succeeded in creating the conditions for a redefinition of the EU's relationship with Ukraine and Moldova, opening up the possibility of integrating both countries into the EU.

In this sense, the Czech PRES has clearly exceeded expectations. At the same time, however, the Slovenian PRES has also brought about a number of significant shifts and improvements in key European policies. This should mean, among other things, there is no significant reason for preconceived doubts that the new Member States are not sufficiently socialised into the EU environment. On the contrary, both analysed PRES have shown a good orientation towards European priorities and at the same time, an ability to link them sensitively with national strategic priorities.

References

Bezpečnostní strategie České republiky (2023). Office of the Government of the Czech Republic. https://mocr.mo.gov.cz/images/id_40001_50000/46088/Bezpecnostni_strategie_Ceske_republiky_2023.pdf.

Bojinović Fenko, A., & Svetličič (2022). Slovenia and the European Union. In: W.R. Thompson (Ed.), *Oxford Research Encyclopedia of Politics*. Oxford, Oxford University Press.

Börzel, T. A., & Risse, T. (2003). Conceptualizing the domestic impact of Europe. *European Integration Online Papers*, *7*, 4, 1–21. https://doi.org/10.1111/1468-5965.00351.

Bureš, O. (2023). A Performance Assessment of the 2022 Czech Presidency of the EU: External Security. *Czech Journal of International Relations*, 58, 1, 113–130.

Cabada, L. (2024). Twenty Years of Czech EU Memberhsip: Between the Pragmatic Bandwagoning and the 'Visegrad-Effect'. *20 years of EU Membership paper Series*, University of Malta, https://www.um.edu.mt/media/um/docs/events/20yearseumembers/LadislavCabada.pdf.

Cabada, L. (2005). *Politický system Slovinska*. Prague: SLON.

Cabada, L., & Hlaváčková, H. (2016). *Zahraniční politika Slovinska*. Prague: LIBRI.

Cabada, L., & Waisová, Š. (2010). Slovenia as an Eu-Member: a Euroenthusiastic Society and Political Elite. In P. Drulák, & Z. Šabič (Eds.), *The Czech and Slovenian EU*

Presidencies in a Comparative Perspective (pp. 37–54). Dordrecht: Republic of Letters Publishing.

Council of the EU (2020). Taking forward the Strategic Agenda. 18-month Programme of the Council (1 July 2020 – 31 December 2021). https://www.eu2020.de/resource/blob/2354332/d2f4bc33ade0af634ae7955206od6332/06-19-pdf-trioprogramme-en-data.pdf.

Council of the EU (2021). Taking forward the Strategic Agenda. 18-month Programme of the Council (1 July 2020 – 31 December 2021). https://data.consilium.europa.eu/doc/document/ST-14441-2021-INIT/en/pdf.

ČT24a (2022). Bojujete I za naše životy a svobodu, vzkázal Fiala. Zelenskyj návštěvu tří premiérů ocenil. *Czech TV*, March 15, 2022. https://ct24.ceskatelevize.cz/clanek/domaci/bojujete-i-za-nase-zivoty-a-svobodu-vzkazal-fiala-zelenskyj-navstevu-tri-premieru-ocenil-23205.

ČT24b (2002). Praha hostile první summit zemí evropského kontinentu. *Czech TV*, October 6, 2022. https://ct24.ceskatelevize.cz/clanek/domaci/praha-hostila-prvni-summit-zemi-evropskeho-kontinentu-prisli-i-rivalove-ze-zakavkazska-a-balkanu-vid-14870.

Digital Slovenia 2030 (2023). Government of the Republic of Slovenia. https://nio.gov.si/products/strategija%2Bdigitalna%2Bslovenija%2B20300.

Drulák, P., & Šabič, Z. (Eds.) (2010). *The Czech and Slovenian EU Presidencies in a Comparative Perspective*. Dordrecht: Republic of Letters Publishing.

Elgström, O. (2003). *European Union Council presidencies: A comparative perspective*. Routledge.

Fink-Hafner, D. (2024). *Party System Changes and Challenges to Democracy*. Cham: Palgrave Macmillan.

Fink-Hafner, D. (2022). Slovenia: Ripe for Autocratisation. In: K. Lynggaard, M. Dagnis Jensen, & Michael Kluth (Eds.), *Governments' Responses to the Covid-19 Pandemic in Europe: Navigating the Perfect Storm* (pp. 209–221). Cham: Springer International Publishing.

Havelka, V. (2023). 2022 Czech Presidency: A Surprising Success. EUROPEUM Policy paper. https://www.europeum.org/en/articles-and-publications/policy-paper-2022-czech-presidency-a-surprising-success/.

Jirušek, M. (2023). The Czech EU Presidency: Strengthening Energy Security Admist the Crisis. *Czech Journal of International Relations*, 58, 1, 147–158.

Kaeding, M., & Udovič, B. (Eds.) (2024). *Slovenia and the EU: 20 Years of Membership in Perspective*. Cham: Springer.

Kaniok, P. (2010). *České předsednictví Rady EU – most přes minulost*. Brno: Masaryk University.

Kaniok, P. (2008). *Předsednictví Rady EU. Příběh půlstoletí*. Brno: Masaryk University.

Kovář, J., & Weiss, T. (2023). The 2022 Czech EU Council Presidency: Performance in the Fields of Security, Energy and Rule of Law. *Czech Journal of International Relations*, 58, 1, 81–99.

Lange, S., & Svetličič, M. (2024). Slovenia's First and Second EU Council Presidency: Lessons Learnt to Enhance the Benefits of EU Membership. *Acta Histriae*, 32, 3, 367–390.

Programme of the Slovenian Presidency of the Council of the European Union, 1 July–31 December 2021 (2021). https://slovenian-presidency.consilium.europa.eu/en/programme/the-programme-of-the-slovenian-presidency-of-the-council-of-the-european-union/.

Schout, A. & Vanhoonacker, S. (2006). Evaluating Presidencies of the Council of the EU: Revisiting Nice. *Journal of Common Market Studies*, 44, 5, 1051–1077.

Sus, M. (2023). Making Use of a Window of Opportunity: The Czech Presidency's Performance in Foreign and Security Policy. *Czech Journal of International Relations*, 58, 1, 101–111.

Svetličič, M., & Cerjak, K. (2015). Small Countries' EU Council Presidency and the Realisation of their National Interests: The Case of Slovenia. *Croatian International Relations Review*, 21, 74, 5–39.

Tallberg, J. (2003). The Agenda-Shaping Powers of the EU Council Presidency. *Journal of European Public Policy*, 10, 1, 1–19.

Tallberg, J. (2006). *Leadership and negotiation in the European Union*. Cambridge University Press.

The Czech EU presidency uses bitter(sweet) pun (2008). *Aktualne.cz*, September 4, 2008. https://zpravy.aktualne.cz/czech-eu-presidency-campaign-uses-bittersweet-pun/r~i:article:615593/.

The Path to Europe' s Digital Decade: The Strategic Plan for Digitalizationof Czechia by 2030 (2023). Office of the Government of the Czech Republic. https://digitalni cesko.gov.cz/media/files/The_Path_to_Europes_Digital_Decade_The_Strategic_Plan_for_the_Digitalization_of_Czechia_by_2030_oOROu6I.pdf.

Tungul, L. (2022). Fighting (Not Only For) Central Europe. an Introduction. In L. Tungul (Ed.), *Shaping EU Presidency Priorities: National Challenges in a European Context* (pp. 9–13), Přerov, Czechia: Jutty.

Udovič, B., & Svetličič, M. (2024). Does the Trio Matter? The Case of Slovenian EU Council Presidency. *Romanian Journal of Political Science*, 12, 2, 4–28.

PART 2

Challenges for European Union Security Policy in a Changing World

CHAPTER 9

Cyber-Enabled Influence Operations during the 2024 Romanian Presidential Elections

Alina Bârgăoanu and Mihaela Pană

Abstract This chapter examines how modern cyber-enabled influence operations weaponize digital platforms affordances to foster emotional vulnerability, undermine critical thinking and satisfy the targeted audience's need to belong to an imagined community. It focuses on a case study from the Romanian digital ecosystem, namely a sprawling network of approximately 75 Facebook pages and 75 clickbait sites pushing content related to "true" traditions, "true" patriotism, "true" religious identity and nostalgia for the Communist period (specifically Ceausescu's time). The findings highlight the *modus operandi* of a long-term, below radar influence campaign carried out with the intention of covertly influencing public opinion and priming it towards embracing more lightly and unknowingly elements of the far right, extremist discourse. It exposes the powerful combination between social media affordances and weaponized nostalgia, that turns in to an offensive capability used in hybrid operations targeting the cognitive domain. At the same time, the case study has the potential to increase awareness of these tactics and to inform public policy responses to counter hybrid threats.

Keywords influence operations – election interference – cognitive domain – weaponized nostalgia – AI slop

1 Introduction

In recent years, hostile actions by state actors directed against other states during peace times have become the subject of both extensive research and government concerns to better understand these activities and to prevent or mitigate their effects. Article 2.4 of the United Nations Charter states that "all Members shall refrain in their international relations from the threat or use of force against the territorial integrity or political independence of any state." Over the past decade, this principle has been interpreted within conceptual

HTTPS://DOI.ORG/10.1163/9789004763623-011

frameworks designed to describe foreign influence and interference such as "hybrid warfare," "gray zone," political and information warfare," or "Foreign Information Manipulation and Interference" (Fridman, 2024, p. 2).

Irrespective of how we may call it, the contemporary political and information warfare is, together with other hybrid tactics, part of a geopolitical influence war taking place *during peace times* and using simultaneously, both in an overt and covert manner, disinformation, propaganda, coordinated influence operations, political violence, election interference, cyberattacks, cyber intelligence, cartographic warfare, algorithmic warfare, data and information harvesting, weaponization of diasporas and of immigration flows, diplomatic pressure, and energy coercion. In fact, what we are witnessing is the emergence of a transnational hybrid industry combining more or less transparent digital advertising, political campaigning, psychological warfare, espionage, cyberattacks, military logics, and logistics in order to achieve several goals: shape the entire information ecosystem; fracture social cohesion; shape social psychology; create paralysis; undermine trust in electoral processes and institutions; undermine reality; prime audiences toward accepting more lightly or unwittingly ideas and interpretations that are detrimental to their interests – both at an individual, community, and societal level.

EU's institutional and regulatory response to such new and emerging threat landscape has been varied, with some notable instruments put in place such as: the Strengthened Code of Practice on Disinformation (since 2022), European Digital Media Observatory (since 2020), Revised Audiovisual Media Services Directive (since 2020), EP resolution (March 2022) on foreign interference, Parliament's new special Committee on Foreign Interference (INGE2, March 2022), Regulation on the transparency and targeting of political advertising (since 2024). A general context of strategic vulnerabilities and reduced tech sovereignty prompted the EU to push several essential pieces of regulation targeting digital identity, cybersecurity, privacy: General Data Protection Regulation and Data Governance Act to ensure control over data and standards, the Digital Services Act for better platform governance, the Digital Markets Act to reduce market concentration and improve economic security with respect to digital markets, EU Cyber Resilience Act introducing mandatory cybersecurity requirements for all digital products sold in Europe, or Cyber Solidarity Act establishing capabilities for threat detection, a cybersecurity emergency mechanism ("cyber shield"), and improved crisis response for critical entities. Taking into account the increasing structural risks of reduced tech sovereignty and strategic dependence, exposure to disruption, interference, or coercion coupled with a rapidly deteriorating geopolitical environment, ongoing debates at the level of EU institutions focus on the importance of implementing the EU's digital rules in full, including in the information field (European Parliament, 2025), as highlighted in the State of the Union Address: "We set our own standards. We set our own regulations. Europe will always decide for itself" (von der Leyen, 2025).

2 Background Information

Despite this rich regulatory, multi-domain, and multi-stakeholder toolbox, the EU information space is still challenged by actions that fall under the general notion of "hybrid warfare during peace times." Dedicated analyses identify several recurring reasons for these persistent challenges: incomplete understanding of the threat environment, the expectation of quick, tool-oriented solutionism, poor or fragmented enforcement (at national or institutional level), incapacity of disrupting capabilities, infrastructures, and incentive systems, fixating the debate on content instead of more substantive things, such the structural levers of modern information systems (algorithms, recommendation and ranking engines, visibility control), thus leaving hosting, amplification, and monetization chains untouched.

The new geopolitical and technological environment and the new regulatory architecture built in response to it make the skills and knowledge in the field of digital, cognitive, and cybersecurity and resilience critical (The Kościuszko Institute, 2024), while the corresponding knowledge gap, especially that at the level of regulators, decision-makers, and high profile political and public figures, is growing into a significant threat to the functioning of EU democracies, economies, and societies.

The Romanian case of annulled Presidential elections provides a case in point of how these unmet challenges, knowledge- and skills gaps, together with other weaknesses, can disrupt high-profile elections, and undermine trust in democratic processes in the long term. In December 2024, the Romanian Constitutional Court took the unprecedented decision to annul the country's presidential election. A runoff was supposed to take place on December 8 between the far-right, fringe candidate Călin Georgescu and the moderate, pro-Western candidate Elena Lasconi. The decision to annul the elections followed the declassification of documents by the Romanian intelligence services that exposed evidence of voting interference and manipulation of public debate through social media platforms, opaque/illegal campaign financing, cyberattacks, and suspected interference from a state actor (namely Russia, as indicated in official documents). The decision has attracted a lot of attention from policy makers and other institutional stakeholders, investigative journalists, analysts, and experts in FIMI, election interference, influence campaigns (Cyfluence Research Center, 2025; Intelligence Online, 2024; Martinescu et al., 2024; Parlato, 2025; Quenel, 2024; VIGINUM, 2025), but the analysis and understanding of the dynamics that have led to Georgescu's sudden rise in online popularity are far from being exhausted. It is our current understanding that this rise in online popularity and its conversion into actual voting patterns were the culmination of a long-term influence campaign. The long-term influence campaign, whose origin can be traced to the pandemic or even earlier, exploited the general dissatisfaction with the political class/the political system and weaponized the digital affordances of various social media platforms in an environment characterized by global trends as the rise of news influencers, the rise of closed, encrypted communication channels, and the blurred line between journalism, political communication, entertainment, and pop culture.

3 Different Narratives Coalescing into on Articulate Worldview

The long-term influence campaign that led to the surprising electoral success of Călin Georgescu in the first round of the presidential elections managed to inject into the information ecosystem (especially the digital, social media-dominated one) narratives such as: "Romania is a colony of the West"; "We are slaves of the decadent West"; "Romanians are losers of the globalization and European integration processes"; "the Western"/globalist elite is constantly lying to the Romanian people, the local politicians are mere lapdogs of EU/NATO/US; "the West is systematically robbing Romania and its citizens of its natural resources and assets"; "the West represents a threat to our national and religious identity."

After Ukraine's full-scale invasion in February 2022, new narratives emerged, reinforcing the previous, pandemic-related ones: "neutrality narratives," "Romania is a sitting duck, fodder cannon, abandoned by its Western rulers in case of a Russian attack"; "the Western elites and their local lapdogs care more about Ukraine than about Romania"; "Ukraine is an enemy state, with which we have major historical, geographical, ethnical and religious disputes"; "Ukrainian refugees and the financial and military aid for Ukraine are a threat to our finances and security."

All these narratives (pandemic and Ukraine-related) coalesced into an articulate worldview according to which "after 1989 and especially after EU accession, Romania lost its dignity"; the Romanian citizens ("the little guys" –

both living in Romania and in diaspora) "are constantly humiliated by the Western/global elites and their local representatives, who seek their national, cultural and religious obliteration." Therefore, "a return to a so-called golden age of historical greatness is a matter of urgency" and Romania should become again a sovereign nation, freed from its "oppressive Western masters."

4 Călin Georgescu's Rhetorical and Digital Footprint

The 2024 presidential electoral campaign took place against this public opinion background, dominated by the generalized feeling that all candidates are more or less the same, that they are equally incompetent/unfit for the job, they represent "the system," and the vote doesn't count because the choice is "rigged" by "Romania's foreign masters." Adding to this were the widespread anti-Ukrainian, anti-Western, lost greatness, peace and neutrality narratives, some of them equally embraced by other candidates.

The label "far-right" applied to candidate Georgescu public persona is justified by the fact that his discourse incorporated a complex rhetorical array of nationalistic, xenophobic, authoritarian, and isolationist overtones, with an all-encompassing narrative, that is, the anti-establishment, anti-system one. He emphasized the trope of the simple, anonymous, powerless citizens who finally have the power to bring down an over-powerful enemy threatening their national, cultural and religious identity. His preferred narratives are organized around strong rhetorical markers ("patriots," "renaissance," "freedom," "garden of Mother Mary," "wake up, Romania," "real Romania," "faith," "apocalypse," "wake up, Diaspora," and "wake up, Dacia" (the name of the current territory of the Romanian state before the Roman occupation at the beginning of the second century AD). Such rhetorical markers contribute to the formation of dense online communities that share a quasi-mystical feeling of belonging. He positioned himself as a "true patriot," "guardian of the Romanian traditions and defender of the Romanian national, cultural and religious identity," "saviour – Messiah who loves and is loved by the masses" engaged in a holy fight against a faceless "them" – the "satanic global elite" and the "banking and financial system which is the head of the snake in the world and which is behind any war." He politically weaponized an emotionally loaded topic – the scale and the condition of the Romanian diaspora living in the more prosperous regions of the European Union, with a focus on its less successful, more disenchanted members feeling "humiliated," "trampled" or simply ignored by their EU counterparts and the Romanian local politicians. Just like the "awakened" Romanians living inside the borders, the Romanian diaspora is deemed as having a quasi-messianic role in saving Romania from its "foreign-induced destruction."

He openly and uninhibitedly glorified the leaders of the WWII legionary movement and the Romanian General Ion Antonescu, who was tried and sentenced as a war criminal at the end of WWII. He made a local reinterpretation of the idea of "lost dignity and lost historical greatness," whose restauration is deemed a matter of national urgency. He remained rather ambiguous about what "golden age period" he had in mind, oscillating between the period between the two World Wars (and expressing praise for General Antonescu as a "strong, patriot man") and the late Communist period (expressing praise for Nicolae Ceaușescu, who is portrayed as having the same qualities). Different from the other political or public figures already endorsing such far right and populist narratives, his discourse has grown more disinhibited in attacking the EU, its leaders, the common currency, the Schengen area of free movement, and Romania's benefits as a NATO and EU member state. His entire political discourse came to be organized around the idea that EU and NATO membership "enslaved" Romania and "brought its tourism, economy, health, education to zero."

He embraced the "peace narrative," positioning himself as the "peace president," fusing this narrative with religious symbolism and emphasizing that "peace" is God's wish. During the last two weeks of the electoral campaign, candidate Georgescu amplified his digital footprint, especially on TikTok, based, among other things, on data-driven micro-targeting, prioritizing or downgrading content based on opaque algorithmic choices, powerful messaging, and hacking of the comments section of any online platform via a combination of real people and automated accounts. At the same time, the campaign and the corresponding alleged foreign interference was cross-platform (TikTok, Telegram, Facebook, Instagram, What's app, YouTube, native advertising platforms). The digital infrastructure underlying seemingly non-political communities (tourism, wellness and nutrition advice, alternative medicine, parenting, cooking) was activated, recycled and repurposed, pushing narratives of "return to a period of historical greatness," "nostalgia," "threatened identity," "declining West," and "traitors and warmongers." These narratives amplified the fear of Romania being drawn into a direct confrontation with Russia, emphasizing that the victory of the centrist candidate Lasconi would lead to "civil war," "offline violence," "insurrection," and "revolution."

The aggressive digital advertising tactics employed by Georgescu's campaign sought to engage disillusioned audiences: inhabitants of heavily de-industrialized areas; "simple" people feeling "left behind" by globalization and EU integration processes; "the little guys" feeling ridiculed for their worldviews by mainstream parties and public figures; their family members living in the diaspora. Among such tactics, the most prominent ones were granular electoral profiling, microtargeted ads, and emotionally charged, highly polarizing content (Martinescu et al., 2024). The pattern of the coordinated inauthentic amplification across various mediums and platforms revealed in fact an increasingly sophisticated digital manipulation threatening European democracies (Mandraud, 2025).

5 Case Study – a Sprawling Network of Clickbait Websites and Facebook Pages

5.1 Methodology

Our research was guided by the questions how the above mentioned narratives were injected into the Romanian public space, and through which specific gateways they managed to manipulate the major terms of the electoral, political, and public conversation. Guided by this question, we identified in Romanian digital environment 72 Facebook pages that promote 75 clickbait websites and that are connected into a sprawling network (see also Bârgăoanu, 2025). The network was identified between January and February 2025 through algorithmic honey potting, using controlled engagement to trigger and systematically capture platform recommendations, consistent with the broader logic of social honeypots for surfacing hard-to-reach online actors and connections (Bardi et al., 2023).

By repeatedly refreshing and accessing pages, the algorithm was tweaked so as to display more pages with similar content, thus increasing access to otherwise opaque content. Although several monitoring tools can help identify content on Facebook, they are limited in detecting connections between pages and clickbait websites, as these connections are mentioned in the comments section (the clickbait sites are strategically referenced in the comments section of the Facebook pages). The promotion of these clickbait sites through the comments section is a strategy used to bypass the platform's algorithm, which can limit the reach of posts containing URL links.

Unlike other monitoring tools and software available for social media content, our search unveiled connections that are visible only through relevant individual, manual search, based on tweaking the social media platform's algorithm. The results were influenced by search history and user (researcher) behavior on the platform. Since the large majority of Facebook pages do not

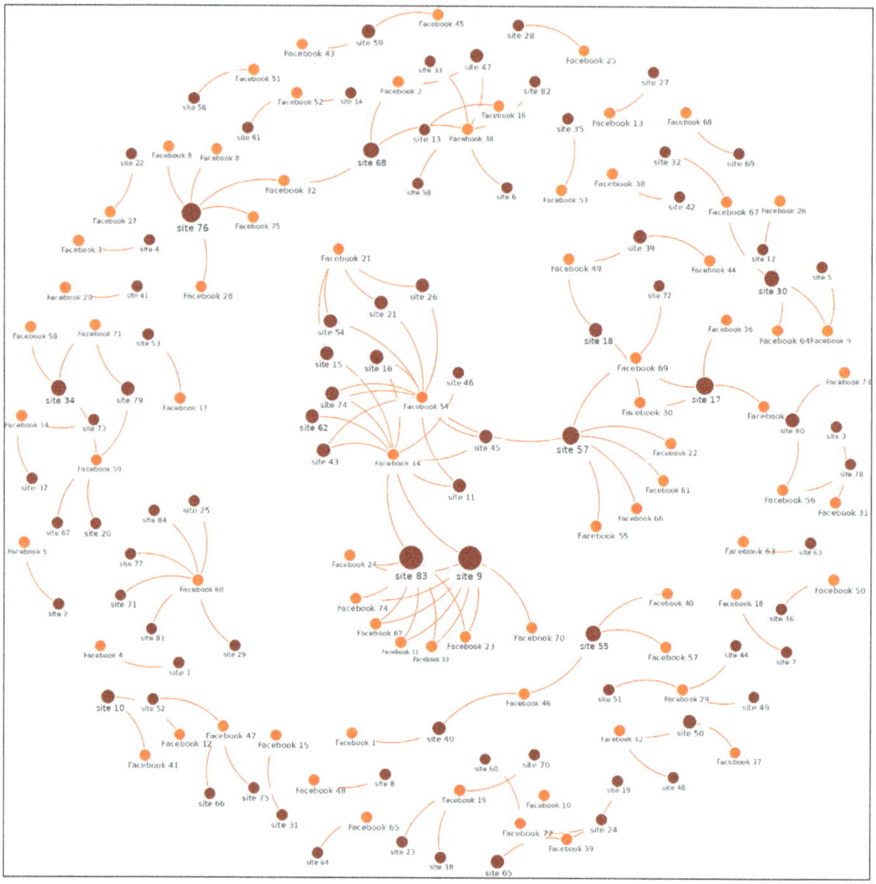

FIGURE 9.1 Network of clickbait websites promoted via comments section of Facebook pages (anonymized)

make any direct connection to the clickbait sites, the only way to discover the latter was through manual search into the comments section.

5.2 Digital Footprint of the Network

The number of followers of the Facebook pages ranges from very low (169) to very high (2,200,000). Although they do not always have unique followers and many of the pages have likely increased the number of followers with unauthentic users and AI-generated audiences, the total number of followers is 28,599,969.00 and counting (the entire Romanian population 19,060,000.00 people, according to the 2023 census) (INS, 2023).

They have in common the efforts to enforce resentment against the current establishment, promote candidate Georgescu, patriotism, and nostalgia over a lost period of historic greatness. The Facebook pages act as distributors, driving users to clickbait sites that host ads (some legitimate, some deceptive), political, social and entertainment content and, in some cases, trigger malware alert. The 75 clickbait websites can be grouped into three categories, based on their domains: news/impersonation of legitimate news outlets; Romania, Romanian traditions, Romanian natural beauties; Orthodox religion. The Facebook pages and the clickbait sites are heavily interconnected – either by sharing the same website, by posting similar or identical content, or by sharing the same administrator.

Figure 9.1 illustrates how several pages belonging to the analysed network evolved from publishing largely non-political lifestyle content to disseminating messages addressing socio-economic, cultural, and religious issues. In the run-up to the elections and during the electoral campaign, these pages increasingly promoted content favourable to Călin Georgescu, suggesting a strategic reorientation of their communication activities and a potential influence on public opinion formation (Toșa, 2025).

Behind these pages, there lies an apparent deliberate intent to redirect the audience to private Messenger and WhatsApp groups, thus making such influence operations even more opaque to public view. A significant number of AI-generated Facebook pages disseminate copy-and-paste scam content, including both images and text messages. Many of them repost the same copy-paste content in Russian and Romanian language pages. This practice is a common strategy employed by scammers to deceive unsuspecting individuals, exploiting automated content generation to enhance the reach and credibility of fraudulent schemes. Another behavioral characteristic is that the clickbait sites that were identified as part of the network and that are used for copy-content distribution is their capacity to seamlessly integrate malicious content across multiple platforms. An example in point is the connection between fraudulent advertisements promoting the BRUA scam, disseminated through different native advertising platforms on clickbait sites belonging to the network and used to amplify pro-Georgescu electoral content (Bloomberg, 2024; Reset Tech, & CheckFirst, 2024b).

5.3 Textual and Visual Footprint of the Network

The network promotes far-right candidate Georgescu and patriotism-related topics as opposed to the current political class/establishment narrated as "corrupt," made out of "traitors," "out-of-touch" with the "real citizens." The identified pages use traditions to create nostalgia for the past and return to "our roots" as opposed to the current European and Western establishment (Lefter, 2025). They appeal to social ("decrease of income," "increase of taxes," "new regulations") and political issues ("lack of professionalism"; "corruption"; "traitors/actions against national interests") to increase resentment and decrease trust in institutions and the political establishment. Many of these narratives are further amplified through Meta paid ads, reels, or sponsored content that are available in their ad libraries (Reset Tech, & Check First, 2024a). Identified pages pertaining to the network include AI-generated content (AI slop), portraying individuals who are sad because they are abandoned, despised, neglected, or not given credit for their merits. This mechanism is designed to trigger emotions and empower the digital community to take action against unidentified attackers who are to be held accountable. While no specific culprits are mentioned, the objective is to create resentment toward those "in power" for neglecting vulnerable yet kind and honest people. Political content is displayed next to non-political one (entertainment, travel to Romania, horoscopes, holiday wishes) for generating engagement.

Most of the Facebook pages have identical images, titles, and content (many posts can be seen on several pages, although they share content from different interconnected websites). Moreover, some pages have clones (Preoteasa, 2025). AI slop encompassed a wide range of visual materials – religious imagery, military personnel experiencing sentiments of isolation or rejection, older people feeling lonely on their birthdays, farmers expressing feeling of neglect for their "hard work." A notable characteristic of these images is the presence of spelling errors and visual anomalies, including the distorted representation of letters or the malformed presentation of flags, which serve as indications of their artificial nature. Beyond visual manipulation, these pages actively promote conspiratorial narratives related to artificial intelligence, non-existent technologies, and so-called miracle cures. They exploit platform algorithms to amplify engagement through alarmist messaging, fear mongering, often prompting users to seek "full" information in the comments section, thereby increasing the visibility of the hosted clickbait sites.

To sum up, the purposes served by this network are at least manifold. First, the interconnected pages and sites represent viralization mechanisms for seemingly non-political content quietly morphing into political and electoral content that taps into socio-economic, cultural, and religious sensitivities. The apparent inauthentic viralization of such content interferes with the free process of opinion formation, since viralization and the implicit popularity signals are interpreted as proxies for "truth" and "legitimacy." Second, they can become vehicles for online scams, exposure to fake or misleading ads and even for malware. Third, they are a vehicle for data harvesting and digital profiling, by collecting the digital footprint of users, registering their political preferences, biases, and online behavior, all to be used in later influence campaign. Finally, the products sold via the clickbait sites can generate revenues that evade fiscal and other legal scrutiny, thus providing for opaque, possibly illegal campaign funding.

6 Detecting Cyber-Enabled Influence Operations before They Get Traction

The 2024 annulled Romanian presidential elections offer a case study into the combination between long-term influence and ad hoc election interference that, under the right socio-political circumstances, can significantly alter the results of high-profile elections. The case is a wake-up call to the scale of cyber-enabled foreign influence operations that allow hostile actors to operate below radar in real time, in contrast to more visible influence campaigns carried out through traditional media.

Our case study builds on previous endeavors and exposes the *modus operandi* of cyber-enabled influence operations targeting the cognitive domain. Exposing such *modus operandi* has the potential to guide further reflection and formulation of adequate responses. Some of the elements of the playbook revealed by this high-profile electoral moment are: injecting the public space with small doses of narratives and interpretations, both political and non-political, patiently and persistently; pushing these narratives from the fringe to the core of the public space (from social media and interpersonal, offline settings to mainstream media and back); amplifying them via hard-to-detect gateways (reels, comment section, sponsored content, clickbait sites, malicious ads, encrypted online groups); quietly priming the public space for the acceptance of such narratives and interpretations by means of ambient cognitive triggers(such as nostalgia over a period of "lost greatness"); activating or repurposing sleeper and pre-positioned accounts; mobilizing influencers, especially micro-influencers ("the little guy" energized to engage politically in the hope of belonging to a community and no longer feeling left behind); amplifying and coordinating cross-platform content; astroturfing and hacking the comments section of social media pages and accounts, thus evading detection mechanisms; camouflaging propaganda, deception, and influence-related narratives into legitimate political and public conversations up to the point when the distinction between the two is completely blurred; weaponizing the dominant sentiments in a society and unaddressed sources of discontent; finally boosting a specific candidate to capture and capitalize on all of the above.

7 Conclusion

Our study has the potential to support further expansion of public policy on countering hybrid threats, with a view to closing the knowledge and skill gaps identified by The Kościuszko Institute and in line with the priorities of the Polish Presidency of the Council of the European Union (2025). It adds to the current conversation around the fact that the gateways through which narratives targeting the cognitive domain are injected into the information space may escape public and institutional scrutiny, Big Data monitoring tools or software available for social media content. Such cyber-enabled influence operations go below radar in unexpected places (clickbait sites hidden from public view in the comment section of social media pages). Without proper early detection mechanisms (including the detection of weak signals, *before* viralization takes place), they gain traction and scale and alter the integrity of public conversations overall, thus decreasing the likelihood for a robust answer. In other words, once they can be observed and quantified and become visible on "radars," it may be already too late for any meaningful response. While Big Data analytics and automated detection systems play a crucial role in countering digital threats, they cannot fully replace manual research and human analysis. The human capacity to detect subtle algorithmic distortions, anomalies, and behavioral inconsistencies is essential in identifying patterns of hostile influence that evade algorithmic detection. This underscores the critical need for interdisciplinary and inter-institutional approaches that combine computational methods with expert-driven analysis to uncover the mechanisms through which these operations shape the digital information ecosystem.

Detecting, attributing, and tackling cyber-enabled influence operations requires a holistic approach that integrates algorithmic scrutiny, manual investigative methods, and regulatory frameworks like the DSA and other instruments that we presented in the beginning of our chapter. Without addressing the structural levers of modern information systems and without disrupting hosting, amplification, and monetization chains enabling such cyber-enabled foreign influence operations, efforts to counter them will remain incomplete, allowing hostile actors to continue to exploit digital vulnerabilities, manipulate public conversations, and undermine democratic processes.

References

Assa, H. (2019). Influencing Public Opinion. In Y. Kuperwasser & D. Siman-Tov (Eds.), *The Cognitive Campaign: Strategic and Intelligence Perspectives* (pp. 25–35). The Institute for National Security Studies and The Institute for the Research of the Methodology of Intelligence.

Bârgăoanu, A. (2025, July 8). The Romanian presidential elections – a "gold mine" still waiting to be mined. Cyfluence Research Center. https://www.cyfluence-research.org/post/the-romanian-presidential-elections-a-gold-mine-still-waiting-to-be-mined

Bardi, S., Conti, M., Pajola, L., & Tricomi, P. P. (2023). Social honeypot for humans: Luring people through self-managed Instagram pages. In M. Tibouchi & X. Wang (Eds.), Applied Cryptography and Network Security: 21st International Conference, ACNS 2023, Proceedings (Lecture Notes in Computer Science, Vol. 13905, pp. 309–336). Springer. https://doi.org/10.1007/978-3-031-33488-7_12

Bloomberg (2024, December 9). *Exposed secret Facebook network targeting Romanian voters*. Cyt. w Agenzia Nova.

Cyfluence Research Center (2025). *The Romanian Presidential Elections 2024: Analysis of Information Operations and Long-term influence Efforts*. https://www.cyfluence-research.org/post/the-romanian-presidential-elections-2024-analysis-of-informa tion-operations-and-long-term-inf.

European Parliament. (2025). Need to enforce the Digital Services Act to protect democracy on social media platforms including against foreign interference and distorted algorithms (2025/2515(RSP)). European Parliament. https://oeil.secure.europarl.europa.eu/oeil/en/procedure-file?reference=2025%2F2515%28RSP%29

Fridman, O. (2024). Defining foreign influence and interference. INSS Special Publication, 1–12. https://www.inss.org.il/publication/influence-and-interference/

Intelligence Online (2024). United States concerned about Russia's possible Facebook operations in Romanian election. *Intelligence Online*. https://www.intelligenceon line.com/government-intelligence/2024/12/19/united-states-concerned-about-rus sia-s-possible-facebook-operations-in-romanian-election,110354031-art.

Kirdemir, B. (2019). *Hostile influence and emerging cognitive threats in cyberspace*. https://www.jstor.org/stable/resrep21052.

Mandraud, I. (2025). The increasingly sophisticated digital manipulation threatening European democracies. *Le Monde*. https://www.lemonde.fr/en/international/art icle/2025/02/22/the-increasingly-sophisticated-digital-manipulation-threatening-european-democracies_6738433_4.html.

Martinescu, A.-L., Stallard, S., Balatchi-Lupascu, A., George Forlafu, M., Kurtov, Y., Bilash, D., Plieshakov, D., & Popov, Y. (2024). *Networks of Influence: Decoding foreign meddling in Romania's elections A collaborative investigation into disinforma-*

tion campaigns and influence operations. https://fpc.org.uk/wp-content/uploads/2024/12/Networks-of-Influence-Decoding-foreign-meddling-in-Romanias-elections-2024.pdf.

National Institute of Statistics (INS). (2023). *Romania's resident population at 1 December 2023 – final data of the 2023 census.* https://insse.ro/cms/en.

Parlato, F. (2025). *Georgescu's Rise Through Russian Cyber Fraud Rocked Romania's Election; Vance Decries His Ouster.* ARTVOICE. https://artvoice.com/2025/02/23/georgescus-rise-through-russian-cyber-fraud-rocked-romanias-election-vance-decries-his-ouster/.

Preoteasa, M. (2025, May 10). *Coordinated inauthentic behaviour surfaces again in the re-run of Romanian elections.* EurActiv (PROMPT project).

Quenel, N. (2024). *AdNow exposed: How Romania became a test lab for Russian interference in the West.* Intelligence Online. https://www.intelligenceonline.com/government-intelligence/2024/12/03/adnow-exposed-how-romania-became-a-test-lab-for-russian-interference-in-the-west,110347109-evl.

Reset Tech, & CheckFirst. (2024a). *Ads, Influence and Democracy: Meta's Role in Romania's 2024 Presidential Election* (Raport).

Reset Tech, & CheckFirst. (2024b). *Secret ad network promoting Georgescu.* Raport cytowany przez Bloomberg i Agenzia Nova.

Seal, T. (2025). Meta Blocked News in Canada. Ads for Scams Are Taking Its Place. *Bloomberg.* https://www.bloomberg.com/news/articles/2025-01-31/meta-blocked-news-in-canada-ads-for-scams-are-taking-its-place.

Segijn, C. M., & van Ooijen, I. (2022). Differences in consumer knowledge and perceptions of personalized advertising: Comparing online behavioural advertising and synced advertising. *Journal of Marketing Communications, 28*(2), 207–226. https://doi.org/10.1080/13527266.2020.1857297.

Ştefan Lefter. (2025, February 24). *Over 1 000 Facebook pages spread disinformation, alarmism, pro-Russian narratives and pseudo-science while promoting candidate Călin Georgescu* – online community sounds the alarm. G4Media.ro.

Toşa, D. (2025). Campanie plătită pro Georgescu pe FB la fel ca pe TikTok. Pagini cu milioane de urmăritori îl susţin. *Newsweek.* https://newsweek.ro/politica/campanie-platita-pro-georgescu-pe-fb-la-fel-ca-pe-tiktok-pagini-cu-milioane-de-urmaritori-il-sustin.

Ursula von der Leyen. (2025, September 10). 2025 State of the Union Address by President von der Leyen. European Commission. https://commission.europa.eu/strategy-and-policy/state-union/state-union-2025_en

VIGINUM. (2025). *Manipulation d'algorithmes et instrumentalisation d'influenceurs.* https://www.sgdsn.gouv.fr/publications/manipulation-dalgorithmes-et-instrumentalisation-dinfluenceurs-enseignements-de.

Votta, F. (2025). Digital Campaign Trail: Unpacking Germany's Political Ad Spending. *Favstats.Eu.* https://www.favstats.eu/post/btws/.

CHAPTER 10

Shaping European Digital Sovereignty: The Council of the EU's Role in Navigating Global and Internal Challenges

Mauro Santaniello and Chiara Spiniello

Abstract As digital technologies increasingly influence global governance, security, and economic competitiveness, the EU sought to assert greater control over its digital infrastructure, data governance, and technological innovation under the framework of "digital sovereignty." The Council of the EU, as a pivotal institution in shaping EU policy, was instrumental in articulating and advancing this concept. This paper examined the role of the Council of the European Union in shaping the concept of European digital sovereignty. By analyzing key Council documents, including regulations, decisions, conclusions, declarations, communications, and press releases, the paper identified recurring themes, policy priorities, and significant shifts in the conceptualization of digital sovereignty over the period 2000–2024. This research contributed to the growing body of literature on European digital sovereignty by providing a nuanced understanding of how the Council of the EU contributed to its development as a strategic policy concept. It underscored the Council's role not only as a policymaker but also as a mediator of competing visions within the EU, offering insights into the institutional and geopolitical dimensions of the EU's digital transformation.

Keywords digital sovereignty – digital governance – Council of the European Union – European Union – strategic autonomy – digital security

1 Introduction

Sovereignty has long been a discursive resource in the domain of digital governance. Assertions of sovereignty over digital assets have characterized a range of political struggles since the 1990s. The principle of sovereignty was initially invoked by the US government to prevent the American technical community from establishing a transnational governance system – independent of US authority and based in Geneva – for the management of IP addresses

 HTTPS://DOI.ORG/10.1163/9789004763623-012

and domain names. The privatization of critical internet resources, alongside continued US oversight of the root zone, was imposed to the world by the Clinton administration through a series of sovereign acts, ultimately leading to the creation of the Internet Corporation for Assigned Names and Numbers (ICANN), a non-profit organization headquartered in California. In response, national governments across the globe began asserting their own sovereignty claims, criticizing US unilateralism and advocating for an intergovernmental model for managing fundamental internet resources. The World Summit on the Information Society (WSIS), held between 2003 and 2005, served as the primary global forum in which sovereignty as a discursive resource for the control of digital assets was deployed, particularly by a coalition of states led by China and Russia. Two decades later, this discourse continues to shape several WSIS outcomes, including the Internet Governance Forum (IGF) and, most notably, the periodic WSIS review processes. Moreover, over the past twenty years, authoritarian regimes have deployed this discourse at the domestic level, formulating doctrines of internet or cyber sovereignty, aimed at consolidating state control over national digital infrastructures and citizens' data.

Beginning in the mid-2010s, however, the discourse on digital sovereignty has expanded significantly. What initially centred on DNS governance has evolved into a cross-cutting issue that spans all layers of the internet stack – ranging from network infrastructure to applications and data. It now encompasses broader technological domains, beyond the internet itself, including microprocessors, artificial intelligence, quantum computing, and other emerging technologies. Also, new actors, including democracies, have embraced the discourse of sovereignty as a guiding principle for their own digital policies. Among these, the European Union is the political entity that has gone furthest in institutionalization of the discourse on digital sovereignty. Through the digital sovereignty discourse, the European Union has completely reframed its digital policy, moving from a set of market-oriented directives to strategic regulation. This study investigates the role of the Council of the European Union in shaping the concept of European digital sovereignty through a qualitative analysis of official documents and press releases drafted between 2020 and 2024. The Council of the EU, as a pivotal institution in shaping EU policy, has been instrumental in articulating and advancing this concept. The chapter analyses how the Council has defined and reframed digital sovereignty in response to evolving global challenges. By analysing key Council documents, including regulations, decisions, conclusions, declarations, communications, and press notes, the paper identifies recurring themes, policy priorities, and significant shifts in the conceptualization of digital sovereignty over the period analysed. The remaining part of the chapter is articulated as follows. The next section resumes main findings from existing research on European digital sovereignty. Then, the chapter outlines research questions and methodology. The following section presents the main findings of the analysis. Finally, the chapter discusses results and concludes with future avenues for research.

2 Digital Sovereignty in the European Union

Among European member states, the concept of digital sovereignty first took root in France and Germany, where, since 2013, it has shaped discussions on digital issues not only among policymakers but also significantly among non-state actors from the private sector, academia, and civil society (Danet & Desforges, 2020). Over time, this discourse transitioned from the national to the European level, where it was formally declared a central objective of the European Union's digital policy agenda (von der Leyen, 2020), and has since become increasingly embedded in the official policy discourse of the Commission (Pohle & Santaniello, 2024). A considerable number of studies have focused on analysing the European discourse on digital sovereignty, mapping a multiplicity of narratives and identifying a complex array of meanings, frameworks, and descriptions enacted by various actors. For example, Pohle (2020), in her analysis of the digital sovereignty discourse in Germany and Europe, identifies three principal dimensions: governmental, economic, and individual. She further connects the emergence of this multifaceted discourse to three key factors: the dominant and centralized role of a small number of powerful technology companies; Snowden's revelations of mass surveillance; and the broader quest for a normative framework that offers an alternative to both the US and Chinese models. A number of other scholars link the digital sovereignty discourse of the EU to an attempt to build a "third way" to digital governance, alternative to the surveillance capitalism of the United States and the technocratic authoritarianism of China and Russia, and to establish a model based on the respect for European norms and values (Giddens, 2020).

Similarly, analysing the German political discourse, Lambach and Oppermann (2023) have found several narratives of digital sovereignty, emphasizing the deep connection of this discourse with the "European way of life" narrative. Moerel and Timmers (2021) identify three primary conceptual approaches to digital sovereignty within the European Union. The first is a risk management perspective, aimed at addressing cybersecurity challenges; the second emphasizes strategic cooperation among like-minded partners to navigate rising geopolitical tensions and reduce reliance on foreign technologies and supply chains; and the third frames digital sovereignty as a global common good, transcending narrowly defined national interests. Situating the EU's digital sovereignty discourse within the broader context of global competition among major powers, Broeders, Cristiano, and Kaminska (2023) describe a process of geopoliticization of EU digital policies. This process entails both the strategic use of trade, competition, and foreign policy instruments to project geopolitical influence, and the emergence of a new generation of deliberately hybrid digital policies that simultaneously address internal market objectives, fundamental rights, and geopolitical considerations. Also, other researchers place European digital sovereignty within the context of emerging geopolitical considerations, highlighting the EU's ambition to fortify its political identity (Monsees & Lambach, 2022; Burwell & Propp, 2020). In the domain of European cybersecurity, Farrand and Carrapico (2022) analyse the EU's shift towards a discourse centred on sovereignty, observing the development of a regulatory mercantilism model in which domestic private sector actors are viewed as reliable regulatory partners, while foreign corporations are perceived as potential risks to cybersecurity policies. Analysing political documents and speeches on platform governance and semiconductors, Lambach and Monsees (2025) argue that the notion of digital sovereignty should be understood more as an overarching economic, societal, and geopolitical project, rather than as a project aimed at achieving any form of political sovereignty or autarky in the digital sector. Roberts, Cowls, Morley, Taddeo, and Floridi (2021) concluded that certain existing policies, particularly those related to data governance, contribute to the achievement of some of the EU's specific objectives regarding digital sovereignty (such as influencing data flows outside the EU), but are more limited in relation to other objectives (such as enhancing the EU's competitiveness and regulating the private sector).

Carver (2024) also focuses on analysing certain specific EU policies in order to understand whether and how the EU has effectively achieved its digital sovereignty objectives. The study finds that in the case of the 5G Toolbox process, the European discourse on digital sovereignty was effective in shaping the policy, while it failed to drive policy changes to the Cyber Diplomacy Toolbox and external CCB assistance initiatives. Adler-Nissen and Eggeling (2024) instead use the GAIA-X cloud computing project as a case study, demonstrating that there is no unified understanding of digital sovereignty within the EU. Vučić (2021) focuses instead on the fields of digital economy, data protection, and artificial intelligence, arguing that the EU's normative and economic power provides it with the best chances of success precisely in these areas.

Most of these studies on European digital sovereignty focus on narratives and policies enacted by the European Commission. The role of other EU institutions is usually overlooked. This study seeks to partially fill this gap of knowledge by assessing the role of the Council of the European Union in shaping the EU's discourse and practice on digital sovereignty. In this study, we draw upon Santaniello's conceptual framework (2025) of the Attributes of Digital Sovereignty to conceptualise digital sovereignty as a discursive construct rather than as a static legal principle or a coherent policy objective. Specifically, Santaniello approaches sovereignty as a dynamic, relational, and performative discourse structured around five key attributes – adversariality, multiversity, latency, instrumentality, and hypocrisy – which enable a systematic analysis of how sovereignty is mobilised, against whom, for what purposes, at which moments, and with which internal contradictions.

3 Research Objectives and Methodology

Previous research has highlighted the significant heterogeneity characterizing the discourse on digital sovereignty. Even within the European Union, there is considerable variation in the motivations and ambitions of the actors invoking digital sovereignty, in the contexts in which this discursive resource is employed, and in the range of policies it is used to legitimize. The aim of this study is to examine how the Council of the European Union has conceptualized and mobilized the notion of digital sovereignty over time. In order to do so, a qualitative analysis was conducted on official documents produced by the Council of the EU between 2019 and 2024. Documents were collected by searching the terms "digital sovereignty" and "technological sovereignty" on official websites (www.consilium.europa.eu and www.eur-lex.europa.eu), both using the internal advanced search tool and Google.com (using queries such as "site: www.consilium.europa.eu "digital sovereignty"). This search yielded a total of 33 valid results, of which 17 were official acts and 16 were official press releases. The collected documents were subsequently analysed to identify conceptualizations of digital sovereignty – including definitions and related concepts – their contexts of use (i.e. digital policy areas such as platform governance, cybersecurity, connectivity, artificial intelligence, etc.), and the policy objectives supported by digital sovereignty claims. The analysis was conducted in two phases. Firstly, conceptualizations, contexts, and objectives were coded through an inductive procedure by the authors independently with high overall inter-coder reliability (0.9). Secondly, results were jointly assessed in order to identify key features of the Council's discourse on digital sovereignty and main trends unfolded over the different presidencies. This approach nonetheless entails certain limitations, which are particularly salient in two respects. First, it does not examine the negotiation dynamics through which each presidency may have shaped – directly or indirectly – the framing and substantive content of the documents analysed. Second, the analysis is confined to official documents produced by a single European body, namely the Council of the European Union (EU); accordingly, the findings should not be assumed to be readily generalizable to other institutional settings or to different documentary *corpora*.

The Council of the European Union was selected as a case study because it plays a pivotal role in the Union's decision-making process. According to Article 16(1) TEU, it exercises "legislative and budgetary functions, [as well as] policymaking and coordinating functions under the conditions laid down in the Treaties". The Council serves as the central hub through which all formal decisions shaping the Union's daily activities are channelled (Adam & Tizzano, 2024). It is, in fact, the principal actor in the exercise of decision-making power at the European level, although it shares responsibility for certain functions with other institutions – for instance, the legislative function with the European Parliament – or specific activities, such as defining the Union's political priorities and general orientations, with the European Council.

The Council is composed of "a representative of each Member State at ministerial level, who may commit the government of the Member State they represent and cast their vote" (Article 16(2) TEU). For this reason, it was referred to as the "Council of Ministers" until 2009 (Wurzel et al., 2020). Each of these representatives takes part in the decision-making process, which, since the entry into force of the Treaty of Lisbon, is based predominantly on qualified majority voting (Lewis, 2016, pp. 138–155), while simple majority voting applies to procedural matters, and unanimity is reserved for the most politically sensitive decisions.

The Presidency of the Council of the EU is held by each Member State in turn, according to a rotating system regulated by a decision of the European Council (Decision 2009/881/EU of 1 December 2009).

This decision specifically provides for a presidency shared by predetermined groups of three States, which, unless otherwise agreed, divide the presidency between them over a period of 18 months, with each State assuming the presidency for six months, supported by the other two, and following a common programme. Each group is composed according to an equal rotation system, considering the diversity of the Member States and the geographical balance of the Union (Puetter, 2022, pp. 78–105).

4 Digital Sovereignty in the Policy Discourse of the Council of the European Union

The term "digital sovereignty" entered the official discourse of the Council of the European Union in 2020, a few weeks prior to Germany assuming the Presidency of the Council. Within the "Conclusions on shaping Europe's digital future", digital sovereignty is described as a fundamental component "to ensure strategic autonomy, global competitiveness and sustainable development, while also promoting common EU values, transparency, human rights and fundamental freedoms on the international stage" (Council of the European Union, 2023a). The document associates digital sovereignty with a wide array of policy issues, including data protection, cybersecurity, cloud computing, connectivity, artificial intelligence, supercomputing, and quantum technologies. Additionally, two primary policy directions come to be clearly identified: public investment and legislation. Public investment serves as an instrument designed to both foster research and development in the public and the private sectors, and to bolster European technological solutions through supply-side policies, so as to achieve European global digital leadership through the rise of European digital champions.

Legislation, on the other hand, functions as a means to both protect the European digital single market, and to safeguard and externally project EU's fundamental values. The German Presidency (July–December 2020) notably expanded upon this nexus between technological autonomy and European values in its "Conclusions on the Charter of Fundamental Rights in the context of Artificial Intelligence and Digital Change". These conclusions advocate a fundamental rights-based approach to artificial intelligence, emphasizing that digital technologies, including AI, are essential for European digital sovereignty, security, innovations and economic development and can contribute significantly to the protection and promotion of fundamental rights and democracy and the rule of law" (Council of the European Union, 2024a). Investments and legislative measures were reaffirmed as crucial instruments to ensure the European Union's capacity to act with self-determination in the digital domain and to enhance the resilience of the EU. Moreover, the EU was encouraged to take a proactive role in the global debate on artificial intelligence, envisioning digital diplomacy as a strategic means of influencing international frameworks and governance mechanisms. Under the German Presidency, the Council also issued two conclusions linking digital sovereignty to openness. The first of these documents, the "Council conclusions on digital education in Europe's knowledge societies", emphasizes the importance of developing open learning platforms and digital educational tools. Here, the discourse on digital sovereignty is a call for public investments in open source software and public platforms for education.

The second document, titled "Council conclusions on the cybersecurity of connected devices," highlights that "the European Union and its Member States need to ensure their digital sovereignty and strategic autonomy, while preserving an open economy" (Council of the European Union, 2024b). Here, openness relates to the market, and the digital sovereignty discourse is framed as a field of tensions between the neoliberal path unfolded in EU's digital policies since their inceptions, and escalating geopolitical transformations. Also, this document links EU ambitions regarding digital sovereignty and strategic autonomy directly with the preservation of core European values, in particular privacy, security, equality, human dignity, rule of law, and open Internet. In December 2020, at the conclusion of the German Presidency, the Council, together with the European Parliament and the European Commission, approved the Policy Objectives and Priorities for 2020–2024, and the EU Legislative Priorities for 2021. The former enlists digital sovereignty as a top policy priority for the EU, and highlights the need to create and shape European digital solutions as a consequence of processes accelerated by the COVID-19 crisis. Digital sovereignty is instrumental to "bolster [EU's] place in the global economy and defend [its] values". Access, data protection, innovation, and infrastructures are identified as the main sectors in need of policy intervention. The issue of openness is framed within the concept of "open strategic autonomy and resilience". In the EU Legislative Priorities for 2021, instead, "Europe's digital sovereignty and innovative capacity" is framed as a key factor "to create a truly functioning single market for digital services within safe and ethical boundaries" (Timmers, 2021).

The discourse on digital sovereignty, as introduced by the German Presidency in the Council of the European Union, was revamped and partially reshaped under the French Presidency (January-June 2022). On 17 March 2022, the Council adopted a general approach on a proposal for a regulation to strengthen EU legislation on batteries and waste batteries. While the document doesn't mention digital sovereignty explicitly, the press release announcing the adoption justifies the need to set up a circular economy sector by targeting all stages of the life cycle of batteries" also in terms of technological sovereignty. In April 2022, the Council issued its conclusions "on building a European Strategy for the Cultural and Creative Industries Ecosystem, where it links digital sovereignty to the objective of maintaining and reaffirming European cultural wealth and diversity in the digital era. In May 2022, the Council entered into negotiations with the Parliament for the establishment of the 2030 Policy Programme "Path to the Digital Decade". The draft document includes digital sovereignty as a principle for the digital transformation of the EU, along with respect of fundamental rights, rule of law, and democracy (Council of the European Union, 2022a). The discourse on digital sovereignty is oriented towards the urgent need to fill the European investment gap, and to reduce technological strategic dependencies while assuring data protection. In a note sent by the Presidency on 23 May 2022, in preparation of the Chips Act, digital sovereignty is used again as just a general principle for EU's digital transformation, with no further connotations or elaboration. At the end of the French mandate, in June 2022, the French Minister for the Economy, Finance and Industrial and Digital Sovereignty, Bruno Le Maire, issued two press notes on behalf of the Council. The first one calls for efficient, competitive, and sovereign space-based communication infrastructures.

The second one announces forthcoming regulation on European crypto-assets aimed at preserving European monetary sovereignty. Under the Czech Presidency (July-December 2022), the Council carries on the negotiations with the other EU institutions on the 2030 Policy Programme, which is approved as a joint decision of the Council and the Parliament on 14 December 2022. In the agreed text, the term digital sovereignty is always specified as "digital sovereignty in an open manner" (Council of the European Union, 2022b). The connection to fundamental rights and European values is reinforced, as well as calls to fill any investment gaps in digital capacities and innovation, and to support European start-ups to scale up and provide European solutions. Digital sovereignty is accordingly described as an enabling factor for the Union's industry and economy, and particular attention is paid to the resilience of value chains. The Czech Presidency ends with the joint Declaration on digital rights and principles signed by the Council, the Commission and the Parliament. The Declaration just recalls the same definition of digital sovereignty in an open manner, as already elaborated in the 2030 Policy Programme. The press note announcing the Declaration, further emphasizes "the EU's ambition is to be digitally sovereign in an open and interconnected world embracing empowered citizens and innovative businesses in a human-centred, inclusive, prosperous, and sustainable digital society" (Council of the European Union, 2022b).

In the spring of 2023, under the Swedish Presidency, Ebba Busch, Swedish Minister for Energy, Business and Industry and Deputy Prime Minister, announced that the Council had reached with the Parliament a provisional political agreement on the Chips Act. The note presents the forthcoming regulation as a way to transform our dependency into market leadership; our vulnerability into sovereignty; our expenditure into investment", and to improve the EU's security of supply, resilience, and technological sovereignty in the field of chips (Council of the European Union, 2023b). In the same period, the Council proposes a regulation to fund a Joint Undertakings under Horizon Europe, aimed at "reinforcing the Union's semiconductor capacity by reducing dependencies, enhancing digital sovereignty, stimulating investment, strengthening the capabilities, security, adaptability, and resilience of the Union's semiconductor supply chain, and increasing cooperation among the Member States and the Commission" (Council of the European Union, 2023c). On 25 July 2023, under the Spanish presidency, the Council approved the Chips Act, reiterating the same wording as in the Joint Undertakings. On 23 November 2023, the Council adopts "Recommendations on the key enabling factors for successful digital education and training", where it calls for "promoting the use of open source, open content or open data solutions and digital commons in general, thus contributing to their development in digital practice and better safeguarding public values, sovereignty and the sustainability of digital resources in education" (Council of the European Union, 2023d). In February 2024, under the Belgian Presidency (January-June 2024), the Council Presidency and European Parliament negotiators reached a provisional agreement on the Strategic Technologies for Europe Platform (STEP), an instrument to mobilise investments in the fields of digital and deep tech, clean tech, and biotech, with the overall aim of strengthening the EU's sovereignty and long-term competitiveness in critical technologies (Council of the European Union, 2024c). The STEP, along with a regulation amending the multiannual financial framework for the period 2021–2027, as well as regulations establishing the Ukraine Fund, was part of a package adopted by the Council in February 2024, aimed at strengthening the EU's long-term budget in a targeted manner and addressing emerging new challenges. In addition, on 29 April 2024, the Council, in conjunction with the European Parliament, adopted a regulation containing "measures to reduce the cost of deploying gigabit electronic communications network". In the introductory part of this act, which amended Regulation (EU) 2015/2120 and repealed Directive 2014/61/EU, it is stated that "high-quality digital infrastructure based on ultra-high-capacity networks [...] is of strategic importance for social and territorial cohesion and, in general, for the competitiveness, resilience, digital sovereignty, and digital leadership of the Union" (Council of the European Union, 2024d). Notably, the principle of European digital sovereignty is reiterated by the Council, also in a press note announcing the conclusions on the future of industrial policy, in May 2024, even though it is not mentioned in the official document.

5 Conclusions

The analysis presented in this chapter demonstrates that the Council of the European Union has played a pivotal – though often overlooked – role in shaping the EU's discourse on digital sovereignty. While much of the existing literature focuses on the European Commission's strategies and proposals, our findings show that the Council has actively articulated, negotiated, and reformulated digital sovereignty across successive presidencies. In so doing, it has both advanced and diversified the discursive framework surrounding this concept, integrating it with broader policy agendas related to fundamental rights, market competitiveness, technological innovation, and global geopolitical pressures.

From Germany's early emphasis on protecting fundamental rights through legislation and investments to France's focus on reducing technological dependencies and fostering strategic autonomy, each presidency contributed specific priorities that further enriched the discourse. The Czech and Swedish presidencies, among others, underscored the tension between "open" approac hes to digital policy – stressing market-driven innovation and external collaboration – and the strategic need to shield critical infrastructures, data flows, and emerging technologies from overreliance on non-EU actors. Notably, the Council's rhetoric consistently positions digital sovereignty as a means of safeguarding not only European interests but also the Union's foundational values of openness, democracy, and rule of law. Moreover, from the outset, the Council's discourse on digital sovereignty has been articulated across a broad range of policy domains: from the digital economy to cybersecurity, from platform governance to artificial intelligence regulation, from connectivity development to data management, from device architectures to the resilience of technological supply chains, from digital taxation to privacy, from semiconductors to the Internet of Things, from quantum computing to algorithms, and from cloud computing to digital rights.

Yet, this evolving discourse is also marked by certain ambiguities. Although the Council aligns digital sovereignty with competitiveness and innovation, the actual implementation of measures – ranging from the Chips Act to various cybersecurity regulations – reveals persistent challenges in ensuring sufficient investment, strengthening supply-chain resilience, and maintaining technological leadership. Furthermore, in the most recent documents, the level of conceptualization of digital sovereignty is quite low, and it seems that rather than being used as a claim to authority and control, it is now a legitimizing principle itself, evoked to justify specific digital policies. This suggests a process of institutionalization of the digital sovereignty discourse, which includes digital sovereignty, among other common principles of the EU's policy-making (such as transparency, democracy, and rule of law). Overall, the Council's evolving engagement with digital sovereignty underscores its significance as a key locus of institutional power, capable of bridging divergent national interests and shaping cohesive policy directions at the EU level. Future research could benefit from a closer examination of how these Council-driven discourses interact with external pressures, such as global market dynamics or interstate competition in emerging sectors like artificial intelligence and quantum computing. In particular, a comparative analysis of the Council's role vis-à-vis the European Parliament and the Commission would offer deeper insights into inter-institutional dynamics in the EU's digital policy domain. By highlighting both the transformative potential and the inherent tensions of digital sovereignty, this chapter contributes to a more nuanced understanding of how the European Union navigates the ever-changing interplay between technological autonomy, political values, and global digital interdependence.

References

Adam, R., & Tizzano, A. (2024). *Manuale di diritto dell'Unione europea*, Torino: Giappichelli Editore.

Adler-Nissen, R., & Eggeling, K. A. (2024). *The Discursive Struggle for Digital Sovereignty: Security, Economy, Rights and the Cloud Project Gaia-X.* JCMS: Journal of Common Market Studies, 62: 993–1011. https://doi.org/10.1111/jcms.13594.

Broeders, D., F. Cristiano, & M. Kaminska. (2023). *In Search of Digital Sovereignty and Strategic Autonomy: Normative Power Europe to the Test of Its Geopolitical Ambitions.* Journal of Common Market Studies 61(5): 1261–80. doi:https://doi.org/10.1111/jcms.13462.

Burwell, F. G., & Propp. K. (2020). T*he European Union and the Search for Digital Sovereignty: Building "Fortress Europe" or Preparing for a New World?*, Issue Brief, Atlantic Council, June 2020. https://www.atlanticcouncil.org/wp-content/uploads/2020/06/The-European-Union-and-the-Search-for-Digital-Sovereignty-Building-Fortress-Europe-or-Preparing-for-a->New-World.pdf.

Carver, J. (2024). *More bark than bite? European digital sovereignty discourse and changes to the European Union's external relations policy.* Journal of European Public Policy, 31(8), 2250–2286. https://doi.org/10.1080/13501763.2023.2295523.

Council of the European Union (2022a). *Negotiations on the Digital Decade policy programme towards 2030: Emphasizing digital sovereignty alongside fundamental rights, rule of law, and democracy*. https://www.consilium.europa.eu/en/policies/digital-decade/.

Council of the European Union (2022b, December 14). *Joint decision of the Council and the European Parliament on the 2030 policy programme*. https://www.consilium.europa.eu/en/policies/digital-decade/.

Council of the European Union (2023a). *Conclusions on shaping Europe's digital future*. https://www.consilium.europa.eu/media/abc123/shaping-europe-digital-future-conclusions.pdf.

Council of the European Union (2023b). *Press note on the provisional political agreement on the Chips Act*. https://www.consilium.europa.eu/en/press/press-releases/.

Council of the European Union (2023c). *Proposal for a regulation on funding joint undertakings under Horizon Europe to strengthen the EU's semiconductor capacity*. https://www.consilium.europa.eu/en/policies/horizon-europe/.

Council of the European Union (2023d). *Adoption of the Chips Act and recommendations on key enabling factors for effective digital education and training*. https://www.consilium.europa.eu/en/press/press-releases/2023/.

Council of the European Union (2024a). *Conclusions on the Charter of Fundamental Rights in the context of artificial intelligence and digital transformation*. https://www.consilium.europa.eu/media/xyz123/charter-fundamental-rights-ai-digital.pdf.

Council of the European Union (2024b). *Council conclusions on the cybersecurity of connected devices*. https://www.consilium.europa.eu/media/xyz123/cybersecurity-connected-devices-conclusions.pdf.

Council of the European Union (2024c, February). *Provisional agreement on the Strategic Technologies for Europe Platform (STEP) and related financial framework amendments*. https://www.consilium.europa.eu/en/press/press-releases/2024/.

Council of the European Union (2024d, April 29). *Regulation on measures to reduce the cost of deploying gigabit electronic communications networks* (Amending Regulation (EU) 2015/2120 and repealing Directive 2014/61/EU). https://www.consilium.europa.eu/en/press/press-releases/.

Danet, D. & Desforges, A. (2020). *Digital sovereignty and strategic autonomy in Europe: From concept to geopolitical reality*. Hérodote, No 177–178 (2), 179–195. https://shs.cairn.info/journal-herodote-2020-2-page-179?lang=en.

Farrand, B., & Carrapico H. (2022). *Digital Sovereignty and Taking Back Control: From Regulatory Capitalism to Regulatory Mercantilism in EU Cybersecurity*. European Security 31(3): 435–53. https://doi.org/10.1080/09662839.2022.2102896.

Giddens, A. (2020). *Foreword*, In Hobbs, C. (ed.) Europe's Digital Sovereignty (London: ECFR). https://ecfr.eu/publication/europe_digital_sovereignty_rulemaker_superpower_age_us_china_rivalry/.

Hummel, P., Braun, M., Tretter, M., & Dabrock, P. (2021). *Data Sovereignty: A Review*. Big Data & Society 8(1). https://doi.org/10.1177/2053951720982012.

Lambach, D., & Monsees, L. (2025). *Beyond sovereignty as authority: the multiplicity of European approaches to digital sovereignty.* Global Political Economy, 4(1), 71–88. https://doi.org/10.1332/26352257Y2024D000000007.

Lambach, D., & Oppermann, K. (2023). *Narratives of digital sovereignty in German political discourse.* Governance, 36(3), 693–709. https://doi.org/10.1111/gove.12690.

Lewis, J. (2016). *the European Council and the Council of the European Union.* In Cini, M., Pérez-Solórzano Borragán, N. (Eds.), *Politics in the European Union*, Oxford: Oxford University Press, 138–54.

Moerel, L., & Timmers, P. (2021). *Reflections on digital sovereignty.* EU cyber direct, Research in Focus series 2021, https://ssrn.com/abstract=3772777.

Monsees, L., & Lambach, D. (2022). *Digital Sovereignty, Geopolitical Imaginaries, and the Reproduction of European Identity.* European Security 31: 377–94. https://doi.org/10.1080/09662839.2022.2101883.

Pohle, J. (2020). *Digital Sovereignty – A New Key Concept of Digital Policy in Germany and Europe.* Berlin: Konrad Adenauer Stiftung. https://www.kas.de/en/single-title/-/content/digital-sovereignty.

Pohle, J., & Santaniello, M. (2024). *From multistakeholderism to digital sovereignty: Toward a new discursive order in internet governance?.* EconStor Open Access Articles and Book Chapters, ZBW – Leibniz Information Centre for Economics, vol. 16 (4), pages 672–691. https://doi.org/10.1002/poi3.426.

Polish Presidency of the Council of the European Union (2025). *Programme of the Presidency.* https://polish-presidency.consilium.europa.eu/en/programme/programme-of-the-presidency/.

Puetter, U. (2022). *the Council of the European Union: Co-Legislator, Coordinator, and Executive Power.* In Hodson, D., Puetter, U., Saurugger, S., Peterson, J., (Eds.), *The institutions of the European Union*, Oxford: Oxford University Press, 78–105.

Roberts, H., Cowls, J., Casolari, F., Morley, J., Taddeo, M. & Floridi, L. (2021). *Safeguarding European values with digital sovereignty: an analysis of statements and policies.* Internet Policy Review, 10(3). https://doi.org/10.14763/2021.3.1575.

Santaniello, M. (2023). *Sovranità digitale e diritti fondamentali: un modello europeo di Internet governance.* Rivista italiana di informatica e diritto 4 (1), 47–51. https://doi.org/10.32091/RIID0058.

Santaniello M. (2025). *Attributes of Digital Sovereignty: A Conceptual Framework.* Geopolitics, 1–22. http://doi.org/10.1080/14650045.2025.2521548.

Thumfart, J. (2021). *The COVID-Crisis as Catalyst for the Norm Development of Digital Sovereignty. Building Barriers or Improving Digital Policies?.* SSRN Electronic Journal. http://dx.doi.org/10.2139/ssrn.3793530.

Timmers, P. (2021). Digital sovereignty as the foundation of the EU's strategic autonomy. In *EU Cyber Direct, Digital Sovereignty: From Narrative to Policy?*

von der Leyen, U. (2020). *State of the Union Address by President von der Leyen at the European Parliament Plenary.* https://ec.europa.eu/commission/presscorner/detail/it/SPEECH_20_1655.

Vučić, M. (2021), *European Union's Quest for Digital Sovereignty: Policy Continuations and Strategy Innovations.* In: Europe in changes: The old continent at a new crossroads. Institute of International Politics and Economics, Belgrade; Faculty of Security Studies at the University of Belgrade, Belgrade, 99–115.

Wurzel, R. K., Liefferink, D., & Di Lullo, M. (2020). *The European Council, the Council and the Member States: Changing Environmental Leadership Dynamics in the European Union. the Future of European Union Environmental Politics and Policy.* In Zito, A.R., Burn, C., Lenschow, A., C., (Eds.), *The Future of European Union Environmental Politics and Policy*, London: Routledge, 62–84.

CHAPTER 11

The Impact of the EU on the Development of the Defense Industry in the Baltic States in the Context of Cybersecurity

Donatas Palavenis

Abstract The article aims to assess the effectiveness of EU and national programs in developing indigenous innovations in cyber defense in the Baltic States (3B). The article utilizes quantitative and qualitative methods, combined with time-series analysis methods, using data from the 2017–2024 August period. The article includes three research questions: 1) Is cyber important in emerging disruptive technologies and future wars? 2) Is the cyber aspect considered the European Defense Fund and precursor programs? 3) What are 3B national cyber policies, cyber-oriented national defense industry support programs, and national champions? The research was not able to verify the hypothesis of whether indigenous entities benefit from participation in the EU and national programs due to inconsistent 3B governmental registers data on the entity's revenue and workforce. Nevertheless, the research concludes that there are clear indications that indigenous entities prefer EU programs over national ones, as participation is higher by at least two times. As a result, 3B could refine priorities within national defense industry support programs while providing opportunities for advancements in the cyber domain, which could benefit the military and society as a whole. Meanwhile, EU defense innovation programs have to include cyber topics with a similar projected rate of grants allocated, as cyber threats are expanding and will become more proliferated due to emerging disruptive technologies.

Keywords Estonia – Latvia – Lithuania – European Defense Fund – Preparatory Action on Defense Research – European Defense Industrial Development Program

1 Introduction

Baltic States (3B) are bordering the Russian Federation and its ally Belorussia, which makes 3B vulnerable through military, economics, diplomacy, and information domains. Russian military engagement in Ukraine and willingness to

HTTPS://DOI.ORG/10.1163/9789004763623-013

use hard power indicate that Russian empyreal ambitions are alive. Therefore, 3B, as former USSR members, feel concerned about their security (Palavenis 2024). As 3B are NATO members, Russia has limited, although broad, abilities to influence developments in 3B while keeping it below the threshold of Article 5, thus avoiding direct military confrontation. This also includes cyber actions that Russia exploits very actively in the region, e.g., Estonia was targeted by distributed denial of service attacks in 2007. As cyber activities for Russia remain a very popular, cheap, and effective tool to effect developments in 3B, the countries of this region have undertaken serious steps to defend themselves in cyberspace. 3B alone and supported by the EU invested in cyber defense capabilities, strengthening the indigenous defense industry that specializes in the cyber field. However, it remains unclear if national and EU defense programs are efficient in strengthening indigenous cyber defense capabilities and what the next steps could be to make 3B cyber defense more innovative and resilient. Furthermore, there is no specific knowledge of indigenous entities acting in the cyber defense area, including their performance within the cyber defense innovation field.

Beyond national security considerations, cyber defense in 3B must be understood within the broader context of the EU's efforts to enhance technological sovereignty through instruments such as the European Defense Fund and its precursor programs. From a theoretical perspective, 3B constitute a particularly relevant empirical case: as small states located on the EU's eastern periphery, they combine high threat perception with limited domestic industrial capacity, making them especially reliant on EU-level security and defense innovation frameworks. Beyond these geopolitical considerations, the three countries share structural characteristics – such as small populations, relatively modest high-tech industrial bases, and strong dependence on EU funding. In this regard, 3B allow for examining how EU initiatives interact with national policies to strengthen indigenous cyber defense innovation.

The aim of the article is to assess the effectiveness of EU and national programs in developing indigenous innovations in cyber defense in the 3B.

The article utilizes quantitative and qualitative methods, combined with time-series analysis methods, using data from the 2017–2024 August period. The research data is collected from the official 3B Ministry of Defense, entities, and the EU Directorate General for Defense Industry and Space websites. Scientific articles used in the text were found using relevant keywords through the Google Scholar website.

The article includes three research questions: 1) Is cyber important in emerging disruptive technologies and future wars? 2) Is the cyber aspect considered the European Defense Fund (EDF) and precursor programs? 3) What are the

3B national cyber policies, cyber-oriented national defense industry support programs, and national champions? The article is structured accordingly, as each part is devoted to a particular research question. The article seeks to verify the hypothesis of whether indigenous cyber-related entities benefit from participation in the EU and national programs in generating revenues, increasing employment, and establishing partnerships. Cyber defense has attracted the attention of many scholars in multiple scientific fields, including technical and social aspects. However, 3B region-specific research in the cyber defense area is less prevalent, including analyses of individual 3B countries. 3B Marek Górka (2023) concludes that bolstering 3B cyber resilience is primarily motivated by the perceived threat from Russia. Māris Andžāns and Reinis Stašulāns (2022) analyzed cyber defense policy developments in 3B and concluded that Latvia has not properly exploited national institutions active in the cyber defense policy framework. Louis Wierenga (2022) notes that 3B achieved many remarkable accomplishments in the cyber defense field, but has to remain steps ahead of adversaries. Dumchykov, Mykhailo, Maryna Utkina, and Olha Bondarenko (2022) assessed the threat to national 3B and Ukraine through the lens of conducted cybercrimes and suggested steps for combating effectively these types of crimes.

Few 3B country-specific researches were conducted. The Estonian case was analyzed by Nick Robinson and Alex Hardy (2021), who noted the state's leadership and expertise in cybersecurity, e-governance, and the National Cyber Security Strategy. Marleen van Ooijen (2020) analyzed the militarization of cybersecurity in Estonia, whereas Matthew Crandall (2024) looked at Estonia's cyber support for Ukraine and concluded that Estonia sees cybersecurity support as secondary to military support and would seek recognition for its cyber expertise. Latvian cyber defense importance within comprehensive defense approach was highlighted by Jānis Bērziņš (2022). Meanwhile, the Lithuanian case study written by Matthew Warren, Darius Štitilis, and Marius Laurinaitis (2023) focused on the key cyber lessons learned in dealing with Killnet and combating disinformation. Darius Štitilis et al. (2017) presented a model of Lithuanian national cyber defense strategy as a result of the best practices of other countries. Ramūnas Vilpišauskas (2024) analyzed the evolution of cyber defense policy in Lithuania due to Russian cyberattacks and concluded that political and institutional change has initially been slow.

Multiple scholars have analyzed cyber defense aspects within the EU and touched on multiple topics. To mention a few, André Barrinha and George Christou (2022) highlighted the sovereignty of EU cyber aspects, noted diplomacy's importance in overall implementation, and highlighted EU ambitions relating to a resilient and safe digital market. Sarah Backman (2023) noted the emerging EU cybersecurity approach, which has threat-based security logic, and highlighted the increasing role of the EU "as a security actor in the wider international cybersecurity landscape." Ana Paula Brandão and Isabel Camisão (2022) concluded that the European Commission is becoming a more relevant player in the EU cybersecurity policy field, which could be explored via "a market–security nexus." Meanwhile, Inga Ulnicane (2022) researched the artificial intelligence aspect in the cyber domain within EU policy, and Pier Giorgio Chiara (2022) analyzed the EU regulatory landscape for cybersecurity and Internet of Things (IoT). Meanwhile, a few scholars speak about the cyber defense aspect in EDF and precursor programs. As a result, the article seeks to enrich knowledge on the EU and 3B national defense innovation programs in cyber defense, as previous analyses have not analyzed this aspect in depth.

2 Cyber Defense in Future Wars

The continuous digital revolution has transformed the way we communicate, access, and exchange information, do business, and live. More than 60% of individuals and organizations in the world exploit advancements in information and communication technology, as quick access to information is critical. As good does not come alone, there are risks within information and communication technologies, which could cause social and economic stability and security challenges. The most prevalent cyber threats include phishing attacks, malware, ransomware, threats against data, social engineering, insider attacks, threats against data availability, disinformation and misinformation, distributed denial of service attacks, and supply chain targeting. Various types of cyberattacks have different impacts in different sectors, e.g., financial, logistics, medical, and military. In the future, artificial intelligence (AI) will provide better opportunities for detecting abnormalities and intrusions and conducting malware detection and analysis. Meanwhile, deep learning tools would be used for threat detection of pirated and malware-infected files within IoT networks, and deep neural networks would be trained to the patterns of normal and malicious applications in IoT devices. Blockchain technology, which is assessed as unhackable, would help store and exchange vital data, including the one stored in a secure cloud. Cyberattacks would become complex, and more advanced techniques would be used to breach cyber defenses, such as multi-vector attacks, zero-day exploits, polymorphic and fileless malware, and advanced persistent attacks. IoT devices would become vulnerable to cyberattacks due to weak default configurations, lack of security upgrades, obsolete firmware/software, insecure data storage, and poor encryption. IoT remains vulnerable through connectivity and could be exploited using eavesdropping, interception, or data modification (Kaur et al. 2023, Admass et al. 2024).

The EU fully understands cyber threats and opportunities from short-, mid-, and long-term perspectives. Halouzka et. al. (2022), while analyzing EU Central European Countries, noted that the most relevant released national documents highlight phishing attacks as they are launched at the beginning of larger cyberattacks. The outcome of phishing activity can lead to the leakage of personal or sensitive data, including payment card details, followed by misuse of the leaked data, and computers infected with malicious codes, which could lead to follow-up exploitations at a later stage. In addition, phishing could be used in hybrid scenarios while influencing people's behavior through fake news towards the desired end state, e.g., to weaken an adversary or strengthen the confidence of one's population.

Furthermore, the EU is seen "as a securitizing actor in the field of cybersecurity," which continues evolving due to the emergence of multiple cyber agencies and voluntary and mandatory legal requirements released by various EU institutions. However, a few challenges in the field remain, such as institutional fragmentation, member states' resistance to particular policies, and the absence of sub-regional cooperation. Analyses of various EU cybersecurity strategies indicate institutional focus shifts, e.g., comparison of 2013 vs. 2020 and reveal differences as more threat-based security logic is being applied in the latest edition. The 2020 cybersecurity strategy entails elements of cyber deterrence, diplomacy actions, and a need for better cyber intelligence cooperation (Backman, 2023). The securitization aspect is crucial to consider while designing responses to future cyber threats, as they would require substantial resources.

Some future challenges in cybersecurity are reflected in the EU's adopted Directive on the security of network and information systems (NIS Directive). The 2022 NIS Directive is more straightforward to implement for Member States and ensures common high-level cybersecurity across the EU. It focuses on managing security risks deriving from the digitalization and the overall surge in cyberattacks (Vandezande, 2024). The Russian war in Ukraine highlighted an important aspect – the EU's cyber-vulnerability linked with broader geopolitical instability, which requires a response. In that regard, the Computer Emergency Response Team for the EU Institutions, Bodies, and Agencies monitors Russia's cyberspace actions. Since the invasion of Ukraine, EU member states have been faced with several types of cyberattacks supported by generative AI. As a result, multiple cyber regulations were released to ensure a sufficient level of cyber resilience in the EU (Carrapico & Farrand

2024). Institution-wise, the EU Military Staff and European Defense Agency become the main actors, focusing on cyber defense initiatives and defining future scenarios that must be considered for upcoming technological development. Some of the foresight technological needs derived through multiple meetings among CapTechs and foresight studies transferred into recommendations for future EDF implementation programs in cyber- and related fields.

In addition to institutional and technical considerations, the implementation of EDF and precursory programs is shaped by the priorities set by the presiding EU member state. Presidencies of the Council of the EU influence which capability areas receive attention, the focus on cyber resilience and defense innovation, and coordination mechanisms across the Union (Håkansson, 2021). The trio presidency system further ensures continuity of strategic priorities across an 18-month cycle, enabling medium-term coordination of EU-wide initiatives. These presidency-driven priorities shape how 3B access EU support, implement innovation programs, and align national policies with broader EU objectives. While individual EU presidencies influence these priorities, a detailed analysis of the actions of each presiding member state is beyond the scope of this chapter, as the focus remains on understanding how EU-level programs interact with national policies in 3B to foster indigenous cyber defense innovation.

3 Cyber Defense within Preparatory Action on Defense Research, European Defense Industrial Development Program, and European Defense Fund

EU uses mainly EDF and precursor programs (Preparatory Action on Defense Research (PADR) and European Defense Industrial Development Program (EDIDP)) to foster innovation in the defense field. Initiatives are available for EU member states and Norway, which has a special agreement. Programs have different implementation timelines and funding volumes. PADR issued calls for participants throughout 2017–2019, with 18 research projects being financed. PADR was designed for research activities and issuing grants with up to 100% funding. Meanwhile, EDIDP, which invited participants throughout 2019–2020, focused on supporting defense capability development. There were 42 projects financed. Depending on the project's technological maturity level, it was financed with a 20–100% rate. PADR had €90 million, and EDIDP had a budget of €500 million. EDF is the most recent European Commission initiative, which combined support for research and development projects in the defense area. Approximately €1 billion is being distributed annually starting 2021.

Support rates depend on the project's technological readiness level. In addition, some incentives are foreseen for projects involving small–medium enterprises, enhancing cross-border cooperation, and being relevant with agreed priorities within the Permanent Structured Cooperation format. Annually, there are around 30–34 thematic call categories within EDF. Most of the financed projects last three years. Not all the programs have a cyber defense topic designed as a separate call.

Between 2019 and 2023, the European Union implemented extensive cyber defense projects under EDIDP and EDF. These projects covered a wide range of thematic areas tailored to the rapidly evolving challenges of digital security.

Key areas of support included situational awareness and the development of defensive capabilities, which were dominant in 2019 and 2020, with a total of five projects carried out. In 2021, the focus shifted to threat analysis and enhanced operational capabilities (two projects), as well as cyber training and exercise capacity (one project). In the following years – 2022 and 2023 – projects concentrated on strengthening cybersecurity to improve system resilience (two projects), situational awareness in emerging computing environments (one project), and the automation of security penetration testing and deployment of autonomous AI agents (two and one projects, respectively, in 2023). The total funding granted for these initiatives over the five-year period amounted to €210.8 million, with a marked increase in investment from €30.4 million in 2019 to over €58 million in 2023 – highlighting the growing prioritization of cyber defense in the EU's security policy (DG DEFIS, 2024).

PADR, the first defense program, had no cyber defense calls, whereas EDIDP and EDF had one call devoted to cyber developments. Grants issued within EDIDP for cyber projects comprised 9% of the total budget, whereas in EDF, this figure was smaller – 5%. Within EDF, one call is devoted to basic research and another to development activities. In most cases, development topics are more attractive as there are normally more projects financed.

Although EU defense programs are popular, not all EU countries are represented equally in the cyber topic. This could be explained by increasing competition for available EU grants, different specializations of participants, and annually changing subtopics. Between 2019 and 2023, European countries engaged numerous entities in the implementation of EU-funded cyber defense projects under the EDIDP and EDF programs (DG DEFIS, 2024). Examples of projects and the number of participating entities by country are presented below:

1. Under the EDIDP program (2019), the ECYSAP project, coordinated by Spain, included six entities from Spain and one each from Italy, Estonia, and France. The PANDORA project (Greece) involved four Greek entities, along with participants from France, Portugal, Austria, Cyprus, Hungary, Spain, and Belgium. The TELUGU project (Greece) included entities from Greece, Cyprus, and Poland. In 2020, the CYBER4DE project (Lithuania) brought together entities from Lithuania, France, Poland, Estonia, Croatia, Italy, and Romania, while the **PRUDENCE** project (Portugal) included participants from Portugal, Austria, Italy, and Spain.
2. Under the EDF program (2021), the P.O. project (Bulgaria) brought together participants from Bulgaria, France, Belgium, Cyprus, the Netherlands, Portugal, Greece, Spain, Ireland, Romania, Italy, Austria, and Germany. The AInception Concept project (Greece) included entities from Greece, Austria, France, Italy, Cyprus, Norway, Bulgaria, Sweden, and the Netherlands. The EU-GUARDIAN project (Spain) involved partners from Spain, France, Poland, Lithuania, Bulgaria, Estonia, and Italy.
3. In 2022, EDF funding supported projects such as the EUCINF Mission (France), which included entities from France, Germany, Austria, Belgium, Italy, Estonia, Hungary, the Netherlands, Cyprus, Spain, Greece, and Latvia, and the FAKT project (Norway), which brought together Norwegian entities and partners from several other European countries. The NEWSROOM project (Austria) involved participants from Austria, Germany, Italy, Estonia, Cyprus, Spain, Finland, Greece, Norway, and France.
4. In 2023, EDF projects such as ECYSAP EYE (Spain), AIDA (Estonia), and TRYTON (Cyprus) engaged numerous entities from various European countries, including Spain, France, Austria, Estonia, Poland, Romania, Greece, Slovakia, Belgium, Norway, Italy, Luxembourg, the Czech Republic, Portugal, and Finland (DG DEFIS, 2024).

Cyber defense projects similar to other types in EDF involve multiple entities from more countries, making project management complex and deliverables not so ambitious. The tendency to work within a particular work package in the project will persist, giving individual participants some freedom. Based on call requirements, bonus system, and willingness to get approval, consortiums tend to attract entities from most EU countries. Multiple countries, within EDIDP and EDF coordinate projects; in most cases, a single country runs one project, with the exception being entities from Greece and Spain, which coordinate three projects each. In the 3B case, Lithuanian and Estonian entities coordinate one project. However, consortium-wise, participants from different countries have different representations. There are 27 entities from Greece, 22 from Spain, and 21 from France, whereas the least represented are the Czech

Republic and Ireland, with one participant. 3B-wise, 13 Estonian entities, 3 Latvian, and 5 Lithuanian are engaged in cyber-related projects.

4 3B National Cyber Policies, Support Programs, and Most Prominent Entities

As it was concluded by Marek Górka, 3B, "cyber resilience [is] primarily motivated by the perceived threat from Russia." However, when discussing cyber policies, scholars mention other factors like the social, economic, and technological environments that can also pressure governments to develop cyber policies that would make countries more competitive globally. In his research, Marek Górka pays attention to the Digital Economy and Society Index, which indicates that in 2022, Estonia was ranked 1st, Latvia 18th, and Lithuania 12th, and these serve as reference points for comparing cyber policies. Furthermore, 3B allocated national and utilized EU funds and private sector investments to reinforce the cyber sector. Advancement was achieved by educating citizens to enhance their digital skills, and investments in advanced digital infrastructure, platforms, and technologies (Górka, 2023). Institutions play a role in defining and implementing cyber policies. Estonia is keen on implementing e-governance solutions in various fields and hosting the Cooperative Cyber Defense Center of Excellence. Latvia is eager to build a resilient society against Russian digital political warfare with a focus on countering propaganda. Lithuania is focused on critical energy infrastructure and cyber through the PESCO initiative, where it runs the Cyber Rapid Response Team and Mutual Assistance in Cyber Security and hosts the biggest cyber exercises in the region (Tschap, 2024). Louis Wierenga echoed similar findings and concluded phenomenally by stating, "although the Baltic states are small states, they carry a big stick when it comes to dealing with cyber and hybrid threats" (Wierenga, 2022). Marek Górka (2023) noted the importance of national defense support programs for cybersecurity within broader cyber policy. All 3B have indigenous support programs and grants available to strengthen their defense forces and industry.

Between 2017 and 2023, countries in the Three Seas region, such as Estonia and Latvia, implemented dedicated defense support programs with a particular focus on cybersecurity. For example, in 2017, Estonia introduced the SecureChat communication system, which was redesigned to operate on the Android operating system. This project received €74,250 in funding from the Estonian Ministry of Defense (EST MON, 2024). In 2021, Latvia concentrated on enhanced electromagnetic protection and cybersecurity in field operations, utilizing innovative shielding technologies and data destruction systems for surveillance equipment. The project, carried out by Riga Technical University, received €465,000 in funding from the Latvian Ministry of Defense (LVA MON, 2024). The lack of support programs for specific cyber defense topics from the MoD and the Armed Forces indicates that other more urgent priorities need to be developed on the battlefield, and lack of particular funding. For example, Estonia, in its indigenous defense industry support programs, is focusing on unmanned aerial vehicles, loitering munition, sights, and radios, rather than on cyber aspects. As 3B entities develop their competencies, expand their workforce, and generate revenues on dual-use items or R&D intellectual property sales, they attract the attention of other entrepreneurs who mimic their behavior and bring innovation into the sector. This is beneficial for cyber defense companies and the ecosystem as a whole. However, background data must also be considered if one assesses the most prominent companies participating in the EU and national companies. Between 2017 and 2023, the Three Seas Initiative countries involved various entities in the implementation of both EU and national defense programs related to cybersecurity (Palavenis, 2025). In Estonia, key participants included institutions such as Uniraja Said, Cybernetica (engaged in EDF and EDIDP programs), GT Cyber Technologies (participating in EDIDP), as well as CR14, Cybexer Technologies, and Advokaadiburoo Sorainen, all active under EDF. In Latvia, an active participant was Riga Technical University, involved in national projects, along with the companies Tilde and Latvijas Mobilais Telefons, both of which cooperated within the EDF framework. Lithuania, in turn, engaged the Baltic Institute of Advanced Technology, participating in both EDF and EDIDP programs, as well as NRD Cyber Security, which operated under the EDIDP (Palavenis, 2025).

As data available in governmental registers on the entity's revenue and workforce is inconsistent across 3B, entities are not scrutinized through this research question, which opens an opportunity for follow-up research. Some 3B entities, e.g., Cybernetica, CR14, and Baltic Institute of Advanced Technology, participated in multiple EU-financed projects, which indicates well-established partnerships and great competencies as coordinators are from various countries. Only one entity – Riga Technical University – is governmental, which attracted additional funds for the project in the cyber domain, which indicates a forward-leaning approach by grant organizers. The number of entities involved in EU programs compared to national programs differs substantially, i.e., 5:1 in the Estonian case, 2:1 in Latvia, and 2:0 in Lithuania, which indicates that national champions favor EU programs.

5 Conclusions

There are clear indications that cyber threats, e.g., state-sponsored attacks, cybercrime, will proliferate rapidly, and state, non-state, commercial actors, and citizens could feel less secure. Cyber defense technologies will expand promptly, and new innovations in the field will emerge. Emerging disruptive technologies, e.g., AI and quantum will influence technological and procedural developments. Meanwhile, the EU indicated a serious approach to cyber aspects and aims to become a securitizing actor. Strong institutional backing, with new cyber-specific players empowered by legal settings, would change the EU landscape, as evidenced by policy paper shifts.

The EU finances defense technologies and innovations through the EDF, EDIDP, and PADR programs, which have calls for cyber projects. Nearly all EU countries and Norway are represented in 14 research and development projects by indigenous entities that form large consortiums, thus creating particular challenges in implementing sophisticated projects in a dynamic cyber environment. Grant-wise, cyber topics receive five percent of the total available budget within annual EDF calls. The sustainability of financing for research and development at this level and above would allow the EU to ensure satisfactory cybersecurity, as cyber threats are increasing and the whole cyber domain is exposed to emerging disruptive technologies.

Although guided by EU legislation and strategies, 3B have slightly different variations on cyber defense priorities. Different views are formed by the threats and expectations in particular countries, e.g., Estonia – which focuses on e-governance; Latvia, which contains propaganda; and Lithuania – which ensures energy security. Few national defense innovation programs focus on cyber defense, although indigenous entities prefer more active engagement in EU programs. Research has not revealed details on entities' participation benefits in national and EU support programs, which was the main focus of the hypothesis, due to inconsistent 3B governmental registers available on entity's revenue and workforce. Nevertheless, there are clear indications that indigenous entities prefer EU programs vs. national ones, as participation is higher in the former by at least two times. As a result, 3B could refine priorities within national defense industry support programs while providing opportunities for advancements in the cyber domain, which could benefit the military and the whole of society.

References

Admass, W.S., Munaye, Y.Y., & Diro, A.A., (2024). Cyber security: State of the art, challenges and future directions. *Cyber Security and Applications*, 2, 100031.

Andžāns, M., & Stašulāns, R., (2022). Cyber security in the Baltic States: Institutionalist analysis of cyber security policy developments (2014–2022). Rīgas Stradiņa universitāte.

Backman, S., (2023). Risk vs. threat-based cybersecurity: the case of the EU. *European Security*, 32 (1), 85–103.

Barrinha, A., & Christou, G., (2022). Speaking sovereignty: the EU in the cyber domain. *European Security*, 31 (3), 356–376.

Bērziņš, J., (2022). Latvia's comprehensive defence approach. *Bezpieczeństwo. Teoria i Praktyka*, 48 (3), 223–235.

Brandão, A.P., & Camisão, I., (2022). Playing the Market Card: The Commission's Strategy to Shape EU Cybersecurity Policy. *JCMS: Journal of Common Market Studies*, 60 (5), 1335–1355.

Carrapico, H., & Farrand, B., (2024). Cybersecurity Trends in the European Union: Regulatory Mercantilism and the Digitalisation of Geopolitics. *JCMS: Journal of Common Market Studies*, 62 (S1), 147–158.

Chiara, P.G., (2022). The IoT and the new EU cybersecurity regulatory landscape. *International Review of Law, Computers & Technology*, 36 (2), 118–137.

Crandall, M., (2024). Understanding Estonia's Cyber Support for Ukraine: Building Resilience, Not Status. *Applied Cybersecurity & Internet Governance*, 3 (1), 78–90.

DG DEFIS, (2024). European Defence Fund (EDF) – Official Webpage of the European Commission. – European Commission. https://defence-industry-space.ec.europa.eu/eu-defence-industry/european-defence-fund-edf-official-webpage-european-commission_en.

Dumchykov, M., Utkina, M., & Bondarenko, O., (2022). Cybercrime as a threat to the national security of the Baltic States and Ukraine: The comparative analysis. *International Journal of Safety and Security Engineering*, 12 (4), 481–490.

EST MoD, (2024). Kaitsetööstuse arendusprojektide konkurss 2024. https://kaitseministeerium.ee/et/eesmargid-tegevused/teadus-ja-arendustegevus/kaitsetoostuse-arendusprojektide-konkurss-2024.

Górka, M., (2023). Baltic States Cyber Security Policy: Development of digital capabilities in 2017–2022. *Stosunki Międzynarodowe–International Relations*, 3 (15), 15.

Håkansson, C., (2021). The European Commission's new role in EU security and defense cooperation: The case of the European Defense Fund. *European Security*, 30 (4), pp. 589–608.

Halouzka, K., Burita, L., Coufalikova, A., Kozak, P., & Frantis, P., (2022). A Comparison of Cyber Security Reports for 2020 of Central European Countries. *Advances in Science, Technology and Engineering Systems Journal*, 7 (4), 105–113.

Kaur, R., Gabrijelčič, D., & Klobučar, T., (2023). Artificial intelligence for cybersecurity: Literature review and future research directions. *Information Fusion*, 97, 101804.

LVA MoD, (2024). Defense Innovation Research Program. https://www.mod.gov.lv/lv/nozares-politika/zinatniski-petnieciska-darbiba/aizsardzibas-inovaciju-petijumu-programma.

Ooijen, M. van, (2020). Cyber securitization or cyberization of conflict?–the militarization of Cyber Security in Estonia. Master's Thesis.

Palavenis, D., (2024). Arms Acquisition in the Baltic States and Transfers to Ukraine: Balancing National Security. *Defense & Security Analysis*, 40 (2), 167–189.

Palavenis, D., (2025). Defence Innovation in the Baltic States: Growth Through National and European Programs. *Comparative Strategy*, 44 (4).

Robinson, N., & Hardy, A., (2021). Estonia: from the "Bronze Night" to Cybersecurity Pioneers. *In*: *Routledge Companion to Global Cyber-Security Strategy*. Routledge, 211–225.

Štitilis, D., Pakutinskas, P., Laurinaitis, M., & de Castel, I.M., (2017). A model for the national cyber security strategy. The Lithuanian case. *Journal of Security & Sustainability Issues*, 6 (3).

Tschap, N., (2024). The New Baltic Way-Baltic Identity amidst the Geopolitics of Digital Transformation. B.S. thesis. University of Twente.

Ulnicane, I., (2022). Artificial Intelligence in the European Union: Policy, Ethics and Regulation. *In*: *The Routledge handbook of European integrations*. Taylor & Francis.

Vandezande, N., (2024). Cybersecurity in the EU: How the NIS2-directive stacks up against its predecessor. *Computer Law & Security Review*, 52, 105890.

Vilpišauskas, R., (2024). Gradually and then suddenly: the effects of Russia's attacks on the evolution of cybersecurity policy in Lithuania. *Policy Studies*, 45 (3–4), 467–488.

Warren, M., Štitilis, D., & Laurinaitis, M., (2023). Cyber Lessons that the World Can Learn from Lithuania. *In*: *European Conference on Cyber Warfare and Security*. 517–524.

Wierenga, L., (2022). Cybersecurity and Hybrid Threats in the Digital Baltic States: a State of the Art. *In*: A. Sprūds, U.A. Bērziņa-Čerenkova, and S. Broka, eds. *Commonalities, Risks and Lessons for Small Democracies: Hybrid Threats in Baltics and Taiwan*. Latvian Institute of International Affairs, 58–67.

CHAPTER 12

Critical Analysis and a Proposed Model for the EU Cyber Resilience Act

Ali Shoker

Abstract The ever-increasing cybersecurity attacks on vital civilian infrastructure and critical defense systems are unprecedented, underscoring the urgent need for cyber-resilient digital systems. The European Union's Cyber Resilience Act (CRA) acknowledges this need by establishing cybersecurity requirements across the entire lifecycle of digital products, aiming to harmonize standards among member states. However, despite its title, the CRA lacks a comprehensive framework for "cyber resilience" implementation, and even fails to explicitly define the term. Our analysis finds that the CRA addresses cyber resilience primarily through general cybersecurity provisions that fall short of embodying foundational resilience principles. In response, we propose a layered cyber resilience model grounded in key principles—decentralization, diversity, agility, degradation, and reformation—tailored to the EU's specific geopolitical and technological context. We conclude by outlining a strategic agenda for embedding these principles into future revisions of the CRA, aiming to transform it into a truly resilience-oriented and effective regulatory instrument.

Keywords Cyber Resilience Act – Cybersecurity – European Union – Regulation

1 Introduction

In an era where everything is connected – from hospitals and energy grids to satellites and smartphones – "everything can be hacked", as President Ursula von der Leyen warned (von der Leyen, 2021). The European Union (EU) has clearly recognized that cybersecurity is no longer a technical afterthought; it is a cornerstone of economic stability, democratic resilience, and geopolitical sovereignty. This is echoed by Vice-President Margaritis Schinas, who asserted that "cyber resilience is no longer a luxury – it is a precondition for a functioning digital economy and democracy" (European Commission, 2022). These

 HTTPS://DOI.ORG/10.1163/9789004763623-014

statements underscore the EU's ambition not only to protect its digital infrastructure but also to lead globally in setting standards for cybersecurity and resilience.

To support this vision, the EU has introduced a series of legislative measures, including the NIS Directive (2016) and its enhanced successor NIS2 (2022), the Cybersecurity Act (2019), and the Cybersecurity Strategy for the Digital Decade (2020). Together, these instruments aim to harmonize cybersecurity across member states, strengthen institutional capacities such as ENISA, and promote a more resilient digital ecosystem (European Commission, 2020).

The Cyber Resilience Act (CRA), adopted in 2024 (Regulation (EU) 2024/2847), represents a critical addition to this framework. It addresses longstanding market failures by establishing horizontal cybersecurity requirements for products with digital elements. Specifically, the CRA mandates that manufacturers adopt security-by-design and -by-default principles, implement effective vulnerability management practices, and provide transparency around product support lifespans. Its objectives are twofold: to reduce the number of insecure digital products entering the market and to empower users with better information for secure product usage (European Commission, 2024).

However, despite its ambitious title, the CRA fails to define "cyber resilience" or present a structured framework for its implementation. This conceptual oversight is more than semantic. It reflects a broader misalignment between the rhetoric of cyber resilience and the Act's actual provisions. In conflating cyber resilience with conventional cybersecurity, the CRA risks neglecting the full scope of capabilities required to ensure continuity of critical functions under adverse cyber conditions. As articulated in the Directive on the Resilience of Critical Entities (Directive (EU) 2022/2557), resilience is defined as "the ability to prevent, protect against, respond to, resist, mitigate, absorb, accommodate and recover from an incident". This broader interpretation is not only absent from the CRA but also inconsistent with its emphasis on pre-market security requirements and limited post-market obligations.

This chapter aims to interrogate this disconnect by analysing the CRA through the lens of established cyber resilience frameworks. Drawing on the NIST Cybersecurity Framework (CSF 2.0) (National Institute of Standards and Technology [NIST], 2023), MITRE's Cyber Resilience Engineering Approach (Ross et al., 2021), and other scholarly and institutional sources (e.g. Wu, 2024; Chiara, 2025), we assess the extent to which the CRA addresses the functional, operational, and systemic aspects of cyber resilience. We argue that while the Act incorporates valuable cybersecurity principles, it misses an opportunity to embed resilience across the full lifecycle of digital products and services.

By identifying this gap, we propose a layered cyber resilience model that integrates principles such as decentralization, diversity, agility, degradation, and reformation – concepts central to modern cyber-resilient architectures. In doing so, we aim to contribute to the policy discourse surrounding digital resilience in Europe and provide a foundation for future amendments to the CRA that align more closely with the EU's broader strategic goals.

2 The Missing Resilience Essence of the CRA

The Cyber Resilience Act (CRA), published in 2024, imposes clear requirements on manufacturers to improve the cybersecurity posture across the EU (European Commission, 2024). Although its title emphasizes "resilience", the content offers neither a definition of cyber resilience nor a practical framework for its implementation. Nevertheless, the Act seeks to address two key issues: (1) the persistently low level of cybersecurity in products with digital components, and (2) the lack of user understanding and access to relevant cybersecurity information.

To address these problems, the CRA establishes a framework for designing secure-by-design and secure-by-default digital products, and for improving transparency about support periods. More detailed requirements (explained in Table 12.1) can be grouped into four categories: Essential Cybersecurity Requirements, Vulnerability Handling Requirements, User Manuals, and Technical Documentation.

The realization of the importance of establishing the Cyber Resilience Act (European Commission, 2024) (CRA) at the EU level is commendable and deserves our attention. Nevertheless, the disappointment quickly arises with the absence of a clear definition of "cyber resilience". This omission is not merely a drafting oversight but a fundamental flaw that undermines the Act's objectives as we explain next. The term "cyber resilience" is often conflated with cybersecurity, particularly in industry, where legacy cybersecurity products are frequently rebranded under the guise of cyber resilience. This ambiguity extends to academia, where the term remains poorly understood, consequently. As one of the key goals of the CRA is to raise awareness, the Act should explicitly define "cyber resilience" upfront to eliminate confusion. The Act even fails to refer to the definition "resilience" used in other EU contexts, e.g. the Resilience of Critical Entities Directive (EU): "resilience" means a critical entity's ability to prevent, protect against, respond to, resist, mitigate, absorb,

TABLE 12.1 The main EU Cyber Resilience Act requirements in 2024 (Source: European Commission, 2024).

Category	Requirements
Essential Cybersecurity Requirements	Manufacturers must ensure products meet cybersecurity requirements (defined by the Act) and address vulnerabilities for the product's lifetime or for five years. Focus on cybersecurity during design, development, and production through up-to-date risk assessments, secure configurations, encryption, and data protection. Provide timely security updates, protect against unauthorized access, reduce attack surfaces, and ensure essential functions remain available and resilient.
Vulnerability Handling Requirements	Identify, document, and remediate vulnerabilities, conduct regular security tests, and publicly disclose fixed vulnerabilities. Emphasize secure update distribution, implement coordinated vulnerability disclosure policies, and provide users with free security updates and advice. Manufacturers have four notification obligations, which may involve informing ENISA, users, or third parties whose components are integrated into the product.
User Manuals and Technical Documentation	Prepare technical documentation showing compliance with cybersecurity requirements before placing a product on the market. Products must come with information (outlined in the Act) to help users install, operate, and use the product securely.
EU Conformity	- Manufacturers must issue an EU declaration of conformity, affix CE marking, conduct conformity assessments, and ensure due diligence when integrating third-party components.

accommodate and recover from an incident"(European Parliament & Council of the European Union, 2022).

The lack of a precise definition has broader implications for the scope and application of the CRA, causing it to be neither comprehensive nor systematic. This could undermine its effectiveness across the EU if not addressed through clearer definitions and guidelines (Wu, 2024). It is uncomprehensive because it sells cybersecurity as cyber resilience; while these concepts have a big intersection, they have different objectives.

To explain this, Figure X depicts a digital service functionality curve, where acceptable service quality is maintained within the Normal Range and should never fall below the Minimum Essential Functionality threshold. The x-axis represents the typical stages of a cyberattack kill chain, illustrating that significant functionality degradation occurs during the Execute stage, when malware is activated. Cybersecurity approaches traditionally emphasize the stages preceding Execute, focusing on vulnerability identification, protection, detection, and initial response measures. In contrast, cyber resilience places greater emphasis on the Execute stage and subsequent phases, aiming to develop advanced response capabilities that enable organizations to mitigate, absorb, accommodate, and recover from cyber incidents (National Institute of Standards and Technology, 2019).

Smarter cyber resilience should also anticipate attacks, to be more proactive, which requires the detection and prevention of cybersecurity techniques, prior to execution. With this vision, the author claims that cyber resilience tends to subsume cybersecurity, although this is not the typical use in practice.

Strategically, the CRA overlooks the advantages of decentralization and diversity within the EU. The diverse political and economic landscape of the EU presents a unique opportunity to strengthen cyber resilience by decentralizing imports and diversifying vendors to mitigate, absorb, accommodate, and recover from an incident. For instance, supply chain requirements under the CRA focus primarily on compliance and sovereignty, without incorporating resilience-enhancing strategies, e.g. requiring to decentralize imports and diversify vendors. Additionally, post-market operations involving critical resources, such as energy and skilled labour, should be decentralized and diversified to bolster resilience.

Furthermore, the Cyber Resilience Act (CRA) lacks a structured implementation strategy at the technical level. Previous research highlights that the cybersecurity requirements outlined in the CRA appear fragmented and inconsistent across different sectors and applications, introducing ambiguity in compliance standards (Burri & Zihlmann, 2023; Chiara, 2025). This issue stems from the absence of a dedicated Cybersecurity or Cyber Resilience Framework and related assessment tools. While the CRA adopts a risk-based approach, the criteria for evaluating and classifying risks are not well defined, which could lead to inconsistent enforcement (Burri & Zihlmann, 2023). Established frameworks, such as the NIST Cybersecurity Framework (National Institute of Standards and Technology- NIST, 2023) (CSF) and MITRE's Cyber Resilience Engineering Approach (Ross et al., 2021), provide structured guidelines and best practices for implementing security controls systematically, which the CRA currently lacks.

Developing a cyber resilience framework is crucial because it is often overlooked by industry, despite being central to the CRA's objectives. The framework should define the minimum cyber resilience requirements through measurable metrics and established resilience techniques. Additionally, a systematic assessment tool is needed to evaluate success and ensure compliance.

3 The Principles of Cyber Resilience for the EU

To address the gaps in the current EU Cyber Resilience Act (CRA), this framework applies the core principles of cyber resilience by design: Decentralization, Diversity, Agility, Automation, Degradation, and Reformation across three critical layers of the cyber ecosystem: Supply Chain, Technical Design, and Operational Execution. These principles are contextualized at the Entity, Member State, and EU levels to build layered, adaptive cyber resilience.

For the sake of understanding our study, it is sufficient at this stage to adopt the cyber resilience definition from: "the ability to prevent, protect against, respond to, resist, mitigate, absorb, accommodate and recover from an incident" (European Parliament & Council of the European Union, 2022). This approach integrates proactive measures, such as threat anticipation and system hardening, with reactive strategies like incident response and recovery planning, to withstand disruptions, minimize damage, and swiftly restore operations, ensuring sustained performance even in the face of evolving cyber threats (Wu, 2024).

3.1 Supply Chain Layer

Decentralization and Diversity: At the entity level, procurement from multiple suppliers using different technologies reduces dependence on any single source. Organizations should secure contracts with multiple manufacturers for the same product, ensuring consistent quality and performance. For example, a cyberattack on a primary supplier should automatically trigger a switch to a backup supplier without service disruption. At the Member State level, governments should establish supply chain deals with vendors in different geopolitical regions. Diversification prevents disruptions from geopolitical instability or economic changes. For example, a supply chain failure in one region should prompt an immediate shift to alternative sources. At the EU level, trade agreements with diverse geopolitical and technological partners create supply chain resilience. Encouraging intra-EU trade and strategic stockpiles enhances redundancy (European Commission, 2020). For instance, a semiconductor shortage in Asia should trigger increased imports from alternative markets within the EU.

Agility and Automation: At the entity level, organizations should source simultaneous components from diverse suppliers to prevent bottlenecks and maintain responsiveness. Automated supply chain management systems should identify performance or geopolitical risks and switch to backup suppliers instantly. For example, a cyberattack on a primary supplier should automatically reroute orders to secondary suppliers, without downtime. At the Member State level, governments should diversify import channels and maintain strategic stockpiles of critical components. Automated systems should monitor economic instability, stock levels, and global risks, triggering immediate reallocation when disruptions occur. For instance, a regional supply chain disruption should activate automated redistribution of resources to alternative suppliers. At the EU level, trade diversification and decentralized automation should enhance resilience by creating a flexible and responsive supply network. A coordinated, decentralized EU platform should monitor global supply chains and adjust sourcing dynamically in response to geopolitical or economic disruptions.

Degradation: At the entity level, organizations should establish tiered service levels to maintain core functions when resources are limited. Automated prioritization systems should shift resources towards critical suppliers and suspend non-essential imports when disruptions occur. For example, a cyberattack disrupting a logistics network should automatically reroute shipments for essential goods while postponing non-essential deliveries. At the Member State level, governments should maintain strategic stockpiles and create contingency plans for supply chain disruptions. Automated systems should assess resource availability and trigger adjustments based on priority (Lin et al., 2019). For instance, a shortage of medical supplies should prioritize healthcare facilities over commercial markets. At the EU level, centralized coordination should enable strategic allocation of resources across member states. Automated resource management platforms should prioritize critical infrastructure and essential services during regional disruptions. For example, a semiconductor shortage should trigger reallocation to defence and healthcare sectors across Member States, deferring less critical production.

Reformation: At the entity level, organizations should integrate post-incident findings into revised security policies and automated feedback loops. For instance, a ransomware attack such as the 2017 *WannaCry* incident should lead to stronger endpoint protection and employee training, as it exposed the risks of unpatched systems across public and private sectors (Lin et al., 2019). At the Member State level, governments should adapt national security frameworks and regulatory standards based on threat intelligence. Cross-border alignment of protocols should strengthen collective defence (Smith & Jones, 2020). For example, a nation-state attack should trigger updates to national cybersecurity strategies. At the EU level, reforms should align member state policies to ensure a unified defence posture. EU directives should mandate the adoption of best practices after major incidents. For example, supply chain vulnerability should lead to revised supplier vetting and security standards across the EU.

3.2 Technical Layer

Decentralization and Diversity: At the entity level, systems should follow fundamental resiliency fundamentals like decentralized and distributed systems, redundancy, active replication, fault avoidance, fault tolerance, failure masking, recovery, self-healing, rejuvenation, elasticity, etc. Decentralization techniques may include blockchain, Conflict-free Replicated Data Types (CRDTs), and cloud infrastructure (Alomari et al., 2020; Demers et al., 1987; Shapiro et al., 2011; Pilkington, 2016; Nakamoto, 2008). These approaches reduce single points of failure and increase adaptability. Edge computing and hybrid cloud models improve performance and resilience (Shi et al., 2016; Marinescu, 2017). Techniques like erasure coding, DAGs, and circuit breakers help ensure consistent functionality and security even under stress (Plank, 1997; Popov, 2018; Nygard, 2007). DIDs enable secure identity verification in decentralized environments (W3C, 2021).

At the Member State level, governments should promote geographically and technologically diverse infrastructures to reduce the risk of regional disruptions and dependence on specific vendors. This includes supporting multiple national cloud providers, varied software stacks, and regionally distributed data centres. The EU should mandate interoperability and incentivize multi-cloud adoption to ensure continuity of service in case of provider failure or cyberattack, enhancing overall digital sovereignty and strategic autonomy (European Commission, 2022).

Agility and Automation: At the entity level, modular architecture using Software-Defined Networks (SDN), containerization, and microservices facilitates rapid response. SDN enables dynamic traffic reconfiguration. Containerization allows lightweight deployment and quick recovery. Microservices support independent scaling and failure isolation. Automated failover and self-healing improve resilience. Failover automation redirects traffic, while self-healing uses monitoring and ML to resolve issues. Infrastructure as Code (IaC) enables consistent, automated deployment. IDPS monitor traffic and isolate threats. A Threat Intelligence Platform (TIP) integrates with Security Information and Event Management (SIEM) or Security Orchestration Automation and Response (SOAR) to automate incident response. AI-driven detection analyses patterns, adapts to threats, and contains zero-day threats (Ahmed et al., 2020). Automated patching and config management ensure security. For example, Kubernetes isolates a breached service and reroutes traffic. SDN controllers adjust traffic to avoid compromised nodes.

At the Member State level, adaptive regulatory frameworks should allow quick modification of standards. Automation facilitates real-time updates to protocols. The EU should integrate security frameworks and protocols using open standards and APIs (Zuech et al., 2015).

Degradation: At the entity level, modular architecture and automated load balancing allow selective shutdowns to preserve critical functions. For example, if a data centre is compromised, the system should isolate the segment and redirect traffic to backups. Services should be prioritized, and lower-priority ones have to be sacrificed in favour of the others. At the Member State level, adaptive infrastructure allows scaling down non-essential services during emergencies. Regulatory frameworks should enable controlled degradation. At the EU level, interoperable systems enable coordinated degradation of non-critical services. Automated resource sharing maintains essential services while scaling down others. For example, a major cyberattack should trigger bandwidth prioritization for emergency services (Anderson & Moore, 2022).

Reformation: At the entity level, organizations should follow agile engineering, like following the DevSecOps model to continuously study and fix identified design and implementation issues. In particular, some modules may have to be redesigned with some resilience principles. Effective applications can be after-market through remote updates. Member States and the EU should establish assessment methods and regulations to vet digital infrastructure and supporting resources, e.g. power, at both intra- and inter-State levels.

3.3 Operational Layer

Decentralization and Diversity: At the entity level, operational resilience requires distributing human and material resources across locations. Having multiple offices in different locations and outsourcing to diverse teams or third-party suppliers improves operational resilience. In addition, having some overlap in the skills of different teams or employees avoids talent shortage or absence. Member States should establish joint task forces and shared infrastructure for cross-border responses. For example, a coordinated cyberattack should activate a unified task force, as demonstrated during the 2007 cyberattacks on Estonia, which led to improved NATO cyber coordination (Smith & Jones, 2020). The EU should centralize real-time threat intelligence and coordinate joint rapid response teams. State and EU governments should manage migration and workforce usage to maintain talent availability despite global crises. The EU may encourage diversifying labours from within the EU as a priority.

Agility and Automation: At the entity level, proactive monitoring and automated threat response systems should adjust security measures instantly. This automation reduces the reliance on human analysts at the entity, Member State, and EU levels to address the global talent shortage. Cross-functional teams should rely on automated playbooks. Member States should use shared infrastructure and automation to enable rapid responses and encourage automated threat intelligence systems across state entities. The EU should enable automated real-time intelligence sharing across Member States. For example, a DDoS attack should trigger an automated and quick EU-wide coordinated response, as modelled in the Cyber Europe exercises conducted by ENISA (European Commission, 2022).

Degradation: At the entity level, an entity may need to reallocate human resources to defend higher critical services. An entity should use economic policies when operational resources (e.g. energy) are compromised. Member States should coordinate degradation responses and follow reallocation to maintain essential infrastructure. The EU should manage real-time degradation strategies using automated systems. For example, a DDoS attack should trigger protection for healthcare and energy while reducing bandwidth for non-essential services. The EU can coordinate reallocations and provisions across the Member States to load balance and absorb crises when they happen.

Reformation: At the entity level, dynamic response playbooks and automated recovery protocols should be updated, talents should be trained, teams should be rotated, and security policies should be revised regularly. Member States and the EU should improve collaboration and intelligence sharing across entities to update their sources, enforce software and patch updates, and coordinate the stopping of rogue software propagation. The EU should oversee cross-border improvements and help Member States to improve and apply reforms.

4 Implementation Agenda

Based on our proposed approach, we now outline an implementation agenda that the EU could adopt to revise the Cyber Resilience Act. Identification and Prioritization: Organize targeted workshops and surveys to identify key digital assets and services across all sectors and member states. Establish dedicated platforms for data collection and real-time tracking of asset status. Prioritize digital assets and services based on their strategic importance and cybersecurity risk levels. Use a structured framework to assess potential impact and vulnerability severity. Analysis and Data Collection: Analyse the cybersecurity and cyber resilience requirements for each asset or service at the entity level, considering supply chain, technical, and operational layers. Start by using established frameworks such as those conducted by the US Department of Homeland Security (DHS) and the Massachusetts Institute of Technology Research and Engineering (MITRE): the Cyber Resilience Engineering Framework (CREF) (Ross et al., 2021), the Cyber Resilience Review (CRR) (US Department of Homeland Security, 2014) self-assessment tool, and Crosswalks (US Department of Homeland Security, 2016), which align various cybersecurity standards and regulations. For example, the Cyber Resilience Review framework provides a high-level structure of the goals, objectives, and techniques required to implement cyber resilience. Such frameworks help organizations align security practices and controls across different standards and guidelines, ensuring consistency and improving overall resilience (National Institute of Standards and Technology, 2019). The techniques are exhaustively explained in the framework and mapped into the famous MITRE ATT&CK cybersecurity framework (MITRE Corporation, n.d.). Nevertheless, ATT&CK is only a good starting point, as it does not solve the zero-day vulnerabilities, and not all threats and techniques apply to all sectors. Collect and consolidate cybersecurity and resilience requirements at the Member State level. Conduct a structured analysis, considering supply chain dependencies, technological vulnerabilities, and operational gaps. At the EU level, aggregate the Member State-level data and conduct a comprehensive analysis, again considering supply chain, technical, and operational factors. Establish ENISA delegations in each Member State to facilitate coordination and information sharing. The first step involves creating a consolidated Strategic Asset Cyber Resilience EU Database (SACRED) to catalogue all identified assets and services, along with their cybersecurity and resilience requirements. It is essential that SACRED is regularly updated and made accessible to support strategic decision-making at the EU level. Next, research projects, workshops, and surveys should be launched to systematically analyse SACRED data in order to identify trends, gaps, and evolving threats, which will help improve resilience strategies. At the same time, frameworks such as CSR, CRR, and CREF need to be refined and adapted to better fit the EU's specific operational and technological environment, ensuring these frameworks remain flexible to accommodate future developments. Finally, the Cyber Resilience Act (CRA) must be amended to incorporate the refined framework, with clear guidelines established for implementation, enforcement, and compliance across entities, Member States, and the EU as a whole. Compliance should be monitored through periodic reviews, and the framework should be adjusted based on real-world incidents and technological advancements.

5 Conclusion

The EU Cyber Resilience Act (CRA) marks an important step in regulating cybersecurity for digital products across the Union. By mandating security-by-design, vulnerability handling, and post-market support, it aims to reduce systemic risks and increase consumer trust. However, this chapter has demonstrated that the Act's treatment of "resilience" is largely rhetorical rather than conceptual. Despite its promising title, the CRA lacks a clear definition of cyber resilience, omits references to established resilience frameworks, and focuses narrowly on pre-market compliance rather than systemic, adaptive, or mission-driven resilience. By mapping the CRA against resilience principles from NIST, MITRE, and EU critical infrastructure directives, we found that while the Act touches on aspects of resilience, it fails to address its core: the ability to anticipate, withstand, recover from, and adapt to cyber disruptions. To address this gap, future amendments should embed cyber resilience as a defined, operational concept. This includes aligning with life cycle-based resilience engineering, strengthening post-market recovery mechanisms, and emphasizing resilience by design.

References

Ahmed, M., Mahmood, A. N., & Hu, J. (2020). A survey of network anomaly detection techniques. *Journal of Network and Computer Applications*, *60*(1), 19–31. https://doi.org/10.1016/j.jnca.2015.11.016.

Alomari, H., Alsanad, A., & Omari, F. A. (2020). Sharding techniques in blockchain systems: A survey. *Journal of Information Security and Applications*, *54*(4), 102580. https://doi.org/10.1016/j.jisa.2020.102580.

Anderson, M., Taylor, R., & Patel, S. (2021). *Cybersecurity resilience: Strategies for defense*. Oxford University Press.

Anderson, R., & Moore, T. (2022). *Security engineering: A guide to building dependable distributed systems* (3rd ed.). Wiley.

Brown, K., Wilson, L., & Martin, D. (2022). *Resilient infrastructure design and response*. MIT Press.

Burri, M., & Zihlmann, Z. (2023). The EU Cyber Resilience Act – An appraisal and contextualization. *Zeitschrift für Europarecht (EuZ)*, 2, B1–B45. https://ssrn.com/abstract=4375552.

Chiara, P. G. (2025). Understanding the regulatory approach of the Cyber Resilience Act: Protection of fundamental rights in disguise? *European Journal of Risk Regulation*, 1–16. https://doi.org/10.1017/err.2025.9.

Demers, A., Greene, D., Hauser, C., Irish, W., Larson, J., Shenker, S., Sturgis, H., Swinehart, D., & Terry, D. (1987). Epidemic algorithms for replicated database maintenance. In *Proceedings of the Sixth Annual ACM Symposium on Principles of Distributed Computing* (pp. 1–12). https://doi.org/10.1145/41840.41841.

European Commission (2020). *EU cybersecurity strategy for the digital decade*. https://ec.europa.eu/digital-strategy/our-policies/cybersecurity-strategy-eu-digital-decade_en.

European Commission (2021). *Strategic autonomy and resilience in critical supply chains*. https://ec.europa.eu.

European Commission (2022). Cybersecurity: Commission proposes a new Cyber Resilience Act to protect Europe's digital products. https://ec.europa.eu/commission/presscorner/detail/en/ip_22_5374.

European Commission (2024). *Regulation (EU) 2024/2847 on horizontal cybersecurity requirements for products with digital elements (Cyber Resilience Act)*. https://eur-lex.europa.eu/eli/reg/2024/2847/oj.

European Parliament & Council of the European Union (2022). *Directive (EU) 2022/2557 on the resilience of critical entities*. https://eur-lex.europa.eu/legal-content/EN/TXT/?uri=CELEX%3A32022L2557.

Lin, J., Yu, W., Zhang, N., Yang, X., & Zhao, W. (2019). A survey on internet of things: Architecture, enabling technologies, security and privacy, and applications. *IEEE Internet of Things Journal*, *7*(5), 4312–4335. https://doi.org/10.1109/JIOT.2019.2957124.

Nakamoto, S. (2008). *Bitcoin: A peer-to-peer electronic cash system*. https://bitcoin.org/bitcoin.pdf.

National Institute of Standards and Technology (2019). *Developing cyber resilient systems: A systems security engineering approach* (NIST Special Publication 800-160 Volume 2). https://doi.org/10.6028/NIST.SP.800-160v2.

National Institute of Standards and Technology (2023). *The NIST Cybersecurity Framework (CSF) 2.0*. https://doi.org/10.6028/NIST.CSWP.29.

Pilkington, M. (2016). Blockchain technology: Principles and applications. In F. X. Olleros & M. Zhegu (Eds.), *Research handbook on digital transformations* (pp. 225–253). Edward Elgar Publishing. https://doi.org/10.4337/9781784717766.00019.

Plank, J. S. (1997). Erasure codes for storage applications. In *Proceedings of the Fourth IEEE International Symposium on High Performance Distributed Computing* (pp. 143–150). https://doi.org/10.1109/HPDC.1997.626459.

Popov, S. (2018). *The Tangle.* https://www.iota.org/research/academic-papers.

Ross, R., Pillitteri, V., Graubart, R., Bodeau, D., & McQuaid, R. (2021). *Developing cyber-resilient systems: A systems security engineering approach* (NIST SP 800-160 Vol. 2 Rev. 1). https://doi.org/10.6028/NIST.SP.800-160v2r1.

Shapiro, M., Preguiça, N., Baquero, C., & Zawirski, M. (2011). Conflict-free replicated data types. In *Proceedings of the 13th International Symposium on Stabilization, Safety, and Security of Distributed Systems* (pp. 386–400). Springer. https://doi.org/10.1007/978-3-642-24550-3_29.

Shi, W., Cao, J., Zhang, Q., Li, Y., & Xu, L. (2016). Edge computing: Vision and challenges. *IEEE Internet of Things Journal, 3*(5), 637–646. https://doi.org/10.1109/JIOT.2016.2579198.

Smith, J., & Jones, P. (2020). *European cybersecurity strategy and defense.* Routledge.

Vogels, W. (2008). Eventually consistent. *Communications of the ACM, 52*(1), 40–44. https://doi.org/10.1145/1435417.1435432.

W3C. (2021). *Decentralized identifiers (DIDs) v1.0.* https://www.w3.org/TR/did-core/.

Wu, J. (2024). Introduction to functional safety and cyber resilience. In *Cyber resilience system engineering empowered by endogenous security* (pp. 203–278). Springer. https://doi.org/10.1007/978-981-97-0116-2_4.

Zuech, R., Khoshgoftaar, T. M., & Wald, R. (2015). Intrusion detection and big heterogeneous data: A survey. *Journal of Big Data, 2*(3). https://doi.org/10.1186/s40537-015-0013-4.

CHAPTER 13

National Priorities and Transnational Security: An Analysis of the Polish Presidency of the Council of the EU through Two-Level Games

José Carlos Hernández-Gutiérrez and Guillermo López-Rodrígsuez

Abstract Twenty years after joining the European Union (EU) and thirteen years after its first presidency of the EU Council, Poland once again assumed the rotating presidency on January 1, presenting an ambitious program focused on security and defense under the slogan "Security, Europe!" Given that organized crime represents one of the main threats to the EU, this chapter examines whether Poland adequately addresses this challenge in its leadership of the EU Council presidency. The chapter has two main objectives: to provide an overview of criminal organizations in the EU and to analyze whether Poland's priorities during its presidency effectively respond to this issue. To achieve this, it describes the main characteristics and activities of the criminal organizations operating in the region. Subsequently, it examines the strategic priorities of the Polish EU Council presidency. While Poland places significant emphasis on security, its internal political context and its predominant focus on border security raise doubts about its ability to comprehensively address European needs in the fight against organized crime.

Keywords Organized crimen – Poland – EU Presidency – Security – Border Policiy

1 Introduction

Thirteen years after holding its first presidency of the Council of the European Union (EU), Poland has once again assumed this role for a six-month term that began on January 1, 2025. Under the slogan "Security, Europe!", the Polish presidency has introduced a program, presented on December 10, 2024, focusing on seven interrelated dimensions of security: external, internal, informational, economic, energy, food, and health. Aimed at ensuring the internal security of Europeans, Poland's approach emphasizes strengthening the pro-

 HTTPS://DOI.ORG/10.1163/9789004763623-015

tection of external borders and developing innovative solutions to address migration, including the establishment of safe spaces in third countries. In this context, it seeks to provide an effective response to hybrid threats, such as the manipulation of migration flows – a challenge that Poland has directly faced (Balázs, 2025).

Addressing threats to European security also requires an analysis of the impact of organized crime. Over recent decades, economic globalization has brought significant social and economic progress but has also introduced substantial security challenges. In this context, geopolitical conflicts in Europe and its neighboring regions have contributed to processes of state weakening and institutional fragility. In several cases, these dynamics have created environments conducive to the expansion of organized criminal networks, particularly in parts of the Western Balkans. These conditions have facilitated the growth of illicit activities such as human trafficking, exacerbated by large migration flows from regions including the Balkans, the Caucasus, the Maghreb, sub-Saharan Africa, the Middle East, Afghanistan, and, more recently, Ukraine (Gandilhon, 2023). Consequently, organized crime, in its various forms, poses not only a global challenge but also a critical threat to the security of the EU (Anguita, 2023).

Despite the transnational nature of organized crime, responses to this challenge continue to be shaped by national threat perceptions and domestic policy priorities. This tension becomes particularly visible during the rotating presidencies of the Council of the EU. Although, in formal terms, the rotating Presidency is expected to act as an honest broker (Priebus, 2025), whose primary objective is to foster consensus among Member States while setting aside its own national interests, domestic political developments nonetheless influence how the country holding the Presidency approaches specific policy issues (Kaczyński, 2025). In the case of Poland, its Presidency coincided with a domestic political context characterized by institutional fragmentation and the holding of presidential elections.

Against this backdrop, this chapter examines how national political circumstances affect the exercise of the Presidency of the Council of the EU in the field of internal security. Its main objective is to assess whether, and in what ways, Poland's domestic political context shapes its agenda-setting role and policy priorities with regard to the fight against organized crime during its Presidency of the Council of the EU. To achieve this objective, the chapter addresses the following research question: To what extent does Poland's national political context influence its approach to organized crime during its Presidency of the Council of the EU, and how does this affect the comprehensiveness of the EU's response to this transnational security threat?

The chapter contributes to ongoing debates on the influence of domestic political dynamics on the shaping of EU-level action (Kassim *et al.*, 2019). Moreover, the Polish Presidency serves as an illustrative case of how internal constraints condition the ability of Member States to act as honest brokers within the institutional framework of the EU.

The remainder of the chapter is structured as follows. The next section develops the theoretical framework, drawing on the concepts of agenda-setting, two-level games, and the politicization of security. This is followed by a detailed account of the methodological strategy. The subsequent section provides empirical data on the main activities carried out by organized criminal groups operating within the EU. The analysis then turns to how domestic political dynamics shape Poland's priorities during its Presidency. Finally, the chapter concludes by highlighting the implications of the findings for EU governance.

2 Theoretical Framework

The rotating Presidency of the Council of the EU possesses significant capacities to shape EU policy, particularly through its influence over agenda-setting. These capacities materialize through different modes of agenda-shaping.

First, agenda-structuring refers to the ability to emphasize or downplay issues that are already present on the Council's agenda (Tallberg, 2003). This form of control is exercised through the organization of informal meetings, the determination of the frequency and timing of formal meetings, and the configuration of the Council's calendar (Vaznonytė, 2022). Through these procedural mechanisms, the Presidency can accelerate or slow down negotiations in line with its political priorities, thereby promoting certain dossiers while sidelining others (Warntjen, 2013). Second, agenda exclusion denotes the capacity to prevent specific issues from being discussed. Such exclusion may be achieved either by refraining from taking a position on politically sensitive matters or by putting forward proposals that are unacceptable to other Member States, thus effectively blocking any substantive progress (Vaznonytė, 2022).

To explain how domestic political contexts condition the exercise of these capacities, this chapter draws on the theory of two-level games (Putnam, 1988). From this perspective, governments negotiate simultaneously at two interdependent levels: the supranational and the domestic. Agreements reached at the European level must be compatible with domestic political constraints, including electoral incentives, institutional configurations, the presence of veto players, and public opinion preferences.

The rotating Presidency constitutes a paradigmatic case of a two-level game. While all Member States operate under domestic constraints in EU negotiations, the Presidency faces additional pressures stemming from its heightened political visibility and its responsibility for agenda-setting (Tallberg, 2003). Decisions taken during the presidential term are therefore more exposed to scrutiny by national media, opposition parties, and the broader public, which increases the political costs of advancing European initiatives that run counter to national priorities.

This dynamic is further intensified in highly politicized policy areas. The literature has emphasized that issues such as migration, border control, and internal security have moved to the center of national political competition. In this context, security policy – and, in particular, the fight against organized crime – occupies a hybrid position between European cooperation and state sovereignty. Although transnational criminal networks require coordinated responses at the EU level, security policies remain closely tied to national authority. As a result, Member States frequently engage in forms of selective Europeanization, supporting EU action in those dimensions that align with their national threat perceptions, while downplaying or redefining others (Börzel & Risse, 2012).

For Council Presidencies, this implies that the definition of "European security" is the outcome of a political construction rather than a neutral process. Depending on their domestic incentives, Presidencies may emphasize certain aspects – such as border control or migration-related crime – while marginalizing other manifestations of organized crime, thereby selectively shaping the Union's security agenda.

3 Methodology

This chapter adopts a qualitative case study research design. The primary methodological approach consists of a systematic review and critical analysis of a documentary corpus comprising three categories of sources: official documents issued by the Polish Presidency, EU institutional sources, and specialized academic literature on organized crime and EU policies. The analytical technique employed combines source triangulation with content analysis. This procedure entails a critical and comparative reading of the documents, with the aim of identifying convergences, divergences, and significant omissions. The main limitation of this methodological approach lies in its reliance on publicly available sources, which may not fully capture informal political negotiation processes, as well as in the exploratory nature of a single-case study design.

4 The Landscape of Organized Crime: Identifying Strategic Threats to EU Internal Security

4.1 Organizations

The EU understands organized crime according to the terms adopted by the UN Palermo Convention about transnational crime. In this sense, any group composed of three or more people that acts in coordination to commit severe infractions to obtain economic profits is considered a criminal organization (Gandilhon, 2023). According to Europol, there are about 821 criminal networks formed by 25,000 members in the EU (Euronews, 2024). While these networks are often analyzed through their operational capacity, their strategic threat lies in their ability to exploit institutional vulnerabilities within and beyond EU borders.

The internationalization of these networks is also reflected in their composition, whose members belong to 112 nationalities, being 68% composed of multinational members and only 32% by national ones. This transnationality challenges the traditional border-centric approach to security. Europol reported that 82% of the most dangerous gangs are led by people located in the main target country or in the country of origin of key members (Europol, 2024a). Regarding their activities, only 18% of criminal groups are involved in multi-criminal markets, and 82% are involved in just a single activity. Drug trafficking is the most relevant sector, as 50% of organizations are profiting from this market and 36% are only involved in narcotic-related activities (Europol, 2024a). Consequently, as the most pervasive threat, drug trafficking represents a benchmark to measure the exhaustiveness of any European security agenda.

Regarding operational methods, 71% use corruption of private and public sectors to improve their activities and avoid state actions (Europol, 2024a). By bribing public officials, criminal gangs are able to access sensitive information and obstruct police and judiciary processes. The systemic nature of this corruption suggests that organized crime is not merely a border issue, but an internal structural threat to democratic institutions.

4.2 Activities

The main criminal activities in Europe include drug trafficking, cyber-criminality, human and migrant trafficking, and fraud related to special taxes (Council of the EU and the European Council, 2024). While these activities represent the objective pillars of criminal profit in the EU, their relative importance in the European political agenda is often subject to the prioritization strategies of rotating Presidencies. In this sense, the main threat to EU security is drug trafficking, as it is the most extended and violent activity (Anguita, 2023).

4.3 Drug Trafficking

The availability of illegal drugs in the EU has increased in recent years, with the increase of cocaine coming from South America (Anguita, 2023) being especially relevant. The infiltration of criminal organizations into major European commercial ports has turned internal logistics hubs into primary security vulnerabilities (Europol, 2024a). Violence and corruption are not limited anymore to regions traditionally associated with organized crime; nowadays, it affects significantly consolidated democracies with strong economies, such as Sweden, France, or the Netherlands (Gandilhon, 2023).

A critical development is the relocation of cocaine production from Latin America to Europe. Criminal networks now export intermediate substances such as coca paste and cocaine base to perform final processes in Europe (Den Held, 2024). This shift requires a transition from border-control policies to deep-rooted internal security cooperation. This displacement of production implies higher sophistication and adaptability of criminal networks, being able to avoid international controls. This fact increases the proliferation of clandestine laboratories in Europe, involving not only challenges to public security but also environmental risks associated with chemicals used for cocaine production (Talking Drugs, 2024). As will be analyzed in Section 13.5, the complexity of this internal production often receives less political attention than more visible threats like irregular migration.

4.4 Cyber-Criminality

The digitalization of services and infrastructures has implied relevant risks for EU security. The dependence on digital networks requires a level of technical cooperation that often transcends national political cycles (Carrapico & Farrand, 2024). The fight against cyber criminality is a priority for police forces and judicial authorities in the EU. The intensive use of communication technologies and connectivity provided by the internet posed several security challenges, with online child exploitation and cyber fraud as the most increasing activities. Technological developments have allowed criminals to expand their networks and amplify their impact (Velasco *et al.*, 2024).

Cyber-criminality not only represents an increasing threat but also has a significantly high scope. The consequences of criminal activities in the cyber domain involve direct effects on individuals, such as economic loss, privacy vulnerabilities, or psychological traumas, as well as structural effects in public and private institutions. The sophistication of digital threats implies risks to the economy and collective security in the EU, which requires the adoption of integral and coordinated measures to reduce their effects (Anguita, 2023). Regarding the role of cybercrime, it can also be used as a proxy by foreign powers to conduct destabilization operations against EU members' critical infrastructures (López-Rodríguez *et al.*, 2024). Securing these critical infrastructures against cyber-proxies is not only a crime-fighting measure but a cornerstone of European technological sovereignty (Barrinha & Christou, 2022).

4.5 Human Trafficking

Human trafficking does not necessarily imply crossing borders, having different exploitation purposes, mainly sexual or working objectives (Eurostat, 2024). The EU defines human trafficking in Article 2, Directive 2011/36/EU as a severe violation of human rights, considering it as a crime against the person with the main purpose of exploiting the victim. The 2022 statistics published by Eurostat (2024) registered 10,093 victims of human trafficking, an increment of 41% compared to the previous year. This surge highlights the vulnerability of displaced populations, such as Ukrainian refugees, and requires a coordinated European social and judicial response.

As in previous years, 63% of victims were women and children. Regarding nationality, 37% of registered victims were EU citizens, which presents as reduction regarding 2021. In addition to this fact, 25% of victims within the EU are citizens of the country in which they were registered, representing a reduction of 19% regarding 2021 (Eurostat, 2024). Furthermore, it is essential to interpret these phenomena in terms of tech-facilitated trafficking. Criminal networks increasingly utilize encrypted platforms and digital tools for the recruitment and exploitation of victims, shifting the locus of the crime from physical public spaces to the digital sphere (Moyo *et al.*, 2025). This technological empowerment of criminal groups requires a shift in the EU response toward digital forensics and cyber-surveillance. Regarding the suspects, the number was reduced to 8,064, showing a gap between victim identification and successful prosecution (Directorate-General for Migration and Home Affairs, 2024).

4.6 Migrant Smuggling

Unlike human trafficking, migrant smuggling is primarily a transnational business based on the unauthorized crossing of borders. In contrast with human trafficking, migrant smuggling is based on personal agreements and has transnational purposes, while human trafficking can happen within national borders (Global Migration Data Portal, 2024). Migrant smuggling provides relevant profits to criminal networks, using several land, air, and sea routes to promote irregular migration to and within the EU, implying several violations of human rights.

Irregular migration is related to the connection of trafficking with terrorism, human trafficking, and money laundering, which increases its relevance for EU security (European Commission, 2024). The reports estimate that most migrants have used criminal networks to arrive in Europe. Near 292,985 migrants arrived in 2023 to Bulgaria, Cyprus, Greece, Italy, Malta, and Spain through the Mediterranean and Atlantic routes (OIN, 2024). The high profitability of these routes (estimated between 320 and 550 million dollars in 2016) makes this activity a highly visible target for political securitization, as it directly involves the integrity of external borders.

4.7 Special Taxes Frauds

Criminal organizations benefit from special tax evasion, which allows them to obtain substantial profits when selling legal products and falsifications at lower prices than their legal equivalents. Fiscal fraud on special taxes is promoted by legislative differences and variable taxes in different EU members. The main fiscal fraud in Europe is related to the smuggling or illegal importation of products subjected to special taxes, illegal manufacturing of products, and distribution of products without paying the tax (Europol, 2024b). This type of fraud represents a direct attack on the fiscal sovereignty of Member States. However, it remains a less visible crime compared to migration or drug-related violence.

The landscape described above reveals a multidimensional threat where financial crimes and drug trafficking constitute the structural backbone of organized crime in the EU. However, the translation of these objective threats into European policy is not automatic; it is mediated by the political incentives and domestic constraints of the Member State holding the Council Presidency. As the next section will explore, this gap between objective data and political agenda-setting is clearly manifested in the case of the Polish Presidency.

5 Polish Presidency of the Council: An Analysis of Two-Level Games and Selective Europeanization

Poland's assumption of the rotating presidency of the Council of the EU in 2025 represented the enactment of a two-level game (Putnam, 1988), enabling it to set the EU agenda while orienting the organization toward its national priorities (Ponczek, 2024). These priorities were strongly shaped by the current geopolitical competition in the international arena, defined by the war in Ukraine, US foreign policy, the migration crisis, and other common security concerns.

Under the motto "Security, Europe!", Poland influenced the structuring of the agenda (agenda-structuring) (Tallberg, 2003). By prioritizing dimensions such as migrant trafficking and border security, the Polish presidency operated under the logic of security politicization. This prioritization responded to Prime Minister Donald Tusk's need to construct a European security definition compatible with Poland's interests. Whereas the previous section identified drug trafficking and tax fraud as some of the main threats to the EU, Poland chose to structure the agenda around migrant trafficking as one of the most pressing challenges (Ministry of Justice, Republic of Poland, 2024). This constitutes a case of selective Europeanization (Börzel & Risse, 2012), as Poland supported EU action and the strengthening of Europol primarily in areas that mitigated its perceived national-level threats (the hybrid threat from Belarus and Russia), while practicing agenda-exclusion in other areas that could generate domestic costs. This selective approach reduces the comprehensiveness of the EU's response, as attention is diverted toward the eastern border, neglecting structural vulnerabilities in other European criminal nodes – such as illicit drug markets or cybercrime described in the previous section – thus fragmenting the cohesion of the common security strategy.

Furthermore, Poland's capacity to act as an honest broker (Priebus, 2025) at the head of the EU Council was significantly eroded by domestic developments. The cohabitation between Prime Minister Donald Tusk and President Andrzej Duda, coupled with the May 2025 presidential elections, which were ultimately won by Karol Nawrocki of the Law and Justice Party (PiS), decisively influenced the role of the Polish Council presidency. The electoral campaign had a significant impact on both Poland's internal political dynamics and the formation of the EU agenda. Donald Tusk faced the challenge of securing another electoral victory to continue effectively implementing his political program, in opposition to PiS (Oliveras, 2025).

The need to circumvent legislative blockage by then-President Duda and competition from PiS compelled Tusk to adopt defensive positions. This translated into an agenda-exclusion strategy in areas likely to produce negative domestic political or economic consequences. This is particularly evident in climate and environmental policies, such as the reform of the Common Agricultural Policy, biodiversity protection, soil restoration, and ambitious CO_2 reduction and economic decarbonization targets (Oliveras, 2025). The political cost of accepting these regulations could be prohibitive at the domestic level, leading Poland to slow down or block these files at the supranational level.

Moreover, to avoid harming foreign relations with the USA, Tusk rejected prioritizing the EU military industry, which conflicted with the positions of Germany, the Netherlands, Austria, Sweden, and Finland, who seek greater defense investments within the EU (Oliveras, 2025). Far from being merely a diplomatic matter, this may have been a maneuver to ensure that the Polish electorate perceived the government as the guarantor of the transatlantic alliance, essential for national security perceptions.

The scenario described illustrates that the complex dynamics of Polish domestic politics significantly influenced the course of the country's Council presidency while simultaneously shaping relations with other Member States. Ultimately, the Polish case reveals that domestic stability and objectives (Level I) were prioritized over the coherence and comprehensiveness of the supranational sphere (Level II).

6 Conclusions

The analysis developed in this chapter allows us to conclude that Poland's presidency of the Council of the EU in 2025 constitutes a paradigmatic case of how domestic dynamics condition supranational governance. It becomes evident that the exercise of the rotating presidency does not operate neutrally; rather, it is embedded in a complex two-level game in which decisions taken in Brussels are ultimately extensions of political competition in Warsaw. The evidence presented demonstrates that Poland's capacity to act as an honest broker was eroded by the need for Donald Tusk's government to navigate a fragmented institutional environment alongside an electoral campaign during the Polish Council presidency. This domestic pressure narrowed the scope of political maneuvering, prioritizing initiatives with lower internal electoral costs while postponing those that, despite their significance, lacked immediate political payoff.

The divergence observed between the objective landscape of organized crime in the European Union and the priorities established under the motto "Security, Europe!" confirms that security is not a static concept but a political construction subject to processes of selective Europeanization. While empirical data indicate that drug trafficking and tax fraud represent the most widespread and costly threats across the Union, the Polish presidency prioritized migration flows and the strengthening of the eastern border. This choice functioned as a mechanism of internal legitimization vis-à-vis domestic opposition. The Polish government leveraged its agenda-structuring capacity to align EU resources with its own national defense objectives.

Finally, this behavior has significant implications for security governance within the EU. The fact that the state, assuming the rotating presidency, can filter the security agenda according to its domestic political cycles poses a challenge to the comprehensiveness and coherence of European collective action. By prioritizing specific dimensions of security aligned with national interests, there is the risk of generating blind spots in the collective response, displacing criminal pressure to other regions or less politically visible types of crime, ultimately weakening the Union's strategic resilience against transnational organized crime.

References

Anguita Osuna, J. E. (2023). La lucha de la Unión Europea contra la delincuencia organizada: la nueva hoja de ruta de 2023. *Anuario de la Facultad de Derecho. Universidad de Extremadura*, *39*, 69–89. http://hdl.handle.net/10662/21518.

Balázs, E. (2025, January 7). Priority dossiers under the Polish EU Council Presidency. *European Parliament*. https://www.europarl.europa.eu/thinktank/es/document/EPRS_BRI(2025)767187.

Barrinha, A., & Christou, G. (2022). Speaking sovereignty: the EU in the cyber domain. *European Security*, *31*(3), 356–376. https://doi.org/10.1080/09662839.2022.2102895.

Börzel, T. A., & Risse, T. (2012). From Europeanisation to Diffusion: Introduction. *West European Politics, 35*(1), 1–19. https://doi.org/10.1080/01402382.2012.631310.

Buras, P. (2025, January 7). New protagonists: Why the EU's future will be decided in central Europe. European Council on Foreign Relations. https://ecfr.eu/article/new-protagonists-why-the-eus-future-will-be-decided-in-central-europe/.

Carrapico, H., & Farrand, B. (2024). Cybersecurity Trends in the European Union: Regulatory Mercantilism and the Digitalisation of Geopolitics. *JCMS: Journal of Common Market Studies, 62*(S1), 147–158. https://doi.org/10.1111/jcms.13654.

Council of the EU and the European Council (2024, July 3). *The EU's fight against organised crime.* https://www.consilium.europa.eu/en/policies/eu-fight-against-crime/.

Den Held, D. (2024, June 17). Drug Traffickers Producing More Cocaine in Europe: EU Report. InSight Crime. https://insightcrime.org/news/drug-traffickers-producing-more-cocaine-in-europe-eu-report/.

Directorate-General for Migration and Home Affairs (2024, February 28). Newly released data show an increase of trafficking in human beings. *European Commis sion.* https://home-affairs.ec.europa.eu/news/newly-released-data-show-increase-trafficking-human-beings-2024-02-28_en.

Euronews (2024, April 5). Europol identifica 821 redes criminales amenazadoras con más de 25.000 miembros en la UE. *Euronews.* https://normas-apa.org/referen cias/citar-periodicos/.

European Commission (2024). *Migrant smuggling.* https://home-affairs.ec.europa.eu/policies/migration-and-asylum/irregular-migration-and-return/migrant-smug gling_en.

Europol (2024a). *Decoding the EU's most threatening criminal networks.* https://www.europol.europa.eu/publication-events/main-reports/decoding-eus-most-threaten ing-criminal-networks#downloads.

Europol (2024b). *Excise fraud.* https://www.europol.europa.eu/crime-areas/economic-crime/excise-fraud#:~:text=Alcohol%2C%20cigarettes%20and%20fuel%20are,pri ces%20than%20their%20licit%20equivalents.

Eurostat (2024). *Trafficking in human beings statistics.* https://ec.europa.eu/eurostat/statistics-explained/index.php?title=Trafficking_in_human_beings_statistics.

Gandilhon, M. (2023). Le crime organisé dans l'Union européenne: une menace stratégique? *Politique étrangère, Été* (2), 163–174. https://doi.org/10.3917/pe.232.0163.

Global Migration Data Portal (2024). *Smuggling of migrants.* https://www.migration dataportal.org/themes/smuggling-migrants.

Kaczyński, P. M. (2025, January 27). What to expect from Poland's EU Council presidency in 2025. Elcano Royal Institute. https://www.realinstitutoelcano.org/en/commentaries/what-to-expect-from-polands-eu-council-presidency-in-2025/.

Kassim, H., Saurugger, S., & Puetter, U. (2019). The Study of National Preference Formation in Times of the Euro Crisis and Beyond. *Political Studies Review, 18*(4), 463–474. https://doi.org/10.1177/1478929919873262.

López-Rodríguez, G., Moreno-López, I., & Hernández-Gutiérrez, J. C. (2023). Cyberwarfare against critical infrastructures: Russia and Iran in the gray zone. *Applied Cybersecurity & Internet Governance*, 2. https://doi.org/10.60097/ACIG/162865.

Ministry of Justice Republic of Poland (2024). Priorities of the Polish Presidency of the Council of the European Union in the Justice Dimension. https://www.gov.pl/web/justice/priorities-of-the-polish-presidency-of-the-council-of-the-european-union-in-the-justice-dimension.

Moyo, T. J., Gunes, O., & Jirotka, M. D. (2025). Investigating Human Trafficking Recruitment Online: A Study of Fraudulent Job Offers on Social Media Platforms. *Proceedings of the ACM on Human-Computer Interaction*, 9(2), 1–31. https://doi.org/10.1145/3711016.

Oliveras, E. (2025, January 3). Las presidenciales polacas coartan la UE. *El Correo Gallego*. https://www.elcorreogallego.es/opinion/2025/01/03/presidenciales-polacas-coartan-ue-113094699.html.

Ponczek, C. B. (2024). *Polish priorities for the presidency of the Council of the European Union*. Trans European Policy Studies Association (TEPSA). https://tepsa.eu/analysis/polish-priorities-for-the-eu-presidency/.

Priebus, S. (2025). The 2024 Belgian and Hungarian Presidencies of the Council of the EU in Comparison: From an Honest Broker to a Dogmatic Disruptor. *Journal of Common Market Studies*, 63(S1), 105–119. https://doi.org/10.1111/jcms.70014.

Putnam, R. D. (1988). Diplomacy and Domestic Politics: The Logic of Two-Level Games. *International Organization*, 42(3), 427–460. http://www.jstor.org/stable/2706785.

TalkingDrugs (2024, January 31). Shifting Geopolitics: How Cocaine Production is Moving into Europe. https://www.talkingdrugs.org/shifting-geopolitics-how-cocaine-production-is-moving-into-europe/.

Tallberg, J. (2003). The agenda-shaping powers of the EU Council Presidency. *Journal of European Public Policy, 10*(1), 1–19. https://ssrn.com/abstract=2180543.

Velasco, C., Garcia Periche, J., Gómez Gómez, J. d. D. & Bueno Benedí, M. (2024). *Inteligencia artificial y crimen organizado*. Programa EL PACCTO 2.0. EU-LAC Partnership on justice and security.

Vaznonytė, A. (2022). *Entrepreneurs of Compromise? The Rotating Presidency of the Council of the EU after Lisbon*. (SIEPS Report No. 2022:3epa). Swedish Institute for European Policy Studies. https://sieps.se/media/oqcgpoza/2022_3epa.pdf.

Warntjen, A. (2013). The elusive goal of continuity? Legislative decision-making and the Council Presidency before and after Lisbon. *West European Politics*, *36*(6), 1239–1255. https://doi.org/10.1080/01402382.2013.826028.

CHAPTER 14

Frontex: From Border Control to People Control

Antonia-Maria Sarantaki

Abstract The European Border and Coast Guard Agency, commonly known as Frontex, is the EU border control agency. It has been created to manage the EU external borders and enhance border security. Yet, Frontex's border control status quo is shifting. The agency seems to be moving its focus from border control to people control. This move from its original mandate brings a shift in Frontex's function that fundamentally shapes the agency's course from border management to a type of policing. This is evidenced through Frontex's new activities for border control and especially debriefing, ETIAS, and lie-detection technology. On that account, we are witnessing Frontex's transformation from a border control agency to a policing agent. The chapter argues that by performing policing, and particularly intelligence-led policing, Frontex has gradually expanded its institutional role that had been set by its creators in its original mission. This transformation, though, raises new accountability considerations, as Frontex is not mandated to control people.

Keywords Frontex – intelligence-led policing – border control – EU governance – EU security – EU agency

1 Introduction

It has become common to link migration to security, and alongside Frontex to security, given that Frontex is the European Union (EU) border control agency tasked to "address migratory challenges and potential future threats at the external borders" (Regulation, 2019). On that account, Frontex's border control activities are often viewed as security practices that contribute to the securitisation of EU borders and migration (Léonard, 2010; Léonard & Kaunert, 2022). But Frontex does not just securitise borders. Neither does it solely focus on border control. Frontex also performs policing, moving its attention to people control.

Frontex, legally, functionally, and institutionally, is not a police force, and neither was it created to perform policing (Regulation, 2004). It partners with

HTTPS://DOI.ORG/10.1163/9789004763623-016

national police organisations and engages in policing activities. That constitutes a deviation from Frontex's founding mandate.

Inspired by critical policing analysis (Maguire, 2000; Bigo & Guild, 2005; Yarwood, 2007), this chapter investigates Frontex's policing role. In light of this, it asks whether Frontex has acquired policing tasks and what it means for Frontex to police. The concept of policing was initially used by police studies to technically evaluate police work (Fyfe, 1995). Yet, in the threat-centric security landscape of the twenty-first century, the functions of police expanded. New forms of control and actors to tackle criminality emerged, renewing the discussion on policing. In this vein, policing started to be employed conceptually to explore the evolving nature of police, including the challenges of police's expansion (Anderson & den Boer, 1994; Benyon, 1994; Loader, 1997). Critical policing analysis points to a critical reflection of the expansion of police functions and powers to ensure social order.

Under this prism, the chapter explores Frontex's shift from a border control agency operating at the EU external borders to a policing agent, controlling people away from the physical borders. Its aim is to reveal what changes have occurred in the agency's function. It argues that Frontex has expanded its operational role and now performs intelligence-led policing (ILP), which does not focus on borders but on people. This indicates a significant change in the agency's function as well as a drift from Frontex's original orientation and legal basis, which raises additional accountability considerations, given that Frontex is not mandated to control people.

This study is based on both secondary and primary data. Secondary data includes scholarly articles, academic books, and policy papers from think tanks. Primary sources comprise EU legislation, Frontex reports, and data drawn from Frontex's Press Team's reply to my written questions (*received on 11 June 2025*).

The chapter starts with a discussion on EU security and borders. The next sections explore the agency's substantial development and its shift from border control to policing. Drawing on intelligence-led policing, the chapter then looks at whether ILP characteristics can be traced in Frontex's function. An ensuing case-study analysis, using mostly qualitative data, is put forward to examine three cases: debriefing, the European Travel Information and Authorisation System (ETIAS), and lie detection technology. These cases have been selected because they constitute new tasks for Frontex. In spite of not being contained in its founding Regulation (2004), these three activities are now part of Frontex's border control repertoire, portrayed by Frontex as effective, wide-ranging, and cutting-edge tools for border control. Besides, these three cases have not been included in previous analyses of Frontex, so their investigation here can reveal new elements about Frontex's role. The chapter concludes by highlighting Frontex's turn towards people control and the implications of policing.

2 EU Security and Borders

Being a peace project in the aftermath of World War II, security has always been at the EU's core, and therefore a driving force for European integration. Increasing the degree of cooperation among its member states and having NATO (North Atlantic Treaty Organisation) as a military guarantor, the EU experienced perpetual peace, without the fear of war in its territory. Even though, since 2022, this fear returned with a vengeance due to the Russo–Ukrainian conflict.

Enjoying long-term stability and peace, the EU gradually grew, both in width and depth – from 6 to 27 member states and from the technical collaboration of the early European Coal and Steel Community to the European Union's acquis communautaire and citizenship. The enhanced interdependence of the EU project, despite promoting peace, is a clear antithesis; it also instigated new security preoccupations. A blatant example illustrating this point is the borders and their control.

The abolition of internal border controls, set in with the 1985 Schengen Agreement, enabled the establishment of an internal market with free flow of people, goods, services, and money. Free circulation within the EU was accompanied by the progressive tightening of external borders in an attempt to protect those living within the EU "walls" from threats posed by outsiders. As clearly put, "the abolition of internal border controls cannot come at the expense of security" (European Commission, 2025). That preoccupation became more pronounced in 2004 with the accession of ten new states to the EU, as it enlarged the EU external border, rendering therefore its protection more challenging, especially for frontline member states.

In parallel, the altered security landscape brought in with the Islamist terrorist attacks of 9/11 in the USA (2001) and, later, on European soil, in Madrid (2004) and London (2005), diverted attention to pre-emptive security. In that spirit, to counter potential threats and to enjoy freedom, the EU border needed to be secured, as reflected in the 2003 European Security Strategy and 2010 Internal Security Strategy. The security model promoted with these documents included a comprehensive approach to security, involving synergies between different EU policies and actors, and an internal/external blurring. Further deterioration of Europe's security environment, with turmoil in

North Africa and the Middle East, led to the so-called 2015 "migration crisis", with over a million people entering the EU irregularly seeking refuge. To manage the crisis and to preserve Schengen, the EU's security became redrawn, with internal borders being re-introduced and external borders being reinforced. The linkage between strong external borders and security is also flagged in the 2020 EU Security Union Strategy, which reproduces the internal/external security nexus and emphasises threat and risk assessment, including the collection of travellers' information prior to their departure and entry into the EU.

The central part of the EU's efforts to protect the EU borders and address security challenges, like irregular migration, while maintaining the freedom of EU intra-mobility, is Frontex. The EU and its member states agreed to set up this border control agency to increase border security and assist EU member states with border control. Over time, Frontex has become an actor in border control (Sarantaki, 2023), with an ever-expanding mandate and competences, and resultantly with a constantly increasing engagement in EU security.

3 Frontex's Expansion

Frontex was established in 2004, having as its purpose to "facilitate the application of existing and future Community measures relating to the management of external borders" (Regulation, 2004). Today, after repeated mandate enhancements, Frontex has evolved into the European Border and Coast Guard Agency, acquiring a wider involvement in EU border control, which transcends external border management.

During the first years of its operation, in implementing the tasks listed in its founding mandate, Frontex showed a clear emphasis on borders and border control, coordinating chiefly joint operations at the EU external air, land, and sea borders. To foster its role at the borders, Frontex's first mandate revision included the formation of Rapid Border Intervention Teams, namely an emergency response mechanism for short-term deployment of border guards at the EU external borders in case of mass migratory pressures (Regulation, 2007).

After its initial organisational phase, Frontex's mandate became both strengthened and expanded with new tasks, such as the development and coordination of the European Border Surveillance System (Eurosur). In 2015, Frontex was involved in "hotspots", namely in border areas facing emergencies, providing expanded migratory assistance. Moreover, it started deploying liaison officers to third countries, enhancing its intelligence function; it also commenced conducting vulnerability assessments for possible security risks at the EU borders (Regulation, 2016a). Next, the agency formed its own uniformed standing corps and acquired a wider involvement in return operations (Regulation, 2019).

In parallel, the agency witnessed a significant expansion in terms of staff, budget, equipment, and resources – from less than 50 staff members in 2005 to 2,500 persons today and from a budget of around €6 million in 2005 to more than €990 million in 2025. What is puzzling though is that this remarkable agency expansion is not accompanied by an equivalent numerical increase in Frontex's joint operations. In 2007, just 2 years after its operational launch, Frontex coordinated 27 joint operations at the EU external borders (Frontex, 2007, p. 17). This number later dropped, amounting, for example, to 16 joint operations in 2019 and 23 joint operations in 2024.

The numerical decrease in its joint operations shows a shift of Frontex's focus from border management towards other activities, not necessarily conducted at the location of the borders, such as risk analysis and proactive planning. Indeed, Frontex started to engage more actively with crime prevention and profiling. In doing so, it deepened its cooperation with Interpol and Europol, which constitute police organisations. The above attest to Frontex's move from border control to people control, that is, an emphasis not on borders but on people.

4 From Border Control to Policing

Border control is "the activity carried out at a border [...] consisting of border checks and border surveillance" (Regulation, 2016b). It includes document checks at concrete border crossing points and surveillance between border crossing points for the prevention of unauthorised entries. In various European countries, like Austria, Belgium, Luxembourg, Norway, and Denmark, border crossing is being managed and controlled by police officers. In others though, such as in Poland, Finland, Iceland, Romania, and Latvia, border control is being conducted by border or coast guards, namely by a distinct border guard department.

Since Frontex's creation, border control can also be carried out by Frontex, namely by Frontex staff deployed at the border during a joint operation. However, Frontex's activities show that its role is not limited to border control. Put differently, Frontex does not carry out solely border checks and border surveillance; it also performs policing activities, particularly ILP.

Policing refers to the "process of preventing and detecting crime and maintaining order" (Mawby, 2008, p. 17). The difference between border control and policing is apparent. Border control emphasises borders, whereas policing emphasises crimes and people, or, put differently, suspected criminals. Even though the police are involved in migration control, it does not inherently mean that border guards have police powers or that they can perform policing, just because they control borders – especially Frontex, which is not a police authority and does not have the power to enforce migration law.

Frontex's link to policing has marginally attracted literature's attention. This literature is mostly focused on humanitarianism, discussing, for instance, how Frontex tries to justify its policing using humanitarian terms (Aas & Gundhus, 2014; Pallister-Wilkins, 2015); how the agency, whilst policing mobility, is being caught between humanitarian and security rationales (Campesi, 2022); or the effects of EU's policing on civil society actors (Carrera et al., 2019). These works do not acknowledge Frontex's turn from border control to people control. Human rights and security considerations describe only a part of Frontex's function. Through policing, and particularly ILP, Frontex has managed to escape from the strict limits of border management, straying therefore away from its founding mandate.

5 Intelligence-Led Policing

Intelligence-led policing (ILP) describes a proactive approach in tackling crimes with the systematic use of intelligence. It started within the UK law enforcement, and later it became integrated into US police departments' guides, like New Jersey's State Police. ILP aims at identifying possible risks, such as individuals that may pose a threat to society based on their profiling (Heaton, 2000, p. 341). The key characteristics of ILP include surveillance, information-sharing, risk assessment, new technologies, use of informants, investigative techniques, emphasis on threats, and proactive action (Ratcliffe & Guidetti, 2008, p. 112; Carter & Carter, 2009). It is striking that every ILP characteristic is traced in Frontex's function. This means that Frontex has adopted and is already performing ILP activities.

Frontex promotes surveillance using sophisticated devices such as radars, satellites, unmanned aircraft, aerostats, and sensors. Its aim is to maintain real-time and 24/7 surveillance at the EU borders "for the disruption of criminal activities" (Frontex, 2021).

Frontex shares information via platforms that it has developed, like the Joint Operations Reporting Application (JORA), which enables users to upload real-time data and then share it with others. Another platform that

Frontex has created involves the collection and then sharing of personal data of "suspected criminals" with Europol (Frontex, 2015).

Frontex has also established various risk assessment products, such as alerts, risk analysis reports, and situational overviews. Actually, Frontex has developed and is implementing its own risk analysis model, which allows the agency to become the evaluator of risks as well as member states' capacity to respond to them effectively (Horii, 2016, p. 247).

On technology, Frontex has a Research and Innovation Unit tasked with promoting and delivering innovation. Furthermore, Frontex has financed, tested, and is promoting various new tools and systems, like lie detection technology (*see next part*).

The use of informants and investigative techniques has gained growing importance in Frontex's activities. It can be linked with the operation of Frontex Liaison Officers (FLOs), both within and outside the EU territory. Through FLOs, Frontex gathers situational data, gets informed about possible developments, and comes into contact, formally or informally, directly or indirectly, with persons who are away from its headquarters. Furthermore, it gets a direct picture about new border security concerns and organised crime groups operating outside the EU soil.

Now, on investigative techniques, debriefing seems like creating a sound correlation with the investigation (*see next part*). After all, debriefing is a form of police interrogation to investigate and extract information.

Moreover, the agency, on its webpage, often uses the terms "security threats" and "prevent(ion)". This terminology illustrates Frontex's emphasis on threats and proactive action for crime, or cross-border crime, prevention.

Frontex's promotion of an ILP approach is also evidenced in personnel selection. Most Frontex staff have a policing background, considering that experience in law enforcement is regularly listed as an essential prerequisite for Frontex staff recruitment, as shown in Frontex job vacancies. This reflects an endogenously provoked shift to Frontex's focus, as these officers become Frontex's staff.

Hence, besides controlling the borders using intelligence, Frontex is also policing, performing ILP. This policing though is not limited to the agency's core mission, namely the management of the borders. Instead, it involves the management of individuals, as ILP is based on the collection, processing, use, exchange, and management of individuals' information through direct or indirect interaction with them. As a result, ILP implementation moves Frontex away from borders and border control and towards a people control direction. This move is already manifested through Frontex's debriefing activities, the implementation of ETIAS, and the promotion of lie detection technology.

5.1 Debriefing

Debriefing constitutes a form of police interrogation aiming to extract information about crimes, for intelligence purposes. This activity takes place during Frontex operations or at the EU hotspots by Frontex officers.

Debriefing concerns the collection of information through oral interviews. Interviewees are irregular border crossers who consent to be interviewed. The information collected is turned into intelligence, so as analysts, to assess risks. Then, intelligence is being shared with national law enforcement authorities and other policing partners, such as Europol (Frontex, 2015).

Indicative of the importance that Frontex attributes to debriefing can be extracted from its operational data. Frontex carried out 5,729 debriefing interviews in 2023 and 4,564 in 2024. Moreover, in 2024, an average of 192 debriefers were deployed per month compared to 84 motor vehicle crime detection officers (*Frontex's reply 2025, June 11*).

Yet, debriefing was not listed as a Frontex task in the agency's initial Regulations. The 2015 European Agenda on Migration first included debriefing in Frontex's function at the EU hotspots, while a year later, debriefing is formally mentioned as one of Frontex's tasks (Regulation, 2016a). However, Frontex was performing debriefing activities since 2008, as evidenced in Frontex reports (2008, p. 31). Hence, debriefing manifests as a shift from Frontex's founding mandate and initial tasks, materialised first by Frontex staff on the field and later formalised in EU legislation.

Debriefing proves an ILP context, which directs operational activities towards people control. The conduct of debriefing, that is, through targeted face-to-face interviewing, turns the attention to individuals, namely irregular border crossers, who give information about their personal lives. Questions asked during debriefing are not limited to extracting a general idea regarding smuggling routes. Instead, they concern the whole life of the person debriefed (FIDH et al., 2014, p. 40). Debriefers are trained even by psychologists to gain interviewees' trust and extract information more easily. Through debriefing, Frontex officers come in direct contact with irregular border crossers, interrogating them even for 3 hours (FIDH et al., 2014, p. 38). That means way more time than the 12 seconds that each border guard has for assessing whether to approve or refuse border crosser's entry to the country (Frontex, 2014). The traits of debriefing analysed above show a tactic employed by Frontex that emphasise on people instead of border management.

5.2 Lie Detection Technology

Since 2012, Frontex has started promoting lie detection technology (Frontex, 2012, p. 25). It has organised and funded simulation exercises, workshops, studies, and trials of the Automated Virtual Agent for Truth Assessments in Real Time (AVATAR) system. AVATAR was developed by the University of Arizona. It constitutes an automated deception detection system based on behavioural indicators. In practice, it is a self-service passport control kiosk with an interactive interface. The electronic interviewer asks questions to the document holder while using remote sensing technology to monitor behavioural indicators of deception, like eye movements, pupil dilation, changes in voice, facial gestures, and muscle rigidity (Frontex, 2013, p. 35).

Parallel to AVATAR, Frontex has supported the Intelligent Portable Control (iBorderCtrl) project, also known as iCROSS. This project was presented at Frontex workshops as well as at the 2019 European Border and Coast Guard Day organised by Frontex. iBorderCtrl was an EU-funded automated border control system combining biometric verification, automated deception detection, document authentication, and risk assessment. It involved a two-stage process – first, pre-travel registration and data collection, including information about social media accounts, as well as an interview with an avatar and second, control at the border. In the whole procedure, the system was designed to follow a risk-based approach, applying lie detection technology, including the analysis of non-verbal expressions (iCROSS, 2020). The system was operationally piloted in Hungary, Latvia, and Greece, while some of its features are included in ETIAS.

Both AVATAR and iBorderCtrl are artificial intelligence lie-detecting systems that carry out traveller behavioural analysis. Hence, their focus is on individuals and not borders, given that they perform surveillance of people instead of border surveillance. Both systems were developed and supported by Frontex, without being included in the agency's initial tasks. Instead, they conform to an ILP risk-based approach that uses new technologies for the maximisation of internal security (RAND, 2021).

5.3 ETIAS

Another activity not included in the agency's original mandate is ETIAS. ETIAS started to be designed in 2016 and is expected to be launched in the last quarter of 2026. ETIAS is an electronic system of pre-travel authorisation for passengers from Schengen visa-free countries. To enter Schengen, these passengers will have to hold both a valid travel document and an ETIAS authorisation (Frontex, 2025).

ETIAS will check the security credentials of travellers intending to visit the Schengen by carrying out a pre-travel screening to assess security and migration risks. This system aims at remedying an "information gap" in relation to non-EU citizens travelling to a Schengen country. Yet, the information required by ETIAS is not limited to basic identity and travel document data. Instead, travellers have to inform about their level of education, occupation, criminal record, and health, as well as answer background questions (European Parliament, 2018).

In practice, ETIAS will gather, store, and update information regarding persons who intend to enter Schengen. It will also crosscheck travellers' identity data against a number of databases, such as Interpol data, the Entry/Exit System, and the Schengen Information System. There will also be a watch list and specific risk indicators, and authorisation can be refused if the traveller is considered a threat to security, or risk to migration or public health. Frontex will have a central role in the implementation and operation of this system. It will host its Central Unit. For this reason, a new Frontex Division was set up for ETIAS' implementation, as shown in Frontex's last organigram. Yet, ETIAS refers to the pre-arrival processing of a traveller. The applicants for an ETIAS travel authorisation will not be at the physical border to be checked while submitting their application but at their house, that is, outside the EU. Furthermore, the collection, management, and storage of individuals' information is matched to a proactive people profiling system; in other words, to an ILP context that uses a tool to assess beforehand whether an individual may pose a threat to EU security. Therefore, ETIAS does not engage with the management of the borders but with the screening of individuals. Undoubtedly, the EU ETIAS has been inspired by the US ESTA paradigm. Yet, ESTA is managed by the US Customs and Border Protection, which constitutes a law enforcement organisation.

6 People Control

Debriefing, ETIAS, and lie detection technology constitute three new Frontex activities introduced and/or promoted by Frontex for the management of borders. Their examination here showed that rather than focusing on borders, they emphasise people. However, people, or people control, goes beyond the scope of border control, as set out in the Schengen Borders Code, namely border checks and border surveillance.

In carrying out these activities, Frontex starts interacting with persons, namely irregular migrants or border crossers. It interrogates them even for hours during debriefing, assesses their facial gestures with lie-detection systems, or proactively profiles them with ETIAS. Simultaneously, Frontex becomes the holder, processor, and analyser of new knowledge that refers to these individuals. It may involve travel intentions derived from ETIAS, or personal experiences extracted with debriefing. This fundamentally alters the character and function of Frontex. From an emphasis on borders and border control, it turns to people and people control, with activities conducted away from the geographic location of borders. Moreover, the use of technologically supported and digitally enhanced tools like lie-debriefing systems and ETIAS boosts a novel techno-policing for border control, tied to people surveillance (Bigo & Guild, 2005) and crime management (Joh, 2017; Brayne, 2021; Campbell, 2024).

A focus on people can also be traced in other new activities, like Frontex's enhanced role in returns. The process of returning a non-EU national to a third country is an administrative procedure taking place after a third-country national's entry to the EU. In other words, it is not part of border control. Likewise, Frontex's provision of reintegration assistance to returnees, such as up to 12 months post-return assistance, does not equate to border control. Certainly, these activities can impact border flows. Returns can dissuade non-EU nationals from irregularly staying in the EU territory, while reintegration assistance can help returnees to remain in their home country. Yet, they are not performed during border control. Moreover, they do not focus on borders. Instead, they are centred on individuals.

The enhancement of Frontex's powers towards this direction is also boosted through EU Presidencies. To achieve a more effective border control and increased EU security, member states back an expanded Frontex role, beyond the limits of border control. Recently, the Polish Presidency (2025) led the discussion for Frontex's empowerment to deal more effectively with evolving threats. The ensuing Danish Presidency followed the same path, as its paper on the agency's future mandate included plans for Frontex to assume more responsibilities to address hybrid threats and attacks to critical infrastructure (Euractiv, 2025). Hence, member states seem on board with Frontex's broadened role.

7 The Implications of Frontex's Policing

The shift from border to people control shows that Frontex is not anymore just a border control agency, it has turned into a policing agent. Yet, Frontex has not been created to police and is not mandated to perform policing. As a result, Frontex's policing function constitutes an unexpected transformation in the agency's role, which raises new considerations about Frontex's role in EU security.

Especially given that policing is more than an activity of preventing crime or enforcing the law; it is considered a "necessity" for the maintenance of social order (Reiner, 2010, p. 4). It reproduces what can, or should, be policed and how. As such, policing acquires social meanings and social implications, while policing agents become social order's "guardians" (Loader, 1997). From this perspective, by carrying out policing, Frontex becomes a guardian of social order that performs social control. So, it is not any more a mere technical border agency that blindly implements its delegated tasks. Policing allows Frontex to increase its authority and symbolic power. Adopting new roles, having access to new information, and creating new knowledge, Frontex's added value is being substantially strengthened, with Frontex becoming a reliable source of information and analysis, beyond its border repertoire. Meanwhile, Frontex's policing has a practical dimension – decisions to be taken, resources to be allocated, operations to be conducted, and tools to be used. Yet, Frontex has not been authorised to act as a policing agent, neither it is treated as such. This raises serious considerations and creates an important policy gap that needs to be addressed in terms of institutional development and accountability. Accordingly, the police serve the public functioning on people's behalf and for people's interest. This triggers the question: who does Frontex serve and whose interest does it safeguard with policing? Moreover, police officers operate in a framework of police oversight and accountability. Within the internal police line of command, police officers are accountable to their chiefs. At a broader spectrum, they are accountable and controlled by the Minister, mostly, of Interior, or the Prime Minister, and the Parliament that ensures public's consent to police actions as well as a democratic accountability and oversight. An analogous system of internal and external checks and balances is missing from Frontex's function. Frontex, as an EU agency, has been designed based on a technical and managerial logic keen for a prompt action. For this reason, Frontex is being governed by a Management Board and its Executive Director, whereas the role of the European Parliament is rather limited and is a common issue of contention among parliamentarians, the Commission, and member states.

To conduct policing, Frontex needs to be restructured. Its legal basis, governance, and institutional architecture need to be adapted to consider its policing function. In parallel, Frontex's mandate should be re-examined and then renewed, as it has not been authorised to perform policing or control people. So, Frontex needs to operate, after having received a fresh mandate and a new consensus from the EU public and authorities. Otherwise, its policing activities need to be considered not just as a significant transformation but a serious misconduct.

8 Conclusion

Policing on borders may not sound so provocative and shocking, given the presence of uniformed police officers in various borders tasked to combat cross-border crime. But what if policing is being conducted by (un)armed Frontex officers operating at the borders or by staff analysing information shown in their computer monitors at Frontex's headquarters? Inspired by these questions, this chapter explored whether Frontex, the EU border control agency, whilst being a non-police authority, performs policing. It has been shown that Frontex's operation goes beyond policing. It follows an ILP approach, which reorients its focus from border control to people control. Analysing the ILP characteristics, it concluded that they can all be traced to Frontex actions manifesting towards people control. This turn is being implemented through Frontex's new activities and especially, debriefing, ETIAS, and lie detection technology. These activities, which were not included in Frontex's initial remit, emphasise on people by collecting, processing, sorting, and analysing individuals' information. They also reflect a strong social context. After all, policing involves social functions determining the subjects and objects of its control. Consequently, through policing, Frontex shapes both the EU security and social order. This unavoidably raises new questions regarding the legitimacy of Frontex's metamorphosis, stressing the need for revisiting both its jurisdiction and accountability framework.

References

Aas, K., & Gundhus, H. (2014). Policing humanitarian borderlands: Frontex, human rights and the precariousness of life. *The British Journal of Criminology*, 55(1), 1–18. https://doi.org/10.1093/bjc/azu086.

Anderson, M., & den Boer, M. (Eds.). (1994). *Policing across national boundaries*. London: Printer.

Benyon J. (1994). Policing the European Union: The changing basis of cooperation on law enforcement. *International Affairs, 70*(3), 497–517. https://doi.org/10.2307/2623708.

Bigo, D., & Guild, E. (2005). Introduction: Policing in the Name of Freedom. *In* D. Bigo & E. Guild (Eds.), *Controlling frontiers: Free movement into and within Europe* (pp. 1–13). Aldershot: Ashgate.

Brayne, S. (2021). *Predict and surveil: Data, discretion, and the future of policing*. Oxford: Oxford University Press.

Campbell, E. (2024). Techno-digital policing and speculative fictions: Towards a criminology of the future. *Crime, Media, Culture: An International Journal, 21*(1), 69–95. https://doi.org/10.1177/17416590241253543.

Campesi, G. (2022). *Policing mobility regimes: Frontex and the production of the European borderscape*. Abingdon: Routledge.

Carrera, S., Mitsilegas, V., Allsopp, J., & Vosyliute, L. (2019). *Policing humanitarianism: EU policies against human smuggling and their impact on civil society*. Oxford: Hart.

Carter, D., & Carter, J. (2009). Intelligence-led policing: Conceptual considerations for public policy. *Criminal Justice Policy Review, 20*(3), 310–325. https://doi.org/10.1177/0887403408327381.

European Commission (2025). *Border crossing*. https://home-affairs.ec.europa.eu/policies/schengen/border-crossing_en.

European Parliament (2018). *Briefing: ETIAS*. https://www.europarl.europa.eu/RegData/etudes/BRIE/2017/599298/EPRS_BRI(2017)599298_EN.pdf.

Euractiv (2025). Member states talk stronger Frontex, test limits on returns. https://www.euractiv.com/news/member-states-talk-stronger-frontex-test-limits-on-returns/.

FIDH, Migreurop, & EMHRN (2014). *Frontex between Greece and Turkey*. https://www.fidh.org/en/region/europe-central-asia/frontex-between-greece-and-turkey-the-border-of-denial-the-deployment.

Frontex (2007). *General report*. Warsaw: Frontex.

Frontex (2008). *General report*. Warsaw: Frontex.

Frontex (2012). *General report*. Warsaw: Frontex.

Frontex (2013). *General report*. Warsaw: Frontex.

Frontex (2014). *12 seconds to decide*. Warsaw: Frontex.

Frontex (2015). *Frontex and Europol agree to exchange information on cross-border crime*. https://www.frontex.europa.eu/media-centre/news/news-release/frontex-and-europol-agree-to-exchange-information-on-cross-border-crime -dEyVpQ.

Frontex (2021). *Frontex tests aerostat systems in Greece for border surveillance.* https://www.frontex.europa.eu/media-centre/news/news-release/frontex-tests-aerostat-systems-in-greece-for-border-surveillance-b5E9I8.

Frontex (2025). *ETIAS.* https://www.frontex.europa.eu/what-we-do/etias-ees/about-etias/.

Fyfe, N. (1995). Policing the city. *Urban Studies, 32*(4–5), 759–778. https://doi.org/10.1080/00420989550012870.

Heaton, R. (2000). The prospects for intelligence-led policing. *Policing and Society, 9*(4), 337–355. https://doi.org/10.1080/10439463.2000.9964822.

Horii, S. (2016). The effect of Frontex's risk analysis on the European border controls. *European Politics and Society, 17*(2), 242–258. https://doi.org/10.1080/23745118.2016.1121002.

iCROSS (2020). *Intelligent portable border control system.* https://cordis.europa.eu/project/id/700626.

Joh, E. (2017). Artificial intelligence and policing: First questions. *Seattle University Law Review, 41*, 1139–1149.

Léonard, S. (2010). EU border security and migration into the European Union: Frontex and securitisation through practices. *European Security, 19*(2), 231–254. https://doi.org/10.1080/09662839.2010.526937.

Léonard, S., & Kaunert, C. (2022). The securitisation of migration in the European Union: Frontex and its evolving security practices. *Journal of Ethnic & Migration Studies, 48*(6), 1417–1429. https://doi.org/10.1080/1369183X.2020.1851469.

Loader, I. (1997). Policing and the social: Questions of symbolic power. *The British Journal of Sociology, 48*(1), 1–18. https://doi.org/10.2307/591907.

Maguire, M. (2000). Policing by risks and targets: Some dimensions and implications of intelligence-led crime control. Policing and Society: An International Journal, 9(4), 315–336.10.1080/10439463.2000.9964821.

Mawby, R.I. (2008). Models of Policing. In T. Newburn (Ed.), *Handbook of policing* (pp. 17–46). Abingdon: Routledge.

Pallister-Wilkins, P. (2015). The humanitarian politics of European border policing: Frontex and border police in Evros. *International Political Sociology, 9*(1), 53–69. https://doi.org/10.1111/ips.12076.

Polish Presidency (2025). Informal Meeting of the Strategic Committee on Immigration, Frontiers and Asylum (SCIFA). https://polish-presidency.consilium.europa.eu/en/news/informal-meeting-of-the-strategic-committee-on-immigration-frontiers-and-asylum-scifa-in-warsaw/.

RAND (2021). *How AI can support the EBCG.* https://www.rand.org/randeurope/research/projects/2021/european-border-coast-guard-artificial-intelligence.html.

Ratcliffe, J., & Guidetti, R. (2008). State police investigative structure and the adoption of intelligence-led policing. *Policing: An international* journal, *31*(1), 109–128. https://doi.org/10.1108/13639510810852602.
Regulation (EC) 2007/2004. (2004). OJ L 349/1.
Regulation (EC) 863/2007. (2007). OJ L 199/30.
Regulation (EU) 2016/1624. (2016a). OJ L 251/1.
Regulation (EU) 2016/399. (2016b). OJ L 77/1.
Regulation (EU) 2019/1896. (2019). OJ L 295/1.
Reiner, R. (2010). *The politics of the police.* Oxford: Oxford University Press.
Sarantaki, A.-M. (2023). *Frontex and the rising of a new border control culture in Europe.* Abingdon: Routledge.
Yarwood, R. (2007). The geographies of policing. *Progress in Human Geography*, *31* (4), 447–465. https://doi.org/10.1177/0309132507079.

CHAPTER 15

Limiting the Right to Asylum in the Post-liberal Era: A Matter of Balancing Security with Universal Rights?

Johannes Thumfart

Abstract This chapter examines recent efforts by Poland, with partial support from the European Commission, to limit the right to asylum in emergency situations under Article 72 TFEU. It contextualizes this development through the tension between security and universal rights, showing how the right to asylum has historically evolved in response to changing political circumstances. The chapter traces the origins of contemporary asylum rights, their development after the 1951 Geneva Convention, and their increasing instrumentalization by authoritarian regimes. The analysis situates current restrictions within broader European developments, from the German Asylkompromiss of 1992 to the migration crises in Hungary (2015) and Greece (2020), and discusses recent initiatives, including Regulation (EU) 2024/1359 and the Commission's interpretation of Article 72 TFEU. It argues that EU law requires a national security rationale to justify suspending asylum rights, often framed through the concept of the "weaponization of migration." However, migration pressures would remain a significant challenge even without such instrumentalization. The chapter concludes that the focus on security obscures a fundamental dilemma of the universal right to asylum: limited resources and administrative overload caused by numerous unfounded claims may ultimately undermine protection for those genuinely in need of international protection.

Keywords Article 72 TFEU – asylum law – Kant – Vitoria – state of emergency – weaponization of migration

1 Introduction

Debates on asylum and migration are polarizing at both the national and supranational levels. On the one hand, there are the individual rights of refugees to various forms of international protection, which are discussed in this chapter under the overarching term "right to asylum." These universal rights

 | HTTPS://DOI.ORG/10.1163/9789004763623-017

are codified in international law, the European Union's Charter of Fundamental Rights, and the constitutions of numerous member states. On the other hand, there is mounting political pressure to manage migratory flows in ways that preserve the stability and security of individual nations and the EU as a whole.

This chapter discusses the conceptual tension between security and universal rights, focusing on limitations on the right to asylum implemented by the following EU member states situated at the external borders: Greece, Hungary, Poland, Lithuania, Latvia, Finland, and Cyprus (Papadimas, 2025; Mikołajczyk, 2025; Mouzourakis, 2025; DW (staff), 2024). Particularly since the onset of the war in Ukraine in 2022, countries on the EU's eastern external borders, such as Finland, Poland, Lithuania, and Latvia, have toughened their asylum policies due to what has been framed as "weaponized migration" (Mikić, 2024). Accordingly, the Polish Council Presidency in the first half of 2025 explicitly prioritized the discussion of "the instrumentalisation of migration" as a "hybrid threat" (Balázs, 2025).

The EU's governing bodies, too, made concessions to the position of the member states located at the external borders by issuing Regulation (EU) 2024/1359 of 14 May 2024, which addresses situations of crisis and force majeure. Additionally, on 11 December 2024, the EU Commission adopted its non-binding *Communication on Countering Hybrid Threats from the Weaponisation of Migration and Strengthening Security at the EU's External Borders* (2024), which has been interpreted as the Commission having "crossed the Rubicon" by legalizing pushbacks (Thym, 2025).

Limitations of the right to asylum stand in stark contrast to the current state of international law. The Geneva Refugee Convention does not include a derogation clause, nor do Articles 18 and 19 of the European Charter of Fundamental Rights. In particular, expulsions to non-EU territories are in conflict with the prohibition of non-refoulement. This conflict between universal rights and security is typical of the "post-liberal era," in which the fundamental concepts of the liberal international order are increasingly at odds with a reality that is characterized by multipolarity, strained resources, and the pressure to act (Thumfart, 2024, p. 282ff.; Thumfart, 2025).

This chapter pursues the methodology of intellectual or conceptual history by contextualizing the current tendency to limit the right to asylum, considering this right's mercurial history. In the second section, I examine the genealogy of the universal rights codified in frameworks like the 1951 Geneva Refugee Convention, starting with their origins in the texts of Vitoria and Kant and in the Cold War context. This *longue durée* perspective illustrates that international agreements regarding refugees were generally shaped by political imperatives arising from specific historical contexts and that Universalist concepts are open to instrumentalization by various actors. Building on this contextualization, in the third section, I discuss European approaches to limiting the right to asylum from the 1990s to the present. In the fourth section, I summarize and discuss the findings.

Overall, this chapter argues that framing the suspension of the right to asylum as an emergency measure represents rather an argumentative move to justify short-term exceptions to the universal right to asylum than a sustainable approach. The current flows of migrants would also be challenging without, in fact, the only insular cases of instrumentalization by authoritarian leaders. The real problem obscured by the "weaponization" discourse is the dilemma at the heart of the individual universal right to asylum. De facto universal access to the right to asylum is morally sound in theory, but untenable in practice. Across Europe, the recognition rate for asylum applications is consistently around 40%, with half of these merely receiving temporary subsidiary protection (Latest Asylum Trends, 2025). Above all, the lack of resources and administrative logjam due to the many unfounded asylum claims threatens the rights of those truly in need of international protection. Instead of leading a "weaponization" discourse, EU governing bodies need to develop sustainable reforms that guarantee that those truly in need of asylum are recognized quickly and that others are rejected quickly and effectively deterred from entering EU territory. Most importantly, the EU must learn the reality lesson that, in an emerging global civil society, its commitment to universal values cannot be determined by legal norms alone but depends on its actual capacities to fulfil this commitment.

2 Instrumentalizing Universal Rights from the *Conquista* to the Cold War

The individual right to asylum is often imagined as a universal value. However, since the beginning of the modern discourse on international law, the right to asylum has been shaped by shifting geopolitical contexts and instrumentalized by various actors. The mercurial history of the right to asylum began with the Spanish theologian Francisco de Vitoria, who is widely regarded as one of the founders of international law. Vitoria formulated an *ius migrandi*, which included the universal "right to migrate" and enjoy the right of hospitality (Augusti, 2025; Di Cesare, 2019, p. 72; Thumfart, 2009; Vitoria, 1991).

It is widely recognized by now that Vitoria is a highly problematic author (Anghie, 1996). However, this knowledge is usually not applied in the discussion of Vitoria in the context of migration law, where references to Vitoria serve to make the point that the right to migrate has a timeless, and hence universal, quality (Augusti, 2025; Di Cesare, 2019). Vitoria's conception of 1532 is anything but timeless and is heavily shaped by geopolitical interests (Thumfart, 2012). In Vitoria's era, the *ius migrandi* served as an ideological tool to justify colonialism in the Americas. In his view, the Spanish could claim their right of hospitality in the Americas, and the American Indians had no right to infringe on this universal right (Anghie, 1996, p. 326). Absurdly enough, Vitoria justified the Spanish *Conquista* as an act of self-defense against an alleged expulsion of the Spanish by the Amerindians.

Vitoria's conception of a universal individual *ius migrandi* was also immediately instrumentalized by rivaling colonial powers. In particular, Hugo Grotius, another founder of international law, based his notion of the Freedom of the Seas on Vitoria's right to travel (Thumfart, 2009). This included Grotius's justification of the use of force by Dutch private merchant vessels if they were hindered from traveling the globe.

One finds a late echo of the widespread abuse of Vitoria's right to international hospitality in Kant's *Towards Perpetual Peace* from 1795, which is commonly understood as promoting cosmopolitan universalism, in particular due to Kant's conception of a universal individual right, the *ius cosmopoliticum* (Kant, 2006). Kant argued that colonizing powers have abused the right of universal hospitality and advocated its limitation to a right to visit (Cavallar, 2002, p. 366). According to Kant, such a right to visit can be exercised only to sojourn in a territory for a limited period, not indefinitely, so that host countries can also turn visitors away if this does not kill them. He wrote:

> If it can be done without causing his death, the stranger can be turned away (…). The right of hospitality, that is, the right of foreign arrivals, pertains (…) only to conditions of the possibility of attempting interaction with the old inhabitants. (…). If one compares with this the inhospitable behavior of the civilized states in our part of the world, especially the commercial ones [i.e. European colonial powers], the injustice that the latter show when visiting foreign lands and peoples (which to them is one and the same as conquering those lands and peoples) takes on terrifying proportions (Kant, 2006, p. 82).

With the rise of the Westphalian order, culminating in the nineteenth-century nation-state, the notion of an unqualified right of hospitality increasingly vanished. It was only partially revived with the League of Nations' Convention relating to the International Status of Refugees negotiated from 1922 onwards and ratified in 1933. In notable contrast to Vitoria's – and partially also Kant's –

Universalist models, this convention was tailored to the specific historical context after the First World War, the Russian Revolution, and the breakup of the Ottoman Empire. The 1933 Convention was explicitly restricted to Russians, Armenians, Assyrians, Assyro-Chaldeans, Syrians, Kurds, and a small number of Turks, and it was only ratified by Belgium, Bulgaria, Egypt, France, Czechoslovakia, Denmark, Norway, Ireland, the UK, and Italy. The text of the 1933 Convention mentioned administrative topics such as identity certificates, labor conditions, welfare, education, and taxation. The core idea was that refugees should receive the same treatment as the most favorable treatment accorded to nationals of a foreign country. Most importantly, in its Article 3, the 1933 Convention introduced the principle of non-refoulement:

> Each of the Contracting Parties undertakes not to remove or keep from its territory by application of police measures, such as expulsions or non-admittance at the frontier [*refoulement*], refugees who have been authorized to reside there regularly, unless the said measures are dictated by reasons of national security or public order. It undertakes in any case not to refuse entry to refugees [*refouler*] at the frontiers of their countries of origin (Convention Relative Au Statut International Des Réfugiés, 1933).

As these references to security in the Convention suggest, the principle of non-refoulement was contentious. Several of the signatory parties explicitly excluded Article 3 of the Convention. The UK, for instance, did not accept the principle of non-refoulement. And neither did Egypt, which reserved itself "the right to expel such refugees at any moment for reasons of public security" (Jaeger, 2001, p. 730). Tragically, the resistance against non-refoulement became only more widespread with the increased persecution of minorities within Germany from 1933 onwards. The *Provisional Arrangement concerning the Status of Refugees coming from Germany* from 1936 even mentioned the possibility to "send people back across the borders of the Reich," as an ultima ratio, and the 1938 *Convention concerning the Status of Refugees coming from Germany* contained similar provisions allowing for pushbacks based on security exceptions (Jaeger, 2001, p. 731). A contemporary article in *Time Magazine* characterized the general mood at the Évian Conference of 1938, discussing the fate of German refugees in the following way:

> All nations present expressed sympathy for the refugees, but few offered to allow them within their boundaries. Britain, France, Belgium pleaded that they had already absorbed their capacity, Australia turned in a flat 'No' to Jews, and the

> US announced that she would (…) admit 27 370 persons ('International Refugees', 1938).

In these years preceding the Second World War, European Jews were often openly rejected by the leading powers of the liberal international order, with spectacular exemplary cases such as the ship St. Louis that, in 1939, was forced to return from the US to Europe with over 900 Jewish refugees on board. Challenged by the massive influx, Switzerland alone turned back tens of thousands of Jewish refugees during the Nazi period (Die Schweiz und die Flüchtlinge zur Zeit des Nationalsozialismus, 2001).

The horrors of the Holocaust inform the 1951 Geneva Refugee Convention. In contrast to its predecessors, it is not limited to a defined set of refugees from certain countries but defines refugees universally as a person who owing to well-founded fear of being persecuted for reasons of race, religion, nationality, membership of a particular social group, or political opinion, is outside the country of his nationality and is unable or, owing to such fear, is unwilling to return to it. Particularly, the reference to the internal psychological state of the "fear of being persecuted" is extremely open to interpretation because it includes the justification of asylum based on circumstances before actual injuries occurred. Simultaneously, Article 33 of the Geneva Convention strengthened the principle of non-refoulement (United Nations High Commissioner for Refugees, 1951).

In addition to being a reaction to the Holocaust, the 1951 Convention's Universalist character served as a Cold War stratagem that embodied the ideology of the liberal "Open Society" (Popper, 2013). By means of their ostentatious openness, the Western signatories distinguished themselves from the closed and totalitarian East (Gatrell, 2013, p. 87). The refugee migration flow was intended to be largely unidirectional from East to West. As a political instrument, the Convention provided an incentive for civil societies to "vote with their feet" and a safety net for dissidents. This Cold War genealogy is crucial to understanding the interest structure behind the Universalist and generous characteristics of asylum law.

However, precisely the universal nature of the right to asylum made it possible for the authoritarian camp to turn it against the West occasionally. This happened on several occasions between 1965 and 2010, when leaders such as Fidel Castro and Gaddafi threatened that they would let emigrants leave their countries if they would not get their will, which included the actual sending of emigrants across the Caribbean (Greenhill, 2010, pp. 75, 331). Comparably, at the height of the refugee crisis in 2016, Turkish President Erdogan demanded €6 billion over two years and addressed European leaders with the following rhetorical question: "So how will you deal with refugees if you don't get a deal? Kill the refugees?" ("Turkey's Erdogan Threatened to Flood Europe with Migrants," 2016)? In 2019, Erdogan used a similar threat to silence European criticism of the Turkish operation in Syria: "Don't call this an invasion (...). If you persist, we will open the gates and send 3.6 million refugees your way (Gatopoulos, 2019)." Putin's opening the borders to refugees, which will be discussed in more detail in the next Section, is just the latest example of attempts to produce "weapons of mass migration" (Leonard, 2021).

In summary, this *longue durée* contextualization demonstrates that with its instrumentalization by authoritarian leaders, the right to asylum came "full circle." Since its beginnings in Vitoria's *ius migrandi*, precisely the universal character of the right of hospitality was instrumentalized by the West, and, of course, during the Cold War, the individual universal right to asylum was part of a strategy to weaken the authoritarian camp. Comparable to other aspects of the international liberal order, the strategic openness of the universal right to asylum made it possible that, in the post-liberal era, authoritarian regimes turned the table and instrumentalized asylum seekers (Thumfart, 2024, p. 282ff.).

3 Adjustments to the Right to Asylum from the 1990s to 2025

The discourse on asylum reached a first turning point around the wars in former Socialist Yugoslavia (1991–1999), where forced migration was part of the toolbox of ethnic cleansing. Particularly Germany, which was hosting the majority of ex-Yugoslav refugees, responded to growing domestic unease by amending its constitution, making asylum rules more restrictive. This amendment ("*Asylkompromiss*") introduced the concept of "safe third countries" into Article 16 of the German Constitution, theoretically preventing everyone entering Germany via the land route from receiving asylum – a forerunner of the Dublin Convention. In reality, this did not change much, particularly as refugee rights, according to the 1951 Geneva Convention, are not affected by this change. However, mainly because the *Asylkompromiss* immediately followed the fall of the Berlin wall in 1989, progressive forces accused the government of building "a new wall around Germany, a wall around Europe" (Bresselau Von Bressensdorf, 2019, p. 121f.).

From 1992 on, "Fortress Europe" became a commonplace trope of, in retrospect, highly exaggerated criticism (Burgdorff, 2023). Over the next decade, the EU introduced or expanded instruments like the 1997 Dublin Convention (later Dublin II and III) and began laying the groundwork for a Common European Asylum System (CEAS) in 1999. In 2004, the European Agency for the

Management of Operational Cooperation at the External Borders (Frontex) was created. Despite these measures, the European migration crisis of 2015/2016 drastically exposed how irregular arrivals could strain member states' political, economic, and social fabrics, generating populist backlashes and challenging European solidarity.

For instance, in response to mass arrivals, Hungary declared a state of emergency, setting up a 175 km-long fence along its Serbian border and transit zones where asylum seekers were effectively detained and in which only a limited number of migrants would be admitted daily (Cantat & Rajaram, 2019). In its judgment of 17 December 2020, the EU Court of Justice found that Hungary's legislation and practice were contrary to EU law and incompatible with the principle of non-refoulement (European Commission v Hungary, 2020).

Another case is Greece. Along the Greek–Turkish border at the Evros River in 2020, Turkey deliberately opened the borders and even bussed migrants there. As a reaction to this exceptional situation, the Greek police undertook pushbacks, reportedly going as far as transporting people from detention facilities land inwards across the border (Karamanidou & Kasparek, 2020). However, unlike Hungary in 2016, the EU Commission supported Greece. Without officially using the metaphor of the "weaponization of migration" yet, the migration issue was reframed as a collective security threat, with President von der Leyen accusing Turkey of seeking "to test Europe's unity" and calling Greece the "European *ασπίδα* [shield]" (President von der Leyen at the press conference in Greece, 2020).

In July 2025, Greece suspended asylum applications for migrants from North Africa, reportedly to signal that "the passage to Greece is closed" (Papadimas, 2025). Although this measure conflicts with human rights, so far, EU institutions have barely reacted to it. The EU Commission's answer to a parliamentary question on this matter from December 2025 informed that the Greek suspension of asylum law had been lifted on 14 October 2025 and that the Commission will "closely monitor the situation" (Answer for Question E-003907/25, 2025). Already in 2024, Greece's neighbor Cyprus has taken similar steps, suspending the processing of asylum applications by Syrian refugees, which is still in effect (DW (staff), 2024).

The tendency towards limiting the right to asylum was framed most clearly as a security measure on the EU's eastern border, starting with Finland. Already in 2016, there were speculations in Finland that the country wasn't facing a refugee crisis but rather an act of Russia's "hybrid warfare" ("Officials See Signs of Hybrid Warfare in Migrant Crisis," 2016). Following the start of the Ukraine War, Finland began constructing a 200 km-long border fence and enhanced border surveillance. In July 2024, the country introduced a law that allows turning asylum seekers away at its eastern border. Human rights organizations have criticized this new law as setting a precedent for the erosion of the principle of non-refoulement (Jochecová, 2024). It must be noted that, in quantitative terms, Finland's actions were not founded in an actual threat but rather on its expectation. Unlike in 2015/2016, there was no significant increase in traffic at the Finnish/Russian border around 2022 (Talleraas & Koch, 2024).

As the route via Poland is far more frequented than the route via Finland, it is no surprise that Poland, too, enacted measures to limit migration and the right to asylum. This discourse peaked in 2024, following an attempt by Tusk's pro-European government to win domestic approval against the extreme right (Buras, 2024). A quantitatively minor but qualitatively extraordinary event accelerated the Polish debate: In June 2024, a Polish soldier patrolling the border was killed by a migrant ("Polish Soldier Dies after Being Stabbed by Migrants on Border with Belarus," 2024). In the wake of this event, Tusk announced in October that Poland would suspend processing asylum applications in cases of the instrumentalization of migration by its authoritarian neighbors. In February 2025, Poland's parliament voted in favor of legislation allowing for suspending the right to asylum for up to 60 days in strained border regions, exempting only particularly vulnerable groups from this suspension (Sieradzka, 2025). The Polish human rights organization SIP doubts that the security threat at Poland's eastern border justifies "such radical steps and violations of fundamental human rights" (Sieradzka, 2025). Similar measures to those adopted in Poland and Finland were adopted in Lithuania and Latvia on similar grounds (Mouzourakis, 2025).

The Polish Council Presidency in the first half of 2025 explicitly prioritized the discussion of "the instrumentalization of migration" as a "hybrid threat" (Balázs, 2025). It is unlikely that Poland's Council Presidency decisively shaped the European debate on this matter, beyond having some rhetorical influence on framing and agenda setting (Vaznonytė, 2020). However, Poland's position evidently resonated with a broader and longer-lasting European development that is unlikely to reverse.

The idea to suspend the right to asylum due to emergency situations was already addressed in Regulation (EU) 2024/1359 of 14 May 2024 relating to conditions of crisis and force majeure. The regulation guarantees European solidarity mechanisms and allows for the derogation of specific aspects of asylum processes in emergency situations, which, however, must be authorized by the Commission and the Council on a case-by-case basis. Possible derogations are codified in its Articles 10 to 13, regarding the obligation to take back applicants, time limits for take charge requests, the asylum border procedure, and the registration of applications.

On 11 December 2024, the EU Commission adopted the non-binding *Communication on Countering Hybrid Threats from the Weaponisation of Migration and Strengthening Security at the EU's External Borders* (2024). In this document, the EU Commission mentions "exceptional situations at the EU's external borders", concluding that "exceptional situations require appropriate measures," referring to Article 72 TFEU, the emergency clause in EU Law. This clause, typical of emergency legislation, can be interpreted as granting far-reaching possibilities to member states and is, at the same time, contentious and not very well defined. Whilst the EU Commission's move might be understood as strengthening subsidiarity and national sovereignty, it factually means that the EU's governing bodies avoid reforming asylum law in a sustainable way by passing the onus to member states that enact temporary and local emergency measures with next to no legal certainty.

So far, it is unclear how far these measures can go and when they are justified. For instance, in 2021, Lithuania adopted a state of emergency on the grounds of Russia's use of migrants as tools of "hybrid warfare" and enacted measures to keep migrants from entering Lithuanian territory to seek asylum and to detain them. In June 2022, the European Court of Justice ruled that Lithuania's invocation of Article 72 TFEU was not convincing because it was based on merely "general considerations (…) which are not (…) capable of justifying the application of that article" (*Valstybės sienos apsaugos tarnyba, Case: C-72/22 PPU*, 2022).

4 Conclusions

This chapter provides a *longue durée* contextualization of the current discussion regarding the limitation of the right to asylum. This contextualization demonstrates that the widespread understanding of the individual right to asylum as a timeless, universal value is questionable. The genealogy of the universal individual right to asylum was always shaped by concrete historical contexts. And within these contexts, it served Western interests, particularly during the Cold War. In the post-liberal era starting around the year 2015, the tables turned, the universal right to asylum was increasingly instrumentalized by authoritarian regimes and, more importantly, it was confronted with an emerging global civil society more mobile and informed than ever before (Thumfart, 2024, p. 282ff.). In this context, the contemporary debate regarding the limitation of the right to asylum is not a unique coincidence but rather the predictable outcome of developments starting with the end of the closed societies of Soviet Communism and the subsequent adjustments of European asylum law to post-Cold War realities. The liberal core promise of global mobility only appeared to be unequivocally desirable until it became a mass phenomenon.

However, the fact that – at least in retrospect – this historical turnaround appears predictable does not mean that suspending the right to asylum constitutes a sound solution. Other than suggested by the metaphor of "weaponization," migrants are no tools without agency that can be "fought back" at the border. Contemporary migration flows are the product of a hyper-connected world in which it is easier than ever before to travel vast distances, to communicate with diaspora members, to get informed about asylum procedures, and to employ various modes of deception, of which fake passports are just the most obvious ones (Bohmer & Shuman, 2018). In part, current migration flows are also a product of the larger tendency that global digital connectivity threatens the bordering capacity of sovereign nation-states, which is addressed in the contemporary digital sovereignty debate (Thumfart, 2024; Augusti, 2025, 4). Smartphones and transnational internet connections are among the key factors enabling mass migration (Leonard, 2021, chapter 2). Since they are the inevitable outcome of the emergence of a global civil society and technological development, contemporary migration flows cannot be stopped by exceptional measures, which are, by definition, temporary and local. Such measures will only result in member states passing the hot potato to one another.

Furthermore, it is highly questionable whether migrants pose a threat to national or international security, which is the *conditio sine qua non* for legitimizing such policies as emergency responses. Violence committed by migrants crossing the border remains far below the level commonly required to characterize an incident as a security threat. Ironically, contemporary political rhetoric attempted to tackle this shortcoming by doubling down and characterizing the influx of migrants as an act of "hybrid warfare." Of course, this only makes the problem bigger. Such rhetoric renders the concept of warfare, which usually requires physically violent acts, dangerously open to interpretation. Despite representing an obvious mismatch on so many levels, the concept of the "weaponization of migration" in the context of "hybrid warfare" is popular among policymakers. The reason for this popularity lies in the fact that this trope suggests that Europe's ongoing migration crisis can be fixed by adopting local and temporary emergency measures without fundamentally changing what seems to be the core problem: the universal individual right to asylum itself, which quickly turns into a form of moral overstretch or simply an empty promise, when confronted with a global citizenry that is well-informed and mobile like never before.

In the post-liberal era, which is the era *after* (in both senses of the word) the triumph of liberalism, fundamental principles of the liberal international order, such as the right to asylum, are confronted with their internal contradictions. The problem the West is facing today is that liberalism has succeeded in creating a global civil society in which everyone feels entitled to pursue their happiness, even across borders. If one trusts the legal procedures, then most asylum applications are unfounded. Across Europe, the recognition rate for different types of refugee protection is consistently around 40% (Latest Asylum Trends, 2025). Distinguishing between the many unfounded and the far fewer well-founded asylum claims places enormous pressure on Europe's administrative bodies and threatens the rights of those truly depending on international protection. A real solution to this dilemma must sustainably reform the legal core of the problem, rather than offering national Band-Aids for emergency situations; it must strengthen European solidarity and provide a more efficient way to help those who are genuinely in need of protection. A good step toward a meaningful reform of EU Asylum Law is Regulation (EU) 2024/1348, which applies from June 2026 on, and which strengthens solidarity among EU member states and accelerates the examination of applications who come from countries in regard to which the share of decisions granting international protection is 20 % or lower (Regulation (EU) 2024/1348 of the European Parliament and of the Council of 14 May 2024 Establishing a Common Procedure for International Protection in the Union and Repealing Directive 2013/32/EU, 2024).

How much of the universality of the right to asylum can be saved under the condition of the post-liberal era depends largely on the EU's ability to design and implement such appropriate, but partially certainly unpopular and morally challenging reforms. In contrast to such endeavors, it is still surprisingly common among intellectuals engaged in normative debates to focus solely on the universality of rights and to criticize the very existence of borders and the associated restrictions on migration (Di Cesare, 2019). However, since human rights are a product of political and intellectual struggles specific to European history, these very rights cannot be interpreted against the existential conditions of European nations without risking the weakening of those rights.

References

Anghie, A. (1996). Francisco de Vitoria and the Colonial Origins of International Law. *Social & Legal Studies*, 5(3), 321–336.

Answer for question E-003907/25. (2025). European Parliament. https://www.europarl.europa.eu/doceo/document/E-10-2025-003907-ASW_EN.html.

Augusti, E. (2025). From Vitoria to Kant: A Genealogy of Ius Migrandi. In L. Nuzzo, M. Pifferi, G. Speciale, & C. Vano, *Legal Responses to Mass Migration* (pp. 3–32). Routledge.

Balázs, E. (2025). *Priority Dossiers under the Polish EU Council Presidency*. European Parliamentary Research Service. https://www.europarl.europa.eu/thinktank/en/document/EPRS_BRI(2025)767187.

Bohmer, C., & Shuman, A. (2018). *Political Asylum Deceptions*. Springer International Publishing.

Bresselau Von Bressensdorf, A. (2019). Diskurse gesellschaftlicher Selbstvergewisserung am Ende des Kalten Kriegs Die Debatte um den Asylkompromiss 1992/93. In R. Etzold, M. Löhnig, & T. Schlemmer (Eds.), *Migration Und Integration in Deutschland Nach 1945* (pp. 113–126). De Gruyter.

Buras, P. (2024, October 25). Tusk's Asylum Coup: Breaking Away from Populist Narratives. *European Council on Foreign Relations*. https://ecfr.eu/article/tusks-asylum-coup-is-a-break-from-populism-possible/.

Burgdorff, C. (2023). A Monument Made of Words? How the Metaphor of "Fortress Europe" Became A Collective Memory (1992–2008). *Journal of European Integration History*, 29(1), 53–70.

Cantat, C., & Rajaram, P. K. (2019). The Politics of the Refugee Crisis in Hungary: Bordering and Ordering the Nation and Its Others. In C. Menjívar, M. Ruiz, & I. Ness (Eds.), *The Oxford Handbook of Migration Crises* (pp. 180–196). Oxford University Press.

Cavallar, G. (2002). *The Rights of Strangers: Theories of International Hospitality, the Global Community and Political Justice since Vitoria*. Ashgate.

Communication from the Commission to the European Parliament and the Council on Countering Hybrid Threats from the Weaponisation of Migration and Strengthening Security at the EU's External Borders (COM(2024) 570 final). (2024). European Commission. https://eur-lex.europa.eu/legal-content/EN/TXT/?uri=COM%3A2024%3A570%3AFIN.

Convention Relative au Statut International Des Réfugiés. (1933, October 28).

Di Cesare, D. (2019). *Resident Foreigners: A Philosophy of Migration*. Polity.

Die Schweiz und die Flüchtlinge zur Zeit des Nationalsozialismus. (2001). Unabhängige Expertenkommission Schweiz.

DW (staff). (2024, April 14). Cyprus Suspends Syrian Asylum Applications. *Deutsche Welle*. https://www.dw.com/en/cyprus-suspends-syrian-asylum-applications/a-68814230.

European Commission (2024). *Communication from the Commission to the European Parliament and the Council on Hybrid Threats and the Instrumentalisation of Migration at the EU's External Borders*. https://doi.org/10.2765/123456.

European Commission v Hungary, Case C-808/18 (ECJ 17 December 2020). https://eur-lex.europa.eu/legal-content/en/TXT/?uri=CELEX:62018CJ0808.

Gatopoulos, D. (2019, October 11). Turkey's Refugee Plan Met with Widespread Skepticism. *AP News*. https://apnews.com/article/3ebfeb637623402e90f775433c04ae6c

Gatrell, P. (2013). *The Making of the Modern Refugee*. Oxford University Press.

Greenhill, K. M. (2010). *Weapons of Mass Migration: Forced Displacement, Coercion, and Foreign Policy*. Cornell University Press.

International Refugees (1938, July 18). *Time Magazine*, 16.

Jaeger, G. (2001). On the History of the International Protection of Refugees. *International Review of the Red Cross*, 83(843), 727–738.

Jochecová, K. (2024, July 22). Finland's Deportation Law Puts EU's Migration Norms to the Test. *POLITICO*. https://www.politico.eu/article/new-finnish-deportation-law-risks-infringing-on-human-rights-eu-migration-norms/.

Kant, I. (2006). Toward Perpetual Peace: A Philosophical Sketch. In P. Kleingeld (Ed.), *Toward Perpetual Peace and Other Writings on Politics, Peace, and History* (pp. 67–109). Yale University Press.

Karamanidou, L., & Kasparek, B. (2020, August 1). What Is Frontex Doing about Illegal Pushbacks in Evros? *Respond Migration*. https://respondmigration.com/blog-1/what-is-frontex-doing-about-illegal-pushbacks-in-evros.

Latest Asylum Trends. (2025). European Union Agency for Asylum. https://euaa.europa.eu/latest-asylum-trends-asylum.

Leonard, M. (2021). *The Age of Unpeace: How Connectivity Causes Conflict*. Bantam Press.

Marder, L. (2018). Refugees are Not Weapons: The "Weapons of Mass Migration" Metaphor and Its Implications. *International Studies Review*.

Mikić, V. (2024). Weaponized migration as a tool of clandestine aggression in contemporary international law. In *Collection Regional Law Review* (Vol. 5, pp. 121–130). Institute of Comparative Law.

Mikołajczyk, B. (2025, July). Polish Law Restricting the Right to Asylum at Borders – EU Immigration and Asylum Law and Policy. *EU Migration Law Blog*. https://www.eumigrationlawblog.eu/polish-law-restricting-the-right-to-asylum-at-borders/.

Mouzourakis, M. (2025, August 21). Asylum Bans in the European Union: How States Break the Law and Get Away with It. *Heinrich Böll Stiftung*. https://gr.boell.org/en/2025/08/21/asylum-bans-european-union-how-states-break-law-and-get-away-it.

Officials see signs of hybrid warfare in migrant crisis (2016, February 14). *Yle News.* https://yle.fi/a/3-8672574.

Papadimas, L. (2025, July 9). Greece Suspends Asylum Applications for Migrants from North Africa. *Reuters.*https://www.reuters.com/world/africa/greece-suspends-asylum-applications-migrants-north-africa-2025-07-09/.

Polish Soldier Dies after Being Stabbed by Migrants on Border with Belarus. (2024, June 6). *Reuters.* https://www.reuters.com/world/europe/polish-soldier-dies-after-being-stabbed-by-migrants-border-with-belarus-2024-06-06/.

Popper, K. R. (2013). *The Open Society and Its Enemies.* Princeton University Press.

President von der Leyen at the Press Conference in Greece. (2020). European Commission. https://ec.europa.eu/commission/presscorner/detail/nl/statement_20_380.

Sieradzka, M. (2025, February 28). *Poland temporarily suspends right to asylum.* Dw. Com. https://www.dw.com/en/poland-temporarily-suspends-right-to-asylum/a-71782793.

Regulation (EU) 2024/1348 of the European Parliament and of the Council of 14 May 2024 Establishing a Common Procedure for International Protection in the Union and Repealing Directive 2013/32/EU (2024). http://data.europa.eu/eli/reg/2024/1348/oj.

Regulation (EU) 2024/1359 of the European Parliament and of the Council of 14 May 2024 Addressing Situations of Crisis and Force Majeure in the Field of Migration and Asylum and Amending Regulation (EU) 2021/1147 (2024). http://data.europa.eu/eli/reg/2024/1359/oj/eng.

Talleraas, C., & Koch, K. (2024). *Border Management in Flux: Comparative Lessons from the Norway- Russia and Finland-Russia Borders (2015–2023)* (1(4); HVL-Policy Brief). Chr. Michelsen institute. https://www.cmi.no/publications/9354-border-management-in-flux.

Thumfart, J. (2012). *Die Begründung der Globalpolitischen Philosophie: Francisco de Vitorias Vorlesung über die Entdeckung Amerikas im ideengeschichtlichen Kontext.* Kulturverlag Kadmos.

Thumfart, J. (2025). Do we have to Choose between Liberalism and Surviving Climate Change? Intertemporal Freedom and the Climate Constitution Quadrilemma. *Law, Innovation and Technology,* 1–31. https://doi.org/10.1080/17579961.2025.2593774.

Thumfart, J. (2009). On Grotius's Mare Liberum and Vitoria's de Indis, following Agamben and Schmitt. *Grotiana,* 30(1), 65–87. https://doi.org/10.1163/016738309X12537002674286.

Thumfart, J. (2024). *The Liberal Internet in the Postliberal Era: Digital Sovereignty, Private Government, and Practices of Neutralization.* Palgrave Macmillan.

Thym, D. (2025). Does the Commission Cross the Rubicon? Legalising 'Pushbacks' on the Basis of Article 72 TFEU. *EU Immigration and Asylum Law and Policy.* https://

eumigrationlawblog.eu/does-the-commission-cross-the-rubicon-legalising-push backs-on-the-basis-of-article-72-tfeu/.

Turkey's Erdogan Threatened to Flood Europe with Migrants: Greek Website. (2016, February 8). *Reuters*. https://www.reuters.com/article/world/turkeys-erdogan-thr eatened-to-flood-europe-with-migrants-greek-website-idUSKCN0VH1R0/.

United Nations High Commissioner for Refugees (1951). *Convention Relating to the Status of Refugees*. United Nations. https://www.unhcr.org/3b66c2aa10.

Valstybės Sienos Apsaugos Tarnyba, Case: C-72/22 PPU (CJEU 30 June 2022). https:// curia.europa.eu/juris/document/document.jsf?text=&docid=261930&pageIndex= 0&doclang=en&mode=lst&dir=&occ=first&part=1&cid=10627695.

Vaznonytė, A. (2020). The Rotating Presidency of the Council of the EU – Still an Agenda-setter? *European Union Politics*, 21(3), 497–518. https://doi.org/10.1177/ 1465116520916557.

Vitoria, F. de. (1991). On the American Indians. In A. Pagden & J. Lawrance (Eds.), *Political Writings* (pp. 231–292). Cambridge University Press.

CHAPTER 16

The Rise of Artificial Intelligence: Threats and Opportunities

Marco Marsili

Abstract This chapter investigates the impact of artificial intelligence (AI) on digital ecosystems, focusing on emerging threats and opportunities within the framework of European regulations (Digital Services Act, AI Act) and Italy's national strategies. Through a legal-policy analysis, the study evaluates regulatory responses to AI-driven risks such as deepfakes, algorithmic bias, and disinformation, addressing how current regulations mitigate these threats, whether innovation can be reconciled with user protection, and the role of international cooperation in harmonizing standards. The central hypothesis posits that mitigating risks requires harmonized legal safeguards, technical countermeasures, and global collaboration.

Findings indicate that the Digital Services Act and AI Act establish important mechanisms, including mandatory deepfake labeling and restrictions on high-risk AI systems. Italy's alignment with EU priorities, through initiatives such as the Italian Institute for Artificial Intelligence (AI4I) and independent fact-checking projects, highlights the importance of ethical AI development in maintaining public trust. However, challenges remain in cross-border enforcement and responses to evolving threats such as real-time deepfake generation. The study emphasizes the need for stronger public-private partnerships, standardized detection tools, and digital literacy programs.

Keywords Artificial Intelligence Governance – Digital Services Act (DSA) – AI Act – Regulatory Frameworks – Italy National AI Strategy – International Cooperation

1 Introduction

The rapid integration of artificial intelligence (AI) into digital platforms has redefined global information ecosystems, offering unprecedented opportunities for innovation while simultaneously introducing systemic risks to democratic integrity. AI-driven tools – ranging from personalized recommendation

 HTTPS://DOI.ORG/10.1163/9789004763623-018

algorithms to generative models like deepfakes – embody this duality: they enhance user engagement, streamline services, and democratize creativity, yet also enable sophisticated misinformation campaigns, cognitive manipulation, and the erosion of public trust. This tension is starkly evident in political arenas, where AI-generated content and algorithmic amplification have been weaponized to distort electoral processes, amplify polarization, and undermine institutional credibility. For instance, the proliferation of deepfakes during the 2020 US elections underscores the urgency of addressing these dualities (Chesney & Citron, 2019).

Historically, AI's evolution – from rule-based systems to today's neural networks and large language models (LLMs) – has been marked by exponential advancements in capability and accessibility. The advent of machine learning (ML) in the 1980s and the rise of generative AI in the 2020s have revolutionized industries but also democratized tools for exploitation, such as synthetic media production (Russell & Norvig, 2021). These technological leaps have outpaced regulatory frameworks, creating a governance vacuum where ethical considerations often lag behind innovation. Against this backdrop, the European Union (EU) has emerged as a global pioneer in AI regulation, enacting landmark frameworks like the Digital Services Act (DSA) and the Artificial Intelligence Act (AI Act). These policies aim to balance innovation with accountability, mandating transparency in algorithmic processes, restricting high-risk AI applications, and imposing stringent penalties for non-compliance (European Commission, 2024). Their influence extends beyond Europe, setting benchmarks for international cooperation and inspiring similar initiatives in regions like North America and Asia (OECD, 2022). This chapter investigates AI's dual impact on digital ecosystems through three research questions:

1. Regulatory Efficacy: How do current EU and national regulations, such as the DSA and AI Act, address AI-driven threats like deepfakes and algorithmic bias?
2. Ethical-Industrial Balance: Can proactive regulatory frameworks reconcile rapid technological advancement with ethical imperatives such as transparency and accountability?
3. Global Governance: What role does international cooperation play in harmonizing standards to counter cross-border AI risks, particularly in jurisdictions with divergent legal traditions?

To answer these questions, the study employs a mixed-method approach, combining legal and policy analysis of EU frameworks and Italy's integration of the AI Act into its National AI Strategy adopted in 2021 and revised in 2024. Italy serves as a critical case study, exemplifying both the potential of aligning national policies with EU standards and the challenges of bridging urban-rural digital divides (OECD, 2022). Central to this work is the hypothesis that mitigating AI's risks requires a threefold strategy:

1. Harmonized legal safeguards: cross-border regulations to ensure platform accountability and consistency.
2. Technical countermeasures: investment in detection tools, such as deepfake identifiers and bias-mitigation algorithms.
3. Strengthened international collaboration: shared norms and enforcement mechanisms to address jurisdictional fragmentation.

While Europe's regulatory frameworks mark significant progress, their long-term success hinges on adaptive governance capable of evolving alongside AI's rapid development. For instance, the DSA's reliance on ex-post content moderation struggles to keep pace with generative AI's real-time content generation, highlighting the need for dynamic, risk-based oversight (Zuboff, 2019).

By critically examining these dimensions, this chapter aims to contribute actionable solutions for policymakers and technologists, emphasizing the urgency of preserving democratic values in an era where AI's transformative potential is inextricably linked to its capacity for harm. The analysis underscores that ethical AI governance is not merely a technical or legal challenge but a societal imperative requiring global solidarity, public engagement, and a redefinition of innovation's boundaries.

2 AI's Evolution, Threats, and Regulatory Challenges

The evolution of AI has transformed digital platforms, beginning with rudimentary rule-based systems in the mid-twentieth century and progressing to today's generative models like GPT-4 and DALL-E. Early AI applications focused on automating simple tasks, such as spam detection and predictive analytics, but lacked adaptability. The advent of ML in the 1980s marked a paradigm shift, enabling systems to learn from data rather than static programming. By the 2010s, deep learning and neural networks revolutionized AI's capabilities, allowing platforms like YouTube and Netflix to personalize content at scale (Russell & Norvig, 2021).

The rise of generative AI in the 2020s introduced unprecedented opportunities – and risks. Models like ChatGPT democratized content creation but also enabled synthetic media (e.g., deepfakes) capable of distorting reality. In Europe, these advancements coincided with growing concerns over algorithmic bias, data privacy breaches, and the weaponization of AI in political campaigns (Marsili, 2023). For instance, during the 2020 elections, AI-generated disinformation targeted Italian voters, exploiting social media algorithms to amplify divisive narratives (Kalpokas & Kalpokienė, 2022).

AI's integration into digital ecosystems has exposed systemic vulnerabilities. Deepfakes, powered by generative adversarial networks (GANs), emerged as a critical threat, enabling hyper-realistic impersonations of public figures. In Italy, fabricated videos of politicians circulated widely during regional elections, undermining trust in democratic processes (Marsili & Wróblewska-Jachna, 2024). Similarly, algorithmic bias in recruitment tools and facial recognition systems perpetuates inequalities, disproportionately affecting marginalized communities (Buolamwini & Gebru, 2018).

Social media platforms exacerbated these risks through engagement-driven recommendation systems. For example, Meta's algorithms prioritized sensationalist content, inadvertently amplifying conspiracy theories during the COVID-19 pandemic. While the EU's Digital Services Act mandated transparency in content moderation, platforms like YouTube struggled to balance free expression with the rapid removal of harmful AI-generated content (Vaccari & Chadwick, 2020). The EU's DSA and AI Act represent landmark efforts to mitigate AI risks. The DSA imposes strict obligations on Very Large Online Platforms (VLOPs), requiring risk assessments, algorithmic transparency, and collaboration with independent fact-checkers. Meanwhile, the AI Act bans "unacceptable risk" systems (e.g., social scoring) and mandates audits for high-risk applications like biometric surveillance. However, enforcement remains fragmented. Compliance with the DSA during the 2024 EU elections revealed persistent challenges: despite deploying AI moderation tools, the platform failed to curb AI-generated disinformation in multiple languages. This case underscores the limitations of relying solely on platform self-regulation and fragmented national oversight. To address such gaps, transnational coalitions – such as the EU-US Trade and Technology Council's proposed database for AI-generated content fingerprints – must operationalize shared enforcement mechanisms. For instance, a coalition-led auditing body could mandate real-time data sharing between platforms and regulators, enabling rapid identification of cross-border disinformation campaigns. Italy's advocacy for harmonized standards within the GPAI exemplifies this approach, linking localized enforcement to global frameworks.

By institutionalizing joint investigations and penalties for non-compliance, such coalitions could bridge jurisdictional divides, ensuring platforms like Meta face consistent accountability across markets. Similarly, the AI Act's focus on pre-market audits overlooks the dynamic nature of generative AI, which evolves faster than regulatory frameworks (Zuboff, 2019).

Italy's alignment with these regulations, through initiatives like General Data Protection Regulation (GDPR)–compliant data governance, highlights progress but also underscores jurisdictional limitations. For instance, AI developers outside the EU often bypass European standards, necessitating stronger international cooperation (OECD, 2022). Europe's regulatory frameworks excel in principle but face practical hurdles, and Italy's experience illustrates both the potential and limitations of these frameworks. While the country adopted GDPR and DSA provisions to combat misinformation, rural-urban digital divides hindered equitable AI deployment. The rise of generative AI raises critical questions in intellectual property. Models like Stable Diffusion or MidJourney, capable of producing digital artworks in seconds, challenge traditional notions of authorship. In 2023, the European Patent Office (EPO) rejected a patent application for an AI-generated invention, asserting that only humans can be recognized as inventors (EPO, 2023). In Italy, the Re:humanism art prize, hosted by Rome's MAXXI Museum, ignited debate: artworks co-created by humans and algorithms raised questions about attribution and compensation. OECD Reports on AI and Intellectual Property (2021, 2023) highlight the need to update EU copyright frameworks to include synthetic content clauses, ensuring fairness between human creators and AI developers.

3 Italy's AI Strategy: Bridging Innovation and Democratic Values

Italy has carved a distinct path in the European AI landscape by intertwining technological ambition with a steadfast commitment to ethical governance. The National Artificial Intelligence Strategy, launched in 2021, revised in 2024, and embedded within the broader Italia Digitale *2026* framework, reflects a dual mandate: to harness AI's transformative potential while safeguarding democratic principles. This approach positions Italy as a laboratory for policies that balance innovation with societal resilience, aligning with the EU's Digital Services Act and AI Act.

At the heart of Italy's strategy lies a commitment to technological sovereignty – a vision that prioritizes homegrown innovation while fostering European solidarity. The government has allocated €1.5 billion through Italia Domani (Italy Tomorrow), the National Recovery and Resilience Plan (NRRP), targeting sectors where AI can yield both economic and social dividends. A cornerstone of this effort is the Italian Institute for Artificial Intelligence (AI4I) in Turin, a hub fostering interdisciplinary research on explainable AI

(XAI) and bias mitigation. By mandating GDPR-compliant data practices, Italy ensures that AI systems in sensitive domains like healthcare adhere to strict privacy standards, setting a benchmark for ethical deployment (OECD, 2022).

Italy's approach diverges from broader EU strategies by emphasizing regional equity. For instance, the NRRP allocates 30% of AI funds to southern regions like Sicily and Calabria, addressing infrastructure gaps that historically hindered technological adoption. Under the National Recovery and Resilience Plan (NRRP), Italy has allocated €1.5 billion to advance AI-driven solutions in agriculture, with a focus on southern regions. A flagship initiative is the Digitalizzazione delle filiere agroalimentari (Digitalization of Agri-Food Chains), which integrates AI and satellite data to optimize irrigation, predict crop diseases, and reduce water waste in drought-prone areas. For example, in Sicily, ML models developed by the National Research Council (CNR) analyze satellite imagery to provide real-time recommendations to farmers, improving yields by 15–20% in pilot zones (NRRP Mission 1). These efforts are part of the broader Agricoltura 4.0 (Agriculture 4.0) strategy, which aligns with the EU's Green Deal objectives. These initiatives underscore Italy's recognition that AI's benefits must extend beyond urban hubs to foster inclusive growth.

Italian small and medium enterprises (SMEs) are emerging as laboratories for responsible innovation. The AI4SMEs project, funded by Italy's Ministry of Economic Development (MISE), supports 200 companies in adopting GDPR- and AI Act-compliant tools, emphasizing algorithmic transparency. For example, GreenAI, a Turin-based startup, uses AI to optimize e-waste recycling, reducing classification errors by 40% (Symbola Report, 2024). Challenges persist: only 7% of small businesses and 15% of medium-sized enterprises are engaged in AI projects (Politecnico di Milano's AI Observatory, 2024). Italy's National Innovation Fund (NRRP Mission 4) allocated €200 million to bridge this gap, fostering university-industry partnerships.

Italy's strategy confronts the existential threat of AI-driven misinformation head-on. During the 2023 regional elections, deepfake videos impersonating political figures flooded social media, threatening to distort public discourse. In response, Italian researchers partnered with the European Digital Media Observatory (EDMO) to develop detection algorithms for synthetic media. While these tools aim to improve content verification, specific accuracy metrics require validation through peer-reviewed studies or official reports. *This initiative complemented* fact-checking projects run by Pagella Politica and Openpolis, which systematically identify and correct misinformation on Italian media platforms. While exact figures are not publicly disclosed, independent reports estimate that similar EU-funded fact-checking initiatives in Italy address thousands of claims annually. Equally critical are efforts to fortify digital literacy: since 2023, Italy's National Plan for Digital Education has introduced pilot programs on AI ethics in secondary schools, aligned with the EU's Digital Education Action Plan.

The COVID-19 pandemic underscored the urgency of countering misinformation. Italian authorities, in collaboration with EU bodies like EDMO and fact-checking organizations such as Pagella Politica, employed AI tools to identify false vaccine narratives on social media. While comprehensive data on removed posts is unavailable for Italy, EU-wide efforts prioritized combating health-related disinformation during the pandemic. Public health campaigns led by the Italian National Institute of Health (Istituto Superiore di Sanità or ISS) contributed to reducing vaccine hesitancy, with national surveys indicating a decline in public resistance to vaccines between 2021 and 2023.

The transformative power of AI is increasingly evident in Italy's public sector. AI tools have been tested in Italy's public sector to process citizen feedback during policy discussions, aligning with the EU's push for transparent governance. Inspired by participatory democracy models like Barcelona's Decidim, Italy has developed digital platforms to enhance citizen engagement. For example, during the 2022 constitutional reform debates, AI-driven tools were employed to analyze public input, though specific figures on submissions require verification from official government communications.

The National Recovery and Resilience Plan prioritizes such innovations under its digital transition goals, aiming to modernize governance through technology. Similarly, the IO App – a government-developed platform used by 16 million Italians – has revolutionized bureaucratic processes. The Italian government's IO app has streamlined tax filings and healthcare appointments, improving efficiency in public services. However, only 18% of Italy's population actively engages in digital governance processes (ISTAT, 2023), underscoring the need for awareness campaigns.

These innovations are not without ethical quandaries. When AI risk-assessment tools at border checkpoints disproportionately flagged non-EU migrants, civil society groups criticized the algorithms' inherent biases. The government responded by revising training datasets and introducing human oversight mechanisms, a move praised by the Council of Europe for aligning with human rights standards (Council of Europe, 2023). Italy's media sector highlights AI's dual-edged potential. Public broadcaster RAI has publicly committed to aligning its digital strategies with the DSA, which mandates transparency in algorithmic processes to counter polarization. While specific metrics on content reduction are not disclosed, RAI's collaboration with initiatives like the European Digital Media Observatory underscores its focus on ethical AI deployment in media. These efforts reflect broader European trends toward responsible AI governance in public broadcasting. Yet smaller media outlets struggle to comply with the DSA's transparency mandates. Italy's commitment to inclusivity extends to education. The AI4T project, an Erasmus+ partnership between 17 EU entities, deploys adaptive learning technologies to address disparities in Italy's education system. Adaptive learning tools, such as those developed by the AI4T project, aim to personalize education in underserved regions. While pilot studies suggest positive outcomes, specific metrics on test score improvements require further validation. Concurrently, Italy's National Recovery and Resilience Plan prioritizes rural digitalization, with significant investments in broadband infrastructure to reduce connectivity gaps (NRRP Mission 1). Despite progress, challenges persist. The rapid evolution of generative AI technologies, such as LLMs, poses regulatory hurdles. Italy's alignment with the AI Act provides a framework for accountability, but enforcement gaps remain, particularly for multinational platforms.

4 Italy's Global Leadership in International Cooperation

Italy's AI governance extends beyond national borders, positioning the country as a pivotal actor in shaping global norms and fostering multilateral collaboration. As a founding member of the Global Partnership on AI, Italy has championed initiatives to counter synthetic media risks while advocating for ethical AI development. In 2023, Italy co-authored GPAI's landmark report Ethical Guidelines for Generative AI, which established protocols for watermarking AI-generated content and mandating transparency in synthetic media deployment. These guidelines, now adopted by 28 member states, draw directly from Italy's domestic experience in combating deepfake-driven disinformation during electoral cycles, offering a model for harmonizing innovation with democratic safeguards.

Italy's commitment to ethical AI aligns closely with UNESCO's Recommendation on the Ethics of AI (2021), which emphasizes inclusivity, cultural preservation, and environmental sustainability. Italy actively collaborates with international bodies like UNESCO to promote ethical AI applications in cultural preservation, aligning with its broader strategy to merge technological innovation with heritage stewardship.

The 2024 EU-US Trade and Technology Council (TTC) meeting in Leuven emphasized transatlantic cooperation on AI governance, including measures to address synthetic media risks. The joint statement endorsed collaborative efforts to improve transparency in AI-generated content, aligning with broader goals of the EU's Digital Services Act and AI Act. Italy's Mediterranean partnerships further underscore its dedication to equitable AI governance. Through EU-funded initiatives such as the PRIMA Programme (Partnership for Research and Innovation in the Mediterranean Area), Italy collaborates with North African nations to deploy AI-driven solutions for sustainable development. For instance, projects under PRIMA focus on optimizing water management in arid regions using ML, enhancing agricultural resilience to climate change. These efforts align with broader EU strategies, such as the Digital4Development (D4D) initiative, which promotes ethical AI deployment in regions vulnerable to environmental challenges. Italy's leadership extends to international forums such as the G7 and OECD, where it advocates for AI policies centered on human rights. In 2020, Italy endorsed the Rome Call for AI Ethics, a global initiative promoting transparency and fairness in algorithmic systems, co-signed by the Vatican, Microsoft, IBM, and UNESCO (RenAIssance Foundation, 2020). Additionally, Italy has aligned with the OECD's revised AI Principles (2023), which emphasize accountability in public-sector AI applications, including measures to counter algorithmic bias. Finally, Italy leverages its cultural diplomacy to advance global AI ethics. The Decidim Italia platform, adapted from Barcelona's open-source participatory democracy model, has been shared with Latin American nations through EU-funded partnerships. In Colombia, Decidim's AI tools have been utilized in citizen-led urban planning initiatives, fostering greater civic engagement (bogotaabierta.co). While these efforts aim to streamline bureaucratic processes, specific metrics on efficiency gains require further validation from official sources. Such efforts reflect Italy's vision of AI as a force for democratic empowerment, ensuring its global governance strategy remains rooted in collaboration, transparency, and equity.

5 Conclusions

AI's dual role as both a catalyst for innovation and a source of societal risk underscores the urgency of balanced governance. This study highlights how AI-driven tools – from generative models to recommendation algorithms – reshape digital ecosystems, enabling unprecedented efficiency and creativity while amplifying threats like deepfake-driven disinformation, algorithmic bias, and erosion of democratic trust. The European Union's Digital Services Act and AI Act exemplify proactive regulatory frameworks, establishing mechanisms for transparency, accountability, and risk mitigation. Italy's alignment with these frameworks, through initiatives like Italia Digitale *2026*, demonstrates a commitment to ethical AI deployment in sectors such as media and public administration. However, gaps persist in cross-border enforcement, adaptive governance, and public awareness, necessitating a threefold strategy: harmonized legal safeguards, technical countermeasures, and global cooperation.

Key recommendations emphasize strengthening enforcement of existing regulations, particularly by equipping oversight bodies with technical expertise to audit AI systems. Public-private partnerships should prioritize ethical AI design, such as developing tools for detecting synthetic media and combating misinformation. Digital literacy programs must empower citizens to critically engage with AI-driven content, while international collaborations – through platforms like the Global Partnership on AI – should harmonize standards to address jurisdictional challenges. Italy's leadership in multilateral initiatives provides a model for bridging regional and global efforts, ensuring inclusivity in governance frameworks.

Future research must address emerging intersections between AI and other emerging and disruptive technologies, which could revolutionize algorithmic capabilities but exacerbate cybersecurity risks. Equally critical is examining AI's societal impacts, including polarization from recommendation systems and strategies to foster digital inclusivity. Ethical AI design, particularly explainable systems, remains paramount to ensuring transparency in high-stakes domains like healthcare and law enforcement. Finally, aligning AI development with human rights principles – such as privacy and non-discrimination – requires interdisciplinary collaboration to anticipate misuse in politically sensitive contexts.

Ultimately, AI's trajectory hinges on reconciling innovation with ethical imperatives. Policymakers, technologists, and civil society must collaborate to build adaptive governance models that safeguard democratic values without stifling progress. By prioritizing transparency, inclusivity, and global solidarity, society can harness AI's transformative potential while mitigating its capacity for harm – a balance essential for fostering trust in an increasingly digital world.

References

Agenzia per l'Italia Digitale (AgID). (2023). *AI in Public Administration: Best Practices*. https://www.agid.gov.it.

AI for and by Teachers (AI4T). (n.d.). https://www.ai4t.eu.

Buolamwini, J., & Gebru, T. (2018). Gender Shades: Intersectional Accuracy Disparities in Commercial Gender Classification. *Proceedings of the 1st Conference on*

Fairness, Accountability, and Transparency, 77–91. https://doi.org/10.1145/3287560.3287593.

Chesney, R., & Citron, D. (2019). Deep Fakes: A Looming Challenge for Privacy, Democracy, and National Security. *California Law Review*, 107(6), 1753–1820. https://doi.org/10.15779/Z38RV0D15J.

Council of Europe (2023). *Ethical Audits of AI Systems in Migration Management.* https://rm.coe.int/ethical-audits-ai-migration/1680a9b7c9.

Decidim Free Software Association (n.d.). *Decidim.* https://www.decidim.org.

European Commission (EC). (2023). *Artificial Intelligence in the EU: Tackling Disinformation.* https://digital-strategy.ec.europa.eu.

European Commission (EC). (2024). *Joint Statement EU-US Trade and Technology Council of 4–5 April 2024 in Leuven, Belgium.* https://ec.europa.eu/commission/presscorner/detail/en/statement_24_1828.

European Patent Office. (2023). Artificial intelligence: Inventorship of AI inventions. European Patent Office. https://www.epo.org/en/news-events/in-focus/ict/artificial-intelligence.

Fondazione Symbola (2024). *GreenItaly 2024.* https://symbola.net/ricerca/greenitaly-2024/.

Global Partnership on AI (GPAI) (2023). *Multistakeholder Expert Group Annual Report 2023.* https://www.gpai.ai/projects/2023-MEG-report.pdf.

European Patent Office (EPO) (2019). *EPO Refuses DABUS Patent Applications Designating a Machine Inventor.* https://www.epo.org/en/news-events/news/epo-refuses-dabus-patent-applications-designating-machine-inventor.

Istituto Nazionale di Statistica ((ISTAT)). (2023). *Cittadini imprese e ICT – Anno 2023.* https://www.istat.it/tag/cittadini-imprese-e-ict/.

Italian Government, Presidency of the Council of Ministers. (2021). *National Recovery and Resilience Plan* (NRRP). https://www.italiadomani.gov.it/content/sogei-ng/it/en/home.html.

Italian Government (2021). *Strategic Program on Artificial Intelligence 2022–2024.* https://assets.innovazione.gov.it/1637777513-strategic-program-aiweb.pdf.

Italian Government (2024). *Italian Strategy for Artificial Intelligence 2024–2026.* https://www.agid.gov.it/sites/agid/files/2024-07/Italian_strategy_for_artificial_intelligence_2024-2026.pdf.

Italian Government, Ministry of Education and Merit (MIM). (n.d.). *AI4T-Artificial Intelligence for and by Teachers.* https://scuolafutura.pubblica.istruzione.it/risorse/intelligenza-artificiale.

Italian Government, Presidency of the Council of Ministers, Department for Digital Transformation (2023). *AI-Driven Responses to Health Misinformation in Italy: 2021–2023.* https://innovazione.gov.it/pubblicazioni/risposte-ai-driven-alla-disinformazione-sanitaria.

Kalpokas, I., & Kalpokiene, J. (2022). *Deepfakes: A Realistic Assessment of Potentials, Risks, and Policy Regulation*. Springer Nature. https://doi.org/10.1007/978-3-030-93802-4.

Marsili, M. (2023). Guerre à la Carte: Cyber, Information, Cognitive Warfare and the Metaverse. *Applied Cybersecurity & Internet Governance* (ACIG), 2(1), 1–11. https://doi.org/10.60097/ACIG/162861.

Marsili, M., & Wróblewska-Jachna, J. (2024). Digital Revolution and Artificial Intelligence as Challenges for Today. *Media i Społeczeństwo*, 20(1), 19–30. https://doi.org/10.5604/01.3001.0054.6506.

Organisation for Economic Co-operation and Development (OECD). (2022). OECD Science, Technology and Innovation Outlook 2023: Enabling Transitions in Times of Disruption. Paris: OECD Publishing. https://www.oecd.org/en/publications/oecd-science-technology-and-innovation-outlook-2023_0b55736e-en.html

OECD. (2022). *Harnessing the Power of AI and Emerging Technologies*. https://www.oecd.org/content/dam/oecd/en/publications/reports/2022/11/harnessing-the-power-of-ai-and-emerging-technologies_c74e020c/f94df8ec-en.pdf.

OECD. (2024). *Recommendation of the Council on Artificial Intelligence*.

OECD. (2024). *G7 Toolkit for Artificial Intelligence in the Public Sector*. https://www.oecd.org/content/dam/oecd/en/publications/reports/2024/10/g7-toolkit-for-artificial-intelligence-in-the-public-sector_f93fb9fb/421c1244-en.pdf.

OECD. (2022). *AI4Med: Climate Resilience in the Mediterranean*. https://www.oecd.org/regional/AI4Med.pdf.

Politecnico di Milano, Osservatorio Artificial Intelligence (2025). *Intelligenza Artificiale, boom del mercato italiano: +58%, 1,2 miliardi di euro*. https://www.osservatori.net/comunicato/artificial-intelligence/intelligenza-artificiale-italia/.

Regulation (EU) 2016/679 of the European Parliament and of the Council of 27 April 2016 on the protection of natural persons with regard to the processing of personal data and on the free movement of such data, and repealing Directive 95/46/EC (*General Data Protection Regulation* or GDPR). *OJ L* 119, *4.5.2016*, p. 1–88 (consolidated version). https://eur-lex.europa.eu/eli/reg/2016/679.

Regulation (EU) 2022/2065 of the European Parliament and of the Council of 19 October 2022 on a Single Market for Digital Services and amending Directive 2000/31/EC (*Digital Services Act*), PE/30/2022/REV/1, *OJ L 277, 27.10.2022*, p. 1–102. http://data.europa.eu/eli/reg/2022/2065/oj.

Regulation (EU) 2024/1689 of the European Parliament and of the Council of 13 June 2024 laying down harmonised rules on artificial intelligence (*Artificial Intelligence Act*) and amending Regulations (EC) No 300/2008, (EU) No 167/2013, (EU) No 168/2013, (EU) 2018/858, (EU) 2018/1139 and (EU) 2019/2144 and Directives 2014/90/EU, (EU) 2016/797 and (EU) 2020/1828 (*Artificial Intelligence Act*), PE/24/2024/REV/1, *OJ L*, 2024/1689, *12.7.2024*, ELI: http://data.europa.eu/eli/reg/2024/1689/oj.

Re:humanism (n.d.). https://www.re-humanism.com/about/.

RenAIssance Foundation (2020). *Call for AI Ethics*. https://www.romecall.org.

Russell, S., & Norvig, P. (2021). *Artificial Intelligence: A Modern Approach* (4th ed.). Pearson.

UNESCO (2021). Recommendation on the Ethics of Artificial Intelligence. Paris: United Nations Educational, Scientific and Cultural Organization. https://unesdoc.unesco.org/ark:/48223/pf0000381137.

United Nations Educational, Scientific and Cultural Organization (UNESCO). (2022). *Recommendation on the Ethics of Artificial Intelligence*, adopted on 23 November 2021, SHS/BIO/PI/2021/1. https://unesdoc.unesco.org/ark:/48223/pf0000381137.

Vaccari, C., & Chadwick, A. (2020). Deepfakes and Disinformation: Exploring the Impact on Democracy. *Social Media + Society*, 6(3). https://doi.org/10.1177/2056305120903408.

Zuboff, S. (2019). *The Age of Surveillance Capitalism: The Fight for a Human Future at the New Frontier of Power*. PublicAffairs.

PART 3

Political Dimensions of Security for States Aspiring to Join the European Union

CHAPTER 17

The Battle for Moldova: Europe's Path Amid Escalated Russian Propaganda

Viktor Denisenko and Kiryl Kascian

Abstract The 2024 Moldovan presidential election and the constitutional referendum on EU membership became key targets of Russian information influence activities. Russia promoted anti-European narratives through disinformation and propaganda, particularly in the pro-Russian regions of Transnistria and Gagauzia, aiming to undermine public support for European integration. This chapter examines the main anti-EU narratives disseminated by pro-Russian actors during the electoral process. The analysis is based on data collected from identified disinformation sources and the EU's EU vs Disinfo database. The findings contribute to understanding Russia's information warfare strategies in the EU's eastern neighborhood.

Keywords Moldova – Russia – EU integration – political communication – disinformation – propaganda

1 Introduction

Moldova is a small, landlocked post-Soviet country in Eastern Europe, situated between Romania and Ukraine. This geographical position, along with its historical experiences, has significantly influenced its political, cultural, and economic development. After gaining independence following the dissolution of the Soviet Union in 1991, Moldova faced complex nation-building challenges exacerbated by internal territorial and ethnic divisions. The issue of national identity has been a central element of Moldovan discourse since independence. As King (1999, p. 229) notes, Moldova is "the only country in Eastern Europe in which major disputes existed among politicians and cultural elites over the fundamentals of national identity." Within Moldova, it is embodied in the division between those who view themselves as fundamentally ethnic Romanians and those who emphasize a distinct Moldovan identity. The former group sees Moldovan identity as an artificial construct, created by the

 HTTPS://DOI.ORG/10.1163/9789004763623-019

Soviet regime, while the latter emphasizes Moldova's unique historical path, sometimes even positioning it as preceding Romanian history (Simionov, 2023, p. 203). Hence, Moldova's complex geopolitical reality is also defined by its deep-rooted cultural and linguistic connections to Romania, alongside the enduring influence of its Soviet history and Russia's focus not only on the country as a whole, but also on specific regions within it, as a notable component of its sphere of influence (Całus & Kosienkowski, 2018, p. 99).

Moldova's route to independence was marked by secessionist movements in its regions of Transnistria and Gagauzia. Three decades later, Transnistria remains a multiethnic, Russophone *de facto* state that employs culturally and linguistically homogenizing policies, with a flexible, eclectic model combining formal and informal institutions (Osipov & Vasilevich, 2019, p. 995). Its continued existence is largely due to direct Russian support, including the presence of a Russian military base (Rutland, 2023, p. 25). In 1995, Gagauzia, a southern Moldovan region with a Turkic Orthodox population, was reintegrated into Moldova as a territorial autonomy. However, Gagauzia and its authorities continuously rely on a "Russian World"-based approach to legitimize their policies (Kosienkowski, 2021). Thus, Russia employs a multi-faceted strategy in Moldova, engaging with the country as a whole while also leveraging its geopolitical influence over Transnistria and Gagauzia to exert broader control over the central authorities in Chișinău.

Moldova's political discourse has largely featured the pursuit of EU integration. The country's 2014 Association Agreement with the EU, which included a Deep and Comprehensive Free Trade Area designed to strengthen political and economic ties, has since been followed by efforts to implement reforms aligned with European standards. However, this process has underlined Moldova's political vulnerability, marked by alternating phases of advancement and regression. Notably, from 2016 to 2020, the EU became concerned with Moldova's implementation of the Association Agreement, particularly issues with the rule of law, democratic standards, and civil society, which resulted in a reduction of financial aid (EaP CSF, 2025, p. 5). This period coincided with a time when Moldova was under the leadership of authorities embedded with Russia, headed by then-president Igor Dodon. This political vulnerability confirms the importance of electoral campaigns for Moldova and its society, as the division of the country's political spectrum into pro-Russian and pro-European political forces shapes domestic policies and the sustainability of its geopolitical choices, all while being subject to external geopolitical dynamics and potential interference.

With the launch of its full-scale aggression against Ukraine in February 2022, Russia has also intensified its hybrid campaign against Moldova, employing tactics ranging from weaponizing energy imports to spreading disinformation to undermine the pro-European government and create tensions in the society, particularly among Russian-speaking Moldovans (International Institute for Strategic Studies, 2023, p. X). In turn, Russia's invasion of Ukraine prompted the EU to reassess its policies toward its Eastern neighborhood, leading Moldova to apply in March and receive EU candidate status by June 2022. Accession negotiations officially began by June 2024, strengthening the country's ties with the Union. To further secure Moldova's path toward European integration, the pro-EU authorities led by Maia Sandu held a referendum to formally include the goal of EU membership in the constitution. This referendum was aligned with the first round of the presidential election, both taking place on October 20, 2024.

The prospective EU membership implies adopting its policies and standards, including those in the digital sphere. These EU digital policies aim to shape societal and economic transformation by promoting innovation, security, and fundamental rights. Thus, they serve as a strict roadmap for Moldova's accession and a means for its democratic and economic modernization. In turn, Russia's diverse hybrid tactics threaten the EU's digital autonomy. Cyberattacks, disinformation campaigns, and interference in elections not only target the EU's security but also erode public trust in its institutions and enlargement plans. Moldova is an important case for the EU in this regard. The puzzle consists of two elements involving both the EU stance and Moldova's domestic developments.

First, an analysis of EU Council presidency programs reflects Moldova's candidate status, which was granted in June 2022 during the French presidency. Since then, all programs treat Moldova accordingly, though with varying degrees of detail. They acknowledge the need to support Moldova in its path toward the EU by implementing comprehensive reforms to foster closer ties with the EU. Digital resilience is a contextually relevant element of this process. For instance, the 2023 Swedish programme suggests the significance of a common reaction to hostile cyber threats and emphasizes the need to more effectively merge cyber diplomacy and the EU's external digital policy with its general foreign policy (Swedish Presidency Programme, 2023, p. 7). The 2024 Belgian presidency aimed to strengthen EU resilience and advance policies related to cyber, hybrid, and digital issues (Belgian Presidency Programme, 2024, p. 14). The 2024 Hungarian EU presidency underscored that digital technologies are key to the competitiveness of the EU and its Member States (Hungarian Presidency Programme, 2024, pp. 23–24). This principle equally applies to candidate countries preparing for EU membership. The 2025 Polish EU presidency addressed both the need to strengthen Moldova's resilience and consistently support the reforms on the country's path to the EU (Polish Presidency Programme, 2025, p. 14).

Second, Moldova's domestic developments reflect an updated National Security Strategy (hereinforth: NSS) from October 2023, adapting to the geopolitical realities in the region. The NSS views the EU as a key security pillar, while authorities recognize Moldova's vulnerability to Russia's "prolonged and high-intensity hybrid war" and consider Russia an "existential threat" to Moldova's sovereignty (Presedinte.md, 2023). Challenges include persistent disinformation and low cyber hygiene among public servants. Nevertheless, Moldova aims to "increasingly act as a de facto member of the European Union," focusing on modernization, economic development, and comprehensive security (Presedinte.md, 2023). The NSS includes actions to boost national resilience, such as strengthening cybersecurity cooperation with the EU and aligning with EU regulations on transformative technologies.

Given Russia's aggression against Ukraine, the Digital Europe program is how the EU responded to weaknesses in its digital supply chains and cybersecurity (European Commission, n.D.). Moldova's case, as an EU candidate, clearly demonstrates that information warfare is an active tool, impacting political processes and thus affecting EU policies and its reputation. The 2024 presidential elections and European integration referendum were the first major campaigns under this new NSS, aiming to strengthen resilience against hybrid threats. This electoral context demonstrates how transposing EU regulatory standards can serve as a countertool against heavy Russian interference, making Moldova's experience crucial for understanding the EU's role as a resilient digital-age actor in preparing candidate countries for membership. The analysis of this campaign thus links the issues of cybersecurity, propaganda, and EU enlargement.

In this chapter, key anti-European narratives promoted by pro-Russian actors in Moldova's political landscape during the electoral process are analyzed. The data was collected from identified disinformation sources using the EUvsDisinfo database, a project of the European Union, to track these narratives. The aim was to outline the dominant anti-European messages in Moldova's information space and shed light on Russia's information warfare strategies in the EU's eastern neighborhood.

2 Propaganda and Manipulation as Tools for Russian Interference in Moldovan Elections

As Galeotti (2022, pp. 160–175) notes, the weaponization of information has become a major global concern, and Russia has effectively leveraged manipulation and disinformation to pursue its national interests against the West. Russian information warfare against other countries is part of the Kremlin's hybrid activities. These actions have certain traditions. It is connected to the conception of the so-called "geopolitics of information." Lithuanian scientist Nerijus Maliukevičius defined this phenomenon as "a policy based on the control of information flows and aggressive methods of psychological impact aimed at establishing dominance in a specific information space affecting the geocultural attitudes of society for own benefit and motivating the political behavior of individual persons" (Maliukevičius, 2008, p. 192).

Moscow has consistently demonstrated significant attempts to affect democratic processes (especially elections) abroad. For instance, at the end of 2024, the Romanian Constitutional Court annulled the results of the first tour of presidential elections owing to manipulations on social media. These circumstances "resulted in the violation of the principle of equal opportunities for all presidential candidates" (Całus, 2024). It was suspected that Russia was behind these activities.

In the case of the Moldovan 2024 presidential election and constitutional referendum, the Kremlin's interference was evident. According to the EU, the situation involved an "unprecedented malign interference by Russia" (Council of the EU, 2024). Russia has directly and indirectly used pro-Russian political actors in Moldova to undermine its pro-European path, employing two key tools: vote-buying and disinformation campaigns. Disinformation campaigns, conducted through Moscow-based platforms promoting anti-EU and pro-Russian narratives, amplified the vote-buying scheme, which is estimated to have targeted over 300,000 citizens and utilized Telegram bots (Marandici, 2025, p. 13).

This highlights the urgent need to analyze the content of the dominant anti-European narratives circulating within Moldova's information space. Consequently, the chapter is structured as follows: it begins by outlining the relevant conceptual framework, proceeds with an analysis that identifies and explains these narratives, and concludes with a summary of the findings.

3 Conceptual Framework

The 2024 presidential elections and referendum on European integration in Moldova should be examined in the context of Russia's efforts to implement information geopolitics. This research analyzes the discourses promoted by Russia, with the support of local media, within Moldova's information environment. The study focuses on identified agents of Kremlin disinformation and propaganda between September 1st and November 10th, 2024, encompassing the period from the start of the autumn political season to one week following the presidential election's second round.

In the context of the electoral process, five key disinformation outlets operating within Moldova's information space were identified through an analysis of data from the "EU vs. DISINFO" database. A detailed examination of this database led the researchers to concentrate on publications originating from Moldova-news.md, Evedomosti.md, Mejdurecie.md, Gagauznews.com, and Tiraspol-news.ru.

Notably, three of these outlets are registered under the Moldovan domain zone ".md." The remaining two outlets, while registered in different domain zones, bear names – Gagauznews and Tiraspol-news – that strongly suggest their affiliation with Gagauzia and Transnistria within Moldova, respectively. Furthermore, the data from the database reveals a significant surge in activity from Russian-backed disinformation sources during the period of the Moldovan election.

This research employs a discourse analysis methodology, examining a total of 140 texts sourced from the selected outlets: 91 from Moldova-news.md, 9 from Evedomosti.md, 16 from Mejdurecie.md, 16 from Gagauznews.com, and 8 from Tiraspol-news.ru.

One limitation of this research stems from its exclusive focus on media outlets publishing (dis)information in Russian. Therefore, narratives disseminated in Romanian, the official language of Moldova (while some continue to use the term "Moldovan"), and languages spoken in Moldova fall outside the scope of this study. Nevertheless, it is hypothesized that Russian-language data adequately captures the central narratives of Kremlin propaganda employed in Moldova during the 2024 electoral period.

A notable characteristic of the selected media outlets is their function, either wholly or partly, as aggregators. The majority of publications chosen for analysis were republished from Russian media sources or social media platforms, such as Telegram channels. The proportion of original articles was comparatively small (26 out of 140). This suggests that while these outlets present themselves as local news sources, they are, in fact, shaping their agendas by mirroring the agenda of the Russian information space.

4 Decoding Anti-European Narratives in Moldova

The analysis revealed clear patterns in how media outlets construct narratives around the election and related topics that align with Kremlin propaganda. A central discourse model emerges, supported by several key sub-discourses, or pillars. Russia's primary goal was to undermine the legitimacy of the election and electoral processes. Even before the election, media publications consistently promoted a narrative suggesting that the upcoming presidential election and referendum would be inherently unfair. This discourse was built upon several distinct narratives. Some narratives exploit genuine points of contention. A key example is the issue of voting access for the Moldovan diaspora. Moldovan citizens residing in Russia have significantly fewer polling places available to them compared to their compatriots in Western Europe. This disparity was used to manipulate public opinion, with a statement from Russian Ministry of Foreign Affairs spokesperson Maria Zakharova being quoted to support claims of "discrimination against Moldovan voters living in Russia by the official Chișinău" (Moldova-news.com, 2024k). Another sub-discourse aimed to establish that only one election outcome was plausible – the one favored by Russia. Moscow signaled its preference for candidate Alexandr Stoianoglo (Gagauz: Aleksandr Stoyanoglu), while expressing disapproval of Maia Sandu's re-election as President of Moldova. The narrative of the authorities' predetermined loss is further illustrated by the following quote: "Nevertheless, despite its political monopoly and complete support by the West, the Sandu regime is powerless in trying to rectify its anti-rating, not only at home but also abroad" (Moldavskie Vedomosti, 2024b). This implied that Sandu's unpopularity should have made her victory impossible. This sub-discourse created a logical trap: any result deviating from the predictions of Kremlin-backed propaganda sources would automatically be deemed illegitimate.

Propaganda outlets also attempted to discredit the pre-electoral processes in other ways. For example, they repeatedly stated that the "pre-electoral campaign in Moldova looked like a farce" (Tiraspol-news.ru, 2024) and that there were many "pre-electoral frauds" on the part of the authorities (Gagauznews.com, 2024b). Moldova's government was accused of using all possible political powers to influence the results of the referendum and presidential election. This emotional statement illustrates the point:

> I declare with all responsibility that Maia Sandu is stealing our elections, the young politician charged. The 2024 election process involves all ministries and government agencies, as well as Parliament, the SIS, the NCBC, the police, the Central Election Commission, the National Bank, the TV and Radio Council, and practically all media outlets (Moldavskie vedomosti, 2024c).

Following the elections, this discourse was reinforced by accusations of election and referendum result falsification. A significant volume of texts amplified the topic of alleged violations during the election. A particular focus for manipulation was Maia Sandu's significant support among Moldovan voters in the diaspora in Europe. Kremlin-backed propaganda exploited this fact in two primary ways. First, they sought to undermine the overall election results by arguing that:

> The opening of polling stations abroad does not guarantee public scrutiny of the voting process. It is physically impossible to monitor the legality and correctness of elections abroad (Moldavskie vedomosti, 2024a).

Second, they repeatedly asserted that Sandu could not be considered a legitimate winner because she lost based solely on the votes cast by voters residing within Moldova, disregarding the diaspora's votes. An example of this sentiment can be found in the following quote: "The so-called 'victory' of Sandu in the elections thanks to the Moldovan diaspora living in the West is cheating because the fate of the country should be decided by those who live in it" (Moldova-news.com, 2024n). Maia Sandu and her political team were frequent targets of propaganda and disinformation, with Sandu herself often demonized through various strategies.

The ruling authorities of Moldova were portrayed as having "betrayed national interests," primarily through "the break of ties with Russia" (Moldova-news.com, 2024c). This narrative was consistently repeated. Publications about the current Moldovan authorities frequently used the terms "regime" or even "nationalistic regime" (Moldova-news.com, 2024d). The following quote illustrates the core message used to frame Moldova's authorities under Sandu's leadership.

Igor Dodon, former president of Moldova and leader of the opposition Party of Socialists, stated that: "I voted in the presidential election to stop the four-year experiment of the current authorities over the country, for a Moldovan head of state who respects the constitution and will stop hatred and xenophobia" (Moldova-news.com, 2024h). A strong narrative emphasized the authorities' supposed crisis. Kremlin-backed media declared that "nowadays authorities are in agony; their time is running out" (Moldova-news.com, 2024c). Claims circulated about the "half-disintegration condition" of Moldova's authorities (Moldova-news.com, 2024l) and that Sandu and her team had no possibility "to improve their anti-ranking" (Moldavskie vedomosti, 2024b). Maia Sandu was personally targeted in the publications. Her administration's policies were labeled "anti-people's" (Gagauznews.com, 2024a). The rhetoric often included emotionally charged statements such as, "no one bullied the people as she [Sandu] did" (Gagauznews.com, 2024d). A recurring narrative claimed that Sandu's governance was "not in the interest of Moldova's people" (Moldova-news.com, 22.10.2024.). In a particularly extreme example of "black communication technologies," Sandu was even accused of suffering from schizophrenia and panic attacks (Gagauznews.com, 2024c). Another tactic used to discredit the Moldovan president and her team involved creating an apocalyptic vision of the consequences of Sandu's re-election and a successful referendum on EU membership. This primarily propagated the threat of (civil) war if Sandu continued her policies (Moldova-news.com, 2024i). Some publications suggested that Moldova was poised to become the next anti-Russia state (Moldova-news.com, 2024e), a particularly significant claim given that a key Russian "argument" for the war against Ukraine was the West's alleged desire to transform it into anti-Russia (Popova & Shevel, 2024). Further fueling fear, some speculated that Sandu's policies could lead to a renewed military conflict in Transnistria (Gagauznews.com, 2024e). Domestic political actors were not the only personalities within these narratives. The discourse surrounding the referendum and presidential election also included two global external actors: the West and Russia. The discourse on the West's role in the context of the referendum and election was negative. Statements mentioned that Maia Sandu had "enormous support from the West" (Moldova-news.com, 2024g) or that foreign politicians were campaigning for her re-election, which would allow her to continue "aggressive anti-Russian and pro-Romanian political course" (Moldova-news.com, 2024b). Moldova was often presented as a puppet under the control of Western powers. Sandu was shown as a political leader ruled "by Brussels and Washington" (Moldova-news.com, 2024j). There were speculations that the president was waiting for the "offer to continue the carrier in the West" (Moldova-news.com, 2024i) or that visits of foreign delegations to Moldova during the pre-election period should be evaluated as Western interference in the electoral process (Moldova-news.com, 2024a).

By contrast, Russia was portrayed in a positive or neutral way. Alexandr Stoianoglo, the primary opponent of Maia Sandu, was presented as a more acceptable candidate from Moscow's point of view. Stoianoglo appeared to be a politician who supported pragmatic relations with Russia, or a smart balance in relations with Russia and Europe, and understood the importance of ties with Moscow in the country's economy (Moldova-news.com, 2024m). The following visible trend in the publications was an attempt to deny any suspicion of Russia's interference in Moldova's election. All these suspicions are represented as "provided without evidence" (Moldova-news.com, 2024f). Manipulations of this topic were sometimes presented as ironic comments (e.g., Moldova-news.com, 21.10.2024f). The general model of this discourse can be described as follows: Europe (equated to the West) consistently accused Russia of intervening in Moldova's election, but Europe was the only actor interfering in it. This was quite a disturbing statement because all narratives presented in this analysis and the communication activities of Kremlin-backed propaganda in the context of Moldova's election prove Russia's attempt to affect the election and its results. Russian propaganda, disseminated by specific outlets in Moldova's information space, chose a side in the 2024 election and referendum. Specifically, Alexandr Stoianoglo was clearly supported during the presidential electoral campaign. In the context of the referendum, these outlets supported an anti-European choice. During the pre-electoral period, these outlets demonized all his possible opponents. This included attacks on Maia Sandu and the acting authorities, as well as broader criticism of Western powers. Moldova was portrayed as a kind of "failed state" under authoritarian and non-independent president Sandu. The possibility of changing the leader of the state to Stoianoglo and strengthening political and economic ties with Russia was presented by these outlets as the only way for Moldova to survive.

5 Conclusion

The 2024 Moldovan presidential election and constitutional referendum on EU membership were at the center of an intense struggle involving competing information campaigns. Russia actively engaged in disinformation and propaganda efforts to sway public opinion. These actions, designed to obstruct Moldova's path toward European integration, capitalized on pre-existing societal divisions and vulnerabilities. By disseminating narratives that cast doubt on the legitimacy of the elections and the advantages of EU membership, Russia, together with its collaborators within Moldova, aimed to destabilize the nation's political environment and preserve its own influence in this country.

The information war went beyond simply opposing EU membership. It was part of a larger plan to discredit Western institutions and promote alternative geopolitical alignments. This strategy involved presenting Russia as a source of stability and safety while simultaneously portraying the West as chaotic and declining. These messages were particularly effective in Moldovan regions with close historical and cultural connections to Russia, such as Transnistria and Gagauzia, where disinformation spread easily.

Overall, the analysis reveals aggressive behavior in Russia's information warfare campaign, which aimed to interfere with the 2024 referendum and presidential election in Moldova. A key element of this propaganda is a stark "black-and-white" worldview. Moldova is portrayed as a territory in a geopolitical struggle between Moscow (the "white" side) and the West (the "black" side). This directly aligns with Russia's broader political view of Europe and political developments there. In the Kremlin's perspective, its "near abroad," which mainly consists of post-Soviet states, is an arena of contention with the West as each seeks to exert influence.

The above analysis of pro-Russia communication indicates that Russia considers itself a stakeholder, and even an active participant, in Moldova's domestic political processes. This also reflects the application of principles of information geopolitics. While this concept is not new, current geopolitical shifts, partially driven by Russia, have compelled Moscow to act in a more openly aggressive manner. It is crucial to recognize that Russia's behavior is a calculated strategy.

References

Belgian Presidency Programme (2024). *Protect, Strengthen, Prepare. Programme of the Belgian Presidency of the Council of the European Union. First half of 2024.* https://wayback.archive-it.org/12710/20241018203746/https://belgian-presidency.consilium.europa.eu/media/3kajw1io/programme_en.pdf.

Całus, K. (2024). *Romania: Constitutional Court annuls the presidential elections.* Centre for Eastern Studies. https://www.osw.waw.pl/en/publikacje/analyses/2024-12-09/romania-constitutional-court-annuls-presidential-election.

Całus, K., & Kosienkowski, M. (2018). Relations between Moldova and the European Union. In P. Flenley & M. Mannin (Eds.), *The European Union and its eastern neighbourhood: Europeanisation and its twenty-first-century contradictions* (pp. 99–113). Manchester: Manchester University Press.

Council of the EU (2024). *Republic of Moldova: Statement by the High Representative on behalf of the European Union on the constitutional referendum and the first round of the presidential elections.* https://www.consilium.europa.eu/en/press/press-releases/2024/10/22/republic-of-moldova-statement-by-the-high-repr

esentative-on-behalf-of-the-european-union-on-the-constitutional-referendum-and-the-first-round-of-the-presidential-elections/.
EaP CSF (2025). *The future of the Eastern Partnership policy, its multilateral track and the role of civil society.* https://eap-csf.eu/content/uploads/2025/03/Policy-Paper-The-Future-of-the-Eastern-Partnership-policy.pdf.
European Commission (n.D.). *The Digital Europe Programme.* https://digital-strategy.ec.europa.eu/en/activities/digital-programme.
Gagauznews.com (2024a). *Molodiozhnyi marsh protiv evrointegratsii sostoyalsia v Komrate.* https://gagauznews.com/117031/molodyozhnyj-marsh-protiv-evrointegratsii-sostoyalsya-v-komrate.html.
Gagauznews.com (2024b). *Politolog: Referendum – predvybornaya makhinatsiya dla privlecheniya storonnikov Sandu.* https://gagauznews.com/117388/politolog-referendum-predvybornaya-mahinatsiya-dlya-privlecheniya-storonnikov-sandu.html.
Gagauznews.com (2024c). *Maia Sandu stradaet shizofreniey i panicheskimi atakami.* https://gagauznews.com/117520/majya-sandu-stradaet-shizofreniej-i-panicheskimi-atakami.html.
Gagauznews.com (2024d). *Rezultaty prezidentskikh vyborov i referenduma vyzyvayut massovoe nedovolstvo.* https://gagauznews.com/117535/rezultaty-prezidentskih-vyborov-i-referenduma-vyzyvayut-massovoe-nedovolstvo.html.
Gagauznews.com (2024e). *Esli Sandu pobedit, v Pridnestrovye nachniotsia boinia.* https://gagauznews.com/117634/esli-sandu-pobedit-v-pridnestrove-nachnyotsya-bojnya.html.
Galeotti, M. (2022). *The Weaponisation of Everything: A Field Guide to the New Way of War.* New Haven: Yale University Press.
Hungarian Presidency Programme (2024). *Programme of the Hungarian Presidency of the Council of the European Union in the Second Half of 2024.* https://wayback.archive-it.org/12090/20250412082714/https://hungarian-presidency.consilium.europa.eu/media/32nhoeop/programme-and-priorities-of-the-hungarian-presidency.pdf.
International Institute for Strategic Studies. (2023). Moldova's domestic and international response to Russia's invasion of Ukraine. Strategic Comments, 29(6), ix–xi. https://doi.org/10.1080/13567888.2023.2252292
King, C. (1999). *The Moldovans: Romania, Russia, and the Politics of Culture.* Stanford: Hoover Institution Press.
Maliukevičius, N. (2008). *Rusijos informacijos geopolitikos potencialas ir sklaida Lietuvoje.* Vilnius: Vilnius University Press.
Marandici, I. (2025). Oligarchs, Euroscepticism, and Russia's strategy in Moldova. *Moldovan Analytical Digest,* 001: 11–19. https://doi.org/10.3929/ethz-b-000715250.

Moldavskie vedomosti (2024a.). *Golosovanie grazhdan Moldovy za rubezhom dolzhno byt otmeneno.* http://www.evedomosti.md/news/golosovanie-grazhdan-moldovy-za-rubezhom-dolzhno-byt-otme.

Moldavskie vedomosti (2024b). *Zakharova: Neuverennost moldavskikh vlastei v sebe tak velika, chto oni vidiat ugrozu dazhe v multfilme "Tri kota."* http://www.evedomosti.md/news/zaharova-neuverennost-moldavskih-vlastej-v-sebe-tak-velika-c.

Moldavskie vedomosti (2024c). "Maia Sandu kradiot nashi vybory": policiya i chinovniki kak elektoralnye agenty PAS. http://www.evedomosti.md/news/majya-sandu-kradet-nashi-vybory-policiya-i-chinovniki-kak-el.

Moldova-news.com (2024a). *Vmeshatel'stvo Velikobritanii v moldavskie vybory.* https://moldova-news.com/world/2024/10/09/56383.html.

Moldova-news.com (2024b). *Vybory i referendum v Moldove: obzor glavnykh narusheniy.* https://moldova-news.com/world/2024/10/20/60805.html.

Moldova-news.com (2024c). *Vlah obvinila moldavskije vlasti v falsifikatsii rezultatov vyborov.* https://moldova-news.com/russia/2024/10/21/61992.html.

Moldova-news.com (2024d). *Vybory v Moldavii stali udarom po partii vlasti, schitaet politolog.* https://moldova-news.com/world/2024/10/21/61614.html.

Moldova-news.com (2024e). *Simonyian: v Moldavii, gde prochodiat "vybory."* https://moldova-news.com/world/2024/10/21/61233.html.

Moldova-news.com (2024f). *Uroven marazma.* https://moldova-news.com/world/2024/10/21/62350.html.

Moldova-news.com (2024g). *Vybory prezidenta Moldavii: rezultaty pervogo tura i prognozy.* https://moldova-news.com/world/2024/10/22/62521.html.

Moldova-news.com (2024h). *Dodon progolosoval na vyborakh prezidenta Moldavii.* https://moldova-news.com/world/2024/11/03/67349.html.

Moldova-news.com (2024i). *Ko vtoroy polovine dnia nakopilas faktura dla analiza khoda moldavskikh vyborov.* https://moldova-news.com/world/2024/11/03/67593.html.

Moldova-news.com (2024j). *Zapadu ne khvataet garantii pobedy Sandu na vyborakh.* https://moldova-news.com/moldova/2024/11/03/67603.html.

Moldova-news.com (2024k). *Zakharova: pobedu Sandu na vyborakh obespechili golosa diaspory v stranakh Zapada.* https://moldova-news.com/russia/2024/11/04/69101.html.

Moldova-news.com (2024l). Politolog Korinenko: Sandu vyigrala vybory tolko blagodaria golosam iz-za rubezha. https://moldova-news.com/world/2024/11/04/68902.html.

Moldova-news.com (2024m). *Ekspert sprognoziroval mrachnoe buduschee Moldavii posle vyborov prezidenta.* https://moldova-news.com/world/2024/11/04/68459.html.

Moldova-news.com (2024n). *Deputat PMR prokommentiroval prezidentskie vybory v Moldove.* https://moldova-news.com/world/2024/11/05/69264.html.

Nantoi, O., Platon, A., & Cristei, A. (2018). *Russian propaganda on "Odnoklassniki" the case of Republic of Moldova.* Chişinău: Institute for Public Policy.
Osipov, A., & Vasilevich, H. (2019). Transnistrian nation-building: a case of effective diversity policies? *Nationalities Papers*, 47(6): 983–999. https://doi.org/10.1017/nps.2018.26.
Kosienkowski, M. (2021). The Russian world as a legitimation strategy outside Russia: the case of Gagauzia. *Eurasian Geography and Economics*, 62(3): 319–346. https://doi.org/10.1080/15387216.2020.1793682.
Polish Presidency Programme (2025). *Security, Europe! Programme of the Polish Presidency of the Council of The European Union. January 1–June 30, 2025.* https://polish-presidency.consilium.europa.eu/media/zkcn0325/programme-of-the-polish-presidency-of-the-council-of-the-european-union.pdf.
Popova, M., & Shevel, O. (2024). Ukraine is Europe: lessons for Europe and for political science. *European Political Science.* https://doi.org/10.1057/s41304-024-00506-4.
Presedinte.md (2023). *National Security Strategy of the Republic of Moldova.* https://presedinte.md/app/webroot/uploaded/Proiect%20SSN_2023_En.pdf.
Rutland, P. (2023). Thirty years of nation-building in the post-Soviet states. *Nationalities Papers*, 51(1): 14–32. https://doi.org/10.1017/nps.2021.94.
Simionov, L. M. (2023). Shifting attitudes towards identity, borders and geopolitical choices: the case of Moldova. *Regional Science Policy & Practice*, 15(1), 200–222. https://doi.org/10.1111/rsp3.12613.
Swedish Presidency Programme (2023). *Swedish Presidency of the Council of the European Union – First Semester 2023.* https://www.consilium.europa.eu/media/65469/2023-jan-jun-se-programme.pdf.
Tiraspol-news.ru (2024). *V PMR prokommentirovali sokraschenie chisla uchastkov na vyborakh v Moldavii.* https://tiraspol-news.ru/society/2024/09/15/137084.html.

CHAPTER 18

Building Cooperation between Ukraine and EU for Strengthening Cybersecurity

Iryna Fyshchuk

Abstract Ukraine has made significant progress in digitalization and remains an important IT development hub despite the ongoing war. At the same time, the country is exposed to numerous cyber threats. This chapter analyzes cooperation between Ukraine and the EU in strengthening cybersecurity at local, national, and international levels. It examines practical measures such as training programs, knowledge exchange, technical support, and resource sharing. A key element of this cooperation is Ukraine's integration into the EU Digital Single Market and participation in programs such as DT4UA, EGOV4Ukraine, and EU4DigitalUA. The study also highlights key challenges, including limited coordination, resource constraints, and the need to further develop digital skills and cyber hygiene.

Keywords cybersecurity – cooperation – EU – digitalization – partnership – Ukraine

1 Introduction

Ukraine has been experiencing a full-scale war for nearly four years, and cybersecurity cooperation with the EU at the national, local, and private levels is a key aspect of strengthening resilience. Cybersecurity in Ukraine remains a critical issue due to ongoing cyberattacks that highlight the vulnerability of state institutions (Fyshchuk et al., 2024). At the same time, cybersecurity is a process (Craigen et al., 2014), not a state. It is not a one – time solution, but a continuous process of improvement, and building partnerships in this field will only make cybersecurity more resilient (Trim et al., 2021). Additionally, cybersecurity is a shared responsibility (Bay, 2016) that requires cooperation at many levels, including national governments, local authorities, private sector stakeholders, and partners from the European Union. In this context, European research on technology and security highlights the interdependence between digital infrastructures, cybersecurity governance, and technological sovereignty, underlining the importance of institutional coordination and increasing resilience within the EU cybersecurity framework (Savaş & Karataş, 2022; Carrapico & Barrinha, 2017; Crespi et al., 2021; ENISA). The Ukrainian case advances EU theory by extending existing analytical frameworks on technological sovereignty, cybersecurity governance, and crisis management.

Cyberattacks increasingly target critical infrastructure (Atkins & Lawson 2021), government agencies, local municipalities (Norris et al., 2024), private sector organizations, and individuals, highlighting the need for a multi-layered approach to strengthening cybersecurity. Indeed, during the time of full-scale Russian invasion, Ukraine has gained unique cyber experience that can be taken as lessons learned to jointly with the EU protect against cyber threats. This is also a war of minds, creativity, cooperativity, unity, and the defense of the values. In March 2022, Ukraine became a contributing member of the NATO Joint Cyber Defense Center of Excellence. The European Center of Excellence for Countering Hybrid Threats as well strengthened its cooperation (Fyshchuk, 2024, p. 219).

The application of technology and the development of technology in the EU countries should stimulate a review of their capabilities and global positioning, especially given China's assertiveness in the digital space and the US President Trump's protectionist policy. Geopolitically, the EU finds itself between two states that are strategically competing in the field of technology, trying to strengthen their positions in the international arena regarding technology by all possible means. Ukrainian cyber experience has the potential to serve as a roadmap for technology development in the EU. This experience, gained through real-world challenges, can contribute to achieving tangible results in the development of cybersecurity. In addition, Ukraine's location in Europe highlights the importance of geography and partnership as fundamental pillars for facilitating and supporting European cooperation in this area.

At the national level, the Ukrainian government has taken steps to align cybersecurity policy with EU standards and improve the legislative framework. However, effective implementation depends on coordination with local authorities and private sector stakeholders. At the local level, municipalities often face resource constraints – financial, technical, and human – with specialized cybersecurity expertise, making them particularly vulnerable to cyber threats, especially now that large numbers of people have left. Strengthening their capacity through inter-municipal cooperation, university curricula, and knowledge-sharing initiatives is essential to building resilience at the grassroots level. At the private level, business plays a vital role in providing digital infrastructure and technological solutions, especially now with Ukrainian cyber expertise. Alongside public-private partnerships, investments in cybersecurity startups, and collaboration with technology companies can strengthen the overall cyber defense not only of Ukraine, but of all of Europe.

This chapter examines how building cooperation between Ukraine and the EU can strengthen and improve cybersecurity between different actors that can create a comprehensive cybersecurity ecosystem. By analyzing existing policies and initiatives, the chapter aims to identify key levels of cooperation for strengthening cybersecurity between Ukraine and the EU.

2 Methods and Study Overview

For this section, the methodology outlines structured approaches: a mixed methods approach as a combination of qualitative analysis with semi-structured interviews with a diverse group of officials from Ukrainian government agencies, EU institutions and cybersecurity experts which were conducted between May 2023 and May 2024, as well as an analysis of documents on EU cyber policy related to Ukraine, Ukrainian cybersecurity strategies, and reports from international organizations including case studies of specific examples of Ukraine-EU cooperation in cybersecurity to assess the impact and lessons learned. The interviews mentioned in this chapter were initiated and organized within the framework of the project "Digital transformation in Ukraine and EU integration", which explored factors that enhance digital security in Ukrainian government agencies, including cooperation with the European Union (EU) and building partnerships. A total of 20 interviews were conducted in the study with a diverse group of experts, including 10 decision-makers, 7 IT managers and specialists, and 3 cybersecurity professionals. In addition, the analysis covered a wide range of documents, including reports, strategies, and programmatic documents from the European Union, published interviews, and materials from the second Kyiv International Cyber Resilience Forum 2025. The study also utilized information obtained from the Telegram channel "State Special Communications Monitoring". Altogether, the documentary materials comprised approximately 380 pages. To investigate the cooperation within digital transformation and cybersecurity during wartime, this chapter includes also a case study approach (Yin, 2009) within the context of the Russia-Ukraine war, with a particular focus on European integration. Being a former Ukrainian national trainer for the EU-funded U-Lead with Europe Program for public administration employees, conducted 20 semi-structured interviews with relevant public authority representatives, decision makers, and cybersecurity specialists. In summary, it is worth highlighting that 50% of respondents have an educational background directly related to the field of cybersecurity, including areas such as computer science, computer engineering, and information security. This group also includes respondents who hold decision-making positions.

Considering the sensitivity of the topic and the current situation in Ukraine, it was quite difficult to organize the interview. Many potential respondents declined, mentioning reasons such as restrictions on their duties as civil servants or related to military law. Other people expressed fear of being interviewed; some simply didn't have the time. In total, 150 potential participants were contacted via email, instant messaging, and phone, and most did not respond to the request. This was combined with a snowball technique (Patton, 2014) to contact relevant people who agreed to be interviewed. Through these interviews and a thematic coding process (Braun & Clarke, 2006), key collaborative initiatives in the digital and cybersphere were identified. The collection, storage, and analysis of interview data were based on ethical standards and the protection of interview participants' rights to voluntariness, anonymity, and confidentiality. The results of the study are described below.

3 Areas of the Cooperation in European Digital Integration Initiatives and Their Impact on Ukraine

The EU is the main partner for Ukraine, especially given the fact that Ukraine received the status of a candidate for EU accession in June 2022 (European Commission, 2022). On 2 February 2023, President von der Leyen and Ukrainian Prime Minister Denys Shmyhal signed the association agreement for Ukraine to join the European Digital Single Market Programme (European Commission, 2023). Furthermore, the European Union emphasizes the development of cooperation between countries in the field of cybersecurity through the creation of a network of cybersecurity centers and the initiative of the Digital Single Market, which provides for the pooling of resources for the development of innovative and safe digital technologies (Lysenko et al., 2024). Regarding the Cybersecurity Strategy of Ukraine until 2021 (President of Ukraine, 2021), it envisaged the need for Ukraine's participation in international cyberspace and the expansion of its own interests to international and European platforms, including the establishment of closer contacts with international stakeholders.

The government CERT-UA team, which operates under the State Service for Special Communications, processed 4,315 cyber incidents in 2024. This is 69.8% more than in 2023, when cybercriminals attacked Ukrainian cyberspace 2,541 times (SSSCIPU, 2025). The entity most vulnerable to cyberattacks is the national government, as confirmed in a special Microsoft report on the situation in Ukraine (Microsoft, 2022). At the same time, the EU has consistently supported Ukraine in its cyber resilience, since the EU Council Presidencies have played a significant role in shaping the digital agenda for Ukraine, particularly in response to the changing security environment following Russia's full-scale invasion (The European Council, 2024). In addition, EU Council Presidencies have influenced the digital agenda toward Ukraine in several distinct phases. In 2022 the Czech Presidency elevated cybersecurity and digital resilience as key priorities in response to Russia's full-scale invasion (Consilium Europa EU, 2022). Moreover, a few days before the Russian invasion, the Lithuanian Ministry of Defense announced the deployment of the EU Cyber Rapid Response Team (CRRT), a project born from EU military cooperation. From 8 to 12 cybersecurity experts from EU countries such as Estonia, Poland, and Romania are helping Ukraine's cyber defense to protect its networks (Omelianenko, 2023). With the Spanish Presidency in 2023, the link between digital transformation, strategic autonomy, and external partnerships was subsequently reinforced, further embedding Ukraine within broader EU digital sovereignty debates (Digital Europe, 2023). And in December 2023, the Tallinn Mechanism – a new multilateral cooperation instrument in the field of cybersecurity – was launched to help Ukraine protect its critical infrastructure (Ministry of Foreign Affairs of Ukraine, 2023). Furthermore, in 2024 the Belgian Presidency emphasized importance of resilience, security, and preparedness, strengthening coordination between cybersecurity, civil protection, and crisis management for Ukraine (Centre for Cybersecurity Belgium, 2023). Building on these achievements, in 2025 the Polish Presidency focused on defining technological security as an integral component of European security, promoting more institutionalized cooperation with Ukraine in the field of cybersecurity, including closer interaction with ENISA, enhanced information exchange, and alignment with the EU framework in the field of cybersecurity, mainly reinforcing Ukraine's role as a contributor to European digital and security resilience (Polish presidency Council of the European Union, 2025).

According to the head of SSSCIP, the exchange of "threat indicators" and joint training for cyber defense specialists are extremely important (Huber, 2022). Importantly, in the Resolution of the Cabinet of Ministers of Ukraine issued in April 2023 on the approval of the Action Plan for the implementation of the Foreign Policy Strategy (Cabinet of Ministers of Ukraine, 2023), Ukraine identifies, among all goals, joining the EU's PESCO projects, conducting a new Cyber Dialogue with the EU, and deepening cooperation with

ENISA in cyberspace. Also, Council working groups, ENISA, and EU coordination mechanisms offer distinct but linked functions – from setting strategic policy and technical guidance (ENISA) to operational implementation and capacity building with Ukraine – forming an essential framework for cohesive EU cybersecurity policy, especially integrating Ukraine's alignment with EU standards (NIS2) through dialogues and joint efforts against shared threats (ENISA, 2023). Despite numerous initiatives and ongoing cooperation efforts, Ukraine remains one of the regions most affected by cyberattacks, with the government sector being the primary target in 2023 (Microsoft, 2023). These findings highlight the need for further strengthening and improvement of cybersecurity cooperation mechanisms.

Another area of strengthening cooperation is the DT4UA – Digital Transformation for Ukraine initiative, which plays a key role in the development of digital infrastructure and cybersecurity in Ukraine. The project is focused on strengthening the protection of state information systems through the modernization of digital services, and developing e-governance and integrating with European cybersecurity standards. This initiative is part of a broader strategy for Ukraine's digital transformation, contributing to its integration into the EU's Digital Single Market and strengthening its position in the field of cybersecurity. The DT4UA and the EU4DigitalUA projects are part of the support of the digitalization of Ukraine by the European Union, which continues the work of the EGOV4Ukraine project. They are aimed at the development of digital government infrastructure, government e-services, personal data protection, and cybersecurity.

Regarding the EGOV4Ukraine initiative that was implemented within the U-LEAD with Europe program and is aimed at the digital transformation of public administration, increasing the efficiency of electronic services, and strengthening cybersecurity. Particularly key aspects of the impact on cybersecurity include such elements as:

1. development of secure electronic registries for local authorities, which reduces the risk of data leakage and cyberattacks;
2. integration of national digital solutions with European cybersecurity standards to increase the level of information protection;
3. strengthening the resilience of Ukrainian communities to cyber threats through the implementation of decentralized and secure e-government systems.

These efforts demonstrate that the EU has recognized the dangers of cyberattacks and has established cooperation with Ukraine to strengthen its resilience in a new type of struggle such as the cyber sphere and has taken measures to eliminate them, but such cooperation still needs to be strengthened.

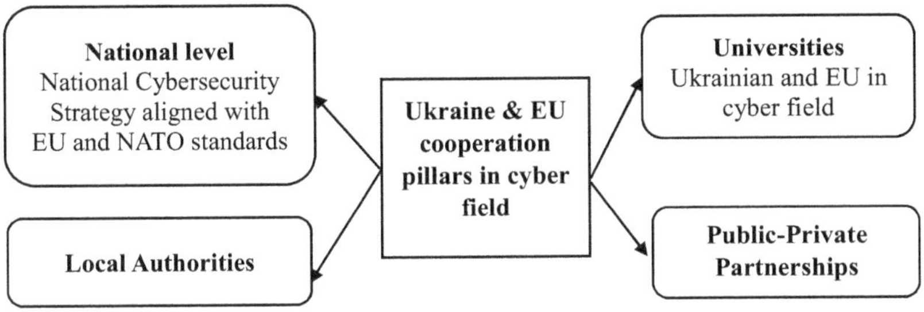

FIGURE 18.1 Pillars of cooperation in cyberspace between Ukraine and the EU (Source: Own elaboration.)

4 Levels of Cooperation for Strengthening Cybersecurity between Ukraine and the EU

Cyber cooperation is essential in determining how the future of a globally interconnected world will be shaped as cyberspace evolves (Radanliev, 2024). Cooperation is critical to successfully responding to incidents and limiting the impact of cyberthreats during cyberattacks. Personal data leaks and large-scale cyberattacks on critical infrastructure can lead to a variety of cyber situations, as in one such case that Ukraine faced – a cyberattack on the Kyivstar mobile operator with 24 million clients (New Voice, 2023), which lost communication possibilities. Additionally, communication is essential during cyber threats, and we need to rely on the reliable and established channels of communication. These channels should include government agencies, private sector organizations, law agencies, and international partners. Cybersecurity is not a sprint; it is rather a marathon, with a long-term building strategy, hence building cooperation in this field is very essential.

These pillars of cooperation between Ukraine and the European Union, with their national, local, university, and public-private levels, demonstrate the interaction and interconnectedness between all participants, while simultaneously providing structure and visual perception for the actors at these levels (Figure 18.1). At each pillar this model shows the importance of involvement in the proactive format of understanding the growing cybersecurity necessity. The examples in the model show that cooperation is taking place, although it still requires attention and systematic operational work in all aspects to strengthen such cooperation to improve cybersecurity in Europe as a whole.

Below, I will describe each component separately to demonstrate the possibilities for strengthening cooperation between Ukraine and the EU in the field of cybersecurity. By starting from the National Cybersecurity Strategy

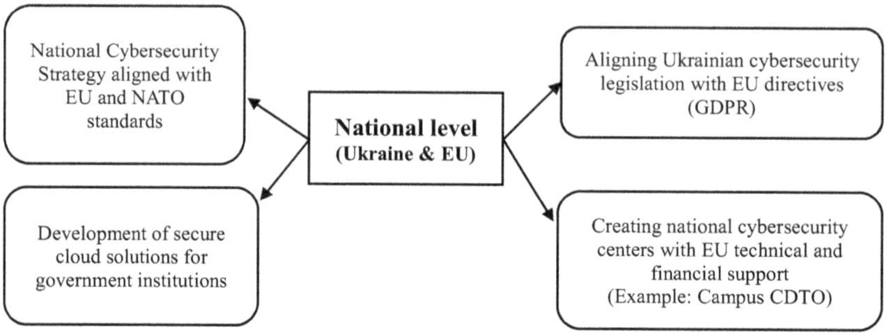

FIGURE 18.2 Cooperation at the national level in cyber field direction (Source: Author's own elaboration.)

aligned with EU and NATO standards, it is worth noting that the decree of the President of Ukraine No. 447/2021 – On the decision of the National Security and Defense Council of Ukraine dated May 14, 2021, "On the Cybersecurity Strategy of Ukraine" was adopted. This strategy defines the priorities of national interests in the field of cybersecurity, goals, and objectives in building a national cybersecurity system for the long term. Also, it is stated about the importance of the development of communication, coordination, and partnership between cybersecurity actors at the national level. Moreover, the development of strategic relations in the field of cybersecurity with key foreign partners, especially with the European Union, the United States, and other NATO member states, and international organizations.

Recently it was held the second Kyiv International Cyber Resilience Forum during 11–12 of March 2025, where as part of the forum, a Memorandum of Cooperation was signed between the European Cybersecurity Competence Centre (ECCC) and the National Cybersecurity Coordination Centre (NCSCC) that will deepen National Cybersecurity Strategy aligned with EU and NATO standards (National Security and Defense Council of Ukraine).

Regarding aligning Ukrainian cybersecurity legislation with EU directives (GDPR) it says that the Ukrainian Law needs to be aligned with the revised European Digital Identity Regulation and Ukraine will need to specify the rules in a number of areas (e.g. compliance with the GDPR or an equivalent level of protection of personal data) and with some provisions of the NIS2 Directive (network and information systems also known as Directive (EU) 2022/2555 of the European Parliament) (Directive, 2022/2555) in order ensure mutual recognition of electronic identification schemes from EU Member States in Ukraine according to European standards (Ukraine 2024 Report, 2024). According to the creation of national cybersecurity centers with EU technical and financial support. It is worth noting that a positive initiative at the national level, the Ministry of Digital Transformation of Ukraine, launched a CDTO Campus in November 2023. CDTO Campus is a national project to train digital transformation leaders, launched in Ukraine with the financial support from EU4DigitalUA; its goal is to provide the state with qualified specialists to implement digital reforms, as well as to train and support civil servants working in the field of digital transformation.

Ukraine is one of the first countries in the world to introduce the position of Chief Digital Transformation Officer (CDTO). Currently, 55 digital transformation deputies are employed across ministries, regional administrations, and city councils, serving as drivers of digital change. The Vice Prime Minister for Innovation, Education, Science, and Technology Development, as well as Minister of Digital Transformation, Mykhailo Fedorov, emphasizes the need for more effective managers capable of accelerating the digitization process both at the national and regional levels. In response to these needs, the CDTO Campus initiative was launched – an educational platform dedicated to training digital leaders. The programs implemented under this project aim to enhance the competencies of current and future CDTOs (CDTO Campus, 2023). CDTO Campus provides programs regarding cybersecurity such as:

1. Leadership Program in Cybersecurity for Executives which aims to equip participants with in-depth knowledge and practical tools to foster a culture of cybersecurity, develop and implement risk management strategies, and respond effectively to incidents.
2. Cybersecurity faculty which focuses on developing cybersecurity skills to ensure the reliable protection of government institutions and organizations (CDTO Campus, 2023).

Development of secure cloud solutions for government institutions is essential for strengthening cybersecurity in government agencies, which face significant challenges during a full-scale invasion. In February 2022, the Ukrainian Parliament adopted the law of Ukraine on Cloud Services (Digital Public Administration Factsheet, 2023). Now the Ukrainian legislation introduces requirements for cloud service providers for public customers; essential terms of the agreement on the provision of cloud services; features of providing cloud services to State authorities and local governments; and personal data processing and information protection when providing cloud services. However, there are still challenges include data breaches, unauthorized access, and compliance with stringent regulatory standards (Abikoye, 2024). By strengthening cloud solution services, it was considered to store government data in a custom-built private cloud, exemplifying the country's strategic alliances and emphasis on data security regarding Ukraine's partnership with Poland; moreover, the server in Poland hosts only Ukrainian data, and it was tested jointly by Ukrainian and Polish officials (Aviv & Ferri, 2023).

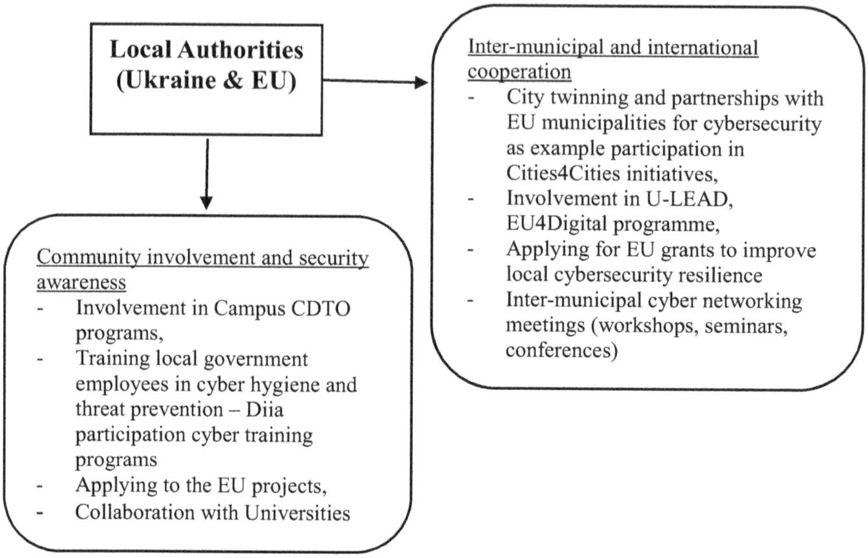

FIGURE 18.3 Cooperation at the local level in cyber field direction (Source: Author's own elaboration.)

When describing the local level represented above, it is important to note the role of inter-municipal and international cooperation with community involvement and security awareness. Where inter-municipal and international cooperation represented with city twinning and partnerships with EU municipalities for cybersecurity as example participation in Cities4Cities initiatives; then involvement in U-LEAD, EU4Digital program; applying for EU grants to improve local cybersecurity resilience; and inter-municipal cyber networking meetings (workshops, seminars, conferences).

Describing city twinning and partnerships with EU municipalities for cybersecurity as an example, it is a participation in the Cities4Cities initiative to exchange cyber experience (Cities4Cities Conference, 2025). This is an online platform, the purpose of which is to facilitate Ukrainian cities in obtaining the necessary assistance of various types, as well as establishing ad hoc or permanent partnerships between Ukrainian cities and cities from other countries belonging to the Council of Europe (Matiaszczyk, 2024). After the Russian aggression against Ukraine, the importance of the U-LEAD with Europe initiative increased (Matiaszczyk, 2024). Within this framework, several cyber hygiene trainings were conducted for Ukrainian communities. Cooperation between municipalities may strengthen resilience and improve cybersecurity at the local level, as public servants have greater access to information and communication channels (Fyshchuk, 2024, p. 2). By tightening the inter-municipal and international cooperation in cyber field, it will benefit cybersecurity competence and cyber skills in times of growing need the growing digitalization and technology development where highly needed digital skills in all sectors (Gervasi et al., 2023). Applying for EU grants to improve local cybersecurity resilience can provide municipalities with critical resources, facilitate knowledge sharing with European partners, and strengthen the overall security infrastructure needed to counter ever-changing cyber threats. However, the EU's support for Ukrainian local authorities and the possibility of applying for different grants, the application procedure for EU projects is often complex (Fyshchuk, 2025), and as mentioned by Smętkowski et al. (2023), there is no experience in applying for external funds in Ukraine. This complexity can be illustrated by the statement by local authorities' representatives that highlighted the significant administrative burden involved in preparing applications, meeting eligibility criteria, and complying with reporting requirements: "We have to deal with recovering physical damage, providing everyday services for our citizens, and at the same time the application for EU-funded projects needs special attention and qualification. We are engaged, but it takes time" (European Court of Auditors, 2021).

The importance of inter-municipal cyber meetings (seminars, workshops, conferences) lies in promoting knowledge exchange, building collective cyber resilience, developing coordinated response strategies, and establishing reliable communication channels, which are crucial for protecting against escalating cyber threats, especially during an ongoing war. As highlighted IT specialist from the western community: "Knowledge exchange is extremely important, and offline meetings are even more crucial, as they enable high-quality interaction, especially in the cybersecurity field" (European Union Agency for Cybersecurity- ENISA, 2022).

Turning to cooperation between universities in Ukraine and the European Union, it is worth emphasizing the opportunities for the exchange of education and knowledge in cyber field specialization presented in this chapter. Below is a table with the ten best Ukrainian universities in the field of cybersecurity. Below is a list of the top ten Ukrainian universities specializing in the field of cybersecurity. This ranking is based on data available on the Edurank.org portal, updated as of March 2, 2025. The highest position is held by V. N. Karazin Kharkiv National University, followed by Lviv Polytechnic National University in second place and Taras Shevchenko National University of Kyiv in third. Among the leading educational institutions are also the National Technical University of Ukraine "Igor Sikorsky Kyiv Polytechnic Institute", Kharkiv National University of Radio Electronics, and the National

Aviation University in Kyiv (Edurank.org, 2025). As noted, a decision maker from the southern region in Ukraine regarding the taken decision: "Since I have an education in computer since from National Technical University of Ukraine Kyiv Polytechnic Institute we have thought ahead and bought technical advanced computer equipment for our community in 2021 with the growing situation of digitalization and therefore cyber threats. The example of Ukraine shows that there is a growing need for education for decision makers, not just specialists in the digital and cyber spheres" (Ministry of Digital Transformation of Ukraine, 2021). Educational cooperation between Ukrainian universities and those of the European Union encompasses a wide range of activities aimed at developing digital competencies and strengthening cyber resilience.

A key component of this collaboration involves twinning programs and dual degree initiatives, which allow students to obtain qualifications recognized in both educational systems. Additionally, short-term study courses are organized to deepen knowledge and skills in specialized fields. Summer schools dedicated to cyber resilience also play an important role. These focus on developing competencies related to risk management and countering digital threats. At the same time, workshops and training sessions are held for local government officials to enhance their qualifications in cybersecurity and effective management of digital transformation processes at the local level. Knowledge sharing is further facilitated through joint research projects, which promote the exchange of experience and the development of innovative solutions in the areas of digitalization and cybersecurity. Best practices in digital governance and European integration processes are jointly developed and disseminated, contributing to the harmonization of standards and procedures. Furthermore, discussion panels featuring experts from various fields are organized to enable in-depth knowledge exchange, analysis of current challenges, and formulation of recommendations for educational and technological policy.

It is a long-term perspective for both sides, enhancing cooperation between businesses and universities in critical sectors (Smętkowski et al., 2023). Moreover, universities' cooperation can contribute to cybersecurity awareness, policy recommendations, and, the most valuable, workforce development. By providing joint study and research programs on cybersecurity between Ukrainian and European universities. Given that Ukraine has been in a situation of full-scale invasion for over three years, and the ATO (Anti-Terrorist Operation) has been going on since 2014, Ukrainian IT specialists have gained unique experience in cyber defense, as well as in restoring information infrastructure and data after cyber attacks. Indeed, this unique cyber experience, competencies, skills, and knowledge are important for cooperation with the EU for joint resilience and countering cyber threats.

In some cases, such cooperation with universities already exists, but it is not systemically. "We cooperate with Denmark on geographic information systems (GIS), also we engage in joint projects with UNICEF to develop general plans, and we have partnership with the Milan Polytechnic University. Additionally, we actively cooperate with the U-Lead program" (European Commission, 2023) – about existing Ukrainian European cooperation mentioned the decision maker from the southern part of Ukraine. Thinking ahead and now, creating and developing partnership programs, double degrees, short-term training courses, cyber resilience summer schools, as well as targeted cybersecurity seminars and trainings between Ukrainian and European universities, will contribute to the exchange of experience and strengthening cross-border academic and practical potential in this field for both the future and the present. It is also important to include knowledge-sharing opportunities within the cooperation framework, such as joint research projects, the development of best practices for digital governance and EU integration, expert panel discussions, and cybersecurity hackathons, all of which foster innovation, mutual learning, and practical collaboration.

Regarding public-private partnership (PPPs), Ukraine and the European Union have established several PPPs in the cybersecurity field, such as: Tallinn Mechanism (2023), ENISA–Ukraine Working Arrangement (2023), ECSO–Ministry of Digital Transformation Memorandum (2023), Cyber Gate Portal Initiative (2024) (CyberGate, 2024), particularly in response to the heightened cyber threats following Russia's invasion. Due to participation in these joint initiatives, it assists in enhancing information exchange, developing cybersecurity expertise and training programs, exchange of best practices, incident response coordination, long-term capacity building efforts, development of cyber hygiene training, shared situational awareness, coordinated mitigation, and recovery efforts. The public–private partnership between Ukraine and the European Union encompasses joint cybersecurity training initiatives aimed at enhancing competencies across both public and private sectors. A fundamental aspect of this collaboration involves the development of knowledge-sharing frameworks intended to improve the effectiveness of preventive and defensive measures in cyberspace. Furthermore, the partnership foresees the implementation of coordinated mechanisms for responding to cyber incidents, enabling timely and effective action in the face of escalating digital threats. Such an integrated approach is essential for strengthening cyber resilience and advancing information security at both national and international levels.

Although there are a range of challenges in PPPs such as different priorities, where public sector often focuses on national security, regulatory compliance, and public service in comparison with private sector which is driven by profit, competitiveness, and customer trust. The next challenge is that trust can limit the sharing of sensitive data, especially when it involves potential vulnerabilities (Axon et al., 2024).

5 Conclusions

Effective cybersecurity measures depend on cooperative information sharing, joint threat mitigation, capacity-building initiatives, and policy alignment by both sides in partnership relations. By building reliable and strong cooperation, organizations and institutions between Ukraine and the EU can increase resilience, develop innovative solutions, and resist emerging cyber threats. Focusing on the four pillars of cooperation in cyberspace between Ukraine and the EU such as national level cooperation, at the local level cooperation in cyber field, university cooperation in cyber sphere, public-private partnerships between Ukraine and the EU could be effective in decision-making processes on protecting citizens' personal data or destroying infrastructure. As digitalization continues to expand and reliance on electricity increases, enhanced cooperation between Ukraine – an EU-bordering country with Ukrainian practical cyber experience – and the EU, with its advanced technologies, can significantly strengthen cyber resilience on both sides.

Ukraine is experiencing real-world challenges from large-scale cyber attacks, and the constant impact of sophisticated cyber and hybrid threats has allowed Ukrainian institutions to develop advanced operational expertise in incident response, critical infrastructure protection, and crisis-driven digital governance. Sharing this expertise enhances the EU's ability to anticipate, detect, and respond to such threats. Moreover, by integrating Ukrainian experience and aligning cybersecurity practices with the EU framework, including cooperation with ENISA, the EU can strengthen its technological sovereignty and collective cyber resilience.

References

Abikoye B.E. (2024). Development of Secure Cloud-Based Government Solutions, *European Journal of Computer Science and Information Technology*, 12(5),17–35.

Atkins, S., & Lawson, C. (2021). An improvised patchwork: success and failure in cybersecurity policy for critical infrastructure. *Public Administration Review*, 81(5), 847–861.

Aviv, I., & Ferri, U. (2023). Russian-Ukraine armed conflict: Lessons learned on the digital ecosystem. *International Journal of Critical Infrastructure Protection*, 43, 100637.

Axon, L., Saunders, J., Esteve-González, P., Carver, J., Dutton, W., Goldsmith, M., & Creese, S. (2024). Private-public initiatives for cybersecurity: the case of Ukraine. *Journal of Cyber Policy*, 9(3), 399–422.

Bay, M. (2016). What is cybersecurity. *French Journal for Media Research*, 6, 1–28.

Braun, V., & Clarke, V. (2006). Using thematic analysis in psychology. *Qualitative research in psychology*, 3(2), 77–101.

Cabinet of Ministers of Ukraine (2023). Resolution On approval of the action plan for the implementation of the Strategy of Foreign Policy Activities of Ukraine. https://zakon.rada.gov.ua/laws/show/327-2023-%D1%80?lang=en#Text.

Carrapico, H., & Barrinha, A. (2017). The EU as a coherent (cyber) security actor?. *JCMS: Journal of Common Market Studies*, 55(6), 1254–1272.

CDTO Campus (2023). Ministry of Digital Transformation of Ukraine. https://thedigital.gov.ua/news/cdto-campus-v-ukraini-startue-natsionalniy-proekt-z-navchannya-lideriv-tsifrovoi-transformatsii.

Centre for Cybersecurity Belgium (2023). Cybersecurity priorities in the upcoming Belgian EU Presidency agenda. https://ccb.belgium.be/news/cybersecurity-priorities-upcoming-belgian-eu-presidency-agenda?utm_source=chatgpt.com.

Cities4Cities Conference | United4Ukraine Initiative. (n.d.). Retrieved August 2, 2025, from. https://cities4cities.eu/about/.

Consilium Europa EU (2022). Priorities of the Czech Presidency of the Council of the European Union in 2022. https://www.consilium.europa.eu/media/57090/2022-jul-dec-cz-priorities.pdf.

Craigen, D., Diakun-Thibault, N., & Purse, R. (2014). Defining cybersecurity. *Technology Innovation Management Review*, 4(10), 13–21.

Crespi, F., Caravella, S., Menghini, M., & Salvatori, C. (2021). European technological sovereignty: an emerging framework for policy strategy. *Intereconomics*, 56(6), 348–354.

Cyber Gate: new Ukraine cybersecurity portal launched with EUAM support, 18.04.2024. https://eufordigital.eu/cyber-gate-new-ukraine-cybersecurity-portal-launched-with-euam-support/.

Digital Europe (2023). 8 priorities for the Spanish EU Presidency. https://www.digitaleurope.org/news/8-priorities-for-the-spanish-presidency/.

Digital Public Administration Factsheet 2023. (2023). European Commission. https://interoperable-europe.ec.europa.eu/sites/default/files/inline-files/DPA_Factsheets_2023_Ukraine_vFINAL_rev.pdf.

Directive (EU) 2022/2555 of the European Parliament. https://eur-lex.europa.eu/legal-content/EN/TXT/?uri=CELEX:32022L2555.

DT4UA official website. https://eu4digitalua.eu/en/dt4ua-en/.

Edurank.org, 2025, https://edurank.org/cs/cybersecurity/ua/.

EGOV4Ukraine official spainwebsite. https://eufordigital.eu/uk/discover-eu/egov4ukraine/.

ENISA (2023). Enhanced EU-Ukraine cooperation in Cybersecurity. https://www.enisa.europa.eu/news/enhanced-eu-ukraine-cooperation-in-cybersecurity.

EU Policy Update (2022). https://www.centr.org/news/eu-updates/june-2022.html?utm_source=chatgpt.com.

European Commission (2022). *Ukraine Membership status: candidate country.* https://enlargement.ec.europa.eu/european-neighbourhood-policy/countries-region/ukraine_en.

European Commission (2023). *Support actions for Ukraine. The single market. What the EU is doing to integrate Ukraine into the single market and promote business cooperation.* https://single-market-economy.ec.europa.eu/support-actions-ukraine_en.

European Commission. (2023). *U-LEAD with Europe: Strengthening local governance in Ukraine.* https://u-lead.org.ua/en.

European Court of Auditors. (2021). *Challenges in the implementation of EU-funded projects by local and regional authorities.* Luxembourg: Publications Office of the European Union.

European Union Agency for Cybersecurity (ENISA). (2022). *Cybersecurity in the public sector: Enhancing cooperation and resilience* (ENISA Report). https://www.enisa.europa.eu/publications/cybersecurity-in-the-public-sector-enhancing-cooperation-and-resilience.

Fyshchuck, I. (2024). Strengthening Municipalities' Resilience against Cyber Attacks in Ukraine, in: Dudzik, S., Kawka, I. and Śliwa, R. (eds). *E-Government. Challenges for Digital Public Services in the EU.* Krakow: Księgarnia Akademicka, 171–188. https://doi.org/10.12797/9788383680255.09.

Fyshchuk, I. (January 2025). Bridging the cyber security gap: Why Ukrainian local authorities need more support for cyber security from the EU. https://hromada.network/ukrainian-hromadas-need-eu-cybersecurity-support/.

Fyshchuk, I. (2024). Stronger together? EU Support for Ukrainian Local Authorities Facing Cyber Attacks (2022–2023). *Applied Cybersecurity & Internet Governance,* 3(1), 204–226. https://doi.org/10.60097/ACIG/190344.

Fyshchuk, I., Noesgaard, M. S., & Nielsen, J. A. (2024). Managing cyberattacks in wartime: The case of Ukraine. Public Administration Review. https://doi.org/10.1111/puar.XXXXXXX

Gervasi, O., Murgante, B., Rocha, A. M. A., Garau, C., Scorza, F., Karaca, Y., & Torre, C. M. (2023). Computational Science and Its Applications – ICCSA 2023 Workshops. In Proceedings of the 23rd International Conference on Computational Science and Its Applications (ICCSA 2023). Springer.

Huber, N. (2022). What Ukraine's cyber defence tactics can teach other nations. *Financial Times*. https://www.ft.com/content/9635c4a0-1f42-44f1-bc9a-503b192f809f.

Lysenko, S., Liubchenko, A., Kozakov V, D. Y., & Krutik, Y. (2024). Global cybersecurity: Harmonising international standards and cooperation. *Multidisciplinary Rev iews*. 2024. Vol. 7,№ e2024spe021. DOI: https://doi.org/10.31893/multirev. 2024 spe021.

Matiaszczyk, N. (2024). City diplomacy as a mechanism of multi-level solidarity and support for Ukraine: A study of the changes following the 2022 Russian invasion. *Journal of Eurasian Studies*, 18793665241254835.

Microsoft Digital Defense Report (2023). *Building and improving cyber resilience*. https://www.microsoft.com/en-us/security/security-insider/microsoft-digital-defense-report-2023.

Microsoft. Special Report: Ukraine (2022). Digital Security Unit. *An Overview of Russia's Cyberat-tack Activity in Ukraine.*

Ministry of Digital Transformation of Ukraine (2021). *Digital transformation and cybersecurity initiatives in Ukraine*. https://thedigital.gov.ua/en.

Ministry of Foreign Affairs of Ukraine (2023). Tallinn Mechanism: Ukraine and international partners launched a new tool for cyber cooperation. https://mfa.gov.ua/en/news/tallinn-mechanism-ukraine-and-international-partners-launched-new-tool-cyber-cooperation.

National Security and Defense Council of Ukraine. https://www.rnbo.gov.ua/en/Diial nist/7142.html.

New Voice. 2023. "*Перші висновки з атаки на Київстар. "Компанія захищена так сильно, як її найслабший співробітник" – інтерв'ю з експертом з кібербезпеки*" [First Conclusions From the Attack on Kyivstar. 'A Company is as Secure as its Weakest Employee'- An Interview with a Cybersecurity Expert]. https://biz.nv.ua/ukr/amp/hakerska-ataka-na-kijivstar-prichini-naslidki-rekomendaciji-50376256.html.

Norris, D. F., Mateczun, L., Hatcher, W., Meares, W. L., & Heslen, J. (2024). Local government cyber insecurity: Causes and recommendations for improvement. *Public Administration Review*, 84(4), 651–659.

Omelianenko, V. (2023). *EU's and Ukraine's approaches to digital diplomacy in the geopolitics of technologies.* Ukrainian Prism. Policy paper. https://prismua.org/wp-content/uploads/2023/05/Digitalization_Ukraine_EU.pdf.

Patton, M. Q. (2014). Qualitative research & evaluation methods: Integrating theory and practice. *Sage Publications.*

Polish presidency Council of the European Union (2025). Achievements in Field of Information Security. https://polish-presidency.consilium.europa.eu/en/news/achievemements-in-field-of-information-security/?utm_source=chatgpt.com.

President of Ukraine (2021). Decree of the President of Ukraine No. 447/2021 On the decision of the National Security and Defense Council of Ukraine dated May 14, 2021 *"On the Cybersecurity Strategy of Ukraine".* https://www.president.gov.ua/documents/4472021-40013.

Radanliev, P. (2024). Cyber diplomacy: defining the opportunities for cybersecurity and risks from Artificial Intelligence, IoT, Blockchains, and Quantum Computing. *Journal of Cyber Security Technology*, 9(1), 28–78. https://doi.org/10.1080/23742917.2024.2312671.

Savaş, S., & Karataş, S. (2022). Cyber governance studies in ensuring cybersecurity: an overview of cybersecurity governance. *International Cybersecurity Law Review*, 3(1), 7–34.

Smętkowski, M., Kniazevych, A., Olechnicka, A., Orchowska, J., & Przekop-Wiszniewska, E. (2023). *Research for REGI Committee – Cooperation between EU cities and regions with their Ukrainian partners.* European Parliament, Policy Department for Structural and Cohesion Policies.

State Service for Special Communications and Information Protection of Ukraine – SSSCIPU (2025). https://cip.gov.ua/ua/news/cert-ua-minulogo-roku-opracyuvala-4315-kiberincidentiv.

The European Council (2024). Overview of dynamics, discussions and decisions. https://www.europarl.europa.eu/RegData/etudes/STUD/2025/774681/EPRS_STU(2025)774681_EN.pdf.

Trim, P. R., & Lee, Y. I. (2021). The global cyber security model: counteracting cyber attacks through a resilient partnership arrangement. *Big Data and Cognitive Computing*, 5(3), 32.

Ukraine 2024 Report. (2024). Communication on EU enlargement policy. https://enlargement.ec.europa.eu/document/download/1924a044-b30f-48a2-99c1-50edeac14da1_en?filename=Ukraine%20Report%202024.pdf.

Yin, Robert K. (2009). Case Study Research: *Design and Methods*, Vol 5. Thousand Oaks, CA: Sage.

CHAPTER 19

Carpathian Ruthenia: A Forgotten Battlefield of Russia's Hybrid War in the Context of the Accession of Ukraine to the European Union

Miroslav Mareš

Abstract This chapter examines the role of Subcarpathian Ukraine/Carpathian Ruthenia in Russia's strategic narratives toward Central Europe. It addresses terminological and geographical issues within the East Central European security complex, including concepts such as primary, subsidiary, and propaganda battlefields in cyberspace. The analysis explores how Russia exploits historical disputes related to this region as part of its hybrid warfare strategy. It identifies various state and non-state actors, as well as methods such as propaganda, cyber activities, terrorism, and organized crime, and highlights their interconnections. The chapter also assesses the impact of Russia's aggression against Ukraine since 2022, including disinformation campaigns targeting Ukrainian military units from the region. Finally, it outlines implications for Ukraine's EU accession and offers policy recommendations for countering Russian influence.

Keywords Carpathian Ruthenia – Hybrid Warfare – Separatism – Russia – Ukraine – EU-Enlargement

1 Introduction

On 23 March 2020, the Polish court convicted three far-right perpetrators for a terrorist act committed in the Ukrainian town of Uzhhorod the previous year. One of the sentenced Polish extremists kept close contact with a German pro-Kremlin journalist. The three Poles had attacked a Hungarian cultural centre under the false flag of Ukrainian nationalists to intensify Hungarian-Ukrainian disputes (Radio Free Europe/Radio Liberty – Ukrainian Service, 2020) – this development is in line with Russian interest in weakening Ukraine and the European Union. The incident is one of many issues related to the territory of Transcarpathia (or Subcarpathian Ruthenia) and nations from the area,

HTTPS://DOI.ORG/10.1163/9789004763623-021

which occurred during the hybrid war between Russia and Ukraine and the West, which has been ongoing since 2014 and intensified after full-scale invasion in 2022.

At first sight, this set of marginal issues can, to summarise, have significant consequences on European politics, including future Ukrainian EU accession. Carpathian Ruthenia demonstrates the complexity of Russia's subversive strategy. This problem is increasingly evident in the online space, which is related to the progress of digitalization. The ability to create profiles on the Internet and social networks that give the impression of strong separatist organizations or 'phantom' combat units in the war in Ukraine or the mobilization of a militant extremist scene against Ukraine and the countries of the Russian Federation contributes to the effectiveness of the contemporary Russian hybrid war. Therefore, it is necessary to emphasize the passages on the connection of hybrid and cyber threats in the European Strategic Compass from 2024: 'At the same time, free and safe access to global strategic domains is more and more contested. Cyberspace has become a field for strategic competition, at a time of growing dependence on digital technologies' (European Union, 2024). The challenge for existing European law is posed by undesirable propaganda campaigns that use illegal content (including hate speech, terrorist content and unlawful discriminatory content), as defined in the Digital Service Act (European Union, 2022).

According to the Programme of the Polish Presidency of the Council of the EU from 2025 'Russia has destroyed the security architecture in Europe – it continues to wage an aggressive war against its neighbour, builds up its offensive military capabilities, and launches hybrid attacks' (Polish Presidency of the Council of the EU, 2025a). The programme further stated: 'The Polish Presidency will strive to strengthen coordination in the fight against disinformation, information manipulation, and to improve the EU's ability to prevent and mitigate the effects of hostile actions in cyberspace' (Polish Presidency of the Council of the EU, 2025a). These activities can also be used against the abuse of separatism and its manifestations in online space as tools of hybrid warfare.

This article's research question examines how a specific peripheral territory, characterized by numerous traditional disputes, can serve the purposes of the leading actors in hybrid warfare. Defining crucial concepts and terms is essential to answering this question. A historical overview of Transcarpathia/ Carpathian Ruthenia is vital to understanding the recent attitudes and behaviour of domestic and external actors. The strategic situation and perspectives of the region will be analysed in the context of hybrid war (including propagandist operations in cyberspace) and the possible EU enlargement. Threats related to cyberspace will be emphasized.

2 Hybrid Warfare: Main and Other Battlefields, and Capitalizing Separatism

US military experts started using the hybrid war concept in the second half of the 1990s. At that time, it was linked to the *mixed* character of actors involved in armed conflicts, that is, conventional and non-conventional forces and tactics (Walker, 1998, pp. 7–8). In this sense, it was latter broadly popularised in literature about the 2006 Israeli–Hezbollah War (Hoffmann, 2007, pp. 35–42). Some authors have researched similar hybrid strategies in various historical conflicts – predominantly insurgencies, counterinsurgencies and other asymmetric wars (Mansoor, 2012; Murray, 2012).

The massive expansion of the concepts hybrid war and hybrid warfare followed the Russian annexation of Crimea, the War in Donbass and the interconnected subversive and information operations against Kremlin adversaries in various countries. According to Ofer Fridman, the aim of the Russian gibridnaya voina (in contrast to the use of military forces in traditional wars) is 'to destroy the political cohesion of an adversary from the inside by employing a carefully crafted hybrid of non-military means and methods that amplify political, ideological, economic and other social polarizations within an adversary's society, thus leading to its internal collapse (Fridman, 2018, p. 96). This political warfare is used against the West, whereas on territories of the region considered Russia's neighbourhood, regular forces (masked as 'little green men') as well as various irregular actors (paramilitaries, terrorists, guerillas, private military companies, etc. under governmental control) serve the Russian strategy. The aggressive Russian behaviour was reasoned by the so-called Gerasimov doctrine – named after the Chief of the General Staff of the Armed Forces of the Russian Federation, General Valery Gerasimov (2013), which was, according to many observers, interpreted as a principle for a new strategic doctrine. However, this view is not generally accepted (Galeotti, 2020; Stojar, 2017). The exploitation of opportunities varies concerning time- and space-based conditions. The traditional concept of hybrid war can be used to label non-linear military operations with the involvement of irregular forces (as in Crimea in 2014). However, the term hybrid warfare is broader, and it interconnects political warfare (including mass disinformation campaigns, trolling, selected elite-capture, etc.) with limited non-military use of physical force (for example, vigilantism, individual cyberattacks, false flag terrorism, etc.). Hybrid campaigns are aimed explicitly at parts of hybrid warfare (to weaken trust of the EU inside European countries, for instance), and attacked states face a spectrum of combined hybrid threats. Hybrid warfare, however, is not necessarily a 'prelude' to a 'hot' hybrid war or even a conventional war, but it can be autonomous, aggressive political activity in and of itself.

The actors and agents who act on behalf of the Kremlin's interests can be directly controlled by Russian governmental control; nevertheless, they can also act autonomously due to their ideological beliefs or economic interests. During hybrid campaigns in cyberspace and the physical world, official Russian policy relies on a wide variety of 'willing helpers', including people who are not intentional allies of the recent Russian regime. They – traditional nationalists, conspiracy theory believers, anti-institutional extremists, etc. – can be manipulated to spread chaos and weaken Western democratic authorities. Hybrid warfare is a complex phenomenon covering many activities in various subfields. The main aim is to break the enemy's ability to resist (Rácz, 2017). The concept of primary and 'other' (subsidiary) battlefields and operations, known from the strategic analysis of traditional armed conflicts (Marble, 2009), can also be used in modified ways within the context of hybrid warfare. Experts Erik Reichborn-Kjennerud and Patrick Cullen state that 'hybrid warfare breaks down the distinction between what is and what is not part of the battlefield (Reichborn-Kjennerud & Cullen, 2016). Hybrid warfare actors can utilize relevant or marginal separatist and irredentist entities (or even initiate their existence) to weaken the enemy and damage its reputation as a functional state with an efficient government in international relations. They can also be an obstacle during the accession process to international and supranational organisations. The capitalizing of separatism as a tool of hybrid warfare is accompanied by intimidation and disinformation campaigns in cyberspace and may be supplemented by cyberattacks.

3 Historical Legacies in Carpathian Ruthenia

The researched territory has an overcomplicated history[1] (Magosci, 2015) which is also reflected in the various names used about this area. Due to the focus on Russian hybrid warfare, the name Carpathian Ruthenia is predomin-

1 An anecdotical joke can illustrate this history. In 2014, some journalists asked a one-hundred-year-old man in a Subcarpathian town, 'Where were you born, Grandpa?' 'Under the Austrian Monarchy', he replied. 'And where did you first go to school? 'In Western Ukraine' he said Where did you receive your diploma?' 'In Czechoslovakia'. 'Where did you serve in military?' 'In Transcarpathia.' 'Where did you work?' asked the journalists. 'In

antly used in this article. Several pro-Kremlin Ruthenians have struggled to establish a quasi-independent state as Subcarpathian Ruthenia (Rus)[2] or Carpathian Ruthenia on what is contemporarily known as the Transcarpathian region or just Transcarpathia, an administrative region in Ukraine. This name will be used in connection with Ukrainian and Hungarian interests in the region throughout this article. In some radical conceptions of what should belong to this new state, even the territories of historical Carpathian Ruthenia (Magosci, 2015, p. 364) in contemporary Poland (the Lemkivshchyna/Lemkovyna region), part of the recently established Prešov region in Slovakia (sometimes called Presevska Ruthenia/Rus) and a very small part of northern Romania (Imaginary Maps, 2019) are considered.

Autonomism and separatism are connected with the Ruthenian (or Rusyn) nation and the Rusyn tongue. In 2012, its members were estimated by political scientist and Ruthenian activist Paul Robert Magosci to number 1,762,000 people (Magosci, 2015, p. 364). Those living in contemporary Ukrainian territory consist of around 853,000 individuals; however, many Ruthenians live in neighbouring European states[3], and there is a diaspora overseas – mostly in North America and Australia (Kokaisl et al. 2020, p. 11). Ruthenians are religiously mostly Greek Catholics or Orthodox Christians.

It is important to mention that only part of the Transcarpathian population of 1,256,800 (Derzhavna sluzhba statistics, 2019, p. 5) identify themselves as Ruthenian, and only a minority support autonomy or the independence of Subcarpathian Ruthenia. The actual number is unknown because the Ukrainian government does not recognise the Ruthenian minority. The majority of people in this territory consider themselves Ukrainian. Moreover, other nationals live there[4].

At the end of the nineteenth century, it was part of the Austrian-Hungarian monarchy under the rule of the Habsburg dynasty (Hořec, 1994, p. 14) in its Hungarian part. However, in the mid-nineteenth century, a Ruthenian

---

the Soviet Union' came his reply. 'And where did you receive your recent pension?' 'In Ukraine.' 'Wow, you have travelled so much!' they said. 'What?! I've never left Mukachevo.'

2 Subcarpathian Rus was also the name of the 1919–45 Czechoslovak administrative area.

3 130,000 in Slovakia; 35,000 in Romania; 30,000 in Poland (a Ruthenian subgroup called the Lemkos in the territory of Lemkovyna); 20,000 in Serbia; 10,000 in the Czech Republic; 6,000 in Hungary and 5,000 in Croatia. Ibid., 1. However, a collective of Czech researchers led by Petr Kokaisl recently estimated much lower numbers in Europe: 33,000 in Slovakia; 11,000 in Serbia; 3,700 in Hungary; 1,800 in Croatia; 1,700 in Montenegro and 800 in the Czech Republic (Kokaisl et al. 2020).

4 Results of the last census in 2001 showed that Hungarians constitute 12.1% of the population of the Transcarpathian region, Romanians form 2.6%, Russians 2.5%, Roma people 1.1%, Slovaks 0.5% and Germans 0,3%. (Świeżak, 2008).

national awakening started with links to Russian helpers and pan-Slavic intellectuals in Bohemia, Poland and other parts of the Eastern European empires. The alleged plot of Subcarpathian Ruthenians to revolt against Hungary in conspiracy with Russia was punished in a 1913 court trial in Marmaros-Sziget (Hořec, 1994, p.15). The stormy development characterised by the impact of the Russian Revolution and post-war conflicts in Eastern Europe caused the existence of short-lived states with territorial claims to the Carpathian area. The Western Ukrainian Peoples Republic supported the Ukrainian national identity of Slavic citizens in 1918–1919. The Russian National Republic (also known as the Lemko Republic) was founded by pro-Moscow Ruthenians in the Lemko region in November 1918, but after several months it was incorporated into Poland (Moklak, 2013, p.31). In general, however, three main streams existed among Slavic elites of the Subcarpathian area at that time. The first struggled for a 'Greater Ukraine' and was connected with Ukrainian nationalism. The two other streams of Ruthenians declared different external preferences – one was oriented towards cooperation with the Czechs and Slovaks, and the other veered in the direction of Russian politics, with ideas of a quasi-independent territory interconnected with Russia. The pro-Russian stream was divided along pro-Soviet and anti-communist (Russian post-Tsarist and democratic opponents of Soviet rule) lines. Hungarians and a small segment of Ruthenians preferred unification with Hungary. The new established Czechoslovakia, with involvement of the Ruthenian overseas diaspora, gained control over the central part of Subcarpathian Ruthenia in 1919 (Wilson, 1996, p.119). In the interwar period, under the Czechoslovak government, brigandage was typical of the territory at the turn of the 1910s and 1920s. The struggle for specific Subcarpathian autonomy in Czechoslovakia went long unfulfilled, but was achieved in 1938 following the Munich agreement under challenging circumstances. The Ukrainian nationalist paramilitary organisation Carpathian Sich (Karpatska Sic) was founded. Even though it was registered according to the Czechoslovak law, their members frequently attacked the Czechoslovak forces. Hungarian (Micianik, 2019, p.122) and Polish (Marek, 2019, pp.289–307) subversive units meanwhile operated in Subcarpathian Ruthenia at the end of 1938 and in the first month of 1939.

On 14 March 1939, the Carpathian Sich started an uprising against Czechoslovakia and proclaimed the independent state of Carpathian Ukraine. However, at the same time, the Hungarian army attacked the territory and, after a short war with both the Sich and Czechoslovak forces, the region fell under Hungarian occupation. Despite only five days in existence, this state still serves as a heroic legacy from the perspective of Ukrainian nationalists. Some members of the Carpathian Sich later fought in the Ukrainian Insurgent Army (Ukrayinska Povstanska Armiya) (Řepa, 2019, pp.158–162).

In 1944, the Red Army liberated the Subcarpathian Ruthenia territory. Ukrainian communists then established a pro-Soviet administration under pressure from Moscow. In 1945, Czechoslovakia was forced to sign an agreement to the loss of Subcarpathian Ruthenia, allowing it to become a definitive part of the Ukrainian Soviet Socialist Republic in 1946 (Rychlík- Rychlíková, 2016, pp. 209–210).

The collapse of the Soviet Union led to the establishment of independent Ukraine on 24 August 1991. On 1 December that year, a referendum was held over the self-governance – the term 'autonomy' was prohibited by Ukrainian authorities – of Transcarpathia within the new state, with 78% of respondents expressing support for it (Hořec, 1994, p. 86). The Ruthenian representation claimed the referendum result as confirmation of Ruthenian identity. However, the Ukrainian government did not recognise the autonomy of the territory. Nor were Ruthenians recognised as a national minority. It was not until 2007, through the local government's decision, and later in 2012, in Law No. 5029-VI on the Principles of the National Language Policy, that this recognition was granted. In February 2014, the parliament abolished the law, but the president did not confirm this (Stefan, 2016, p. 228). This long-term tension between Ruthenians and the Ukrainian government has led to its misuse by pro-Kremlin forces within the context of hybrid warfare. Moreover, many other actors from Ukraine and neighbouring states are today intentionally or unintentionally affiliated with this issue.

3.1 Ruthenian Separatism before and after the Start of the Ukrainian Crisis

As in many ethno-nationalist movements, among the Ruthenians, however, more and less radical streams exist – non-violent. In 1993, the self-proclaimed 'Provisional Government of the Association of Subcarpathian Ruthenia' was established under Professor Ivan Turyanitsa's leadership. One year later, Ukrainian authorities declared it unlawful (Żochowski & Iwański, 2015). This attempt at autonomy was probably without Russian support; however, in the mid-2000s, a new wave of Ruthenian separatism began that was connected with Russian interference, similar to various pro-Kremlin autonomist and separatist movements in other parts of Ukraine (including Donbass). Leader of the Congress of Subcarpathian Ruthenians (Sojm Pydkarpatskych Rusynev), Dmitry Sydor, was also a priest in the Orthodox Church of the Moscow Patriarchate. The congress declared the independence of the Republic of Carpathian Ruthenia in 2008, following the Russo-Georgian War, which was branded a Kremlin plot by the Ukrainian government (Roth, 2015, p.165). Activist Pyotr Getsko became the self-proclaimed prime minister of the Government of the Republic of Subcarpathian Ruthenia (Pravitelstvo Respubliky Podkarpatskaa Rus), which was initially seated in Minsk, Belarus (Gusarov, 2016). Later Sydor stated the right to armed resistance against the Ukrainian authorities and he was sentenced to 'three years of imprisonment with a respite of two years for the encroachment on the territorial integrity of Ukraine' (Religious Information Service of Ukraine, 2012).

After the Maidan demonstrations, the annexation of Crimea and the start of the War in Donbass, the fringes of Ruthenian separatism became more militant. The main goal of the separatists in eastern Ukraine was to establish the new state of Novorossiya, consisting of Lugansk, Donetsk, Odessa, Zaporizhye and other regions in the east. However, separatist movements in other regions were also used to weaken governmental authority (Roth, 2015, pp.161–165). After the start of the Ukrainian crisis in the winter of 2013/2014, Ruthenian separatism, under Getsko's leadership, became more visible on pro-Kremlin propagandists' channels, mostly in various discussions on social media networks and servers (Facebook, Vkontakte, Livejourna (Getsko, 2014)). The Vkontakte profile of the Government of the Republic Subcarpathian Ruthenia actively supported separatism in Novorossiya and tried to interconnect Ruthenian separatism with them. Calls for 'little green men', the nickname given to members of Russian special forces in Crimea and Donbass wearing uniforms without insignia, were propagated in Subcarpathian Ukraine as a part of the media strategy to spread fear. Pyotr Getsko refused to exclude the use of force against Kyiv and asked for Russian help (Goble, 2014).

In late summer 2014, during heavy fighting in Donbass, information about the alleged existence of the battalion I. G. Kundri in the Lugansk People's Republic was widespread on online social networks. The battalion included 387 Ruthenian fighters from Subcarpathian Ruthenia, Slovakia, Poland (Lemko region) and Hungary. The shared support for the independence of Subcarpathian Ruthenians and Novorossiya was expressed in a propagandist statement that included the flag of both separatist states (Respublika Podkarptskaya Rus, 2014). However, the actual existence of the unit is disputable due to an inability to ascertain reliable data about its further activities.

The wave of separatist successes in Eastern Ukraine caused new growth in provocations in Subcarpathian Ruthenia. In February 2015 Prime Minister Pyotr Getsko warned in the Russian press that the war could spread from Ukraine to Europe. Accusations that the Ukrainian government was fascist and Nazi were typical of his rhetoric (Sudakov, 2015). On 13 March 2015, he again demanded autonomy and national independence, allegedly in the name of the Coordination Council of the Rusyn Organisations of Transcarpathia (TASS, 2015). However, Yevhen Zhupan, the head of the People's Council of Ruthenians in Transcarpathia (an official organisation recognised by Ukrainian authorities), rejected this step, denying the existence of Getko's council and criticising his role in pro-Russian propaganda (Coynash, 2015).

The Ukrainian authorities, including the secret security service SBU, started Getsko's prosecution; in 2019, he was sentenced in absentia to 12 years in prison for disrupting Ukrainian territorial integrity (Kovalenko, 2019). Getsko lives abroad and is active in new organisations, mostly the International Centre Mother of Ruthenians (Mezhdunarodny Center Matica Rusinov). He is still promoting separatism and anti-Ukrainian politics. In 2025, Getsko supported. His organisation participated in February 2025 at the Ruthenian congress in Moscow. In its declaration, it supported the goals of Putin´s Special Military Operations, demanded the de-nazification of Ukraine, and the EU and NATO were defamed (Mezhdunarodny Center Matic Rusynov, 2025).

3.2 Controversies about Subcarpathian Ruthenia in Central European Countries

In several countries in Central Europe, various links to Subcarpathian issues exist. The Czech Republic is an external actor not only with historical ties to the territory of Subcarpathian Ruthenia but also one which has had very recent impact on Ruthenian separatism. The main reason for this is the headquarters of some Ruthenian organisations in the Czech Republic and the existence of a relatively strong pro-Kremlin segment of the Czech political spectrum. Cooperation between these two political entities contributed to tensions with Ukrainian authorities.

In addition to these strong Czech links to the situation in Carpathian Ruthenia, there is also marginal Czech irredentism with territorial claims to this territory; however, it can be assessed more as political curiosity than a real security issue, as demands of the Czech far-right party, the Association for the Republic – Republican Party of Czechoslovakia (SPR-RSČ) struggled for unification with Subcarpathian Ukraine before and even after the divorce of Czechoslovakia (its chairman Miroslav Sládek also threatened the use of military force) (Gjuričová, 2012, pp. 337–339), or the similar demands of the obscure National Bolshevik Party of Czechoslovakia at the end of 2010s (Národně-bolševická strana Československa, 2010). In 2014, another marginal far-right party, National Democracy (Národní Demokracie), together with the World Council of Subcarpathian Ruthenians (small pro-Kremlin organisation registered in the Czech Republic) and the association Subcarpathian Ruthenians (Podkarpatskí Rusíny) signed the Dašice Declaration, demanding autonomy for Subcarpathian Ruthenia and suggesting separatist methods for its future development should Ukraine not accept its autonomy (Podkarpatskí Rusíny, 2014). Soon a more critical partner was found in the Czech political spectrum for the World Council of Subcarpathian Ruthenians, namely, the head of the Czech Republic, President Miloš Zeman, who promoted Russian interests (Orenstein, 2019, pp. 208–209). Two members of this organisation, Micyhalo Tyasko and Valery Zhugan, were invited by Zeman to the Independent Czechoslovak State Day ceremonies on 28 October 2019. They were accompanied by members of a pro-Kremlin, Crimean Tatars organisation, which recognises the Russian annexation of Crimea. According to their statement, Zeman recognised the unification of Crimea and Russia. However, Zeman denied it, and a council representative apologised to him for the misunderstanding. Ukrainian officials criticised strongly the whole incident (JH, 2019).

While the Czech role is relatively limited directly to the territory of Subcarpathian Ruthenia, the Hungarian card is used for a more important game. Hungarians in Transcarpathia are within the focus of the government in Budapest as part of its care for Hungarian minorities abroad. Radical Hungarian politicians even declare irredentist claims, and this fact is negatively perceived from Kyiv. This tension is welcomed from the Russian side because it can weaken the Ukrainian position. Some Ruthenians prefer cooperation with Hungary, or, at the very least, they try to use Hungarian support for their own autonomist or separatist demands. Hungary promotes strong support for the Hungarian minority in education, culture and business. During the Ukrainian crisis, Viktor Orbán's government (with friendly ties to Putin's regime) declared its protective role for Hungarians and Ruthenians in Ukraine. In March 2015, the Hungarian government 'announced that it distributed Hungarian citizenship papers to 94 000 people in Transcarpathia' (Bugajski, Assenova, 2016, p. 163). The politician Bela Kovacs from the party Jobbik was very active in Transcarpathia with irredentist goals in the first half of the 2010s (Sinitsyn, 2017). Later, in 2022, he was sentenced in Hungary in absentia due to espionage for Russia. He lives in Moscow (Hetzmann, 2024).

The alleged Hungarian unit – the Legion of Saint Stephen (Szent István Légió/Legiója) consisting mainly of Hungarians from Transcarpathia – had to fight as part of the Donetsk People's Republic military forces for Novorossiya and after incorporation of the DNR/LNR into the Russian Federation forces, full-scale invasion by the Russian troops in 2022. This was propagated in far-left as well as in far-right internet media channels, mostly in 2014–16 (Slava Novorossii) (Donetskaya Respublika, 2015). Its interconnection with

Russian volunteer units was declared, and the legion was supported by both the Jobbik party and Getsko's separatist Ruthenian government (Anons Zakarpatya, 2014). The number of fighters and their actual activity was negligible. According to several sources, the whole existence of the legion was a hoax (Pulemoytov, 2014). However, some activities of the Legion of St. Stephen are documented. The main goal of this 'phantom unit' was to strengthen Hungarian-Ukrainian tensions (Mareš, 2017, p. 36).

The same goal but different modus operandi was at the root of another act: an arson attack on the headquarters of the Transcarpathian Hungarian Cultural Association (Kárpátaljai Magyar Kulturális Szövetség) on 4 February 2018. In connection with the attack, Nazi symbols were sprayed on the building. Despite suggestions that Ukrainian nationalists committed the attack, the real perpetrators came from the pro-Russian Polish sphere. The two perpetrators, Tomasz Szymkowiak and Adrian Marglewski, under the command of 28-year-old Michał Prokopowicz, both had links to the pro-Kremlin party Change (Zmiana) and the militant organisation Falanga (one of the few Polish groups supporting separatism in Eastern Ukraine) (Wyrwał, 2019). Prokopowicz stated in his testimony that he was paid 1,500 euros 'to organize the attack by Manuel Ochsenreiter, a German journalist with multiple links to Russia, who works as an adviser to AfD MP Markus Frohnm' (Walker, Davies, Schulteis, 2019). Even though Ochsenreiter denied these allegations (Walker, Davies, Schulteis, 2019), the attack is considered in Poland to be part of the Russian hybrid war (Zaryn, 2019). All three Polish perpetrators were sentenced by a court of first instance in Krakow on 23 March 2020; Prokopowicz was sentenced to three years in prison, Szymkowiak to two years and Marglewski to two years of supervised release (Radio Free Europe/Radio Liberty – Ukrainian Service, 2020).

In 2025, relations between Ukraine and Hungary deteriorated. Ukrainian intelligence reported the discovery of a spy network in Transcarpathia. Two spies were tasked with gathering information about Ukraine's military capabilities and preparing the territory for the potential entry of Hungarian troops (Sluzhba bezpeky Ukrainy, 2025). Subsequently, Budapest expelled two Ukrainian diplomats accused of spying (Kyiv Post, 2025). Russia also exploited the situation. In July 2025, Russian Foreign Minister Lavrov urged Hungary to cooperate in protecting minorities in Ukraine (The Ministry of Foreign Affairs of the Russian Federation, 2025).

3.3 Pro-Kremlin Neo-Nazi On-Line White Separatism in Zakarpathia

In Western countries, before and after full-scale invasion in 2022, the impact of hard-line Ukrainian nationalists and organized crime (Zapałowski, 2017) from the area remains a closely watched security threat from Transcarpathia. The global neo-Nazi scene maintains contacts, in particular, with the revived Carpathian Sich (Tarasiuk & Umland, 2021). This fact was misused in Kremlin propaganda to show Ukraine as a nazi-friendly state (Ukr Leaks, 2025).

However, as a counterweight to this influence in the neo-Nazi milieu, the pro-Kremlin transnational neo-Nazi organization The Base, led by the American activist Rinaldo Nazzaro operating from St. Petersburg, announced its project. The Base was founded in 2018 and promotes a violent method of struggle. In the spring of 2025, its White Phoenix project began (Makuch, 2025). Its goal is to create an ethnically pure white supremacist state. White neo-Nazis would have a base there. The organization also declares war 'on the Liberal Globalist regime in Kiev' (White Phoenix, 2025) and collects money via cryptocurrencies to start a guerrilla campaign. The group stated, 'We will continue to target infrastructure, military vehicles, and government buildings until the Ukrainian government officially grants independence to the Republic of Carpathia, currently known as Zakarpattia Oblast (White Phoenix, 2025). The capital should be "Eihwaz (formerly Uzhhorod)" (White Phoenix, 2025)'.

The group is allegedly responsible for one urban attack in Odessa (White Phoenix, 2025). However, the project's primary focus to date has been propaganda. The Russian regime is trying to use neo-Nazi militants to destabilize the territorial integrity of Ukraine. Although after the campaign in media, The Base stated that Ukrainian members initiated the Project Phoenix, the Russian secret services are very likely involved in such a project of capitalisation online anti-Ukrainian neo-Nazi propaganda to win impact on militant accelerationism.

4 Conclusion

Russian hybrid warfare aimed at the territory of Subcarpathian Ruthenia has predominantly had a subsidiary character to the main battlefields in Crimea and Donbass and after invasion in the whole Eastern Ukrainian territory. However, it has also consisted of Russian influence operations in Central European countries. The conditions for establishing a parastate according to the model of Transnistria, Southern Ossetia or the Donetsk People's Republic and Lugansk People's Republic were not fulfilled in Subcarpathian Ukraine, mostly due to the lack of direct connection with the Russian territory during the 'hot' conflict. However, the region's unstable situation could change that condition if pro-Russian forces in Slovak and Hungarian politics were to grow. In case of the EU-enlargement, Russia.

The strategies from this 'subsidiary' battlefield of Russian hybrid warfare in Subcarpathian Ruthenia have been identified in the following points:

1. Weakening the territorial integrity, governmental integrity and international reputation of the enemy by supporting multiple claims to a territory (Ruthenian autonomism and separatism, Hungarian and Czech irredentism).
2. Fuelling historical prejudices and hate through the capitalization historical legacies.
3. Using proxies (the Czech or Hungarian politicians acting in line with Russian interests).
4. Conducting false flag terrorist operations (arson attack in Uzhhorod).
5. Propagandising 'phantom units' and pro-Kremlin neo-Nazi militants predominantly in on-line dimension.
6. Threatening military action in the background.

Can Russia intensify online and offline hybrid operations to strengthen Ruthenian separatism and Hungarian irredentism and pro-Kremlin neo-Nazism related to the territory of the Subcarpathian area in the case of the negotiations about the Ukrainian EU accession? Probably yes, including targets within the EU territory (false-flag attacks are not excluded, including in cyberspace). In view of calls to intensify the campaign in Transcarpathian Ukraine, further escalation of threats in cyberspace is to be expected amid ongoing digitalisation and the advancement of artificial intelligence, including new forms of cyber attacks and hacktivism. Russia's war of aggression against Ukraine is considered by experts at ENISA to be one of the most important driving factors of contemporary hacktivism (Bakatsis & Boutemeur, 2025).

It is therefore commendable that the Warsaw Call Declaration was adopted during the Polish Presidency in 2025, which is a 'key reference point for the EU's future efforts to protect the digital space and enhance resilience amid growing geopolitical challenges. The document sets out thirteen recommendations, endorsed by ministers and representatives of the member states' (Polish Presidency of the Council of the European Union, 2025b). Among other points, it includes the following calls: 'Developing a roadmap for emerging technologies, boosting EU expertise in strategic cybersecurity forecasting, and harmonising efforts to increase cybersecurity investments' and 'Reinforcing the horizontal impact of the NIS 2 Directive, focusing on harmonised, innovation-friendly implementation, simplifications, reduced regulatory burdens, and strengthening EU-wide cybersecurity risk assessments' (Polish Presidency of the Council of the European Union, 2025b).

To avoid unwonted results related to Russian utilisation of Subcarpathian issues, it is essential from the point of view of the EU: to be able to recognize the Russian influence operations in hybrid warfare (including hidden support of separatist and irredentist entities); to support Ukraine in its struggle to keep territorial integrity (of course, with respect to minority rights); to eliminate networks of separatist and irredentist anti-Ukrainian entities on the EU territory and (if it is possible) worldwide. The main part of this paper was written as part of the research project 'Manipulative Techniques of Propaganda in the Age of the Internet' (MUNI/G/0872/2016) of the Masaryk University Grant Agency.

References

Anons Zakarpatya (2014). Getsko vmeste s fashistami „Jobbika" sazdayut terrorististicheskuyu organizacyu „Legion svyatogo Iszhtvana". http://anons-zak.com.ua/kriminal/17010-gecko-vmeste-s-fashistami-yobbika-sozdayut-terroristicheskuyu-organizaciyu-legion-svyatogo-ishtvana.html.

Bakatsis, I. & Boutemeur, J. (2025). ENISA Sectoral Threat Landscape. Chalandri: European Union Agency for Cybersecurity. https://www.enisa.europa.eu/sites/default/files/2026-01/ENISA%20Public%20Administration%20TL%202024%20-%20v1.2.pdf.

Bugajski, J. & Assenova, M. (2016). *Eurasian Disunion, Russia´s Vulnerable Flanks.* Washington: The Jamestown Foundation.

Coynash, H (2015). Moscow Replays Rusin 'Separatist' Card in Ukraine's Zakarpattia". *Kharkiv Human Rights Protection Group.* http://khpg.org/en/index.php?id=1426415683.

Derzhavna sluzhba statistiky (2019). *Pasport Zakarpatskoi oblasti. Uzhhorod.* Uzhorod: Derzhavna sluzhba statistiky.

Donetskaya Respublika (2015). *Slava Novorossii, Pravye Otryadi na Strazhe Ruskoy Vesny.* https://vk.com/donbass_res.

European Commission (2025). State of the Digital Decade 2025: Keep building the EU's sovereignty and digital future. https://digital-strategy.ec.europa.eu/en/library/state-digital-decade-2025-report.

European Union (2022). Regulation (EU) 2022/2065 of the European Parliament and of the Council of 19 October 2022 on a Single Market For Digital Services and amending Directive 2000/31/EC (Digital Services Act) (Text with EEA relevance).

Document 32022R2065 Regulation – 2022/2065 – EN – DSA – EUR-Lex Accessed 12 January 2026.

European Union (2024). A Strategic Compass for Security and Defence. For a European Union that protects its citizens, values and interests and contributes to international peace and security. https://www.eeas.europa.eu/sites/default/files/documents/strategic_compass_en3_web.pdf.

Fridman, O. (2018). *Russian Hybrid Warfare. Resurgence and Politicisation.* London: Hurst & Company.

Galeotti, M. (2020). The Gerasimov Doctrine. *Berlin Policy Journal.* https://berlinpolicyjournal.com/the-gerasimov-doctrine/.

Gerasimov, V. (2013). Cennost Nauki v Previdenii. *Voyenno-promyshlennyi kurer,* 8. https://www.vpk-news.ru/articles/14632.

Getsko, P. (2014). Rusiny dali start narodnoy revolucii v Zakarpatye. *Livejournal.* https://getsko-p.livejournal.com/2014/04/07/.

Gjuričová, A., M. Kopeček, P. Roubal, J. Suk & T. Zahradníček (2012). *Rozděleni minulostí. Vytváření politických identit v České republice po roce 1989.* Praha: Ústav pro soudobé dějiny Akademie věd.

Goble, P. (2014). Transcarpathian Rusin Leader 'Does Not Exclude' Using Force Against Kyiv. *Eurasia Daily Monitor,* 11 (109). https://jamestown.org/program/transcarpathian-rusin-leader-does-not-exclude-using-force-against-kyiv/#.VQTv9eE7eUk.

Gusarov, I. V. (2016). European bison: from nursery to freely breeding population. In Habitat condition and fauna of hunting animals in Russia and adjacent territories (Proceedings of the 2nd International and 7th All-Russian Scientific-Practical Conference), pp. 117–121. Balashikha.

Hetzmann, M. (2024). Hungarian refugee in Russia, Béla Kovács, accuses Western media of misleading public. Daily News Hungary. https://dailynewshungary.com/bela-kovacs-political-refugee-russia/.

Hoffmann, F.G. (2007). *Conflict in the 21st Century: The Rise of Hybrid Wars.* Arlington: Potomac Institute for Policy Studies. https://www.potomacinstitute.org/images/stories/publications/potomac_hybridwar_0108.pdf.

Hořec, J. (1994). *Podkarpatská Rus – země neznámá.* Jinočany: H&H.

Imaginary Maps (2019). *Republic of Carpathian Ruthenia* 2019. https://www.reddit.com/r/imaginarymaps/comments/daw08h/republic_of_carpathian_ruthenia_2019/.

JH (2019). Podkarpatští Rusíni se omluvili prezidentu Zemanovi kvůli Krymu. *Česká televize,* https://ct24.ceskatelevize.cz/domaci/2972459-podkarpatsi-rusini-se-omluvili-prezidentu-zemanovi-kvuli-krymu.

Kokaisl, P., A. Štolfová, P. Fajfrlíková, V. Němcová, J. Zychová, I. Cejpová et al. (2020). P*o stopách Rusínů v Evropě.* Praha: Nostalgie.

Kovalenko, J. (2019). SMI: Premera 'Podkarpatskoy Rusi' prigovarili k 12 godam zaklyutschenia, *KP v Ukraine*, https://kp.ua/incidents/648591-smy-premera-podkarpatskoi-rusy-pryhovoryly-k-12-hodam-zakluichenyia.

Kuzio, T. (2005). The Rusyn question in Ukraine: Sorting out fact from fiction. *Canadian Review of Studies in Nationalism*, 32 (1): 1 –15.

Kyiv Post (2025). Budapest Expels Ukrainian Diplomats After Ukraine Exposes Hungarian Spy Ring. Kyiv Post. https://www.kyivpost.com/post/52366.

Latmier, J. (2005). *Burma. The Forgotten War*. London: John Murray.

Mansoor, P. R. (2012). Introduction: 'Hybrid Warfare in History'. In W. Murray and P. R. Mansoor (eds): *Hybrid Warfare: Fighting Complex Opponents from the Ancient World to the Present*, (pp. 1–17): Cambridge: Cambridge University Press.

Marable, S. (2009). Offensive versus subsidiary attacks, 1916–1918: The British Expeditionary Force balancing its options. *Journal of the Society for Army Historical Research*, 87 (351), 273–287.

Marek, J. (2019). *Válka o červeného medvěda. Zapomenutí čeští vojáci na Podkarpatské Rusi v březnu 1939*. Cheb: Svět křídel.

Mareš, M. (2017). Foreign fighters in Ukraine: Risk analysis from the point of view of NATO. In K. Rekawek (ed.): *Not Only Syria? The Phenomenon of Foreign Fighters in a Comparative Perspective* (pp. 31–39). Amsterdam: IOS Press.

Magosci, P. R. (2015). *With Their Backs to the Mountains: A History of Carpathian Rus' and Carpatho-Rusyns*. Budapest: Central European University Press.

Makuch, B. (2025). US neo-Nazi group with Russia-based leader calls for targeted Ukraine attacks. The Guardian. https://www.theguardian.com/us-news/2025/apr/05/the-base-neo-nazi-russia-ukraine.

Mezhdunarodny Center Matic Rusynov (2025). Itogovoe reshenie 40 konferencii Suverenov, Titulyarov Rusinov, Rusiniv-Rusov. https://rusmatica.org/materials/40krrm.html.

Mičianik, P. (2019). *Malá vojna Maďarska proti Slovensku 1938*. Bratislava: DALI-BB.

Moklak, J. (2013). *The Lemko Region in the Second Polish Republic, Political and Interdenominational Issues 1918–1938*. Kraków: Jagiellonian University Press.

Murray, W. (2012). Conclusion: What the past suggests. In W. Murray and P. R. Mansoor (eds.): *Hybrid Warfare: Fighting Complex Opponents from the Ancient World to the Present*, (pp. 289–307). Cambridge: Cambridge University Press.

Národně-bolševická strana Československa (2010). *Černobílý manifest. Program Československých národních bolševiků*. Brno: NBSČ (author´s archive).

Orenstein, M. A. (2019). *The Lands in Between. Russia vs. the West and the New Politics of Hybrid War*. Oxford: Oxford University Press.

Pieniążek, P. (2018). How to set up your own para-state. *The New Eastern Europe*, 8 (3–4): 22–26.

Plachý, J. (2020). Oběti "železné opony" na Podkarpatské Rusi po sovětské anexi, *Historie a vojenství*, 69 (3), 60–75.

Podkapatskí Rusíny, Světová rada Podkarpatských Rusínů and Ne Bruselu – Národní demokracie (2014). *Dašická deklarace*. Dašice.

Polish Presidency of the Council of the EU (2025a). Programme of Polish Presidency of the Council of the EU. Security, Europe. https://polish-presidency.consilium.europa.eu/en/programme/programme-of-the-presidency/.

Polish Presidency of the Council of the EU (2025b). Warsaw Call Declaration adopted at the informal TTE Telecom Council on cybersecurity. Warsaw Call Declaration adopted at the informal TTE Telecom Council on cybersecurity.

Prchal, L., & M. Švec, M. (2018). Zemanovi ve štábu fandili i Rusíni bojující za autonomii. Na Hrad je zve často, vzal je i do Ruska. *Aktualne.cz*. https://zpravy.aktualne.cz/domaci/zeman-se-styka-s-bojovniky-za-autonomii-rusinu-podporuji-rus/r~6f07a67006aa11e880d30cc47ab5f122/.

Pulemyotov (2014). Vnimanye! Feyk! Legion svyatogo Iszhtvana. *Livejournal.com*. https://pulemjotovlivejournal.com/319486.html.

Rácz, A. (2015). *Russia's Hybrid War in Ukraine. Breaking the Enemy's Ability to Resist*. Helsinki: The Finnish Institute of International Affairs. https://www.fiia.fi/wp-content/uploads/2017/01/fiiareport43.pdf.

Radio Free Europe/Radio Liberty – Ukrainian Service, (2020). Polish Court Convicts Three Men For Torching Hungarian Center In Ukraine. *Radio Free Europe/Radio Liberty* 24 March. https://www.rferl.org/a/polish-court-convicts-three-men-for-torching-hungarian-center-in-ukraine/30505582.html.

Reichborn-Kjennerud, E., & Cullen, P. (2016). *What Is Hybrid Warfare?* Oslo: Norwegian Institute of International Affairs. https://core.ac.uk/download/pdf/52131503.pdf.

Religious Information Service of Ukraine (2012). Suspended Prison Sentence Passed Upon Separatist Priest Dymytrii Sydor. https://risu.ua/en/suspended-prison-sentence-passed-upon-separatist-priest-dymytrii-sydor_n55418.

Respublika Podkarpatska (Podkarptskaya) Rus (2014). Rusyny-doborvolcy na zaschite Novorossiyi, *Vkontakte*, https://vk.com/public67712004.

Roth, Ch. F. (2015). *Let's Split! A Complete Guide to Separatist Movements and Aspirant Nations, from Abkhazia to Zanzibar*. Sacramento: Litwin Books.

Řepa, T. (2019). *Banderovci. Politické souvislosti, následky zneužití tématu komunistickou propagandou, návaznosti na hybridní konflikt současnosti*. Praha: Academia.

Rychlík, J., & Rychlíková, M. (2016). *Podkarpatská Rus v dějinách Československa 1918–1946*. Praha: Vyšehrad.

Sinitsyn, M. (2017). Separatism in Transcarpathia: From Kremlin's hand to hand of Budapest", *112.ua*. https://112.international/opinion/separatism-in-transcarpathia-from-kremlins-hand-to-hand-of-budapest-21821.html.

Sluzhba bezpeky Ukrainy (2025). CBU vpversche v istrorii Ukraini vikrila agenturnu merazhu voyennoy rozvydki Ugorshini. Facebook. https://www.facebook.com/SecurSerUkraine/videos/585341354575606/?t=15.

Stefan, S. (2016). Multiple Identities in a Unitary State: Tracing the Origins of the Ukrainian Crisis Back. In M. Nicolini, F. Palermo & E. Milano (eds.): *Law, Territory and Conflict Resolution: Law as a Problem and Law as a Solution* (pp. 219–243). Leiden; Boston: Martinus Nijhoff Publishers.

Stojar, R. (2017). Vývoj a proměna konceptu hybridní války. *Vojenské rozhledy*. 26(2), 44–55. https//doi.org/10.3849/2336-2995.26.2017.02.044-055.

Sudakov, D. (2015). War in Ukraine will continue in Europe Kyiv. *Pravda.ru*. https//doi.org/https://english.pravda.ru/world/129908-war_ukraine_europe/.

Świeżak, P. (2008). Znaczenie ruchu rusińskiego na Ukrainie. *Bezpieczeństwo narodowe*, 3 (7–8), 248–264.

TASS (2015). Rusiny Zakarpatya trebuyut prinanya svoyei nacionalnosti i avtonomii kraya cherez dialog. https://tass.ru/mezhdunarodnaya-panorama/1828915.

Tarasiuk, T., & Umland, A. (2021). Unexpected Friendships: Cooperation of Ukrainian Ultra-Nationalists with Russian and Pro-Kremlin Actors. Illiberalism – GW's Institute for European, Russian and Eurasian Studies (IERES). https://www.illiberalism.org/unexpected-friendships-cooperation-of-ukrainian-ultra-nationalists-with-russian-and-pro-kremlin-actors/.

The Ministry of Foreign Affairs of the Russian Federation (2025). Foreign Minister Sergey Lavrov's interview with the Hungarian newspaper Magyar Nemzet. The Ministry of Foreign Affairs of the Russian Federation. https://mid.ru/en/foreign_policy/news/2034494/.

Ukr Leaks (2025). From America and Europe with a Swastika. How a Gang of Foreign Neonazis Carried Out Killings of Russian Prisoners of War. Telegra.ph. https://telegra.ph/FROM-AMERICA-AND-EUROPE-WITH-A-SWASTIKA-HOW-FOREIGN-NEONAZIS-CARRIED-OUT-KILLINGS-OF-RUSSIAN-PRISONERS-OF-WAR-06-21.

Walker, R. G. (1998). *The United States Marine Corps and Special Operations*. Monterey: Naval Postgraduate School. https://apps.dtic.mil/dtic/tr/fulltext/u2/a359694.pdf.

Walker, S., Davies, C., & Schulteis, E., (2019). Polish far-right trial raises spectre of „false-flag" tactics, *The Guardian*. https://www.theguardian.com/world/2019/jan/27/polish-far-right-trial-raises-spectre-of-false-flag-tactics-german-journalist-russia-ukraine-fire-court.

White Phoenix (2025). White Phoenix. TGstat. https://tgstat.com/channel/@op_W_P.

Wilson, A. (1996). Ukraine. between Eurasia and the West. In S. Dunn and T.G. Fraser (eds.): *Europe and Ethnicity. The First World War and contemporary ethnic conflict* (pp. 110–137). London: Routledge.

Wyrwał, M. (2019). Łączy nas Rosja. Polscy terroryści, komuniści, nacjonaliści. *Wiadomości w Onet,*. https://wiadomosci.onet.pl/tylko-w-onecie/laczy-nas-rosja-polscy-terrorysci-komunisci-nacjonalisci/ybohxsb.

Zapałowski, A. (2017). Zagrożenia związane z separatyzmem na obszarze Zakarpacia. *Przegląd Geopolityczny*, 9 (20), 9–27.

Żaryn, S. (2019). Russia's hybrid warfare toolkit has more to offer than propaganda. *Defence News*. https://www.defensenews.com/opinion/commentary/2019/08/09/russias-hybrid-warfare-toolkit-has-more-to-offer-than-propaganda/.

Żochowski, P. Iwański, T. (2015). Zakarpattia – together, but separated. Warsaw: Centre for Eastern Studies, 2015). https://www.osw.waw.pl/sites/default/files/commentary_185.pdf.

CHAPTER 20

Norway and the European Union's Security and Defence Policies: Ever-Closer Partners, but Why?

Bjørn Olav Knutsen and Elisabeth Pettersen

Abstract This chapter examines the EU–Norway security and defence partnership signed in 2024, covering areas such as crisis management, defence industry, space, critical infrastructure, and hybrid threats. It analyses the empirical mechanisms behind this cooperation and explores how European integration influences Norway's security and defence policies. The study asks why Norway partners with the EU and what causal mechanisms drive this development. It applies a neo-functionalist approach, arguing that integration reshapes actors' expectations, roles, interests, and identities. The chapter also compares key Norwegian government reports from 2012, 2023, and 2024, highlighting how the worsening European security environment has increased the need for cooperation with the EU. Finally, it discusses paradoxes in Norway's position and argues that theories of EU integration should account for non-member states.

Keywords EU – Norway relations – security and defence cooperation – European integration – neo-functionalism – hybrid threats – foreign policy

1 Introduction

1.1 Analytical Approach and Research Questions

Norway's relationship with the European Union (EU) is the most contentious one in Norwegian politics, ever since Norway applied for membership for the first time in what was then called the European Economic Communities (EEC) in April 1962 (Norwegian Government, 1962). The membership issue has split Norwegian parties along the left-right conflict line as well as the urban-rural and cultural lines of conflict known from the comparative politics literature (Norderval Means, 1974; Saglie, 2000; Grindheim, Heidar & Strøm,

HTTPS://DOI.ORG/10.1163/9789004763623-022

2017, pp. 328–337). Norway is the only country in Europe that has applied for membership three times and rejected membership in two referenda, in 1972 and in 1994, respectively. However, even though Norway has rejected membership, over the years, the country has developed a very close partnership with the EU, covering all aspects of policies. This includes agreements stretching from the European Economic Area (EEA) to association with the Schengen Agreement and over 100 other agreements (NOU 2024: 7; I Europa, 2025), including what we are to discuss in this chapter: Norway's relationship with the EU in security and defence affairs.

In the EU's first defence strategy – the Strategic Compass made public in March 2022 – the EU states, "We will deepen our constructive relations with Norway, as our most closely associated partner through the Agreement on the European Economic Area (EEA)" (EEAS, 2022, p. 56). By building such a partnership, the aim is to contribute to upholding a "rules-based international order and effective multilateralism" to set international norms and standards based upon the principles laid down in the Charter of the United Nations (UN) (EEAS, 2022, p. 53).

The research questions we ask in this chapter are why Norway becomes an ever-closer partner with the EU in security and defence affairs, and, furthermore, what kind of causal mechanisms drive this Norwegian approach to a closer partnership. Our main argument is that the EU today is the most important framework for Norwegian foreign- and security policies, while NATO and the United States are still Norway's most important military ally. By answering these research questions, we aim to speak to different parts of the research literature in International Relations (IR). First, we aim to analyse how a non-member of the EU develops an ever-closer partnership with the union. In so doing, we apply a neo-functionalist approach where we aim to fill a lacuna in the integration theory research literature on how a non-member adapts to and takes part in the European integration process. We henceforth seek to explain the most important mechanisms behind this process of adaptation by Norway and how we can adapt neo-functionalist theory to include non-members in the integration theory research literature. In this regard, we aim to demonstrate how Norway cooperates and takes part in the EU process by pursuing three different forms of strategies, which we label as adaptation, being an external resource, and shielding towards the EU (Kjølberg, 2006, p. 14; Knutsen, 2021; see also Svendsen, 2022). We argue that all these strategies are causal mechanisms that explain Norway's relations with the EU's security- and defence policies. The argument is that these strategies become even more difficult to pursue since Norway stands outside the EU's political governance structures, priorities, and flow of information that takes place inside the EU's political bodies. This is a demanding development and can increase uncertainty in crisis situations and geopolitically tense situations. A challenge for Norway is that the EU is a hybrid security actor, where security challenges are built into other policy areas. Measures aimed at improving the EU's ability to implement policy in vital areas complicate Norway's relationship with the EU.

Second, we apply this theoretical framework to explain the empirical developments in Norway's relationship with the EU, comparing the findings in four governmental reports on Norway's relationship with the EU: one from 2012, two from 2023, and one from 2024. The aim of this comparison is to analyse how Norway relates to the EU process when the security political environment changes dramatically, and where more security considerations are built into EU policies. These reports were chosen because they represent the intersection between research and politics and therefore provide a balanced expression of Norwegian approaches to European integration. We organise this chapter in the following way. In the next part, we describe our theoretical approach based on neo-functionalism and how we can include European non-members in the analysis. We describe this development as part of a wider "externalisation" of the integration process. Thereafter, we describe the development of Norway's relationship with the EU from 2012 onwards, when the first report was issued, focusing on some of the most important paradoxes in Norway's relations with the union. Based on our theoretical foundation, we finally explain the main mechanisms in Norway's relations with the EU in the security- and defence spheres and seek to identify the main drivers in this development.

1.2 Methodological Considerations

In our research for this chapter, we have conducted both rigorous document analysis and interviews, which include a triangulation of both primary and secondary sources. Our aim was to search for the most relevant documents, research articles, and books related to our research questions. The interviews were face-to-face, semi-structured interviews with experts working within the field of Norway's relations with the EU, covering security and defence issues. These interviews were conducted in Brussels in July and December 2024 in the Norwegian delegation to the EU, in the European Commission, in the Norwegian delegation to NATO, at the Vrije Universiteit, and at Egmont Institute. The interviews were transcribed while they took place, which gave us the opportunity to check if we had understood the answers correctly.

2 Neo-functionalism and Non-EU Members

Theories of European integration abound, stretching from regarding the EU as a "tightly coupled pluralistic security community" (Deutsch, 1957/2004; Adler & Barnett, 1998) to stressing that the EU process is based upon neo-functionalist logics (Haas, 1958; Haas, 1964; Niemann, Lefkofridi & Schmitter, 2019). Other theories emphasise the importance of the Member States, which underpins the intergovernmentalist approaches to European integration (Hoffmann, 1964; Moravcsik & Schimmelfennig, 2019). Intergovernmentalism can furthermore be divided into a liberal version (Moravcsik, 1998) and a so-called "new intergovernmentalism", in which intergovernmental policy coordination has become the dominant mode of governance, where supranational institutions have been complicit in it (Bickerton, Hodson & Puetter, 2015; see also Fiott, 2023, p. 448). Other theoretical approaches can be included, of course, like social constructivism (Risse, 2019), governance approaches (Börzel, 2019), and historical institutionalism (Pollack, 2019).

In this chapter, we utilise core insights from neo-functionalism and highlight the dynamics between the intergovernmental level of cooperation and supranationalism. Neo-functionalism is beyond doubt the most famous – and most criticised – approach but is also the most sophisticated theory of European integration (Hooge & Marks, 2019; Schimmelfennig, 2024). It has a high explanatory power, exceeding far beyond the harsh criticism it has received from its most ardent opponents. As Niemann, Lefkofridi and Schmitter (2019, p. 50) underline, "... many of the criticisms levelled against neofunctionalism misrepresented its claims, distorted its arguments, or interpreted the approach selectively".

Neo-functionalism has resurfaced during recent years, including in studies of the evolution of the EU as a security and defence actor (see e.g. Bergmann, 2019; Haroche, 2020). Nevertheless, when studying neo-functionalist theory, we must move beyond the traditional understanding of integration laid down in Ernst B. Haas' book "The Uniting of Europe" (1958/2004), where he defines integration as a process through spill-over mechanisms where the end result is a political community. The determinism and normativism embedded in this theoretical approach lost its explanatory power early since it underestimated nationalism and national sovereignty as important factors in European integration (Hoffmann, 1964; Moravcsik, 1998).

Therefore, Haas′ book "Beyond the Nation-State" (1964) seems to be a more fruitful approach in studying European integration, including how non-members relate to this process. In this book, Haas studies the International Labour Organisation (ILO), but the theory claims to have general validity, also when it comes to regional integration processes (Haas, 1964, pp. vii–viii). Instead of regarding integration as a transfer of loyalty from the national to the supranational level, where one bypasses the nation-state, we can apply this theoretical approach to study integration through the EU Member States, where these states delegate parts of their sovereignty to supranational institutions. This takes place in the form of convergencies of the integration strategies by the Member States through the common institutions laid down in the EU treaties. In his book, Haas labels this development in general terms as a more "universal type of system" (Haas, 1964, p. 79).

In addition, newer research contributions, too, underline the transformative character of the integration process emphasised by neo-functionalists, since it changes the objectives of the actors involved, including the institutions being developed by the actors themselves (Niemann, Lefkofridi & Schmitter, 2019, p. 59). The core argument is that since integration is transformative in nature, the concept of "learning" must be included, as introduced by Haas (Haas, 1964, p. 81). Learning is a key concept since it is through learning that the actors can reformulate their interests, resulting in a higher level of integration if the interests of the actors converge in that direction. Haas furthermore introduces the concept "powers", which he defines as the actors' abilities to learn. He describes it as the "art of revising their demands and seeking new ways of satisfying them" (ibid.).

We now come to the point where the status of the EU's closest non-member partners must be discussed further. The Russian attack on Ukraine in 2014 and onwards, and the full-scale invasion in 2022, have made the EU even more important for Norwegian security and defence. Norway is neither an actor in the European integration process nor a country that can learn from previous integration steps taken by the EU. Norway has consequently fallen into an "integration trap" (Eriksen, 2022), where new integration steps from the EU's side can fall both within and outside of the agreements Norway has with the EU. As a non-member, Norway has no influence over the decisions and actions taken by the EU Member States or by the supranational institutions like the European Commission and the European Parliament. They are, however, heavily influenced by the EU integration process.

In 1992, Norway signed the EEA Agreement, which entered into force in 1994 (Norwegian Government, 1992). Later, Norway joined the Schengen Agreement on passport-free travel inside the Schengen area. This illustrates the externalisation effects of EU integration through "… deepening and widening in terms of actively influencing and reshaping the international environment so as to make it more responsive to the agreed-on aims and common policies of the regional entity" (Sæter, 1998, p. 38). Hence, EU integration will transform the union's geographical environment through a process that changes non-members' status in European integration and security.

It is therefore obvious that integration theorists have not taken the external effects of the EU process sufficiently into account. The geopolitical circumstances are changing, namely the Russian war on Ukraine and an increasingly uncertain role for the United States in European security, and thereby also a changing role for NATO. It is thus imperative to analyse further how non-members relate to EU integration that now also encompasses security and defence issues. This is beyond doubt a challenge for scholars taking a more theoretical approach to European integration, as it is for other scholars and policymakers as well. Especially within the field of security and defence, we have witnessed the most interesting developments in European integration in recent years. Ideas like developing the European Commission into a "geo-political Commission", possibly leading towards a "European defence union", are examples of the transformative effects of European integration. Earlier research has also shown how European integration in security and defence has gone beyond intergovernmentalism as scholars study the role of the European Commission (Haroche, 2020; Brøgger, 2024). Even Norway's delegation to NATO emphasises the EU's ability to "rise to the occasion" when it comes to the further development of the EU as a security and defence actor (Interview, Norway's delegation to NATO, 4 July 2024).

This constitutes a need for a more comprehensive approach in analysing the EU–Norwegian relationship. Norway takes part in the integration process, but in a very special way. Since Norway is not an actor in the EU and thereby has no ability to "learn" from previous integration steps, we suggest an approach to study how Norway approaches the EU's security and defence policies. We also lay down some criteria for how successful such criteria might be. We argue that the most common approach for Norway is *adaptation*, which we observe in the EEA Agreement, where the principle of homogeneity applies to ensure that the same rules apply to EFTA states (that is, non-EU, EEA member states) as to EU countries. Adaptation therefore implies a strategy whereby Norway aims to align its security- and defence policy with EU priorities (Knutsen, 2021, p. 89). The second Norwegian strategy is to act as an *external resource*, which implies that Norway can further develop its role as a respected partner in international military operations and as a significant military power taking responsibility for European security through NATO. Today, this also implies assisting Ukraine in its defence efforts and taking part in different EU initiatives, like the European Defence Fund (EDF), the Permanent Structured Cooperation (PESCO), and contributions to the European Peace Facility (EPF). The last strategy is *shielding,* which is aimed at limiting EU influence on Norwegian security. As we will discuss, the weakness with these strategies is that the Norwegian approach towards the EU takes NATO as its point of departure. Norway currently experiences that NATO is not the institutional framework for EU security and defence policies, even though a closer interinstitutional partnership between these institutions is evolving (NATO, 2024). This is a phenomenon that can be explained by neofunctionalist theorising and the externalisation effects of European integration.

3 To Build a Security and Defence Partnership with the EU without Membership

3.1 Four Governmental Committee Reports

Since Norway rejected membership in 1994, foreign- and security policy has traditionally been a "blind-spot" in the country's relationship with the EU. From a Norwegian perspective, the perception has been that the EU's foreign- and security policies are too inefficient, and also potentially harmful to NATO and the transatlantic relationship. Norway's traditional Atlantic security outlook has therefore overshadowed Norway's relationship with the EU (Archer, 2004). This is very much still the case, but Norway's views on the EU as a foreign policy and security provider have changed due to the rising importance of the EU's foreign- and security policies, especially after 2000 (Knutsen & Tvetbråten, 2022; Rieker & Riddervold, 2021; Riddervold & Rieker, 2023). Norway has, as part of this development, signed several agreements with the EU on these areas. The most recent one (per March 2025) is from May 2024 on a partnership agreement on a broad set of issues from security and defence, but also space, critical infrastructure, economic security, and hybrid threats (Norwegian Government, 2024a).

To describe this development, we will analyse four governmental committee reports that cover Norway and the EU as a security provider. We focus on two dimensions: The first one is how the working relationship between NATO and the EU is described, which is of interest due to Norway's traditional Atlanticist foreign policy outlook. The second is how effective the EU's security policies are from a Norwegian viewpoint, with emphasis on how the EU handles the war in Ukraine and Norway's response to EU policies.

The first report, "Outside and Inside" from 2012, covered Norway's agreements with the EU from 1994 until 2012. It discussed all aspects of Norway's relations with the EU (NOU 2012: 2, 2012). The second report, "This is serious – Prepared for an uncertain future" from 2023, discusses Norway's total preparedness from a societal security perspective centring on a discussion of the overall risk, vulnerability, and threat picture that Norway at present faces (NOU 2023b: 17, 2023). The third report, also from 2023, is the Defence Commission's, "Defence for Peace and Freedom" (NOU 2023: 14, 2023). This report assesses which potential security and defence policy choices and priorities Norway can take to best safeguard Norwegian security in a 10–20-year perspective. The last report, from 2024, is a follow-up to the 2012 report. It is entitled "Norway and the EEA: Development and experiences" (NOU 2024: 7, 2024). It underscores the major and dramatic changes occurring in and around Europe over the past decade and how the EU has responded to these changes, not least the changes in geopolitics.

3.2 Findings

The first report, from 2012, provides a general assessment of how Norwegian politicians have traditionally assessed the role of the EU in security and defence. That assessment is one of ambivalence and scepticism based on Norwegian Atlanticism and the alleged presumptions that all security arrangements outside the Atlantic frameworks are ineffective:

> Since the early 1990s, Norwegian authorities have sought cooperation with the EU. At the same time, the Norwegian authorities' view of the development of EU foreign and security policy cooperation has been ambivalent. On the one hand, they have sought to obtain as many agreements and as close a connection as possible to the EU. On the other hand, they have both hoped and believed that the EU would not develop an independent foreign, security, and defence policy (NOU 2012: 2, 2012, p. 723).

Even though this report was published two years before the first Russian attack on Ukraine, this Norwegian scepticism endures, albeit in a different form. Common for all reports, but mostly so for the three from 2023 and 2024, is that they emphasise that the EU has become not only a more effective security actor over the years, but also a comprehensive one, with a broad approach to security. All three reports highlight the importance of the war in Ukraine as important for the development of the EU as a security- and defence actor. Norway therefore needs a closer partnership with the union. Nevertheless, the reports still define the EU's role within an Atlantic security framework, which implies that NATO and the transatlantic relationship are fundamental to European security. This approach is most prominent in the report from the Defence Commission. The Commission emphasises that Norwegian authorities have traditionally been clear that the EU must not develop into a military alternative or a competitor to NATO. Norwegian authorities have warned against ambitions to establish EU forces, which could weaken NATO's internal cohesion and military effectiveness (NOU 2023: 14, 2023, p. 269). For example, the EU's decision in 2022 in the Strategic Compass (EEAS, 2022) to establish a Rapid Deployment Capacity (RDC) of some 5,000 soldiers to be operative in 2025, was not considered a positive proposal, neither from Norwegian authorities nor from the Norwegian Secretary General of NATO at the time, Jens Stoltenberg (VG, 30 November 2021). In comparison, the present Secretary General Mark Rutte has a more positive approach to EU security and defence initiatives than his predecessor (Interview, Norwegian NATO delegation, 4 July 2024).

The Strategic Compass is the steering guide for the development of the EU's security- and defence policies. The aim is to make the EU a stronger and more capable security provider (EEAS, 2022). The challenge in this regard is to develop collective action within a field dominated by the Member States, through common defence planning, exercises, and headquarters (Interview, European Commission, 3 July 2024). This also includes the European defence industry, where the European Commission works on a European Defence Industry Strategy and Programme (EDIS/EDIP) (European Commission, 2025). In addition, the European Defence Fund (EDF) and the Permanent Structured Cooperation (PESCO) are important tools in the further development of EU policies within this field. The overarching aim is to reduce capability gaps and enhance European deterrence (Interviews, European Commission, 3 July 2024 and the Norwegian EU delegation, 13 December 2024).

This development of the European integration process takes place at the same time as European security is at a turning point. The war in Ukraine and the new Trump administration's approach to international cooperation and security undermine the basic foundations for the institutionalised European security order, which has NATO as its core. Several European politicians are therefore now stating that Europe can no longer outsource security to others, and they underline the need for a "European defence union" and "European strategic autonomy" (see e.g. Adler, 2025; Scazzieri, 2025). Such autonomy is a precondition for the build-up of a European pillar inside NATO and a more sustainable sharing of burden. For such a pillar to be operative, it must include the development of European force packages to respond to external conflicts and crises and thereby European deterrence. A prerequisite for deterrence, however, is reducing capability shortfalls in European defence, which is thoroughly discussed in the Niinistö-report (European Commission, 2024). One of the conclusions in this report is that for the EU to be a fully fledged security actor, the EU "... must address external threats from a position of strength together with partners across the globe in ways that uphold and strengthen the rules-based international order" (European Commission, 2024, p. 6). All these developments are well reflected and discussed in the different governmental reports. When comparing the 2012 and the 2024 reports, it is obvious that the EEA Agreement over the years has become even more important to Norway (NOU 2: 2012; NOU 2024: 7, 2024). Two of the main reasons are the war in Ukraine and the experiences from the COVID-19 pandemic. Norway has therefore connected more closely to the EU through civil emergency preparedness and health security cooperation, which is discussed in the 2023 report on societal security (NOU 2023: 17, 2023). As the 2024 report discusses, today's EU has developed into a hybrid security actor, which implies a much broader approach to security than NATO (NOU 2024: 7, pp. 304–306). Furthermore, this report discusses these developments and argues that the EEA Agreement can be a framework that can be expanded within certain areas and sectors of foreign- and security policy. The main reason is that the EU is now more prepared and has closer cooperation with NATO. This is a significant difference from what was the case in 2012.

In this regard, the reports stress that the EU should be complementary to NATO, and they therefore argue for a closer partnership between these two institutions. When comparing the Defence Commission's report with the report on societal security, it is interesting to observe the differences between them. While the Defence Commission has a more traditional approach to security, the report on societal security stresses the comprehensiveness of the EU's approach and stresses the differences in decision-making between NATO and the EU. Decisions made within the EU framework oblige the Member States to a greater extent than decisions in NATO (NOU 2023: 17, p. 106). As decisions made in the EU are legally binding to a higher degree and its security actorness is more comprehensive in nature, it is especially interesting to assess the EU's role in providing for the Member States´ total defence. Of special importance is the civil protection mechanism (UCPM). As the 2023 total defence report stresses (NOU 2023: 17, 2023), the EU cooperation on resilience and preparedness is developing rapidly, and the work to strengthen preparedness is taking place in parallel with ongoing crisis management. As the same report further underscores, as one of the EU's closest cooperation partners in the war in Ukraine, Norway is largely included. Norway is also, through association with the EU through the EEA Agreement and several other agreements, covered by EU directives and regulations in several sectoral areas

(NOU 2023: 17, p. 104). This illustrates how Norway is steadily developing a partnership with EU security, one which takes on a more comprehensive approach. Such a comprehensive societal security approach is therefore a necessary condition for building European security after the Russian attack on Ukraine. Simultaneously, European resilience must be built while facing a Russian threat which is far wider than just the military threat.

The new security situation after the Russian attack on Ukraine has consequently been a turning point in Norwegian security. Although NATO is still the framework for Norway's defence policies, the EU has become the most important framework for Norwegian foreign- and security policies. This is clearly reflected in the partnership agreement signed by Norway and the EU in May 2024. Besides stating that Norway is "the EU's most closely associated partner, including in the area of security and defence", the aim of the partnership is to "establish and implement a tailor-made, mutually beneficial security and Defence Partnership that will frame their overall cooperation across the security and defence spectrum" (Norwegian Government, 2024a). This agreement must furthermore be understood in the perspective of the conclusions drawn in the three reports from 2023 and 2024. The 2024 report states that the government should follow up on the EU's ambition for a strategic partnership, with a view to clarifying in more detail how such cooperation can be designed. The three reports also emphasise the seriousness of the security situation, which has deteriorated significantly. Therefore, they highlight the need for Norway to change the pace of its cooperation with the EU to ensure Norwegian security by further strengthening ties with the country's European allies and neighbours.

4 How to Explain Norway's Relations with the EU's Security- and Defence Policies

The changing character of the European security order has increased the need for Norway to connect its foreign- and security policies closer to the EU. Norway's partnership with the union is therefore an expression of a larger EU process of deepening and widening, and thereby influencing and reshaping the international environment. Through this form of externalisation, the aim is to defend a European security order based on European integration and to support Ukraine.

Norway's approach is to build a closer partnership with the EU, but without the rights and obligations that follow from membership. Instead, Norway follows the path of adaptation, where it, in some cases, also aims to act as an external resource. The key to this development is the Russian full-scale attack on Ukraine, which has transformed the EU into a more coherent and comprehensive security- and defence actor. But even though this development is of fundamental importance to European security, several challenges and problems remain.

The main problem in European defence is obviously fragmentation and duplication. Europe still depends on US strategic enablers, and this limits the extent of European strategic autonomy. As emphasised by Biscop (2024, p. 516), Member States should "... integrate national capabilities into large multinational formations, where NATO's new force model could provide an extra stimulus to that end". Therefore, Europe needs to be able to deter its opponents and defend itself. This should happen within the framework of the EU and in close cooperation with NATO, where the United Kingdom plays a fundamental part in Europe's defence. Norway's challenge is that the country's association with the EU is a static and adaptive one, while the EU's security policy dimension has a dynamic character based upon neo-functionalist logics. That is, the Member States, in their interactions with each other and with the European Commission, develop an ability to learn from previous experiences. This enables them to further develop new integration steps, like moving beyond intergovernmentalism in European security and defence, where the European Commission is an actor today. We also observe this development when analysing EU policymaking: it becomes cross-sectoral, and areas like justice, emergency preparedness, trade, and foreign aid are included in EU security- and defence policy. This is an important part in developing strategic autonomy, where a total defence approach to the autonomy concept must be added. A far more demanding geopolitical situation in Europe explains this development. As the Norwegian 2024-report underscores: "In the wake of the COVID-19 pandemic, the spotlight was turned on reducing economic dependence on foreign supply chains, and since 2021, strategic autonomy has encompassed the vast majority of EU policy areas" (NOU 7, 2024, p. 297).

It is thus a paradox that the processes in the EU that are intended to simplify the institutions and streamline cooperation will present a challenge for Norway in the times to come. This illustrates well that the three mechanisms pursued by Norway towards the EU – adaptation, acting as an external resource, and shielding – will become more difficult to pursue. A changing geopolitical environment with a reduced role for the US in European security shakes the foundations for Norway's approach to European security and integration. The forthcoming European pillar of NATO will be built around the EU and the United Kingdom. This inevitably amplifies the asymmetry of Norway's relationship with its other European allies. Such a development is also described in the long-term plan for the Norwegian defence in light of intensified great power competition. Here, the Norwegian government states that "Norway's room for manoeuvre is under pressure. A main challenge for Norwegian security is the increased risk of being marginalised, meaning that our interests or security are not taken into account" (Norwegian Government, 2024b, p. 36).

As Eriksen (2022) argues, Norway has fallen into an integration trap. His argument is that Norway faces a significant democracy problem that is difficult to address within the existing frameworks. Our argument is that the same is the case when it comes to European security. Norway risks falling on the outside of the new European security structures that now evolve centred around the EU in partnership with the United Kingdom.

From this discussion, it is evident that the EU plays a fundamental part in the European security framework and that the interaction between intergovernmental and supranational decision-making should in no way be considered dichotomous. This is also emphasised by Haas (1964) as he regards integration in terms of convergence of the integration strategies among the participating governmental actors. How the EU further develops as a security actor is nevertheless determined by how the US will relate to Europe and to NATO going forward. Norway is a country with a strong Atlantic security orientation. This Norwegian Atlanticism will last for the foreseeable future. Moreover, Norway's relationship with the EU has been stable since 1994, when the country last had a referendum on membership. We expect this to remain so for the foreseeable future.

5 Conclusions

We started out this chapter by asking why Norway becomes an ever-closer partner with the EU in security and defence. We also asked what kind of causal mechanisms lie behind this Norwegian approach to a closer partnership. The neo-functionalist framework helps us to understand why Norway takes part in this process, not by being an actor in the EU, but through externalisation of the European integration process. Today, this implies a recast of the European security order where Europe must take on more responsibility for its own security. The Norwegian approach towards the EU is one of adaptation, acting as an external resource and shielding. As this chapter has shown, the Norwegian approach towards the EU is a static and adaptive one, while the EU's security policy dimension has a dynamic character. It is therefore paradoxical that processes in the EU that intend to simplify institutions and streamline cooperation present a challenge for Norway. However, this Norwegian status is not likely to change. Continued Norwegian adaptation to the EU, with the partnership agreement from May 2024 as a framework, will therefore steer the EU–Norwegian relationship on security and defence.

One of the largest challenges in today's European security order is that of the transatlantic relationship. NATO's future is at stake. It is this institution that lays the framework for Norway's security- and defence political relationship with its European allies and with the US. An uncertain role for NATO will also make Norway's status in European security more unpredictable.

References

Adler, E., & Barnett, M. (Eds.). (1998). Security communities. Cambridge University Press.

Adler, K. (2025, February 24). New German leader signals seismic shift in transatlantic relations. BBC News. https://www.bbc.com/news/articles/clyz4j5z4npo

Archer, C. (2004). Norway outside the European Union: Norway and European integration from 1994 to 2004. Routledge.

Bergmann, J. (2019). Neofunctionalism and EU external policy integration: The case of capacity building in support of security and development (CBSD). Journal of European Public Policy, 9(4), 1253–1272. https://doi.org/10.1080/13501763.2018.1526204.

Bickerton, C., Hodson, D., & Puetter, U. (Eds.). (2015). The new intergovernmentalism: States and supranational actors in the post-Maastricht era. Oxford University Press.

Biscop, S. (2024). European defence: No Zeitenwende yet. Defence and Peace Economics, 35(4), 516–520. https://doi.org/10.1080/10242694.2023.2201739.

Brøgger, T. E. (2024). A 'Europe of defence'? The establishment of binding commitments and supranational governance in European security and defence. Journal of European Integration. https://doi.org/10.1080/07036337.2024.2379421.

Börzel, T. (2019). Governance Approaches to European Integration. In A. Wiener, T. A. Börzel, & T. Risse (Eds.), European integration theory (pp. 107–128). Oxford University Press.

Deutsch, K. (1957/2004). Political Community and the North Atlantic Area. In B. F. Nelsen & A. Stubb (Eds.), Readings on the theory and practice of European integration (pp. xx–xx). Lynne Rienner Publishers. https://www.lsu.edu/faculty/lray2/teaching/7971_1s2009/deutsch1957.pdf.

EEAS. (2022). A strategic compass for security and defence. European External Action Service. https://www.eeas.europa.eu/sites/default/files/documents/strategic_compass_en3_web.pdf.

Eriksen, E. O. (2022). Demokratisk overskudd i Europa – hva med Norge? Nytt Norsk Tidsskrift, 39(3), 187–200. https://doi.org/10.18261/nnt.39.3.3.

European Commission (2024). Safer together: Strengthening Europe's civilian and military preparedness and readiness. https://ec.xeuropa.eu/info/publications/safer-together-strengthening-europes-civilian-and-military-preparedness-and-readiness_en.

European Commission. (2025). The European defence industry programme at a glance. European Commission.

Fiott, D. (2023). In every crisis an opportunity? European Union integration in defence and the war on Ukraine. Journal of European Integration, 45(3), 447–462. https://doi.org/10.1080/07036337.2023.2183395.

Grindheim, J. E., Heidar, K., & Strøm, K. W. (2017). Norsk politikk. Universitetsforlaget.

Haroche, P. (2020). Supranationalism strikes back: A neofunctionalist account of the European Defence Fund. Journal of European Public Policy, 27(6), 853–872. https://doi.org/10.1080/13501763.2019.1609570.

Hoffmann, S. (1964). The European process at Atlantic cross-purposes. Journal of Common Market Studies, 3(1), 85–101. https://doi.org/10.1111/j.1468-5965.1964.tb01096.x.

Hooge, L., & Marks, G. (2019). Grand theories of European integration in the twenty-first century. Journal of European Public Policy, 26(8), 1113–1133. https://doi.org/10.1080/13501763.2019.1569711.

Håkansson, C. (2023). The Ukraine war and the emergence of the European Commission as a geopolitical actor. Journal of European Integration, 46(1), 25–45. https://doi.org/10.1080/07036337.2023.2239998.

Haas, E. (1958/2004). Uniting of Europe: Political, social and economic forces, 1950–1957. University of Notre Dame Press.

Haas, E. (1964). Beyond the nation-state: Functionalism and international organization. Stanford University Press.

I Europa. (2025). Norges avtaler med EU. https://ieuropa.no/norges-avtaler-med-eu

Kjølberg, A. (2006). EUs sikkerhetslogikk (FFI-rapport 2006–01857). FFI. https://ffi-publikasjoner.archive.knowledgearc.net/bitstream/handle/20.500.12242/1745/06-01857.pdf.

Knutsen, B. O. (2021). German–Norwegian relations in security and defence: What kind of partnership? European View, 20(1), 88–96. https://doi.org/10.1177/17816858211001228.

Knutsen, B. O., & Tvetbråten, K. (2022). A reluctant European: How Norway responds to the EU's quest for strategic autonomy. European View, 21(2), 196–205. https://doi.org/10.1177/17816858221095897.

Moravcsik, A. (1998). The choice for Europe: Social purpose and state power from Messina to Maastricht. Cornell University Press.

Moravcsik, A., & Schimmelfennig, F. (2019). Liberal Intergovernmentalism. In A. Wiener, T. A. Börzel, & T. Risse (Eds.), European integration theory (pp. 67–89). Oxford University Press.

NATO (2024). Relations with the European Union. https://www.nato.int/cps/ru/natohq/topics_49217.htm?selectedLocale=en.

Niemann, A., Lefkofridi, Z., & Schmitter, P. C. (2019). Neofunctionalism. In A. Wiener, T. A. Börzel, & T. Risse (Eds.), European integration theory (pp. xx–xx). Oxford University Press.

Norderval Means, I. (1974). Norway and the EEC. Scandinavian Studies, 46(4), 352–369. https://www.jstor.org/stable/40917474.

Norwegian Government (1962, April 30). Norway's application for membership to the EEC. https://www.cvce.eu/en/obj/norway_s_application_for_membership_to_the_eec_oslo_30_april_1962-en-8a1ef152-25ca-4108-9eb2-be4599f39287.html.

Norwegian Government (1992). St.prp. nr. 100 (1991–92) Om samtykke til ratifikasjon av Avtale om Det europeiske økonomiske samarbeidsområde (EØS), undertegnet i Oporto 2. mai 1992. https://www.regjeringen.no/no/dokumenter/stprp100_1991/id627296/.

Norwegian Government (2024a, June 6). Norway and EU sign Security and Defence Partnership [Press release]. https://www.regjeringen.no/en/aktuelt/norway-and-eu-sign-security-and-defence-partnership/id3040958/.

Norwegian Government (2024b). Prop. 87 S (2023–2024), 2024: Forsvarsløftet – for Norges trygghet. Langtidsplan for forsvarssektoren 2025–2036 [The Norwegian Defence Pledge. Long-term Defence Plan 2025–2036]. https://www.regjeringen.no/no/dokumenter/prop.-87-s-20232024/id3032217/.

NOU (2012). Utenfor og innenfor. Norges avtaler med EU [Outside and inside: Norway's agreements with the EU] (NOU 2012: 2). https://www.regjeringen.no/contentassets/5d3982d042a2472eb1b20639cd8b2341/no/pdfs/nou201220120002000dddpdfs.pdf.

NOU (2023a). Forsvarskommisjonen av 2021. Forsvar for fred og frihet [The Defence Commission of 2021. For peace and freedom] (NOU 2023: 14). https://www.regjeringen.no/no/dokumenter/nou-2023-14/id2974821/.

NOU (2023b). Nå er det alvor. Rustet for en usikker fremtid [This is serious – Prepared for an uncertain future] (NOU 2023: 17). https://www.regjeringen.no/no/dokumenter/nou-2023-17/id2982767/.

NOU (2024). Norge og EØS: Utvikling og erfaringer [Norway and the EEA: Development and experiences] (NOU 2024: 7). https://www.regjeringen.no/no/dokumenter/nou-2024-7/id3033576/.

Pollack, M. A. (2019). Rational Choice and Historical Institutionalism. In A. Wiener, T. A. Börzel, & T. Risse (Eds.), European integration theory (pp. 45–66). Oxford University Press.

Riddervold, M., & Rieker, P. (2023). Finally coming of age? EU foreign and security policy after Ukraine. European Security, 33(3), 497–516. https://doi.org/10.1080/09662839.2023.2209587.

Rieker, P., & Riddervold, M. (2021). Not so unique after all? Urgency and norms in EU foreign and security policy. Journal of European Integration. https://doi.org/10.1080/07036337.2021.1977293.

Risse, T. (2019). Social Constructivism and European Integration. In A. Wiener, T. A. Börzel, & T. Risse (Eds.), European integration theory (pp. xx–xx). Oxford University Press.

Saglie, J. (2000). Values, perceptions and European integration: The case of the Norwegian 1994 referendum. European Union Politics, 1(2), 227–249. https://doi.org/10.1177/1465116500001002004.

Scazzieri, L. (2025, January). Towards an EU "defence union"? [Policy brief]. Centre for European Reform. https://www.cer.eu/publications/archive/policy-brief/2025/towards-eu-defence-union.

Schimmelfennig, F. (2024). Crisis and polity formation in the European Union. Journal of European Public Policy, 31(10), 3396–3420. https://doi.org/10.1080/13501763.2024.2313107.

Svendsen, Ø. (2022). The politics of third countries in EU security and defence: Norway, Brexit. Springer.

Sæter, M. (1998). Comprehensive neo-functionalism: Bridging realism and liberalism in the study of European integration. NUPI.

VG. (2021, November 30). NATO-sjefen advarer mot EU-plan: Kan ikke ha en allianse i alliansen [NATO-chief warns against EU plan: Cannot have an alliance within the alliance]. VG.

CHAPTER 21

The Importance of Türkiye in the Area of European Union Security in the Programs and Priorities of the Presidency of the Council of the EU in the Years 2016–2024

Dominika Liszkowska

Abstract This chapter examines the role of Türkiye as a strategic partner for the European Union in key areas such as migration, counter-terrorism, and economic cooperation. It discusses the persistent lack of EU membership despite long-term integration efforts and highlights factors affecting the accession process. The study focuses on Türkiye's importance in EU security in the context of the priorities of successive Presidencies of the Council of the EU between 2016 and 2024. It addresses how member states perceive Türkiye's role in EU security, why some downplay its importance, and which countries view it as a key partner. The analysis is based on qualitative research of EU Council Presidency programmes and priorities, using an agenda-setting framework. It argues that the positioning of Türkiye in EU security policy is shaped by strategic political choices at the supranational level.

Keywords Türkiye – European Union – security – foreign policy – Presidency of the EU Council

1 Introduction

Creating security throughout Europe was the driving force behind the establishment of the European Union (EU). During the Cold War, cooperation between European states evolved towards creating a powerful economic bloc through the European Community (EC), as well as ensuring collective defence against external threats through the North Atlantic Treaty Organization (NATO) (Desai, 2005, p. 369). For years, the policy of enlargement of the EC and then the EU has been a strong anchor for peace, democracy, prosperity, security, and stability on the continent (Stanicek, Przetacznik, & Albala-

 | HTTPS://DOI.ORG/10.1163/9789004763623-023

dejo Roman, 2023, p. 2). The influence of EU values and principles on new members and candidates has significantly reduced the likelihood of instability and conflict in Europe (Missiroli, 2004, p. 16). Additional member states have expanded the platform for exchange with new interests and opportunities, increasing the scope of common policies and strengthening the EU as an international actor (Desai, 2005, p. 366).

Although Türkiye has become an active player in the European security and defence architecture since the beginning of the Cold War (Doğan, 2003, p. 1), despite many years of efforts, it has not managed to become a member of the EU. However, the state plays an important role in cooperation within the framework of the EU's common foreign and security policy. Over the years, it has actively participated in EU missions (mainly within the framework of the NATO-EU cooperation scheme), including nine of the 30 operations led exclusively by the community. Türkiye's contribution is one of the largest after France, Germany, and the United Kingdom (Szymański, 2019, p. 181).

This chapter highlights Türkiye's importance in the area of EU security in the context of the priorities of the upcoming presidencies of the Council of the EU. The analysis aims to verify the following research hypotheses:

H_1: For the next presidencies of the Council of the EU, the most significant areas of security in relations with Türkiye are the management of migration crises, the fight against terrorism, and ensuring stability in the Middle East region.

H_2: Some Member States holding the Presidency of the Council of the EU in 2016–2024 ignored Türkiye's role in ensuring EU security due to the political situation in that country and problems with democracy and the rule of law.

H_3: The perception of Türkiye's role in European security in the programmes of the next presidencies of the Council of the EU reflects the attitude of individual Member States towards this state's accession to the EU.

In order to verify the above research hypotheses, an analysis of existing sources was conducted, and the desk research technique was used. An important source for this research was the programs of the individual EU Council presidencies (2016–2024). This study adopts the theoretical framework of agenda-setting. Its premise is that EU agenda-setting is a strategic choice made by political actors to elevate issues to the supranational level. This is because European states differ significantly in terms of their political institutions and the prevailing understanding of specific issues (Princen, 2017, p. 545). Therefore, the topics of EU presidency programs need not be limited solely to "truly" European or supranational issues, but also to the interests of individual states. National actors' motives may lead to an emphasis on addressing issues relevant to a given Member State, rather than the EU as a whole (Princen, 2007, p. 35). This approach therefore helps explain why some states, in their six-month presidency programs, focus on issues related to Türkiye's role in the security of the entire EU, for example, in combating crime, terrorism, and illegal migration. Other states, on the other hand, completely omit Türkiye from their six-month presidency plans, revealing particular motives and national interests. This type of agenda selectivity can be interpreted as an indirect signal of disputes or tensions in bilateral relations. Therefore, applying the agenda-setting approach allows for the analysis not only of the priorities declared by the states holding the presidency but also reveals the importance of omitted topics, which equally reflect political and relational conditions. This corresponds to the classic approach of agenda-setting theory used in policy analysis at the national level (Cobb & Elder, 1971, pp. 907–908).

2 The Importance of Türkiye in the European Security System

Türkiye's importance in the European security system stems from three important factors: NATO membership, military capabilities, and geostrategic position (Müftüler-Bac, 2000, p. 490). Türkiye became a member of the North Atlantic Alliance on 18 February 1952, as a result of the perceived threat from the Soviet Union. However, in recent years, Türkiye's relations with several NATO states, including the United States (US), have been marked by numerous crises of trust. The mechanisms of political consultation between the parties have been undermined by, among other things, the negative perception of Türkiye by other member states (Shkurti Özdemir, 2022, p. 5), as well as the decision of the Turkish authorities to purchase the S-400 air defence system from Russia. Significant differences between Türkiye and US have also been revealed by the issue of policy in the Middle East, when the US began to pragmatically cooperate with the Kurdish Democratic Union Party in the fight against the self-proclaimed Islamic State (Ülgen, Besch, & Toygür, 2024, p. 2). In turn, since the failed coup attempt in Türkiye in July 2016, the narrative of the ruling Justice and Development Party (AKP) has begun to emphasize the alleged involvement of the US in the attempt to destabilize the state. Moreover, from Türkiye's perspective, the reduction in the role of the US and the EU in some regions has become the basis for the disorientation of the global order. A significant expression of this was the withdrawal of NATO forces from Afghanistan, initiated by the US in 2021 (Ülgen, Besch, & Toygür, 2024, p. 2). All this has influenced an increasingly disappointed and frustrated Turkish society, which perceives US actions as disregarding Türkiye's security concerns. According to a 2024 Pew Research Center study, only 18% of Turks had a positive assessment of the current US actions in the international arena (Clancy, Poushter, & Hernandez Ramones, 2024).

Türkiye's geopolitical significance is considered crucial for European security (Doğan, 2003, p. 1). The state's strategic location between Europe, the Middle East, and Central Asia makes it the focal point of Eurasian geopolitics (Colibasanu, 2024, p. 5). However, Türkiye is surrounded by several unstable states (Carley, 1995, pp. 23–24). Therefore, despite being a NATO member, Türkiye is seeking to expand its list of non-Western partners, including Russia and China. At the same time, the state's actions reflect the broader context of the domestic political narrative, portraying it as "a strong, independent regional leader, unafraid of confrontation with both old friends and rivals" (Colibasanu, 2024, p. 16). In line with this vision, Türkiye asserts its influence in conflicts in Syria, Libya, and the Caucasus, demonstrating a strategic autonomy that oscillates between global powers (Ülgen, Besch, & Toygür, 2024, p. 3).

The atmosphere of regional rivalry implies the dynamics of power and international alliances, and also affects numerous security threats, including those of an asymmetric nature (Kumelska-Koniecko, 2023, p. 104). This is why Türkiye continues to seek deeper involvement in the EU's Common Security and Defence Policy and in EU defence initiatives. It is an important participant in EU missions and operations and also the largest power, outside the Member States, of the military contingent for Operation ALTHEA in Bosnia and Herzegovina, through EU-led forces (European Commission, 2023a).

However, despite emerging calls for a greater European role in transatlantic security, Brussels has yet to recognize Türkiye as a "closer partner in defence" (Coşkun, 2024). Even in the face of Russia's aggression against Ukraine, the community has not decided to engage Türkiye in any significant way in the field of defence and security. Opposition from Greece, Cyprus, and France has blocked, for example, financing from the European Peace Fund (Defence Security Asia, 2024) for the supply of Turkish Bayraktar TB2 drones and artillery missiles to Ukraine (Ioannidis, 2024). This state of affairs is influenced by disputes with Cyprus, the conflict with Greece in the Aegean Sea, and the national positions of some EU members (Coşkun, 2024).

According to Münevver Cebeci, however, the creation of a common European security culture without Türkiye's involvement is incomplete (Cebeci, 2014, p. 26). This reflects a fundamental difference in the understanding of security culture, as Türkiye's approach remains rooted in traditional concerns about state sovereignty. In turn, for the EU and some member states, until recently, it was the result of a postmodern and post-military approach to interpreting security (Bagcı & Gaudıno, 2020, pp. 12–13). However, in recent years, the European paradigm has changed as a result of two important events. The first is Russia's full-scale aggression against Ukraine, which destroyed the security architecture. The date of 24 February 2022 is even considered to be "a historical turning point that forced the EU to become a full-fledged security actor in the largest geopolitical conflict in Europe since World War II" (Alpan, 2024, p. 9).

Another significant issue has been the change in US priorities and the announcements of this state's reduced involvement in the security sphere on the old continent (Ülgen, Besch, & Toygür, 2024, p. 6). After Donald Trump took office as president in 2025, the US administration clearly emphasized the medium- and long-term perspective. This aims to reduce US military participation in Europe and (in the short term) partially withdraw support for the war in Ukraine in connection with the negotiations with Russia (OSW, 2025). This state of affairs has led to the adoption of a number of political initiatives to increase the level of security in Europe, including an increase in defence spending in the years 2021–2024, amounting to over 30% (European Council, 2025). Taking into account the above premises, Türkiye seems to be a natural partner in every dimension of strengthening the European defence pillar. This is influenced not only by its long-standing activity in the Western security architecture, but also by the importance of the developing defence industry, as well as the risks and uncertainties coming from the south of the continent (Coşkun, 2024).

The debate on Europe's greater responsibility for defence further highlights the EU's arms industry deficit and the need to collaborate with partners outside the EU that offer technological flexibility and rapid production capabilities. In this context, Türkiye stands out as a particularly valuable partner. Its strong defence industry makes it a strategic partner for some member states, as evidenced by the signing of numerous bilateral and multilateral agreements in this area in recent years (Musiienko, 2025). For example, in late 2024, Italy approved the sale of domestic aircraft manufacturer Piaggio Aerospace to Turkish company Baykar, known for its unmanned aerial vehicles (UAVs). In October 2025, the Spanish government approved a €3.12-billion deal for the purchase of 45 Turkish Hürjet trainer aircraft (Kostidis, 2025). Cooperation in the field of purchases from the Turkish defence industry also applies to Central and Eastern European states such as Romania and Poland, in particular in the context of the Bayraktar TB2 system (Özdilek, 2025, p. 131).

It's worth emphasizing that the technological dimension of EU-Türkiye relations doesn't end with defence. The technological solutions designed by this state are increasingly addressing areas such as border management and migration control. Türkiye was the seventh state in the world to develop its own biometric system and implemented it in 2022 (Karakaya Polat, 2024, p. 69). To prevent irregular flows, the state uses biometric identification systems, such as fingerprint and iris scanning, to register and track individuals crossing the border. Since 2023, mobile vehicles equipped with biometric technology have been used for random checks in urban areas with the highest migrant density (Karakaya Polat, 2024, p. 67). New solutions are also being introduced in the context of cooperation with the EU in combating terrorism, organized crime, and cybercrime. For example, the EU, in cooperation with Germany, financed a €1.3 million project in Türkiye that supported the state in adapting its national internet regulations to European standards. In addition, the Turkish National Police has gained a 24/7 contact point in its cybercrime department (Delegation of the European Union to Türkiye, 2022), facilitating cooperation and information exchange with European services in the fight against cybercrime.

3 EU Member States and the Presidency of the Council of the EU on Cooperation with Türkiye in Managing the Migration Crisis (2016–2018)

The issue of migration has become particularly important for the entire EU during the period of increased migration flows that led to the crisis in 2015–2016. In the programme of the presidency of the Council of the EU, the importance of cooperation with Türkiye in overcoming it was already highlighted in the first months of 2016 by the Netherlands. This state considered effective control of Europe's external borders, improving the initial reception of refugees on the continent and in the region, as a priority, as well as fair distribution of the burden of migration (EUNL2016, 2016, p. 9).

During the Dutch presidency, an agreement was concluded on 18 March 2016 between Türkiye and the EU, the aim of which was to reduce the influx of illegal migrants into Europe. Under the agreement, Türkiye decided to accept the rapid return of all immigrants not requiring international protection who entered Greece from its territory. It also declared that it would accept back all irregular migrants who were detained in its territorial waters. In turn, the EU, in addition to financial support for Syrian refugees in Türkiye, announced the liberalisation of the visa system and the lifting of visas for citizens of that state by the end of June 2016. Both parties also agreed to further strengthen measures aimed at combating migrant smugglers (Council of the EU, 2016).

The statement was seen as the embodiment of a new stage in relations between the two parties (European Commission, 2016). In reality, however, it was a confirmation of the EU's strategic dependence on cooperation with Türkiye in the management of irregular migration flows (Turhan & Wessels, 2021, p. 198) and the "transactional turn" (Alpan, 2024, p. 12) in Turkish–EU relations. Despite criticism from the EU in the context of the halting of democratization processes, Türkiye gained significant assurances of taking action to re-energize the accession process, including by opening another chapter of accession negotiations: (33) Financial and budgetary provisions.

The role of Türkiye as a key partner in overcoming the largest migration crisis in Europe since World War II was therefore confirmed in the programmes by subsequent presidencies of the EU Council: Slovakia, Malta, Bulgaria, and Romania. Referring to the priorities of the Slovak presidency, effective cooperation in the field of returns and readmission with Türkiye contributed to "a significant decrease in migration flows in the Aegean Sea and stabilisation of the situation on the migration route through the Western Balkans" (SKEU2016, 2016, p. 9). As it turned out, joint actions by Türkiye and the Member States in counteracting the effects of the migration crisis almost immediately blocked the mass flow of migrants (Liszkowska, 2020, p. 66). In turn, Malta emphasized the important role of the strategic partnership with Türkiye in maintaining peace and stability in the region. The state also stressed the need to appreciate the Turkish contribution to reducing irregular migrant flows and resolving the crisis as a result of the EU-Türkiye statement (MaltaEU2017, 2017, p. 16). The Bulgarian EU Council Presidency also expressed its support for the full implementation of the Statement (EU2018BG.BG, 2018, p. 21), considering it crucial for strengthening cooperation on migration issues with key countries of origin and transit in Asia. For Bulgaria, its long common border means that any shift in Türkiye's migration policy could expose it to significant pressure, particularly if refugees were encouraged to move further west, which is why Bulgaria remains one of Türkiye's closest allies in the EU and the only Member State that decided to deport alleged supporters of the preacher Fethullah Gülen to his homeland following the 2016 coup d'état. In response, Türkiye reciprocated by accepting the return of illegal migrants from Bulgaria (Gotev, 2017).

4 Türkiye's Actions in the Eastern Mediterranean as a New Element of the Security Agenda in Türkiye-EU Relations (2018–2020)

Since the signing of the EU-Türkiye statement on migration, Ankara has repeatedly threatened to terminate the agreement due to the EU's failure to meet its financial obligations. In addition, the EU side has not fulfilled the arrangements for visa liberalization for Turkish citizens. Therefore, one of the topics of wide discussion during the meeting scheduled for 28 March 2018 in Varna was once again the issue of managing migration flows (Leviev-Sawyer, 2018). During the meeting, EU representatives confirmed their commitment to provide further support aimed at improving the situation of over three million refugees that Türkiye has accepted.

However, the agenda of the talks also covered other important security issues, on which the position of EU states was not in line with Türkiye's line of action. The first issue concerned Turkish involvement in Syria and the capture of the Kurdish-controlled city of Afrin. EU leaders expressed concern about the actions of the Turkish army and its allies. The President of the Council of Europe, Donald Tusk, even noted that "the escalation of violence and displacement must be avoided in an already tragic and unstable situation" (The Sofia Globe Staff, 2018).

Another problem in bilateral relations was the issue of Türkiye's actions in the Eastern Mediterranean and the Aegean Sea. The EU's "illegal" Turkish attempts to extract gas and oil in the Eastern Mediterranean, as well as the dispute over the course of the maritime border, caused a serious crisis in Türkiye's relations with Greece and Cyprus. Both Türkiye and the Republic of Cyprus claim jurisdiction over oil and gas exploration in Cyprus's maritime zones (an island located close to the coasts of Türkiye, Syria, and Lebanon). The EU clearly sided with Cyprus in this respect. At the same time, it recalled "Türkiye's obligation to respect good neighbourly relations and normalise relations with all EU Member States", which is "a condition for the development of EU-Türkiye relations, including the accession process" (Council of the EU, 2018).

In the following months, the situation in EU-Türkiye relations was not improved by Türkiye's continued activity in the Mediterranean Sea, nor by mutual accusations of violating the air and sea borders between Türkiye and Greece (in the context of the decades-long dispute over the division of the continental shelf) (Chudziak, 2020). In June 2018, the Council of the EU stated for the first time that Türkiye was increasingly moving away from the EU and

References

Alpan, B. (2024). Türkiye-EU Relations and the European Political Community in the Aftermath of Russian Invasion: Is the Transactional Turn Here to Stay?. *EuroMeSCo Paper*, 69, 7–22.

AVİM. (2018). *Romanian PM: Cooperation between Romania and Turkey is very high.* https://avim.org.tr/en/Bulten/ROMANIAN-PM-COOPERATION-BETWEEN-ROMANIA-AND-TURKEY-IS-VERY-HIGH.

Bagcı, H., & Gaudıno, U. (2020). Involving Turkey in EU Common Foreign, Security and Defence Policies. *Eurasian Research Journal*, 2(1), 7–27.

BBC (2018). *Macron tells Erdogan: No chance of Turkey joining EU.* https://www.bbc.com/news/world-europe-42586108.

Carley, P. (1995). *A Conference Report: Turkey's Role in the Middle East.* https://www.usip.org/sites/default/files/pwks1.pdf.

Cebeci, E. M. (2014). Turkish and EU Security Cultures in Perspective: Why do we need a holistic approach to European Security?. *Marmara Üniversitesi Avrupa Araştırmaları Enstitüsü Avrupa Araştırmaları Dergisi*, 22(2), 25–62.

Cengiz, S. (2024). *Why Turkiye is overlooking the European Political Community.* https://www.arabnews.com/node/2552646.

Chudziak, M. (2020). *Turcja: konfrontacyjna polityka na Morzu Śródziemnym.* https://www.osw.waw.pl/pl/publikacje/analizy/2020-08-12/turcja-konfrontacyjna-polityka-na-morzu-srodziemnym.

Clancy, L. & Poushter, J. & Hernandez Ramones, S. (2024). *How People in Turkey View International Affairs.* https://www.pewresearch.org/global/2024/10/16/how-people-in-turkey-view-international-affairs/.

Cobb, R.W. & Elder, Ch. D. (1971). The Politics of Agenda-Building: An Alternative Perspective for Modern Democratic Theory. *The Journal of Politics*, 33(4), 892–915.

Colibasanu, A. (2024). *Turkey's Evolving Geopolitical Strategy in the Black Sea.* https://www.fpri.org/wp-content/uploads/2024/12/turkeys-evolving-geopolitical-strategy2.pdf.

Coşkun, A. (2024). *Turkey as Europe's Defense Pillar.* https://www.gisreportsonline.com/r/turkey-europe-defense-pillar/.

Council of the EU (2016). *EU-Turkey statement, 18 March 2016.*

Council of the EU (2018). *EU-Turkey leaders' meeting, 26 March 2018.*

Council of the EU (2019). *Turkish drilling activities in the Eastern Mediterranean: Council adopts conclusions.*

Council of the EU (2022). *EU-Turkey leaders' meeting, 6 October 2022.*

Council of the European Union (2018). Presidency Programme of the Council of the European Union: 1 January – 30 June 2018. Brussels.

Council of the European Union. (2019). *Nielegalne odwierty prowadzone przez Turcję we wschodniej części Morza Śródziemnego: Rada przyjmuje ramy sankcji, 11 listopada 2019.*

Council of the European Union. (2020). *Morze Śródziemne: Rada przedłuża o rok sankcje za nielegalne odwierty, 6 listopada 2020.*

Defence Security Asia (2024). *Greece, Cyprus, France Block EU from Purchasing Turkish Drones, Missiles for Ukraine.* https://defencesecurityasia.com/en/greece-cyprus-france-block-eu-from-purchasing-turkish-drones-missiles-for-ukraine/.

Delegation of the European Union to Türkiye (2022). *Supporting Türkiye for a Safe Cyberspace,* https://www.eeas.europa.eu/delegations/t%C3%BCrkiye/supporting-t%C3%BCrkiye-safe-cyberspace_en.

Desai, S. (2005). Turkey in the European Union: A Security Perspective – Risk or Opportunity?. *Defence Studies,* 5(3), 366–393.

Doğan, E. (2003). Turkey in the New European Security and Defence Architecture. *Perceptions: Journal of International Affairs,* 8(1), 1–18.

EU2018BG.BG. (2018). *Programme of the Republic Of Bulgaria for the Presidency.*

EU2019.FI. (2019). *Sustainable Europe: Finland's Presidency Programme, Presidency of the Council of the European Union 1 July – 31 December 2019.*

EU23. (2023). *Programme Spanish presidency of the Council of the European Union.*

EUNL2016. (2016). *Programme of the Netherlands Presidency of the Council of the European Union 1 January – 30 June 2016.*

European Council. (2024). *Konkluzje Rady Europejskiej w sprawie Ukrainy, Bliskiego Wschodu i Turcji, 17 kwietnia 2024.*

European Council. (2025). *Bezpieczeństwo i obrona.* https://www.consilium.europa.eu/pl/topics/security-and-defence/.

European Council. (n.d.). *Turcja.* https://www.consilium.europa.eu/pl/policies/turkey-enlargement/.

European Commission (2016). Komunikat Komisji do Parlamentu Europejskiego, Rady Europejskiej i Rady: Pierwsze sprawozdanie z postępów w zakresie realizacji postanowień oświadczenia UE-Turcja. COM(2016) 231 final.

European Commission (2023a). *Wspólny komunikat do Rady Europejskiej: Stan stosunków politycznych, gospodarczych i handlowych między UE a Turcją.* JOIN (2023) 50 final.

Gotev, G. (2017). *Estonian presidency passes the Turkey decision to Bulgaria.* Euractiv. https://www.euractiv.com/section/enlargement/news/estonian-presidency-passes-the-turkey-decision-to-bulgaria/.

HU24EU. (2024). Programme of the Hungarian Presidency of the Council of the European Union in the Second Half of 2024.

İKV. (2018). *The Austrian EU Presidency and Prospects for Turkey-EU Relation.* https://www.ikv.org.tr/images/files/newsletter-41.pdf.

Ioannidis, S. (2024). *Greece, Cyprus, France Block Bayraktar Financing*. https://www.ekathimerini.com/news/1231979/greece-cyprus-france-block-bayraktar-financing/.

Kaczan-Winiarska, E. (2020). Austrian Veto against European Union Membership Negotiations with Turkey, *Polish Review of International and European Law*, 8(1), 103–122.

Karakaya Polat, R. (2024). Biometric Technology and Policy Narratives on Migration Control: The Case of Turkey, *Social Review of Technology and Change*, 2(1), 65–84.

Kostidis, M. (2025). *Spain approves €3.12 bln deal for 45 Turkish-built Hurjet aircraft*. https://www.ekathimerini.com/politics/foreign-policy/1285178/spain-approves-e3-12-bln-deal-for-45-turkish-built-hurjet-aircraft/.

Kumelska-Koniecko, M. (2023). Ewolucja tureckiej polityki bezpieczeństwa. *Przegląd Geopolityczny*, 45, 104–118.

Leviev-Sawyer, C. (2018) *Great, or not so great, expectations for EU – Turkey summit in Varna*. https://sofiaglobe.com/2018/03/25/great-or-not-so-great-expectations-for-eu-turkey-summit-in-varna/.

Liszkowska, D. (2020). Migration Crisis and Its Impact on the Internal Situation in Turkey. *Studia Europejskie – Studies in European Affairs*, 3, 63–83.

MaltaEU2017 (2017). *Programme of the Maltese Presidency of the Council of the European Union 1 January–30 June 2017*.

Missiroli, A. (2004). The EU and Its Changing Neighbourhood. In R. Dannreuther (Ed.), *European Union Foreign and Security Policy Towards a Neighbourhood Policy* (pp.12–26). London: Routledge.

Müftüler-Bac, M. (2000). Turkey's Role in the EU's Security and Foreign Policies. *Security Dialogue*, 31(4), 489–502.

Musiienko, V. (2025). *From Defense Cooperation to European Integration: Can Security Dynamics Revitalize Turkey–EU Relations?* https://tdcenter.org/2025/12/18/from-defense-cooperation-to-european-integration-can-security-dynamics-revitalize-turkey-eu-relations/.

Osiewicz, P. (2009). Turcja na drodze do członkostwa w UE: uwarunkowania, przebieg oraz perspektywy negocjacji akcesyjnych. *Rocznik integracji europejskiej*, 3, 181–210.

OSW. (2025). *Co po Monachium? Perspektywa zmniejszenia obecności USA w Europie*. https://www.osw.waw.pl/pl/publikacje/analizy/2025-02-17/co-po-monachium-perspektywa-zmniejszenia-obecnosci-usa-w-europie.

Özdilek, E. (2025). Cooperation between the European Union and Turkey in the Context of Security. *Florya Chronicles of Political Economy*, 11(2), 111–149.

Princen, S. (2007). Agenda-setting in the European Union: a theoretical exploration and agenda for research. *Journal of European Public Policy*, 14(1), 21–38.

Princen, S. (2017). Agenda-Setting and Framing in Europe. In E. Ongaro & S. van Thiel (Eds.), *The Palgrave Handbook of Public Administration and Management in Europe* (pp. 535–551). London: Palgrave Macmillan.

Ricard, P. (2022). *Macron's project for a 'European political community' stumbles over Turkey.* https://www.lemonde.fr/en/international/article/2022/09/07/macron-s-project-for-a-european-political-community-stumbles-on-turkey_5996056_4.html.

Romania2019.EU (2019). *Programme of the Romanian Presidency of the Council of the European Union 1 January – 30 June 2019.*

Shkurti Özdemir, G (Ed.). (2022). *70-Years of Turkey-NATO Relations.* Istanbul: SETA.

SKEU2016 (2016). *Programme of the Slovak Presidency of the Council of the European Union 1 July – 31 December 2016.* Bratislava, Slovakia: Ministry of Foreign Affairs.

Stanicek, B., & Przetacznik, J. & Albaladejo Roman, A. (2023). *Enlargement policy: Reforms and challenges ahead.* https://www.europarl.europa.eu/RegData/etudes/BRIE/2023/757575/EPRS_BRI(2023757575_EN.pdf.

Sweden2023.EU (2023). *Swedish Presidency of the Council of the European Union.*

Szymański, A. (2019). Differentiated Integration and EU-Turkey Relations: A Chance for Cooperation in Foreign and Security Policies?. *Sprawy Międzynarodowe*, 1, 175–196.

The Sofia Globe staff (2018). *EU – Turkey leaders' meeting in Varna: No agreements, but 'readiness to keep up dialogue'.* https://sofiaglobe.com/2018/03/26/eu-turkey-leaders-meeting-in-varna-no-agreements-but-readiness-to-keep-up-dialogue/.

Turhan, E. (2016). Europe's Crises, Germany's Leadership and Turkey's EU Accession Process. *CESifo Forum*, 2, 25–29.

Turhan, E., & Wessels, W. (2021). The European Council as a Key Driver of EU–Turkey Relations: Central Functions, Internal Dynamics, and Evolving Preferences. In W. Reiners & E. Turhan (Eds.), *EU-Turkey Relations* (pp. 185–217). Cham: Palgrave Macmillan.

Ülgen, S. & Besch, S. & Toygür, İ. (2024). *Strategic Autonomy as a Dynamic of Convergence in Türkiye-EU Relations.* https://carnegie-production-assets.s3.amazonaws.com/static/files/Ulgen_Besch_Toygur_Turkiye%20EU-final.pdf.

CHAPTER 22

The European Union's Contribution to the Development of Post-Dayton Bosnia and Herzegovina

Marko Aćić and Miloš Grujić

Abstract This chapter examines cooperation between Bosnia and Herzegovina and the European Union in the post-Dayton period. It focuses on the role of the EU Special Representative in supporting Bosnia and Herzegovina's European integration process. The analysis highlights EU regional initiatives in Southeast Europe as instruments aimed at improving state functionality within Bosnia and Herzegovina's complex institutional structure. It also discusses the implementation of the Stabilization and Association Agreement and the challenges related to delayed EU candidate status. Further attention is given to monetary stability under the currency board system and cooperation with the European Border and Coast Guard Agency (Frontex) in the field of border management.

Keywords European Union – Bosnia and Herzegovina – EU integration – Special Representative of the EU – Stabilization and Association Agreement – Frontex – currency board system

1 Introduction

The role of the European Union (hereinafter referred to as the EU) in post-Dayton Bosnia and Herzegovina (hereinafter referred to as BiH) can be examined through several segments. The first segment relates to its role in ending the war conflicts that took place on the territory of BiH, encompassing two phases: the first, where the EU unsuccessfully attempted to end the war through support or as an initiator of peace plans, and the second, where it witnessed the signing of the Dayton Peace Agreement. The second segment in the relations between the EU and BiH needs to be viewed through the broader prism of relations with the Western Balkan countries, for which the Stabilization and Association Agreement was specifically designed and imple-

 | HTTPS://DOI.ORG/10.1163/9789004763623-024

mented. This agreement serves as a framework through which comprehensive reforms should be carried out to enable these countries to meet the defined conditions and acquire candidate status for full EU membership. In addition to these segments, the EU has approved a significant portion of financial resources through projects and by supporting reforms, such as the IPA funds, to achieve the ultimate goal defined in the aforementioned second segment.

Notwithstanding the above, it is necessary to emphasize that although the EU has been a partner in numerous reforms implemented in BiH, it has also applied double standards towards it and other Western Balkan countries, leaving them to wait excessively long for EU candidate status, while continuously reiterating in its official statements that the future of these countries lies within the EU. Conversely, countries such as Ukraine were granted candidate status for purely political reasons, despite the situation within the country being far from meeting the actual conditions for EU membership candidacy. Regardless of these significant political decisions made at the EU level, this paper aims to specifically highlight that the EU has invested over 1,045,760.00 KM in BiH to date and that only partial goals have been achieved, despite the turnover of eight EU Special Representatives in BiH across various mandates (Aćić & Grujić, 2023, p. 31). The degree of success of specific projects and their impact on BiH's path to full EU membership will be discussed further in the following text. The aim of this paper is to analyze the role of the EU in the development and stabilization of BiH from the Dayton Peace Agreement to the present, with a particular emphasis on key initiatives, programs, and challenges on the path to European integration. The main research question of this paper is: How has the EU influenced the post-war reconstruction and stabilization of BiH, and what are the remaining challenges on the path to full EU membership?

This paper employs a qualitative methodology, including the analysis of documentary sources, official EU reports, and relevant academic literature, to examine the EU's contributions in various segments, such as peace initiatives, economic reforms, institutional strengthening, and migration control. The research questions are focused on how the EU has influenced the post-war reconstruction of BiH, which are the most significant reforms implemented with its support, and what are the remaining challenges on the path to full EU membership. The main thesis of the paper is that, although the EU has played a significant role in the stabilization and development of BiH, the existence of double standards and the slow pace of the integration process represent significant obstacles to achieving full membership. Therefore, in order to gain a deeper theoretical and contextual understanding of the mechanisms of EU influence and the dynamics of enlargement, the paper is expanded by analyzing key theories of Europeanization and digital policies, as well as the specific role of the rotating Presidency of the Council of the EU and the digital dimension in the context of the Western Balkans. The paper will focus on three main segments: (1) the EU's role in ending the war and implementing the Dayton Agreement, (2) economic and political reforms supported through programs such as IPA funds and the Stabilization and Association Agreement, and (3) contemporary challenges in cooperation, including migration control through cooperation with Frontex and military stabilization through the EUFOR Althea mission. Through the analysis of these segments, the paper seeks to highlight the contradiction between the significant invested resources and the slow progress of BiH on the path to the EU, while emphasizing the need for further reforms and strengthening the political will to overcome existing challenges.

There is still no single, general, and comprehensive theory of Europeanization (Sošić, 2006). Instead, it is a dynamic, multidimensional, and often controversial field of study within political science and European studies, with numerous competing theoretical approaches and perspectives that explain how the EU and its institutions, policies, and values spread and influence national states, societies, and policies, yet none cover all aspects and processes (Olsen, 2002; Jambrovic & Maresic, 2020; Bodur Ün & Arıkan, 2022). In analyzing the EU's influence on Bosnia and Herzegovina, approaches relying on institutionalism (focus on the conditions set by the EU) and sociological constructivism (focus on changing identities, discourse, and values) are particularly relevant (Radaelli, 2023; Schimmelfennig & Sedelmeier, 2005). Europeanization here is viewed as a process of external imposition and diffusion (out-of -the-gate), where the Enlargement Policy serves as the main tool for transformation. The context of the Council of the EU Presidency further shapes this process, giving Member States with historical and geographical ties to the Balkans (such as Germany, Austria, or Croatia) a special role in transmitting standards and defining priorities for stabilization and digital cooperation in the region (Vachudova, 2005).

The study of the EU's external influence on the Western Balkans is predominantly framed within the literature on external Europeanization, with the enlargement policy being considered its most powerful instrument. Two complementary theoretical approaches are most commonly used: rationalist institutionalism and sociological constructivism. The rationalist approach emphasizes conditionality, where the EU uses market access and financial aid as leverage, and the Balkan states, as rational actors, adopt reforms to achieve membership (Schimmelfennig & Sedelmeier, 2005; Vachudova, 2005). Conversely, constructivism focuses on norm diffusion and socialization, analyzing how the EU promotes democratic, administrative, and legal values, influencing the change in identities and discourse among elites (Lavenex, 2004). However, in the Balkan region, the effectiveness of conditionality often declines due to the problem of weak absorptive capacity within states and an unclear membership perspective, leading to selective or passive Europeanization (Todorova, 2000). These mechanisms are further intertwined with the role of Member States which, through the EU Council Presidency, directly shape regional priorities, particularly in the domains of digital policy and security, with emphasis placed on the role of newer members that have geographical and historical ties to the region (Dimitrova, 2010).

The theoretical understanding of digital policies, especially in the context of the EU and its external influence on the Balkans, relies on several perspectives that are interwoven with traditional theories of international relations. Network institutionalism emphasizes the role of transnational actors, such as large technology companies and technical regulatory bodies, in shaping global digital governance, often outside the direct control of states (Drezner, 2014). On the other hand, digital geopolitics analyzes the digital sphere as a new arena for the struggle for power, where states and blocs (such as the EU, the USA, and China) use standardization, infrastructure, and data control as strategic tools (Nye, 2021). The theory of digital sovereignty has become particularly important in the EU, focusing on the need for Europe to build resilience and autonomy in the digital space. The external influence of EU digital policies on the Western Balkans often occurs through a process of regulatory diffusion, where the EU, with the help of the Council Presidency, seeks to impose its standards (such as the GDPR or the Digital Services Act) in the region, which is crucial for the security of digital infrastructure and technological transformation.

Despite the fact that the Council of the EU Presidency formally has the role of a mediator and administrator, its political and agenda-setting power is particularly pronounced in the context of foreign policy and enlargement, where the Presidency can significantly influence the focus and dynamics of negotiations (Kočan & Lovec, 2021). Member States taking on the rotating Presidency use this opportunity for the strategic promotion of their national interests and regional priorities, particularly towards their geographical neighbors (Balfour, 2016). In the case of the Western Balkans, the role of the Presidency takes on additional significance; for example, the German and Austrian Presidencies have traditionally used their influence to maintain stability and economic support (Aleksandra, 2020), while the Slovenian and Croatian Presidencies, as newer members from the immediate neighborhood, have sought to keep the Western Balkans high on the EU agenda (Kočan & Lovec, 2021).

Their engagement is directed not only toward political stabilization but also toward themes such as digital transformation and the security of digital infrastructure in the region, which is crucial for alignment with the ambitions of "Digital Europe." Through the Presidency, the priorities and intensity of the application of the conditionality policy in the Stabilization and Association Process are thus directly influenced (Gafuri & Muftuler-Bac, 2021; Đorđević et al., 2021).

2 The Role of the EU in Ending the War Conflicts in Bosnia and Herzegovina

Since the onset of the war in Bosnia and Herzegovina (BiH), the EU has actively mediated an end to one of the largest conflicts on European soil since World War II. Some of its key members, including Germany, provided diplomatic and other support to the separatist republics of the former Yugoslavia, primarily Slovenia and Croatia, and later to one of the warring factions involved in the conflict within BiH. Acting as a diplomatic mediator, the EU sponsored several significant initiatives to halt the horrors of war in BiH. Following the failures of the Cutileiro, Vance-Owen, and Owen-Stoltenberg peace plans, the EU initiated further negotiations based on previously reached agreements. On November 18, 1993, under the Franco-German initiative led by foreign ministers Juppé and Kinkel, separate meetings were conducted with Boban, Karadžić, and Silajdžić, representing the three equal nations. This plan aimed to broaden negotiations to encompass the entire former Yugoslavia rather than just the territory of Bosnia and Herzegovina. All efforts to endorse this plan were directed towards the Bosniaks and their elected representatives due to the ongoing inability to reach a consensus with them throughout the war (Tokić-Marić). Unfortunately, both this plan and earlier peace proposals failed because the warring parties presented maximalist demands that had support from segments of the international community. The Dayton Peace Agreement ultimately concluded the conflicts in BiH, with the EU, United Nations, as well as representatives from France, Germany, Great Britain, the USA, and the Russian Federation, witnessing the signing of this agreement as noted in Article XI.

After the war's conclusion, the EU launched a wide array of programs, projects, and initiatives in BiH. Among the most significant activities of the EU, which will be further discussed in this paper, is the establishment of the EU Special Representative in BiH. In addition to the Special Representative, Lopandić emphasizes that BiH, as a Southeast European country, has also engaged with the EU to participate in initiatives such as the Central European Initiative (CEI), Central European Free Trade Area (CEFTA), Organization for Black Sea Economic Cooperation (BSEC), multilateral cooperation in Southeast Europe (PSJEI), the Royaumont Process for good neighborliness and cooperation in Southeast Europe (recently integrated into the Stability Pact), Southeast European Cooperative Initiative (SECI), Adriatic-Ionian Initiative (JJI), Stability Pact for Southeast Europe (SP), and Stabilization and Association Agreements (SAA) (Lopandić, 2001, pp. 20–23).

2.1 Special Representative of the EU

The EU and BiH have developed and institutionalized cooperation that has lasted for more than thirty years[1]. The EU decided, in addition to various pro-

1 We can follow the mentioned cooperation through the following most significant dates, observing them chronologically backward: 21.03.2024. – The Council of the EU made a decision to open accession negotiations with Bosnia and Herzegovina, 15.12.2022. – The European Council made a decision to grant Bosnia and Herzegovina candidate status for membership in the EU, 20.08.2016. – The General Affairs Council of the EU accepted the application of Bosnia and Herzegovina for membership in the EU and instructed the European Commission to prepare a questionnaire for BiH. 23.04.2015. – Both chambers of the Parliamentary Assembly of Bosnia and Herzegovina adopted the Decision on the confirmation and adoption of the Declaration of the Presidency of BiH on the commitment of institutions at all levels of government in Bosnia and Herzegovina to implement the necessary reforms within the process of accession to the EU at special sessions. By confirming the Declaration in the Parliamentary Assembly of BiH, the key condition of the EU for the entry into force of the Stabilization and Association Agreement between BiH and the EU was fulfilled, 21.04.2015. – The EU Council adopted the Decision on the conclusion of the Stabilization and Association Agreement between BiH and the EU, 1.6.2015. – The Stabilization and Association Agreement between Bosnia and Herzegovina and the EU, which aims to prepare our country for EU membership, entered into force, from 15.12.2010. – The EU Council for Justice and Home Affairs adopted a decision to abolish the visa regime for citizens of Bosnia and Herzegovina. In June 2008. – The Stabilization and Association Agreement was signed, at the beginning of 2007. – The Instrument for Pre-accession Assistance (IPA) was established, intended for all pre-accession activities financed by the European Commission, January 2006. – The Council of the EU adopted the second European Partnership with Bosnia and Herzegovina, 2004 – The EU accepted the European Partnership with BiH. 2000 – Publication of the EU Road Map. This document established 18 key conditions that Bosnia and Herzegovina needs to fulfill in order to start the Feasibility Study for the start of negotiations on the Stabilization and Association Agreement, which enabled the introduction of duty-free access for products from Bosnia and Herzegovina to the EU market. 1999 – The Stabilization and Association Process (SAP) began. SAP offered a clear possibility of integration for Bosnia and Herzegovina, as well as for the other five countries of the Western Balkans region into the EU, 1997 – The Council of Ministers of the EU set political and economic conditions for the development of bilateral relations. Bosnia and

grams and projects that it would implement, to open an office of the EU Special Representative (EUSR) for BiH (hereinafter referred to as the EUSR). The concept of the EU Special Representative finds its true basis in the Maastricht Treaty and began operating as early as 1996. It is necessary to note that such practice is not new within the common European foreign and security policy but originated before it. So far, it has appointed missions in Afghanistan, the African Union, Central Asia, the Middle East, the South Caucasus, Serbia-Kosovo and Metohija, Macedonia, Moldova, Sudan, and the Great African Lakes. The complete definition of the future role of the EU Special Representative is attributed to Article 30 of the Lisbon Treaty, which, among other things, states that these representatives will be under the auspices of the European External Action Service (EEAS) and will have to cooperate closely with the EU diplomatic missions in the host countries. The final decision on approving this engagement on behalf of the EU is given by the EU Council of Ministers, upon the recommendation of the High Representative for Foreign Affairs and Security Policy, to whom they are directly accountable for the work they perform.

The specificity of their engagement in BiH is reflected in the fact that since the first appointment in 2002, the High Representatives themselves, as heads of the Office of the High Representative in BiH, have also performed the function of the EUSR. In that capacity, Paddy Ashdown acted from 2002; Christian Schwarz-Schilling from 2006; Miroslav Lajčák from 2007; and Valentin Inzko from 2009 to 2011 (Aćić, 2022, p. 328). Until 2011, the EUSR office was integrated into the Office of the High Representative for BiH and was held by one person. That practice was changed by the decision of the EU Council of Ministers on July 11, 2011, which specifically defines that in the future, the EUSR will perform the function independently of the Office of the High Representative in BiH. This decision also emphasizes that: "The Union will further strengthen its policy and presence on the ground through a unified and strengthened Union representative who will take the lead in supporting Bosnia and Herzegovina on Union-related issues to support that country's progress towards integration with the Union, including a broad and balanced set of instruments" (Council of the European Union, 2011).

So far, 4 EUSRs have operated independently in BiH: Peter Sørensen with a mandate from 2011 to 2015, Lars-Gunnar Wigemark from 2015 to 2019, Johann Sattler from 2019 to 2024, and Luigi Soreca from 2024. Within its mandate, the

Herzegovina enjoys autonomous trade preferences. 1995 – The Dayton (Paris) Peace Agreement stopped the war and established two entities in Bosnia and Herzegovina: the Federation of BiH and Republika Srpska.

EUSR has so far acted by encouraging reform processes that should bring BiH to the status of a full EU member in several areas: a commission was appointed to deal with defense reform issues, a commission was formed to reform the intelligence and security sector, and a Commission for Police Reform was appointed. In addition to the above, in its regular reports to the EU bodies, it has repeatedly encouraged the process of European integration in BiH with the aim of reversing it from the phase of stagnation and obstruction.

2.2 Other Programs and Projects of the EU in BiH

The European Commission actively participated in financing numerous projects in Bosnia and Herzegovina. In projects financed by EU grants and loans, both the private and public sectors were treated equally[2].

2 The aforementioned cooperation can be traced through the following significant dates, presented in reverse chronological order: On March 21, 2024, the Council of the EU decided to open accession negotiations with Bosnia and Herzegovina. On December 15, 2022, the European Council decided to grant Bosnia and Herzegovina candidate status for membership in the EU. On August 20, 2016, the General Affairs Council of the EU accepted Bosnia and Herzegovina's application for EU membership and instructed the European Commission to prepare a questionnaire for BiH. On April 23, 2015, both chambers of the Parliamentary Assembly of Bosnia and Herzegovina, during special sessions, adopted the Decision on the confirmation and adoption of the Declaration of the Presidency of BiH on the commitment of institutions at all levels of government in Bosnia and Herzegovina to implement the necessary reforms within 1 the process of accession to the EU; the confirmation of the Declaration in the Parliamentary Assembly of BiH fulfilled the key condition of the EU for the entry into force of the Stabilization and Association Agreement between BiH and the EU. On April 21, 2015, the EU Council adopted the Decision on the conclusion of the Stabilization and Association Agreement between BiH and the EU. On June 1, 2015, the Stabilization and Association Agreement between Bosnia and Herzegovina and the EU, aimed at preparing the country for EU membership, entered into force. On December 15, 2010, the Council of the EU for Justice and Home Affairs adopted a decision to abolish the visa regime for citizens of Bosnia and Herzegovina. In June 2008, the Stabilization and Association Agreement was signed. In early 2007, the Instrument for Pre-accession Assistance (IPA) was established, intended for all pre-accession activities financed by the European Commission. In January 2006, the Council of the EU adopted the second European Partnership with Bosnia and Herzegovina. In 2004, the EU adopted the European Partnership with BiH. In 2000, the EU Road Map was published, establishing 18 key conditions that Bosnia and Herzegovina needed to fulfill to proceed with the Feasibility Study for the start of negotiations on the Stabilization and Association Agreement, enabling duty-free access for products from Bosnia and Herzegovina to the EU market. In 1999, the Stabilization and Association Process (SAP) was initiated, offering a clear path to integration for Bosnia and Herzegovina, as well as for the other five countries of the Western Balkans region into the EU. In 1997, the Council of Ministers of the EU set political and economic conditions for the development of bilateral relations, and Bosnia and Herzegovina enjoyed autonomous trade

Some of the most significant projects financed in the private sector are often related to issues in the economic sphere – specifically projects related to sustainable economic development, support for employment, improvement of the competitiveness of domestic businesses, as well as agricultural competitiveness, and projects related to quality improvement with a special emphasis on rural development.

On the other hand, in the public sector, the financed projects were related to: strengthening joint institutions at the BiH level, including public finance and judicial reforms, while at the entity level, priorities were given, among other things, to projects related to the social protection system, harmonization of domestic legislation with the EU, and those specifically related to support in fulfilling obligations under the Stabilization and Association Agreement[3].

The Stabilization and Association Agreement (SAA) represents an important step in the European integration process of Bosnia and Herzegovina. This agreement, signed on June 16, 2008, in Luxembourg, entered into force on June 1, 2015, and replaced the Interim Agreement on Trade and Trade-related Matters. The SAA aims to establish a close and lasting relationship with the EU, enabling Bosnia and Herzegovina to become a full member of the EU in a transitional period of six to ten years (Glintić, 2013, 105). The basic elements of the SAA include political, economic, and security stabilization. One of the essential requirements for signatory countries is the gradual harmonization

preferences. In 1995, the Dayton (Paris) Peace Agreement ended the war and established two entities in Bosnia and Herzegovina: the Federation of BiH and Republika Srpska. These milestones collectively illustrate the progressive trajectory of cooperation between the EU and Bosnia and Herzegovina, highlighting the incremental steps taken towards deeper integration and eventual EU membership. The chronological arrangement of these events underscores the sustained commitment of both parties to fostering a stable and prosperous future for Bosnia and Herzegovina within the European framework.

3 In the domains of transport and energy, initiatives have included: the implementation of the Law on Railways of BiH, the establishment and operation of the State Regulatory Body for Railways, assistance to the energy sector in fulfilling obligations arising from the Treaty Establishing the Energy Community of Southeast Europe (EnC) and other international commitments, reform of the energy sector in BiH, integration of civil air traffic into EU civil aviation, construction of a hydroelectric power plant on the river Bosna (Doboj), support for the implementation of the General Transport Network project in Southeast Europe and assistance for priority investments in the transport sector identified in sub-sector strategies, and the establishment and strengthening of cooperation between state and entity electricity companies. For further information on sectoral projects and IPA programs, refer to: EU projects in BiH http://europa.ba/?page_id=558, http://europa.ba/?page_id=562; http://ec.europa.eu/enlargement/instruments/funding-by-country/bosnia-herzegovina/index_en.htm, accessed December 19, 2024.

of legislation with the EU's legal acquis, known as the acquis communautaire (Nezirović et al., 2022). A free trade area is also established, which opens new opportunities but also challenges for domestic producers in terms of product quality and standards. From an economic perspective, the SAA has a significant impact on Bosnia and Herzegovina's trade relations with the EU. Open borders and reduced tariffs facilitate access to the EU market for domestic producers, enabling them to compete in a broader market. This particularly applies to sectors such as agriculture, textiles, and the wood industry, where BiH has natural resources and a tradition of production. However, market access requires domestic products to meet strict quality and safety standards, which can pose a challenge for entrepreneurs who must adapt (Vranješ et al., 2017). The SAA focuses on sectoral cooperation in areas such as the judicial system, environment, transport, and energy policy. Within the SAA, contributions are made to strengthening institutional capacities and training personnel to ensure the effective management and implementation of policies related to EU legislation. In these areas, Bosnia and Herzegovina can benefit from European funds and programs, which can further enhance the capacities of public institutions.

In its provisions, the SAA also contains provisions that regulate human rights, minority rights, and democratization. These provisions are of particular importance for Bosnia and Herzegovina, given its complex ethnic composition and past experiences with wars and conflicts. Through the implementation of the agreement, it was expected to strengthen democracy and the rule of law, which should reduce tensions and build trust among different ethnic groups (Džidić, 2014, p. 86). In addition to the above, it is important to emphasize that cooperation between BiH and the EU has also taken place through programs: PHARE, ECHO, Reconstruction, CARDS, IPA, as well as less significant projects that accompanied them, with an emphasis on specific areas in which EU projects also led to progress. After the successful implementation of these programs, a completely new and unique IPA program appeared in further cooperation between BiH and the EU with five components[4]: assistance in transition and institution building, regional and cross-border cooperation, rural development, regional development, and projects related to capacity building. "For the period from 2007 to 2014, the IPA program provided 11.468 million euros to each of the countries that can use its funds. The program for the distribution of funds is created based on a multi-annual indicative framework for a planned period of three years, after which proposals for distribution

4 Among the other agreements that have been available to BiH in the past period, we mention: TAIEX, P2P, LAF, IfS.

by countries and the mentioned segments (components) are made. Analyzing the planned assistance for BiH through this program, we can conclude that in the observed period of analysis, for BiH from 2007 to 2012, the following was planned: 58.1 million euros in 2007, 69.8 million euros in 2008, 83.8 million euros in 2009, 100.6 million euros in 2010, 102.6 million euros in 2011, and 104.7 million euros in 2012" (Aćić, pp. 342–343).

In addition to the above, the EU Delegation in BiH, with the support and coordination of the EUSR, was also engaged in projects of building BiH institutions, some of which were imposed by the Office of the High Representative for BiH, contrary to the Dayton Peace Agreement, namely:

1. Structural dialogue in the field of justice, aimed at resolving issues related to the work of the Court and Prosecutor's Office, which were imposed as bodies at the BiH level by the decision of the Office of the High Representative for BiH. Regarding this reform, Blagojević says, "During the period of authoritarian rule of the country, state judicial institutions were established, the Court, the Prosecutor's Office of BiH, the High Judicial and Prosecutorial Council, and the reform of procedural legislation was carried out as a prerequisite for the functioning of the judicial system" (Blagojević, p. 80).
2. Implementation of the public administration reform program.
3. Support for the development of the statistical system.
4. Support for the development of customs and taxation systems.
5. Projects related to the work of the High Judicial and Prosecutorial Council of BiH.

Although all the mentioned and other projects that the EU has implemented in BiH so far have partially led to improvements in the areas they covered, there are still many challenges and problems that BiH faces, which the highest EU bodies have also noted in their conclusions. The European Parliament, in its 2015 Resolution, notes: "We regret to note that, 20 years after the end of the war and the establishment of the General Framework Agreement, emphasizing the main aspects of the peace agreement and the future form of the country, successive governments have failed to build a fully functional and self-sustainable state" (European Parliament, 2015).

Since then, regardless of all projects and implemented reforms, EU bodies in their resolutions and decisions similarly conclude that BiH has not yet fulfilled all the prescribed conditions to become a full member and appeal for more active work on the started reforms through the Stabilization and Association Agreement and other active programs and initiatives to reach the status of a full EU member.

2.3 The Berlin Process: A Decade of Cooperation and Challenges on the Way to European Integration

The Berlin Process was initiated in 2014 as a platform for stronger cooperation between the Western Balkan countries and the EU. This initiative arose in response to the need for enhancing regional stability, connectivity, and economic development in the context of European integration, and at its outset aimed to facilitate the path towards EU membership.

On October 14, 2024, a major summit was organized in Berlin, held as part of the tenth anniversary of the Berlin Process. This event brought together heads of state and government from six Western Balkan countries: Albania, Bosnia and Herzegovina, North Macedonia, Montenegro, and Serbia, along with representatives from the German government and other European host countries. Although agreements on mutual recognition of personal documents and academic qualifications were reached in 2022, the implementation of these agreements has not yet been fully realized. Many promises remain unfulfilled, creating barriers to further cooperation. The Berlin Process was initiated at the behest of then German Chancellor Angela Merkel (Đukanović & Krstić, 2016, p. 625). Prior to this, European Commission President Jean-Claude Juncker announced a "pause in the enlargement" of the Union, raising concerns among the countries of the Western Balkans. In the context of the 2014 example, Merkel gathered representatives from Balkan countries and ten EU member states to support the integration process (Mulaosmanović et al., 2018).

Over the past years, annual summits have been held in various European cities such as Vienna, London, and Paris, with new priorities added each time covering topics such as economic cooperation, digitization, and sustainable development. Within the framework of the Berlin Process, ministerial conferences have addressed security issues, youth exchanges, agriculture, as well as the rights of Roma communities and gender equality.

Regular high-level meetings contribute to preserving the EU perspective for the countries of the Western Balkans, but it is noted that the processes themselves have not permanently stabilized regional relations. Serbian policy toward Kosovo continues to create tensions, complicating progress. Critics point out that the Berlin Process operates with much improvisation, despite numerous meeting formats and initial agreements. Musliu notes that the existing political dimension of cooperation is often neglected, and the lack of concrete results remains a significant problem. Moreover, many regional tensions, such as the situation in Serbia and its relationship with Kosovo, further complicate the implementation of agreements. Additional criticisms are directed at various unimplemented laws and reforms concerning democracy and media freedom (Marciacq, 2017, p. 22). Nevertheless, the Berlin Process can boast of certain successful projects, such as the gradual abolition of roaming fees in the region and the Regional Youth Cooperation Office (RYCO), which has gathered a significant number of young people from the Balkan countries. The EU has provided substantial funds for economic development and investment in the region, but many agreements on mobility remain unfulfilled (Delegation of the EU to Bosnia and Herzegovina, 2023). On the occasion of the tenth anniversary, new agreements are planned, particularly within CEFTA (the Regional Free Trade Agreement), which would further strengthen the ties between the Western Balkan countries and the EU internal market, aimed at enhancing regional stability and cooperation.

2.4 Treaty Establishing the Energy Community: Overview and Significance for Bosnia and Herzegovina

Within the European energy sector, the requirements of the EU and Bosnia and Herzegovina's (BiH) international obligations are significantly enshrined in the provisions of the Treaty Establishing the Energy Community (Decision on the Ratification of the Treaty Establishing the Energy Community, "Official Gazette of BiH – International Agreements", No. 9/06). This document provides a framework for integrating energy markets in Southeast Europe and establishes a basis for alignment with EU standards.

Bosnia and Herzegovina initiated its engagement in the regional electricity market through the Athens Process, which focused on establishing a regional energy market in Southeast Europe. By signing two Memoranda of Understanding on this market in 2002 and 2003, BiH demonstrated a clear commitment to energy integration. In this context, on October 25, 2005, in Athens, the first multilateral and legally binding international treaty related to energy in this part of Europe was signed – the Treaty Establishing the Energy Community, which entered into force on July 1, 2006. This treaty has created the largest internal market for electricity and gas in the world, encompassing the EU and eight other Contracting Parties: Albania, Bosnia and Herzegovina, Montenegro, Kosovo, North Macedonia, Moldova, Serbia, and Ukraine. In addition to these states, the work of the Energy Community bodies includes 19 EU member states as participants, while Armenia, Georgia, Norway, and Turkey participate as observers. The objectives of the Energy Community include creating a stable and unified regulatory framework and market space that enables reliable energy supply and attracts investments in the electricity and natural gas sectors. Furthermore, the importance of developing alternative gas supply routes, improving environmental conditions, promoting energy efficiency, and using renewable energy sources is emphasized.

By concluding this Treaty, the Contracting Parties commit to establishing a common electricity and gas market that will function according to EU standards, applying the acquis communautaire, including relevant directives and regulations in the fields of electricity, gas, environmental protection, competition, renewable energy sources, energy efficiency, oil, and statistics. It is important to note that the Ministerial Council of the Energy Community unanimously decided on October 24, 2013, to extend the Treaty, which was initially concluded for ten years, for an additional ten years. The Ministry of Foreign Trade and Economic Relations of BiH is responsible for fulfilling the obligations undertaken by signing the Treaty and, together with the relevant entity ministries, actively cooperates with the Energy Community bodies. This treaty not only represents the legal framework for the energy sector in BiH but also a fundamental instrument for further integration into European energy structures, thereby creating a foundation for enhancing energy policies and security of supply throughout the region.

2.5 The Althea Mission: A Legal and Operational Framework for EU Involvement in Bosnia and Herzegovina

Notwithstanding its role in ending the war conflicts in BiH and signing the Dayton Peace Agreement, the EU made additional efforts to consolidate the achievements of the peace settlement, particularly concerning the implementation of the military aspects of this Agreement through the Althea mission. The mission, which commenced immediately after the conclusion of the SFOR mission, provided an opportunity for the EU to develop its "first serious military operation, and SFOR was replaced by EU forces – EUFOR" (Ivanić, 2005, p. 280). Viewed in the "context of European integration, this mission was a sign that BiH had indirectly progressed towards the EU, as clearly indicated by its "different goal compared to SFOR and IFOR" (Vukičević, 2013, p. 172). Despite being a mission whose commencement would be the responsibility of the EU, "the legal basis for its establishment can be found in UN Security Council Resolution 1551, adopted on July 9, 2004, which generally welcomed the intention to launch this mission. Further positions of the UN Security Council can be traced through the adopted Resolution 1575, adopted on November 22, 2004, which, among other things, emphasizes that the UN welcomes the EU's intention to engage more concretely in the military aspects of the implementation of the DPA, as well as the fact that, as a result, an EU military operation in BiH is launched. This resolution confirms the authorization for EU member states, and those cooperating with them in this mission, which they take over from NATO, to establish multinational stabilization forces – EUFOR, which will act as the legal successor to SFOR" (Aćić, 2017, p. 415).

In December 1995, the UN Security Council approved the establishment of IFOR, a multinational "Implementation Force" under NATO command, comprising 60,000 personnel with a one-year mandate. IFOR's specific role was to implement the military annexes of the Dayton Agreement. When IFOR was succeeded by NATO's Stabilization Force (SFOR) on December 21, 1996, the number of troops was reduced to approximately 30,000. Based on the decision of the Council of the EU adopted on December 2, 2004, SFOR, a NATO-led mission, was replaced by the EU Force (EUFOR). Initially, Operation "Althea" involved 7,000 troops. Ten years later, and after continuous restructuring, that number was reduced to 600 troops. However, this troop level is flexible and can be adjusted whenever the situation demands it. The decision on whether an international military presence in BiH is still necessary is made by the UN Security Council, which renews the mandate of international forces annually, most recently on November 11, 2014.

"Its composition has so far included 21 countries, including four non-EU members (Albania, North Macedonia, Turkey, and Chile). It began in 2004 and at its peak comprised 7,000 troops, deployed in three areas: the Multinational Task Force North, headquartered in Tuzla, with 1,300 troops; the Multinational Task Force Northwest, headquartered in Banja Luka, with a contingent of approximately 1,000 troops; and the Multinational Task Force Southeast, headquartered in Mostar, with human resources numbering approximately 1,400 troops." The mission underwent two troop reductions: the first reduction occurred in 2007 when the Althea mission was reduced to about 1,600 troops, and finally, in 2012, that number reached only 600 troops, deployed in liaison and observation teams, known as LOT teams (Liaison and Observation Teams) (Aćić, 2017). LOT teams currently operate at 20 locations across BiH, with the LOT Coordination Center located at the "Butmir" Military Base in Sarajevo.

2.6 The Role of Frontex in Addressing Irregular Migration in Bosnia and Herzegovina

The European Border and Coast Guard Agency, known as Frontex, is becoming an important partner for Bosnia and Herzegovina through an agreement that will enable the deployment of officers in this area. The agreement was initialed on December 18 in Sarajevo and establishes the basis for better control of irregular migration and more efficient border management.

The external dimension of Frontex's work, namely cooperation with third countries, has become increasingly important due to the blurring of lines between internal and external security following the introduction of the Schengen Area and growing concerns about terrorism (Kalkman, 2021, p . 165–181). Cooperation with non-EU countries has become a priority. Meanwhile, Frontex has been tasked with fostering cooperation between EU member states and non-EU countries in the field of external border management. The agency has gradually strengthened its capacities (Meissner, 2017, p. 155). From an initial budget of 6,280,199 euros in 2005, the budget has grown to over 250 million euros. The number of employees has also increased significantly, from 45 in 2005 to 450. These data indicate a significant strengthening of Frontex over the years.

The agreement is expected to be officially signed in late February or early March, during a visit by the EU Commissioner for Home Affairs and Migration. This cooperation represents a significant step in the fight against organized crime and smuggling, as well as in strengthening relations between BiH and the EU. Bosnia and Herzegovina is thus positioned as a serious partner of the EU, which helps it to face the challenges of irregular migration. According to official statements, about 90% of illegal border crossings on the "Balkan route" occur on the border between BiH and Croatia. In addition, the visa-free regime with BiH is being abused by individuals from countries with which Bosnia has such agreements. The agreement with Frontex has the potential to improve the situation on all borders (Neal, 2009, p. 349), which will enable BiH to better manage migrants and increase its security. During the drafting of the agreement, full consensus was reached between all parties, and Frontex resources will be available to BiH. These resources include the deployment of officers, training of border police, and data exchange with border services in Europe. Bosnia and Herzegovina, which until now did not have an operational agreement with Frontex, now receives important tools to improve its border police.

The agreement emphasizes that officers deployed by Frontex will work under the command of their colleagues from BiH. This includes cooperation not only with the state Border Police but also with police at the entity and cantonal levels. In previous negotiations, disagreements arose due to the demands of representatives of Republika Srpska who requested that Frontex officers be under the command of the entity police. Finally, an agreement was reached that allows for a unified command over all officers who will be engaged. In the past, Bosnia and Frontex initialed an agreement that excluded cooperation with entity and cantonal police, but that agreement fell through due to a veto. Today, thanks to new agreements reached through months of negotiations, Bosnia and Herzegovina can expect better conditions for protecting its borders and more effectively managing migrants. Bosnia and Herzegovina currently has about 1,800 border guards to guard more than 1,500 kilometers of border with Croatia, Serbia, and Montenegro. According to data, one border guard secures approximately one and a half kilometers of border. About 600 border police officers are expected to retire by 2026, which creates additional pressure on already existing resources and capacities. According to records, in the first eleven months of 2023, 24,523 illegal migrants were recorded in BiH. Most migrants come from Morocco, Afghanistan, and Syria, with the majority being men trying to continue their journey to Croatia and further into Western Europe. At this moment, about 800 migrants are housed in temporary centers in BiH, while the country needs additional support to adequately cope with the current migration challenges.

The agreement with Frontex provides technological and operational support and is a symbolic step towards integrating BiH into European institutions, thereby strengthening its place in the wider regional context. The progress of these negotiations, in cooperation with the EU, creates the basis for long-term reforms in the sectors of security, migration, and justice necessary to achieve the standards required for joining the EU.

2.7 The Monetary Framework of Bosnia and Herzegovina: Stability through a Currency Board System

Bosnia and Herzegovina, as a post-conflict state consolidated through the Dayton Peace Agreement, maintains its sovereignty as an independent nation (Aćić, 2019, p. 25; Hartwell, 2019, p. 44). A cornerstone of its economic stability is the currency board arrangement, which anchors the domestic currency to foreign reserves. The Central Bank of Bosnia and Herzegovina (CBBH) commenced operations on August 11, 1997, under Article 7 of the Dayton Peace Agreement (Vitomir & Lazić, 2021, p. 206; Vranješ, 2020, p. 81). Establishing the CBBH under a currency board model was instrumental in fostering monetary order, restoring confidence in the post-war economy, and achieving macroeconomic stability. A fundamental tenet of this system is the commitment to a fixed exchange rate (Gregović & Hodžić, 2017, p. 112). The CBBH operates under stringent regulatory constraints, precluding the creation of unbacked currency, the extension of credit to the government, or the function of a "lender of last resort" for commercial banks (Aćić & Grujić, 2022). Furthermore, it is prohibited from engaging in open market operations or manipulating exchange rates to address domestic economic challenges (Jović, 2020, p. 9).

The implementation of the currency board in 1999 yielded substantial positive outcomes, notably in curbing inflation and stabilizing the financial sector, thereby mitigating the risks associated with economic transition (Grujić, 2018, p. 205). This system mandates a fixed exchange rate, necessitating adequate foreign exchange reserves to fully collateralize the domestic currency in circulation while ensuring unrestricted internal and external convertibility (Gregović & Hodžić, 2017, p. 114). While the CBBH's regulatory framework, which prohibits the creation of uncollateralized money, mitigates the risk of hyperinflation, it also constrains its ability to manage the domestic money supply. The money supply is predominantly determined by foreign exchange inflows, which are susceptible to fluctuations due to external factors. Nevertheless, the CBBH transcends the simplistic role of a "currency exchange." Within the parameters of the currency board, it can execute a degree of sovereign, discretionary monetary policy, complemented by macroprudential regulation.

Economic policymakers in BiH must critically assess the implications of monetary policy, refine its instruments, and ensure the CBBH's accountability for its impact on the national economy. Given the CBBH's capacity to influence both real and financial variables through monetary policy tools, a synergistic understanding and proactive approach are imperative. Furthermore, the significant role of international financial institutions warrants recognition. Their financial assistance, including credits, funds, and project initiatives, has been pivotal in supporting the stable functioning of the BiH economy in the post-Dayton era. This collaborative engagement underscores the importance of external support in cultivating a resilient economic framework within a complex post-conflict context. In addition to its stability-enhancing features, the currency board system in Bosnia and Herzegovina presents challenges to long-term growth, as the rigidity of the fixed exchange rate can impede competitiveness in the global market. Furthermore, while the maintenance of substantial foreign reserves is essential, it also exposes the economy to external shocks, such as fluctuations in global commodity prices or international demand. Finally, the interplay between the currency board and political considerations necessitates ongoing dialogue among policymakers to ensure that monetary stability aligns with broader economic reforms and developmental objectives.

2.8 Extension Elements

The Presidency of the Council of the European Union, while formally serving as a facilitator, plays a key role in setting the agenda and directing the Union's attention to specific foreign policy priorities (Bunse, 2009). The activities of the Presidency concerning the Western Balkans are often defined through thematic summits, ministerial meetings, and informal working groups formed during the six-month mandate, allowing the presiding state to intensify certain processes or issues in the region (Balfour, 2016). The role of the Presidency is to ensure the coherence and continuity of enlargement policy, mediating between the interests of Member States and the aspirations of candidate countries. In the context of external Europeanization, the Presidency reinforces the EU's influence through mechanisms of conditionality and technical assistance, thereby directly impacting reforms in the areas of rule of law and economic stability (Malo, 2017).

Regional proximity significantly influences the priorities of individual Presidencies. In this regard, the role of the German and Austrian Presidencies has often been strategic and focused on stabilization. Germany, with its Berlin Process initiative, continuously utilizes its Presidency period to strengthen regional cooperation and economic integration of the Western Balkans, offering a strong political impetus. Austria, given its historical and economic ties with the region, traditionally focuses its mandate on migration and security issues, and on supporting the European perspective (Džihić, 2014). This strong and structural support is crucial for maintaining the momentum in the Stabilization and Association Process, especially when support among other EU Member States is inconsistent (Vachudova, 2005).

The influence of the Presidency is particularly visible through the engagement of Croatia and Slovenia, states that share an immediate border with the Western Balkans. The Croatian Presidency (e.g., 2020) strongly emphasized the new enlargement methodology and sought to increase the region's presence on the Brussels agenda, acting as an advocate and bridge between the EU and the Western Balkans (Dimitrova, 2010). Slovenia, as a proponent of enlargement, utilized its mandates to organize key summits, confirming its commitment to accelerating the integration process and resolving bilateral issues (Kočan & Lovec, 2021). These activities are vital for the diffusion of EU standards and for the operationalization of enlargement policy on the ground (Lavenex, 2004).

Finally, the impact of the Presidency on security policies in the region encompasses a wide spectrum of areas, from the fight against organized crime to the security of digital infrastructure. The presiding state directs the work of the Political and Security Committee (PSC) and can push for specific topics, such as cyber-security, which has become an imperative for Europe (Nye, 2021). Actively promoting the harmonization of legislation in the field of digital security and data protection with the EU acquis is also part of the Presidency's mandate, especially for those with a strong digital agenda (Balfour, 2016). This ensures that security challenges in the Western Balkans are treated as an integral part of the wider European security architecture.

2.9 The Digital Dimension in the Context of the Western Balkans

Technological transformation represents a key opportunity for the Western Balkans (WB) to address long-standing economic and structural challenges, yet insufficient implementation threatens to exacerbate problems such as brain drain and economic divergence from the EU (Stević & Grujić, 2017, p. 55). Although digitalization is defined as a top priority, the countries in the region still lag the EU average in adopting digital technologies, despite recorded convergence (Mrdović, 2023) and a high correlation between digital skills and productivity (Levkov & Kitanovikj, 2024). The development of the 5G network, supported by the Memorandum of Understanding for the Digital Transformation Roadmap, is crucial for future economic growth and opens the door for advanced services like smart cities (Broz et al., 2020). Despite regional initiatives, digital transformation remains uneven; for instance, Montenegro shows above-average results in terms of basic digital skills and accessibility, while other countries significantly lag behind.

The security of digital infrastructure is becoming an imperative in the region, particularly after significant cyber incidents targeting government IT systems in Albania and Montenegro in 2022, which underscored vulnerability. Cybercrime, including ransomware and DDoS attacks, poses a major security threat (Pawlak, 2025). Existing institutional and legal frameworks for cybersecurity are often weak or non-existent, and technical response capabilities are insufficiently equipped (Tahirović, 2024). Regional cooperation, in line with EU standards, is crucial for building resilience, with the potential application of artificial intelligence being explored to enhance cybersecurity techniques in the context of good governance (Sollaku, 2023).

The EU's digital policy plays a transformative role through the so-called "Brussels Effect," seeking to project its norms and regulations onto the WB (Asderaki et al., 2021). The European integration process requires alignment with the acquis, including regulations such as the General Data Protection Regulation (GDPR) and, increasingly, the Digital Services Act (DSA) and electronic communications laws. Alignment with the acquis in the domain of cybersecurity (e.g., the NIS 2 Directive) directly improves the protection of critical infrastructure. Additionally, the EU insists on the development of national cybersecurity strategies and operational capacities as prerequisites for accession.

Technological geopolitics in the Balkans is reflected in systemic competition, given that digital transformation does not occur in a vacuum. While the EU strives for harmonization and integration, China, through the Digital Silk Road initiative, offers alternative models of financing and building digital infrastructure. This Chinese approach, which implies control over critical technological supply chains, imposes dilemmas for WB countries regarding standardization, security, and long-term economic and political dependence (Malkin, 2022). Geopolitical tensions are exacerbated by poor governance and internal geopolitical disputes within the region itself.

The role of the Presidency (of the Council of the EU) in these processes is significant because it uses its mandate to set the Union's priorities. The Presidency actively promotes the harmonization of laws in the field of digital security and data protection with the EU acquis, thereby directly influencing the reform agenda in the WB. States taking on the Presidency, especially neighboring ones, often use this position to keep the WB high on the EU agenda, focusing on digital transformation and security, thereby strengthening the Union's influence through conditionality and technical assistance (Kočan & Lovec, 2021; Aleksandra, 2020).

3 Conclusion

This study examined the role of the European Union (EU) in the development and stabilization of Bosnia and Herzegovina (BiH) from the Dayton Peace Agreement to the present, focusing on key initiatives, programs, and challenges in the pursuit of European integration. Following the research questions posed, it can be asserted that the EU played a pivotal role in ending the war conflicts through support for peace initiatives, followed by the implementation of the Dayton Peace Agreement. Although numerous peace plans were unsuccessful due to the divergent demands of the warring parties, the Dayton Agreement facilitated the cessation of hostilities, with the active participation of the EU as a signatory witness.

The mechanism of this external influence is best understood through the lens of Europeanization theories, particularly the rationalist framework of conditionality and the sociological approach of norm diffusion. BiH's slow adoption rate highlights the limitations of EU leverage when faced with deep-seated domestic resistance and weak absorption capacity. Furthermore, the rotational Presidency of the Council of the EU plays a critical, often underestimated, role. Presidencies, especially those held by regionally connected Member States – such as Germany, Austria, Slovenia, and Croatia – strategically use their mandate to keep the Western Balkans high on the EU agenda, thus injecting political momentum and defining specific regional priorities.

Crucially, the EU's focus on engagement is shifting to encompass modern strategic dimensions. The growing emphasis on the "Digital Europe" agenda means that the EU's influence is now extending into new security policies, namely the security of digital infrastructure, technological transformation, and digital geopolitics. The success of the EU's stabilizing mission in BiH will increasingly depend on its ability to effectively diffuse its emerging digital policies and standards across the region. This requires BiH to rapidly enhance its institutional capacity to manage cyber threats and participate in the EU's pursuit of digital sovereignty, ensuring that the country's future trajectory is fully integrated into the wider European security and technological architecture. Moving forward, the effectiveness of the EU's contribution will hinge on maintaining consistent, high-level political engagement – often facilitated by the Council Presidency – and successfully aligning BiH's complex institutional landscape with the demands of the modern digital acquis.

Since the onset of war conflicts in Bosnia and Herzegovina, the EU assumed an active mediating role, initiating a series of peace efforts, including failed plans such as the Cutileiro, Vance-Owen, and Geneva plans of 1993. Despite persistent efforts, peace plans were unsuccessful due to the divergent demands of the warring parties, and the war was only concluded with the Dayton Peace Agreement in 1995, with the involvement of the EU and other international actors. Post-conflict, the EU continued to provide support to BiH through various programs and initiatives, including the institution of the EU Special Representative and the integration of the country into regional economic and political initiatives.

The EU and Bosnia and Herzegovina maintain a longstanding institutionalized partnership, within which the EU established the Office of the EU Special Representative (EUSR) for BiH, intending to support reforms and European integration. From 2002 to 2011, the function of the EUSR was performed by the High Representatives for BiH, after which an independent EUSR position was established, whose holders were instrumental in promoting reforms in the defense, intelligence-security, and police sectors. The EU has significantly contributed to the development of Bosnia and Herzegovina through various programs and projects, such as PHARE, CARDS, IPA, and others, focusing on economic development, judicial reform, institutional strengthening, and harmonization of legislation with EU standards. Despite numerous reforms and support, BiH continues to face challenges in building a functional and self-sustaining state, as frequently emphasized in resolutions and statements by EU bodies. Continued active cooperation through the Stabilization and Association Agreement and other initiatives is crucial for further progress of BiH on its path to full EU membership. Following the Dayton Peace Agreement, the EU assumed a role in maintaining peace in Bosnia and Herzegovina through the EUFOR Althea military mission, which replaced NATO forces and became a key factor in maintaining stability. Concurrently, through cooperation with Frontex, the EU enabled BiH to more effectively control migration and strengthen border security, representing an important step towards European integration. Despite the progress achieved, BiH continues to face challenges in managing migration and implementing reforms, but cooperation with the EU remains crucial for achieving stability and preparing for future Union membership.

While the Stabilization and Association Agreement (SAA) significantly contributed to the development of Bosnia and Herzegovina, the path to EU membership remains challenging. Successful completion of the integration process necessitates addressing key issues such as strengthening the rule of law, combating corruption, and reforming public administration. Additionally, greater political will and willingness to compromise are required to overcome deep social divisions. Given the complexity of the topics covered by the SAA, the expert and wider public must follow these issues and contribute to reforms and the removal of obstacles on the path to the EU. While the Berlin Process promotes the ambition for integration and regional cooperation, remaining problems and challenges, including unresolved conflicts, the need for reforms, and political instability, continue to affect its significance and success. Effective implementation of agreements and continuous dialogue between member states are essential for the future development and stability of the region. Economic stability in BiH is ensured by a currency board, which links the domestic currency to foreign reserves. The Central Bank of BiH, established in 1997, plays a crucial role in maintaining monetary stability, preventing hyperinflation, and ensuring macroeconomic balance, although the rigidity of the currency board poses challenges for long-term growth and competitiveness. Despite limitations, the CBBH has scope for implementing discretionary monetary policy, with the support of international financial institutions, which is crucial for further economic reforms and the country's development.

References

Aćić, M. (2017). The role of international organizations in the emergence of the Dayton Bosnia and Herzegovina established by the Dayton Peace Agreement – doctoral dissertation, University of Belgrade, 415–416.

Aćić, M. (2019). The evolution and practice of international administration on states and territories. *Acta Politica Polonica*, 1(47), 15–27. DOI: 10.18276/ap.2019.47-02.

Aćić, M. (2022). The role of international organizations in the creation and functioning of the Dayton BiH monograph, NUBL, 328, 342–343.

Aćić, M., Grujić, M. (2022). The role of selected financial organizations in the reconstruction of Bosnia and Herzegovina after the Dayton Peace Agreement. *Acta Politica Polonica*, 53, 15–28. DOI: 10.18276/ap.2022.53-02.

Aćić, M., & Grujić, M. (2023). The role of pre-accession assistance in the process of joining the European Union. *Acta Politica Polonica*, 56, 25–37.

Aleksandra, S. (2020). The role of the Presidency of the EU Council based on the example of the EU Enlargement Policy towards the Western Balkans–subject matter and methodological approach. *Przegląd Europejski*, 3(26), 97–110.

Asderaki, F., Panfil, G., Molnár, A., & Paile-Calvo, S. (Eds.). (2021). *Cybersecurity in the Western Balkans: Is the 'Brussels Effect' in Play?* Publications Office of the European Union. DOI: 10.2855/3056225.

Balfour, R. (2016). Change and continuity: A decade of evolution of EU Foreign Policy and the creation of the European External Action Service. In R. Balfour, C. Carta & K. Raik (Eds.), The European External Action Service and National Foreign Ministries (pp. 31–44). Routledge.

Banning, T. (2014). The Bonn powers of the high representative in Bosnia Herzegovina: Tracing a legal figment. *Goettingen J. Int'l L.*, 6, 259.

Beglerović, N. (2022). The role, expectations and challenges of high representative in Bosnia and Herzegovina. *Politics in Central Europe*, 18(3), 397–432.

Blagovčanin, S. (2016). The European Union and Bosnia and Herzegovina: State building through the process of European integration. In R. Balfour, C. Carta & K. Raik (Eds.), The European External Action Service and National Foreign Ministries (pp. 31–44). Routledge.

Bodur Ün, M., & Arıkan, H. (2022). Europeanization and de-europeanization of Turkey's gender equality policy: The case of the Istanbul convention. *JCMS: Journal of Common Market Studies*, 60(4), 945–962.

Broz, T., Buturac, G., & Parežanin, M. (2020). Digital transformation and economic cooperation: The case of Western Balkan countries. *Zbornik radova Ekonomskog fakulteta u Rijeci: časopis za ekonomsku teoriju i praksu*, 38(2), 697–722.

Bunse, S. (2009). Small states and EU governance: The role of the Council Presidency in agenda-setting and policy leadership.

Council of the European Union (2011). *Council Decision 2011/426/CFSP of 18 July 2011 appointing the EU Special Representative in Bosnia and Herzegovina.* Official Journal of the European Union, L 188/30, p. 15.

Delegation of the European Union to Bosnia and Herzegovina (2023, March 14). *EU4BiH – financial assistance: EU projects in BiH.* European External Action Service. https://www.eeas.europa.eu/delegations/bosnia-and-herzegovina/eu-projects-bih-0_en?s=219.

Dijkstra, G. S., Raadschelders, J. C. (2022). The high representative in Bosnia-Herzegovina: The unusual institutional arrangement of a non-authoritarian, yet controlled, democracy. *World Affairs*, 185(2), 285–311.

Dimitrova, A. L. (2010). The new member states of the EU in the aftermath of enlargement: Do new European rules remain empty shells?. *Journal of European public policy*, 17(1), 137–148.

Drezner, D. W. (2014). *The System Worked: How the World Stopped Another Great Depression.* Oxford University Press.

Džidić, M. (2014). The application of the right of establishment in context of the Stabilisation and Association Agreement for Bosnia and Herzegovina. Collection Papers Actualities of Civil & Commercial Law and Legal Practice, pp. 80–95.

Đorđević, V., Turcsányi, R. Q., & Vučković, V. (2021). Beyond the EU as the 'Only Game in Town': the Europeanisation of the Western Balkans and the role of China. *Eastern Journal of European Studies*, 12(2).

Đukanović, D., & Krstić, M. (2016). Berlinski proces–Nemačka „Zapadnobalkanska" inicijativa. *Srpska politička misao*, 54(4), 169–185.

Eufor (2025). https://www.euforbih.org/index.php/background.

EUFORBIH (2015). *EUFOR across Bosnia and Herzegovina.* https://www.euforbih.org/index.php/eufor-houses.

European Commission (2023). EU projects in BiH. https://www.eeas.europa.eu/delegations/bosnia-and-herzegovina/eu-projects-bih-0_en?s=219 Accessed February 20, 2024.

European Commission (2025). *Funding by county – Bosnia and Hezegovina.* http://ec.europa.eu/enlargement/instruments/funding-by-country/bosnia-herzegovina/index_en.htm.

European Commission (2025). http://europa.ba/?page_id=558;http://europa.ba/?page_id=562.

European Commission (2025). http://www.europarl.eu/sides/getDoc.do?pubRef=-//EP//TEXT++P8-TA-2015-0471+0+DOC+XML+V0//EN.

European Parliament (2015). *European Parliament resolution of 17 December 2015 on the 20th anniversary of the Dayton Peace Agreement* (*2015/2979*(*RSP*)) [P8_TA(2015)0471]. http://www.europarl.eu/sides/getDoc.do?pubRef=-//EP//TEXT++P8-TA-2015-0471+0+DOC+XML+V0//EN.

Gafuri, A., & Muftuler-Bac, M. (2021). Caught between stability and democracy in the Western Balkans: a comparative analysis of paths of accession to the European Union. *East European Politics*, 37(2), 267–291.

Glintić, M. B. (2013). Sporazum o stabilizaciji i pridruživanju pred Evropskim sudom pravde. *Strani pravni život*, (3), 105–119.

Gregović, N., & Hodžić, N. (2017). Perspectives of currency board in Bosnia and Herzegovina. *Ekonomski Izazovi*, 11, 108–125. https://scindeks-clanci.ceon.rs/data/pdf/2217-8821/2017/2217-88211711108G.pdf.

Grujić, M. (2018). Endogeni i egzogeni faktori neuspjeha procesa tranzicije na Zapadnom Balkanu. *Poslovne Studije*, 10(19–20), 209–218. DOI: 10.7251/POS18199G.

Hartwell, L. (2019). Conflict esolution: Lessons from the Dayton Peace Process. *Negotiation Journal*, 35(4), 443–469. https://doi.org/10.1111/nejo.12300.

Ivanic, M. (2005). The international community and Bosnia and Herzegovina. *Cambridge Review of International Affairs*, 18(2), p. 280.

Jambrovic, D. D., & Maresic, M. (2020). The subnational dimension of Europeanization. *Rev. Eur. & Comp. L.*, 42, 7.

Jović, D. (2020). Currency board, monetary regulation, and macroprudential regulation in Bosnia and Herzegovina. *Bankarstvo*, 49(1), 8–36. https://scindeks-clanci.ceon.rs/data/pdf/1451-4354/2020/1451-43542001008X.pdf.

Kalkman, J. P. (2021). Frontex: A literature review. *International Migration*, 59(1), 165–181. https://doi.org/10.1111/imig.12729.

Kočan, F., & Lovec, M. (2021). *What Can the Slovenian Presidency of the Council of the European Union Do for the Western Balkans?*. Austrian Society for European Politics (ÖGfE).

Lavenex, S. (2004). EU external governance in "wider Europe". *Journal of European Public Policy*, 11(4), 680–700.

Levkov, N., & Kitanovikj, B. (2024). The importance of digital skills for the Western Balkans-comparative analysis between the Western Balkans and the European Union. Economy, Business and Development: *An International Journal*, 5(1), 28–43.

Lopandić, D. (2001). *Regionalne inicijative u Jugoistočnoj Evropi*, Institut za međunarodnu politiku i privredu, Evropski pokret u Srbiji. Beograd: Prometej, Beograd.

Malkin, A. (2022). The made in China challenge to US structural power: Industrial policy, intellectual property and multinational corporations. *Review of Interna tional Political Economy*, 29(2), 538–570.

Marciacq, F. (2017). The EU and the Western Balkans after the Berlin Process: Reflecting on the EU enlargement in times of uncertainty. Friedrich-Ebert-Stiftung, Dialogue Southeast Europe, Sarajevo.

Meissner, V. (2017). The European Border and Coast Guard Agency Frontex beyond borders. TARN Working Paper Series 16/2017, Technische Universität München.

Mrdović, P. (2023). The role of digitalisation in transforming Western Balkan societies. ÖGfE Policy Brief 14'2023, Austrian Society for European Politics, Vienna.
Mulaosmanović, A., Karić, M., Kulanić, A., Knezović, E., Mustafić, A., & Preljević, H. (2018). Bosna i Hercegovina i Berlinski proces. *Analiza stanja ključnih procesa u BiH pred londonsku konferenciju.*
Musliu, V. (2021). The Berlin Process for the Western Balkans. What is in a name?. The Journal of Cross-Regional Dialogues/La Revue de dialogues inter-régionaux. DOI: 10.25518/2593-9483.172.
Neal, A. W. (2009). Securitization and risk at the EU border: The origins of FRONTEX. *JCMS: Journal of Common Market Studies*, 47(2), 333–356.
Nezirović, S., Živko, A., Durmišević, B., & Hodžić, A. (2022). Stabilisation and association agreement between the Western Balkan countries and the European Union. *Journal of Geography, Politics and Society*, 12(2), 36–50.
Nye, J. S. (2021). The new geopolitics of technology. *Foreign Affairs*, 100(2), 108–117. https://moderndiplomacy.eu/2025/09/23/the-new-geopolitics-of-technology/.
Olsen, J. P. (2002). The many faces of Europeanization. *JCMS: Journal of Common Market Studies*, 40(5), 921–952.
Paile, S., Asderaki, F., Panfil, G., & Molnar, A. (2025). The Common Security and Defence Policy: New Developments and Challenges. 2529–3362.
Pawlak, P. (2025). Forged in crises: Learning and adaptation in the European Union's Cyber Diplomacy. *The Hague Journal of Diplomacy*, 20(4), 765–780.
Radaelli, C. M. (2023). Europeanization in public policy. In A. Wiener, T. A. Börzel, & T. Risse (Eds.), Encyclopedia of public policy (pp. 1–5). Cham: Springer International Publishing.
Resolution of the European Parliament of 17 December 2015 On the 20th anniversary of the Dayton Peace Agreement (2015/2979(RSP), http://www.europarl.eu/sides/getDoc.do?pubRef=-//EP//TEXT++P8-TA-2015-0471+0+DOC+XML+V0//EN, accessed 20.02.2025.
RTRS (2015). Pendeš i Mogerini potpisale sporazum o učešću BiH u operacijama EU, 19.09.2015. https://www.rtrs.tv/vijesti/vijest.php?id=168008.
Schimmelfennig, F., & Sedelmeier, U. (Eds.). (2005). *The Europeanization of Central and Eastern Europe*. Cornell university press.
Sollaku, O. (2023). AI for Good Governance and Cybersecurity in the Western Balkans: Opportunities and Challenges. DCAF. https://www.dcaf.ch/sites/default/files/imce/ECA/OrnelaSollaku_Young-faces2023.pdf.
Sošić, M. (2006). Europeizacija nacionalnih politika: koncept i istraživački pristup. *Anali hrvatskog politološkog društva*, (03), 229–246.
Stević, L., & Grujić, M. (2017). Bosna i Hercegovina na Novom putu svile: Strategija, stanje, perspektive i pozicioniranje u skladu sa novom kineskom strategijom u zemljama CIE. Bosnia and Herzegovina on the New Silk Road: Strategy, State,

Perspectives and Positioning in Line With The New Chinese Strategy in CEE Countries], *Politeia*, 5(10), 53–86. DOI: https://doi.org/10.7251/POL1610053S.
Tahirović, M. (2024). The main security challenges and threats in the Western Balkans and possible model for solving them in the process of European integration. *Studia Europejskie*, 28(3), 27–55.
The effect of the agency's external dimension. TARN Working Paper Series 16/2017, December 2017. https://papers.ssrn.com/sol3/papers.cfm?abstract_id=3085529.
Tibor, Ö. (2023). Digitalisation in the Western Balkans. *AARMS–Academic and App lied Research in Military and Public Management Science*, 22(3), 91–107.
Todorova, M. N. (2000). The Balkans: From invention to intervention. In *Kosovo: Contending Voices on Balkan Interventions* (pp. 259–172). Wm. B. Eerdmans Publishing.
Tokić-Marić, S. (2018). Mirovni planovi i sporazumi za Bosnu i Hercegovinu, https://core.ac.uk/download/pdf/287773466.pdf.
Vachudova, M. A. (2005). *Europe Undivided: Democracy, Leverage, and Integration After Communism*. Oxford University Press.
Vitomir, J., & Lazić, Đ. (2021). Orthodox currency board and currency boards of the second generation: Operational aspects. *Baština*, 53, 205–229. doi: 10.5937/bastina31-31454.
Vranješ, M. (2017). Financial Legal Aspects of the Stablization and Association Agreement between the European Communities and Their Member States and Bosnia and Herzegovina. In *Collection Papers from Conf. Org. on Occasion Day Fac. L.*
Vranješ, N. (2020). Dayton and post-Dayton concept of the administrative power in Bosnia and Herzegovina: Twenty five years after Dayton Peace Agreement. *Politeia*, 10(20), 81–93. DOI: 10.5937/politeia0-29197.
Vukčević, D. (2013). *Evropska unija kao strateški akter*. Teorija i praksa bezbednosne i odbrambene politike, Beograd: Institut za političke studije.

CHAPTER 23

Türkiye and the European Union Relationship in the Digital Age: A Perspective for Cybersecurity Cooperation

Nezir Akyesilmen, Fuad Jomma and Kamil Tarhan

Abstract The chapter examines the prospects and challenges of cybersecurity cooperation between Türkiye and the European Union (EU) in the context of increasing digitalisation and growing cyber threats. It aims to assess the degree of convergence between policies, institutions, and legal frameworks, and to identify key opportunities and barriers to cooperation. The study is based on qualitative analysis of literature, policy documents, and official reports, complemented by quantitative data where appropriate. The findings show that both the EU and Türkiye consider cybersecurity a strategic priority and have developed extensive legal and institutional frameworks in this field. Significant areas of convergence include critical infrastructure protection, capacity building, and international cooperation. However, differences in governance models, institutional structures, and regulatory approaches continue to limit full harmonisation. Existing cooperation mechanisms, including NATO, ENISA, and CERT-EU, provide a strong foundation for further collaboration, but require enhanced trust, regulatory convergence, and improved information sharing.

Keywords Türkiye – cybersecurity cooperation – cyber governance – digital security

1 Introduction

Türkiye–EU relationships, which have historically encompassed issues such as political, economic, and cultural reliability, have entered a new phase in the digitalization age. The new opportunities and threats brought about by the digitally enriched world are forcing these two actors to cooperate. Major global cyber-attacks that emerged in the 2000s, including the DDoS attacks against Estonia in 2007, the hybrid war that took place in Georgia in 2008, the Stuxnet attack on the Iranian nuclear infrastructure in 2010 (Tarhan, 2022, p.404), and finally, the explosion of Hezbollah's pagers in 2024, have made cybersecurity a critical component of national and international security (Arquilla & Ronfeldt, 1993, p. 155). And it has been proven in many studies that the most effective way to ensure cybersecurity is international cooperation. Therefore, it has become an inevitable fact that Türkiye and the European Union, which face common threats, need to cooperate. However, this cooperation gives rise to both challenges and opportunities. For instance, in contrast to comprehensive, inclusive, and holistic frameworks, regulations, institutions, and strategies developed by the European Union regarding cybersecurity, Türkiye has developed an understanding that focuses more on the protection of critical infrastructures and national security.

This chapter employs a governance-oriented and functional integration approach to examine cybersecurity cooperation between the European Union and Türkiye. The primary analytical tool applied for the analysis is the International Telecommunication Union (ITU)'s five-pillar framework, namely legal regulations, technical and institutional setup, capacity building, and cooperation (ITU, 2024, pp. 8–12). Within the broader context of cyber sovereignty, cyber geopolitics, and strategic autonomy, this approach depicts how the EU goes on with its cyber norms, rules, and practices beyond its borders, while Türkiye selectively aligns with EU norms in line with common cyber threats and interdependence. Therefore, the chapter provides an in-depth analysis of the limits of EU-Türkiye cybersecurity cooperation in the digital age.

1.1 Research Methodology and the Research Question

Although the major methodology of this research is qualitative, quantitative tools will also be utilized when necessary. This mixed method will be more efficient in understanding the complicated cybersecurity relationship between the two actors. Therefore, utilizing primarily from ITU's five-pillar framework, the research method also focuses on literature review of cybersecurity cooperation studies between the two actors, document analysis, policy analysis, and statistics will be applied when needed.

The central question of the research is what are the opportunities and difficulties of cybersecurity cooperation between Turkey and the European Union? The results of the research to be conducted in this domain will guide both the EU and Türkiye to converge their policies and practices in the cybersecurity domain. For that purpose, the proper hypotheses of the research are as follows. The first hypothesis is that the different cybersecurity understandings of Türkiye and the European Union weaken the possibility of effective cooperation. The second hypothesis is that the serious cyber-attacks faced by Türkiye and the European Union reveal cooperation opportunities. And the third, Türkiye has some difficulty in adapting its own legislation to the comprehensive institutional and regulatory framework developed by the European Union.

This section investigates the possible cooperation dynamics between Türkiye and the European Union in cybersecurity. In this context, cooperation between the two actors will be a critical dynamic in eliminating the cyber threats encountered. A notorious synergy will emerge due to the convergence of cybersecurity policies, practices, institutions, and strategies developed by the two actors in the last two decades. The more cybersecurity cooperation between these two actors, the more benefits it will provide for both parties. However, it should not be forgotten that the differences in the approaches of the two parties to cybersecurity constitute a serious obstacle to cooperation.

2 The State of Cybersecurity in Türkiye and the EU

Türkiye and the European Union, as two distinct international actors, have developed strong infrastructures and frameworks and continue to spend efforts in cybersecurity, but it is a fact that both sides will benefit from cooperation. While Türkiye as a state benefits greatly from the EU cybersecurity ecosystem, where the experiences and knowledge of the EU countries are accumulated, the EU will also benefit from the strategies developed by a candidate country.

2.1 The Evolution of Turkey's Cybersecurity Strategy

Turkey has taken critical steps in the last decade to strengthen its cybersecurity strategies, regulations, and institutional structure. These include legal and administrative regulations, technical measures, institutional infrastructure, capacity development, and cooperation at national and international levels. The national cybersecurity strategy documents, which hold a critical place among these regulations, have shaped Türkiye's priorities, practices, and institutional structure. In addition, Türkiye's newly established Cybersecurity Presidency aims to centralize the entire national cybersecurity ecosystem in Türkiye. Despite the progress Türkiye has made in cybersecurity in the last decade, the works of international institutions illustrate the fact that it needs to do more.

Türkiye published its first cybersecurity strategy document and action plan in 2013, and has developed four different versions roughly in three-year periods so far. These documents not only determine the country's goals and principles regarding cybersecurity, but also bring together its policies, regulations, institutions, cooperation strategies, and capacity-building initiatives. This framework determines all sorts of cyber threats to the country and develops measures to be taken at various levels.

The goals of the National Cybersecurity Strategy and Action Plans are to increase cybersecurity awareness and strengthen institutions' security measures. Furthermore, public- and private-sector cooperation, protection of critical infrastructures, and priority use of national resources are emphasized. The National Cyber Incident Response Center (USOM – Turkish abbreviation) and Sectoral and Institutional Cyber Incident Response Teams (SOME – Turkish abbreviation) have been established, legal regulations have been made, human resources have been developed, and emphasis has been placed on national technologies in these strategy documents (Resmi Gazete, 2013). Türkiye further developed its new strategy in the second version of the Document, targeting to protect all systems and stakeholders by emphasizing that cybersecurity is an integral part of national security. The plan, which covers all national cyberspace, including public information systems, critical infrastructures, and SMEs, envisages taking administrative and technological measures against cyber threats (T.C. Ulaştırma ve Altyapı Bakanlığı, 2016). In 2020, the third cybersecurity strategy document did not focus only on national security and the protection of critical infrastructures, but also highlighted eight different strategies and goals. These are protecting critical infrastructures and increasing resilience, developing national capacity, establishing an organic cybersecurity network, ensuring the security of new-generation technologies, developing the fight against cybercrime, developing and supporting domestic and national technologies, integrating cybersecurity into national security, and developing international cooperation (T.C. Ulaştırma ve Altyapı Bakanlığı, 2020).

Apart from Strategy documents and administrative documents, Türkiye has also developed legal regulations for strengthening its position in cybersecurity. The government first started to issue its legal regulations in the 1990s. The most comprehensive regulation on national cybersecurity initiatives is the Council of Ministers Decision No. 2012/3842, issued in June 2012 (Akyeşilmen, 2022, p. 124). The government decided to establish the Cybersecurity Board, which includes participants from different ministries and institutions (Resmi Gazete, 2012). Yet the law with significant outputs was adopted in 2007. Law No. 5651 provided precautions against online crimes and empowered the authorities to intervene in issues such as data breaches and unauthorized access. It also obliged internet providers to protect user data and cooperate with authorized bodies. These regulations, supported by the criminal provisions in the Turkish Penal Code, create a legal framework against illegal digital activities (T.C. Cumhurbaşkanlığı Mevzuat Bilgi Sistemi, 2007). In addition, the Personal Data Protection Law was issued in 2016 (T.C. Cumhurbaşkanlığı Mevzuat Bilgi Sistemi, 2016). This law is designed particularly to be compatible with the EU's relevant regulations. Other legal regulations are the Electronic Communications Law (No. 5809), the Electronic Signature Law (No. 5070), the Law on the Regulation of E-Commerce (No. 6563), and changes made in the Turkish Penal Code (No. 5237) and the Turkish Commercial Code (No. 6102) (Akyeşilmen, 2022, p. 124).

Another critical step by the government has been the establishment of a strong institutional setting. For this purpose, the Cybersecurity Presidency was established very recently by the Presidential Decree on January 8, 2025. This institution aims to gather and carry out nationwide cyber activities directly under one institutional umbrella (Resmi Gazete, 2025).

The efforts by the government of Türkiye have been evaluated by several international organizations through various indices. A combination of all these studies illustrates that the average success is 75 out of 100, as shown in Figure 23.1.[1]

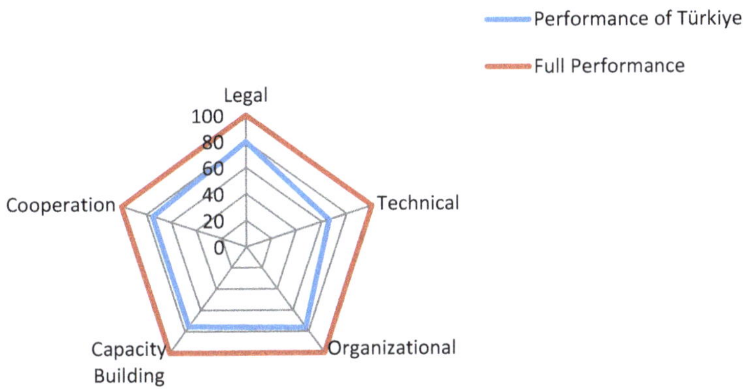

FIGURE 23.1 Türkiye's cybersecurity performance (Source: Akyeşilmen, N. (2022). Türkiye in the Global Cybersecurity Arena: Strategies in Theory and Practice. *Insight Turkey*. Summer 2022, Vol. 24, No.3. https://www.insightturkey.com/file/1483/turkiye-in-the-global-cybersecurity-arena-strategies-in-theory-and-practice.)

1 Figure 23.1 is based on the findings of the ITU's Global Cybersecurity Index (GCI), which has been published four times (2015, 2017, 2018, and 2020), as well as the National Cyber Security Index (NCSI) and the National Cyber Power Index (NCPI). These indices collectively help evaluate whether national cybersecurity strategies are effectively translated into action.

As shown in the graph, Türkiye has been able to achieve a rate of approximately 75 to 100. That represents a commendable degree of performance, though not flawless. To enhance certain indications across all five pillars, the most vulnerable aspect appears to be the technical pillar, while the most robust performance is demonstrated in legal measures. Among the remaining three pillars, the technical one appears to be the most vulnerable, while the other pillars are intermediate but require enhancement to bolster Türkiye's national performance (Akyeşilmen, 2022, p. 123).

Türkiye has taken serious steps in recent years to develop its cybersecurity strategies and institutional infrastructure. National cybersecurity strategy documents have strengthened the country's security measures against cyberattacks, the protection of critical infrastructures, and international cooperation. Furthermore, numerous legal and administrative regulations have been made, institutional infrastructures have been developed, and the law on combating cybercrimes has been brought to a higher level. The establishment of a Cybersecurity Presidency is critical in terms of centralizing cybersecurity-related institutions that will both coordinate and create a safer cyber ecosystem in Türkiye.

2.2 Cybersecurity in the EU

The EU's economy, democracy, and society are more dependent than ever on digital tools and connectivity (European Commission, 2020). Therefore, cybersecurity has become one of the cornerstones of building a resilient, sustainable, and digital Europe. Thus, the EU has developed a comprehensive and holistic approach to cybersecurity that includes regulations, policies, strategies, institutions, and applications.

Cybersecurity issues have been the subject of EU policies with the eEurope initiative in 2002. It was followed by the establishment of the EU's dedicated cybersecurity strategy agency, the European Network and Information Security Agency (ENISA), in 2004. ENISA, as the EU's cybersecurity agency, aims to standardize cybersecurity across Europe. ENISA's primary objectives are to enhance Europe's digital security and safeguard critical infrastructures via information exchange, capacity building, and awareness promotion. The EU implemented its inaugural comprehensive cybersecurity plan in 2013 and subsequently amended it in 2017. The Cybersecurity Strategy listed five priority domains: Creating new institutions, enhancing detection and protection capabilities, promoting training efforts, combating cybercrime, and defending the EU economy. A key element of this approach was its focus on maintaining fundamental rights, democracy, and the rule of law in a digitally enriched world, guaranteeing that the norms regulating the physical realm are valid in the digital sphere.

In addition to strategy documents, the EU has issued critical legislative measures to regulate and enhance cybersecurity. The 2016 Network and Information Systems (NIS) Directive (Directive 2016/1148) was the first comprehensive legal framework addressing cybersecurity at the EU level (EUR-Lex, 2016). This directive was followed by Regulation (EU) 2019/881, known as the first Cybersecurity Act, in 2019. The EU further strengthened the role of ENISA and established an EU-wide cybersecurity certification framework for digital products, services, and processes. It aims to increase resilience against cyber threats, increase trust in digital technologies, and provide a coordinated EU approach to cybersecurity challenges (EUR-Lex, 2019). The Cybersecurity Act was followed by the Data Governance Act (COM/2020/767), which aims to create a framework for secure and reliable data sharing within the EU in 2020 (EUR-Lex, 2020a). In the same year, two other critical laws were adopted within the EU. These are the Digital Markets Act (EUR-Lex, 2020b) and the Digital Services Act (EUR-Lex, 2020c). They have been accompanied by NIS2, adopted by the EU Council in 2022 (EUR-Lex, 2022). Finally, the Cyber Solidarity Act (European Commission, 2024) and the Cyber Resilience Act (Council of the European Union, 2024) have been adopted by the EU Council in 2024 to further strengthen the EU cybersecurity legislation.

Another critical initiative is that the EU has developed a strong institutional structure, such as Security Operations Centers (SOCs) and a Joint Cyber Unit, to strengthen cooperation between member states. Within this framework, the sovereignty of the EU and its member states and their technological competition are also underlined (European Commission, 2020).

The EU has positioned itself as a leading actor in regional and global cybersecurity governance through legislation, regulations, directives, applications, comprehensive projects, and competent institutions. Thus, the policies, directives, and laws created by the EU are defined as best practices, promoting international cooperation and setting international standards on a global scale by the United Nations Institute for Disarmament Research (UNIDIR) (UNIDIR Resources, 2019). This illustrates the EU's proactive role in shaping cybersecurity policies and strategies both regionally and globally, reinforcing its commitment to a safe, resilient, and free digital domain.

The EU Council presidency and EU leadership in general play an agenda-setting and coordinating role in cybersecurity and digital strategic autonomy. Europe's digital strategic autonomy and the Council's influence in this process are shaped by the strategic positioning of cybersecurity, the shaping of digital strategic autonomy, and the impact of EU presidencies and leaders (Escriche, 2022).

In particular, strategic autonomy aims to increase Europe's capacity to make its own choices by reducing its over-reliance on microprocessors and telecommunications infrastructure (such as 5G). According to the then-President of the EU Council, Charles Michel, achieving overall strategic autonomy is impossible without achieving digital sovereignty. He stated that this process is not only about defence but also about the ethical and human-centred use of data as an economic resource (Council of European Union, Press releases, 2021). In general, the influence of EU presidencies and leadership aims to transform Europe into a digital powerhouse. In this context, the EU budget has begun to be used in these areas.

3 Common Ground and Differences in Their Cybersecurity Strategies

Türkiye and the EU have developed similar in some respect but also distinct cybersecurity strategies, regulations, applications, and institutions. While the EU, as a supranational organization, has developed effective cybersecurity strategies that affect its member states, Türkiye, as a nation-state, has pursued cybersecurity policies in line with both global initiatives and the EU's cyber regulations.

3.1 Common Ground

A fundamental commonality between the two actors is the prioritization of cybersecurity as national and regional security concern. The EU has been integrating cybersecurity in its cyber initiatives in the last two decades (Carrapico & Barrinha, 2018). Similarly, cybersecurity dominates discourse of politicians and language of cyber activities in Türkiye (Çakır & Arınmış Uzun, 2021, p. 367).

Cybersecurity initiatives of both Türkiye and the EU have been in line with the framework prepared by ITU that is based on five basic pillars: Legal measures, technical and procedural measures, organizational structures, capacity building, and international cooperation.

The institutional structure for cybersecurity of both actors has also been a shared priority. While the EU tries to ensure the safety of its cyber ecosystem with ENISA, Türkiye engages in cybersecurity governance through the Cyber Security Board and, more recently, in the Cybersecurity Presidency.

Conceptualization of both cybersecurity and cyber risks and threats of both actors illustrates notable similarities. Both Türkiye and the EU define cybersecurity as the protection of digital systems, networks, programs, information, and interconnected infrastructures from all types of cyber threats. Their policy frameworks focus on the core principles of confidentiality, integrity, and availability of critical digital assets, underscoring a shared understanding of cybersecurity issues (Resmi Gazete, 2013).

3.2 Differences

Türkiye and the EU have commonalities, but they also have some notable distinctions in their cybersecurity strategies, particularly in their cybersecurity governance model, their global influence, and institutional structure.

The main difference between the EU and Türkiye is that the EU is a leading global cybersecurity actor (Backman, 2023), while Türkiye's cybersecurity strategies are much more limited in scope. Another critical difference between these two actors' cybersecurity policies is their cybersecurity governance model. The EU's model is based on an approach that advocates a multi-stakeholder model in internet governance, emphasizing the principle of shared responsibility. This model aims to increase resilience by spreading security responsibility from central authorities to local governments and the private sector (Dunn Cavelty, 2018, p. 313). In this sense, it is aligned with the freedom of the internet and fundamental human rights on a global scale. Türkiye, on the other hand, has produced strategies for developing cyber capacity more through national security, with a centralized understanding dominated by governmental bodies.

The scope and implementation of cybersecurity policies also diverge. The EU, as a supranational organization, has developed comprehensive common standards and frameworks for its member states. Decision-making and implementation processes are multilevel and complicated in nature. Yet on the other hand, Türkiye as a single sovereign entity has more flexible decision-making and implementation procedures based on national security priorities (Altın, 2023, p. 495).

Furthermore, there are distinctions between the identification and protection of critical infrastructures of both actors. The EU has a comprehensive legal and administrative regulation and institutional framework for protecting critical infrastructures based on initiatives and NIS directives of 2016. However, there has been no similar definition, clarification, or legal and institutional structure for critical infrastructures protection in Türkiye, despite cybersecurity being considered a fundamental component of national security.

Although both actors prioritize cybersecurity as a strategic standpoint and have developed similar initiatives and frameworks, they have different approaches regarding governance models, global influence, and institutional structures. The EU has developed a comprehensive cybersecurity regulation for member states, positioning itself as a global leading standard-setter in cybersecurity and fostering a multi-stakeholder approach in implementation. While Türkiye, enhancing its capacity in cybersecurity, has a more centralized and government-dominated institutional framework. Türkiye also tries to align its cybersecurity policies and legal regulations with the EU's standards.

4 Türkiye: EU Cybersecurity Cooperation: Historical Context

Turkey's cybersecurity policies are shaped through bilateral cooperation with international institutions and organizations, including the European Union (EU), being a primary partner in this regard.

Türkiye has a long history of cultural, diplomatic, strategic, and economic relations with the EU and its member states for decades. Although there have been fluctuations in bilateral relations from time to time, relations between the two parties have reached a sustainable point. It has always been considered as a notable ally by Europe due to its energy transit route, a special state in EU-Russia relations, a candidate country for EU membership, an important factor in the refugee crisis in recent years, its bilateral relations with Muslim countries in the Middle East (Ünalp Çepel, 2023) and more recently cooperation in cybersecurity, alongside with other security concerns.

The first historical situation in terms of cooperation between Türkiye and the EU in the field of cybersecurity is the Budapest Convention on Cybercrime in 2001. Although it is an initiative of the Council of Europe, Türkiye is a party to this convention due to its membership in the Council of Europe and its relationship with the EU that has lasted more than half a century. The EU did not participate in the convention directly as an organization but participated through the individual participation of member states. Türkiye signed this agreement in 2010 and put it into effect in 2014 (Council of Europe, 2001).

4.1 Multi-lateral Forums on Türkiye-EU Cybersecurity Cooperation

Türkiye, a NATO member, maintains close relationships with countries within the organization. Today's global system is faced with multi-dimensional and complex threats including cyber conflicts. Overcoming these threats requires close cooperation with international organizations that share common values, similar security concerns, and crisis management, such as NATO and the EU, that play leading roles in global economic and political security. The fact that 22 countries are common to both organizations and the existence of a complex interdependence between them further increase the significance of cooperation for security matters. For instance, the EU's use of NATO's resources for crisis management in the past years illustrates the complex interdependence of these two structures (Aghniashvili, 2016, p. 68). Naturally, Türkiye, a NATO member and an EU candidate country, is a part of this cooperation.

Both actors cooperate in cyber defense and cybersecurity issues within the NATO cybersecurity initiatives, such as the Cooperative Cyber Defense Center of Excellence (CCDCOE), that aims to foster cyber defense capabilities, encourage information sharing, and create a collective defense strategy against cyber threats. It also contributes to the development of cybersecurity policies and practices by conducting studies on technical, strategic, operational, and legal dimensions (CDOCOE, 2025).

Cybersecurity cooperation between NATO and the EU has expanded in recent years. In 2016, NATO and the EU signed a technical arrangement in the field of cyber defense to better prevent cyber threats and respond more quickly to possible attacks. With this arrangement, the NATO Cyber Security Center and the EU's Computer Emergency Response Team (CERT-EU) created a framework for information exchange and sharing of best practices (NATO, 2024). The same year, they also agreed on more than 40 measures to create a more stable, safer, and free digital environment (NATO, 2016).

4.2 Areas of Collaboration

Although there is no formal cooperation on certain matters, Türkiye tries to harmonize its cybersecurity regulations, practices, and institutions with the EU ones. The Information and Communication Technologies Authority (BTK) has initiated efforts to align its regulatory framework with EU cybersecurity standards. Furthermore, legal regulations such as the Law on Personal Data Protection (KVKK) and legislation on payment systems have been structured in accordance with EU directives (Beghe Sönmez & Karakaşlar, 2023). Türkiye also cooperates with ENISA for best practices in risk assessment and cyber resilience strategies. In addition, USOM continues to cooperate with the CERT-EU to improve threat detection and response (Özker, 2022, p. 25).

Capacity-building initiatives represent another key area of engagement. In the conclusions of the EU Council on June 26, 2018, it was proposed to establish an EU External Cyber Capacity Building Network to collectively include the expertise of EU Member States and neighboring countries including Türkiye (EU Cybernet, 2025). In line with this program, Türkiye has received both financial support and entered into cooperation in the field of strengthening capacity against cybercrime (European Commission, 2021, pp. 60–62).

On the other hand, the EU's pursuit of strategic autonomy means not merely turning inward, but also having the capacity to make independent choices in an open global environment. Türkiye's participation in the Digital Europe Programme is both a concrete step directly aligned with the EU's goals of strategic autonomy and digital sovereignty, and a positive step toward strengthening cooperation efforts in the cyber domain (European Commission, 2023). In 2023, the year Türkiye joined the Digital Europe program, the EU Council presidencies were held by Sweden and Spain. During this period, cybersecurity was emphasized within the framework of the EU's internal security and digital resilience. While the Swedish Presidency focused on the implementation of the NIS2 Directive and the protection of critical infrastructure, the Spanish Presidency linked cybersecurity to the goals of digital sovereignty and strategic autonomy.

The 2023 EU Council presidencies addressed cybersecurity within the context of strategic autonomy and resilience, while prioritizing cooperation with third countries. This approach limited the scope of EU-Türkiye cyber cooperation, as normative and regulatory harmonization took precedence, despite continued technical cooperation in the cyber field with candidate countries like Türkiye. This was also noted in the Türkiye Report published by the Commission in 2024. In particular, the increased efforts on digital services legislation with the EU, the need to further develop normative values through personal data protection legislation, and freedom of expression were highlighted (European Commission, 2024).

5 Conclusion

Türkiye and the European Union's relationship, whose political, diplomatic, and economic history has been shaped by various factors, has continued for a very long time, dating back to the 1960s. Even though there have been tensions, uncertainties, and crises from time to time in these bilateral relations, both parties have managed to overcome these problems and have always maintained their relations. Although Türkiye–European Union relations are multi-dimensional, security issues have always been at the center. In this context, in the last decade, cybersecurity has tried to become an important component in these relations. Cyber threats, mostly international in nature, have necessitated international and regional cooperation and, in this context, forced Türkiye and the European Union to closer cooperation.

As emphasized from the beginning, this research explores the possibilities and challenges of cooperation between Türkiye and the European Union in the context of cybersecurity based on ITU's five-pillar framework. In this context, the primary research question guiding the study was: What are the opportunities and difficulties of cybersecurity cooperation between Türkiye and the European Union? Research findings allow us to understand the dynamics of this relationship. And it proves that it is possible to establish much deeper cooperation between Türkiye and the European Union; there are many opportunities in this context, but there are also some challenges.

If the three hypotheses put forward in the introduction part of the study are examined according to the study findings, the following conclusions are inferred. The first hypothesis suggested that differing cybersecurity perceptions between Türkiye and the EU weaken the prospects for effective cooperation. This hypothesis was partially confirmed because Turkey's understanding of cybersecurity is based more on national security priorities, while the European Union envisages a multi-stakeholder model. This model adopts an approach that centers on human rights and democratic governance. This difference in approach between the two actors creates some difficulties for cooperation. The second hypothesis was as follows: major cyber-attacks faced by Türkiye and the EU could create opportunities for cooperation. To the extent that the research findings confirm this hypothesis, both actors have common concerns about being exposed to serious cyber threats and attacks and seem very willing to cooperate. In this context, Turkey's participation in some cybersecurity projects of the European Union, its development of some collaborations with the cybersecurity agency of the European Union, ENISA, and its development of some partnerships with NATO's cybersecurity center of excellence are clear indicators of this cooperation.

The final hypothesis examines whether Turkey encounters difficulties in adapting its legal and institutional framework to the EU's comprehensive cybersecurity regulations. This hypothesis is also validated by research findings. Because Turkey is making efforts to harmonize its cybersecurity legislation with the European Union legislation as it encounters serious cyber attacks and threats. However, despite Turkey making significant progress on this path and making critical advancements in its institutions, cybersecurity framework, and applications, it still faces some difficulties in protecting critical infrastructures and developing a more compatible structure with Europe.

Although Türkiye's European Union membership is increasingly moving toward an equation with many unknowns, cooperation in the field of cybersecurity is developing day by day. Within this framework, different cyber activities such as technical and financial assistance, joint projects, and joint security initiatives are being developed for Turkey. Of course, Turkey's NATO membership offers the opportunity for this cooperation to progress and develop to a higher level. In order for this cooperation to deepen and develop further, both parties need to develop a more comprehensive and pragmatic approach. Harmonization of legal and administrative regulations, development of trust between the parties, reinforcement of common institutional and application areas, and sharing of good examples will further advance cooperation in the field of cybersecurity. In the future, more information sharing, sharing of good practices, joint training and exercises, development of coordinated policies, and protection of critical infrastructures will create a safer, freer, and more stable digital domain for the two actors in the face of serious cybersecurity threats.

References

Aghniashvili, T. (2016). Towards More Effective Cooperation? The Role of States in Shaping NATO-EU Interaction and Cooperation. *Connections: The Quarterly Journal*, 15(4), 67–90. https://doi.org/10.11610/Connections.15.4.04.

Akyeşilmen, N. (2022). Türkiye in the Global Cybersecurity Arena. *Insight Türkiye*, 24(3), 109–134. https://doi.org/10.25253/99.2022243.8.

Altın, O. (2023). AB'nin Siber Güvenlik Alanındaki Politikalarının ve Uygulamalarının Etkinliği: Bir Siber Güvenlik Temsilcisi Olarak AB'nin Yeterliliği. *Çankırı Karatekin Üniversitesi İktisadi ve İdari Bilimler Fakültesi Dergisi*, 13(2), 482–507. https://doi.org/10.18074/ckuiibfd.1276923.

Arquilla J., & Ronfeldt D. (1993). Cyberwar Is Coming!. *Comparative Strategy*, 12, 141–165.

Backman, S. (2023). Risk vs. Threat-Based Cybersecurity: The Case of the EU. *European Security*, 32(1), 85–103. https://doi.org/10.1080/09662839.2022.2069464.

Beghe Sönmez, S. & Karakaşlar, M. (2023). Cybersecurity Türkiye, Lexology. 2023 Law Business Research.

Çakır, H. & Arınmış Uzun, S. (2021). Türkiye'nin Siber Güvenlik Eylem Planlarının Değerlendirilmesi. *Ekonomi İşletme Siyaset ve Uluslararası İlişkiler Dergisi*, 7(2), 353–379.

Carrapico, H. & Barrinha, A. (2018). European Union Cyber Security as an Emerging Research and Policy Field. *European Politics and Society*, 19(3), 299–303. https://doi.org/10.1080/23745118.2018.1430712.

CDOCOE (2025). https://ccdcoe.org/.

Council of Europe (2001). *Details of Treaty No.185*. https://www.coe.int/en/web/conventions/full-list?module=treaty-detail&treatynum=185.

Council of Europe (2024). *Cyber Resilience Act: Council Adopts New Law on Security Requirements for Digital Products*. https://www.consilium.europa.eu/en/press/press-releases/2024/10/10/cyber-resilience-act-council-adopts-new-law-on-security-requirements-for-digital-products/.

Dunn Cavelty, M. (2018). Europe's Cyber-Power. *European Politics and Society*, 19(3), 304–320. https://doi.org/10.1080/23745118.2018.1430718.

EU CyberNet (2025). *About Project*. https://www.eucybernet.eu/about-project/.

EUR-Lex (2016). *Directive (EU) 2016/1148 of the European Parliament and of the Council of 6 July 2016 Concerning Measures for a High Common Level of Security of Network and Information Systems Across the Union*. https://eur-lex.europa.eu/legal-content/EN/TXT/?uri=CELEX%3A32016L1148&qid=1736710804290.

EUR-Lex (2019). *Regulation (EU) 2019/881 of the European Parliament and of the Council of 17 April 2019 on ENISA (the European Union Agency for Cybersecurity) and on Information and Communications Technology Cybersecurity Certification and Repealing Regulation (EU) No 526/2013 (Cybersecurity Act) (Text with EEA relevance)*. https://eur-lex.europa.eu/eli/reg/2019/881.

EUR-Lex (2020a). *Proposal for a Regulation of the European Parliament and of the Council on European Data Governance (Data Governance Act)*. https://eur-lex.europa.eu/legal-content/EN/TXT/?uri=CELEX:52020PC0767.

EUR-Lex (2020b). *Proposal for a Regulation of the European Parliament and of the Council on Contestable and Fair Markets in the Digital Sector (Digital Markets Act)*. https://eur-lex.europa.eu/legal-content/en/ALL/?uri=COM:2020:842:FIN.

EUR-Lex (2020c). *Proposal for a Regulation of the European Parliament and of the Council on a Single Market for Digital Services (Digital Services Act) and Amending Directive 2000/31/EC*. https://eur-lex.europa.eu/legal-content/EN/TXT/?uri=CELEX:52020PC0825.

EUR-Lex (2022). *Directive (EU) 2022/2555 of the European Parliament and of the Council of 14 December 2022 on Measures for a High Common Level of Cybersecurity Across the Union, Amending Regulation (EU) No 910/2014 and Directive (EU) 2018/1972, and Repealing Directive (EU) 2016/1148 (NIS 2 Directive) (Text with EEA relevance)*. https://eur-lex.europa.eu/legal-content/EN/TXT/?uri=CELEX%3A32022L2555&qid=1736710804290.

European Council (2021). *Digital Sovereignty Is Central to European Strategic Autonomy – Speech by President Charles Michel at "Masters of Digital 2021" Online Event*. https://www.consilium.europa.eu/en/press/press-releases/2021/02/03/speech-by-president-charles-michel-at-the-digitaleurope-masters-of-digital-online-event/.

European Commission (2020). *Joint Communication to the European Parliament and the Council: The EU's Cybersecurity Strategy for the Digital Decade.* https://eur-lex.europa.eu/legal-content/EN/TXT/PDF/?uri=CELEX:52020JC0018&from=EN.
European Commission (2021). *International Cyber Capacity Building: Global Trends and Scenarıos.* Luxembourg: Publications Office of the European Union.
European Commission (2023). Commission Opens Digital Europe Programme to Türkiye. https://enlargement.ec.europa.eu/news/commission-opens-digital-europe-programme-turkiye-2023-09-01_en.
European Commission (2024). *Commission Welcomes Political Agreement on Cyber Solidarity Act.* https://ec.europa.eu/commission/presscorner/detail/en/ip_24_1332.
European Commission (2024). *Commission Staff Working Document: Türkiye 2024 Report.* https://enlargement.ec.europa.eu/document/download/8010c4db-6ef8-4c85-aa06-814408921c89_en.
Escriche, I. Arco. (2022). *How Can the European Union Achieve Digital Strategic Autonomy? Views From Future Leaders, CIDOB Briefings.* https://www.cidob.org/sites/default/files/2024-09/42_CIDOB%20BRIEFINGS_ANG.pdf.
International Telecommunication Union (ITU) (2024) Global Cybersecurity Index 2024. Geneva: International Telecommunication Union. https://www.itu.int/itu-d/reports/statistics/global-cybersecurity-index-2024/
NATO (2024). *Cyber Defence.* https://www.nato.int/cps/en/natohq/topics_78170.htm.
NATO (2016). *NATO and the EU Deepen Ties as Ministerial Meeting Wraps Up.* https://www.nato.int/cps/en/natohq/news_139112.htm.
Özker, U. (2022). *Türkiye'de Kritik Altyapı ve Siber Güvenlik,* Konrad-Adenauer-Stiftung Türkiye.
Resmi Gazete (2012). *Karar Sayısı: 2012/3842.* https://www.resmigazete.gov.tr/eskiler/2012/10/20121020-18.htm.
Resmi Gazete (2013). *Ulusal Siber Güvenlik Stratejisi ve 2013–2014 Eylem Plan.* https://www.resmigazete.gov.tr/eskiler/2013/06/20130620-1.htm.
Resmi Gazete (2025). *Siber Güvenlik Başkanlığı Hakkında Cumhurbaşkanlığı Kararnamesi (Kararname Numarası: 177).* https://www.resmigazete.gov.tr/fihrist?tarih=2025-01-08.
T.C. Cumhurbaşkanlığı Mevzuat Bilgi Sistemi. (2007). *İnternet Ortaminda Yapilan Yayinlarin Düzenlenmesi ve Bu Yayinlar Yoluyla İşlenen Suçlarla Mücadele Edilmesi Hakkinda Kanun (Kanun 5651).* https://www.mevzuat.gov.tr/mevzuat?MevzuatNo=5651&MevzuatTur=1&MevzuatTertip=5.
T.C. Cumhurbaşkanlığı Mevzuat Bilgi Sistemi. (2016). *Kişisel Verilerin Korunması Kanunu (Kanun 6698).* https://www.mevzuat.gov.tr/mevzuat?MevzuatNo=6698&MevzuatTur=1&MevzuatTertip=5.
T.C. Ulaştırma ve Altyapı Bakanlığı. (2016). *2016–2019 Ulusal Siber Güvenlik Stratejisi.* https://hgm.uab.gov.tr/strateji-eylem-planlari.

T.C. Ulaştırma ve Altyapı Bakanlığı. (2020). *Ulusal Siber Güvenlik Stratejisi 2020–2023*. https://hgm.uab.gov.tr/strateji-eylem-planlari.

Tarhan, K. (2022). Historical Development of Cybersecurity Studies: A Literature Review and Its Place in Security Studies. *Przegląd Strategiczny*, 15, 393–414. https://doi.org/10.14746/ps.2022.1.23.

Ünalp Çepel, Z. (2023). European Union Global Strategy and Türkiye: An Analysis of the Term 2016–2021. *TESAM Akademi Dergisi*, 10(1), 357–368. https://doi.org/1030626/tesamakademi.1075694.

UNIDIR Resources (2019). The Role of Regional Organizations in Strengthening Cybersecurity and Stability: Experiences and Opportunities. *Report of the 2nd International Security Cyber Workshop Series Geneva. Switzerland.*

PART 4

Economic Policy in the Area of Security and New Technologies

CHAPTER 24

Poland's Policy toward the European Union's Decisions to Combat Disinformation in the Energy and Climate Sector

Małgorzata Kamola-Cieślik

Abstract The chapter analyzes the actions of the European Union and Poland in countering disinformation in the energy and climate sector in the digital space between 2022 and 2025. It aims to identify the main disinformation narratives and assess policies and measures implemented at both EU and national levels, including actions taken by governmental and non-governmental institutions. The study is based on document analysis, comparative data analysis, and decision-making analysis. The findings indicate that despite intensive efforts at both EU and national levels, the effectiveness of countering disinformation has remained limited. The main challenge has been coordinated disinformation campaigns by the Russian Federation on social media, which have influenced public opinion and undermined trust in institutions responsible for energy transition and climate policy.

Keywords disinformation – energy transition – climate policy – European Union – Poland

1 Introduction

Before the outbreak of war in Ukraine in 2022, the European Union (EU) imported an average of 60% of its annual energy needs from the Russian Federation. The European Union condemned Russia's attack on Ukraine and gradually introduced energy sanctions against the country, which was expected to weaken Russia's economy and reduce its ability to conduct warfare. On April 8, 2022, the EU adopted a fifth package of sanctions banning the purchase, import, and transfer of coal and other solid fossil fuels to the EU originating in or exported from Russia. In the same year, the union's embargo on the import of Russian oil by sea took effect. Despite the sanctions imposed on Russian oil, there were some exceptions in the supply of this commodity

HTTPS://DOI.ORG/10.1163/9789004763623-026

through the "Friendship" pipeline to, among others, Hungary, Slovakia, and the Czech Republic. Regarding Russian gas, the EU has not imposed a full embargo on this raw material. Still, the countries of the union import liquefied natural gas (LNG) from Russia, as opposed to pipeline gas supplies, which have been severely curtailed. In 2024, Russia supplied the EU market with 10% of oil and 33% of gas of the EU's total imports. Russian coal has ceased to be imported by the Union States. As a result of EU sanctions on energy raw materials, Russia changed its export direction. Most of the raw materials began to go to the Asian market (China, India), which resulted in financial losses for the Russian Federation due to the lower prices it had to offer to remain competitive in the new market.

The Russian Federation's loss of its main market for energy raw materials made it undertake disinformation activities in the energy and climate sectors of EU Member States. These actions were aimed at destabilizing the EU energy market and weakening solidarity among member states. Poland was exposed to Russian disinformation activities because of the public and military support given to Ukraine and the abandonment of Russian energy resources. For both the EU and the Polish government, combating disinformation in the energy and climate area was important in order to protect citizens from manipulation that could contribute to changes in energy and climate policies. (Skalamera, p. 9; Vakulchuk, Overland & Scholten, 2020).

The aim of the chapter is to analyze the EU's decisions to combat disinformation in the energy and climate sector in the digital space and the actions taken by the Polish government and non-governmental institutions to counter disinformation in the energy and renewable energy sources (RES) area. In pursuit of this goal, the following thesis was formulated: Despite the decisions of the EU and Poland to combat disinformation in the energy and climate sector, their effectiveness was limited due to the intensive use of social media by the Russian Federation as a propaganda tool.

To verify the above thesis, the following questions were sought to be answered: 1. What issues related to the Polish energy and climate sector were most often the subject of disinformation on the Internet in 2022–2025? 2. What decisions has the EU taken to combat disinformation in the energy and climate sector? 3. What measures has Poland taken to provide the public with reliable information in the energy-climate area? In order to answer the research questions, the methods used were desk research analysis, comparative data analysis, and decision-making analysis. The use of the method of desk research analysis consisted of analyzing documents produced by the EU, the Polish Parliament, and non-governmental institutions, taking into account their impact on combating false information in the energy and climate sector. The comparative analysis made it possible to compile differences and similarities in the activities of the EU and non-governmental institutions in Poland in countering disinformation on energy-climate issues. The decision-making method, in turn, was used to assess the effectiveness of decisions taken by the EU, the Polish government, and non-governmental institutions in the context of limiting the impact of disinformation on society and political-economic processes in the energy-climate sector.

In an effort to understand the mechanisms of disinformation in the energy-climate sector and the reactions of the EU, the Polish government, and non-governmental institutions, the theory of propaganda by Edward Herman and Noam Chomsky was used (Herman & Chomsky, 2002). Through the application of this theory, it was explained how the Russian Federation used the media to shape public opinion and influence policy in the context of the energy transition, among other things. The agenda-setting theory of Maxwell E. McCombs and Donald L. Shaw was also applied, allowing us to understand how the media shape public discussions of energy crises and climate change, and how they influence Polish and EU policy. In addition to normative acts, reports and articles were used in the conducted analysis. The time caesura adopted in the thesis chapter covers the years 2022–2025. The first marginal date is associated with the start of the war in Ukraine, which resulted in an increase in disinformation activities in the energy and climate sector in Poland and other EU countries. The second marginal date is June 2025, associated with the end of Poland's presidency of the EU Council. The chapter is divided into sections which relate to: 1. disinformation in the Polish energy sector in the digital space and the response of the Polish government; 2. EU efforts to combat disinformation in the energy sector; 3. countering disinformation as part of building information resilience in the energy sector in Poland.

2 Disinformation in the Polish Energy Sector in the Digital Space and the Response of the Polish Government

The possibility of using the Internet quickly and freely through various communicators has made it an important channel of information for people around the world. For several years, the digital space has begun to serve as the main source of information on global events for most Poles, as evidenced by the results of a survey conducted by Instytut Badań Internetu i Mediów Społecznościowych (IBIMS) (the Institute for Internet and Social Media Studies) and Fundacja Instytut Badań Rynkowych i Społecznych (IBRIS) (the Institute for Market and Social Research Foundation) in November 2021. According to the report by these institutions, 62% of respondents said their main source of information was online portals such as Wirtualna Polska or Onet. Just under 40% admitted that they get their knowledge from TVN's "Facts" reports at 7 p.m. In third place as a source of knowledge, 38.5% of people pointed to social media such as Facebook and Twitter (IBIMS IBRIS, 2021). At the same time, since the beginning of the war in Ukraine, Internet users have been increasingly exposed to false information on energy-climate security of Poland and the EU, among other issues.

The Digital Poland Foundation's 2022 report, "Disinformation through the Eyes of Poles," indicated that the most susceptible to fake news were young people who have difficulty distinguishing between real information and fake information. The condition was influenced by fast and cursory reading involving scrolling a smartphone screen. Of the respondents to the survey, 81% had encountered disinformation, and 82% had encountered fake news. The report, "Disinformation Through the Eyes of Poles," noted that the energy and climate area was significantly susceptible to the spread of false information.

In light of the propaganda theory, it should be noted that the Russian Federation has used the digital space to shape public opinion by posting messages with the following content: 1. The nuclear power plant poses a threat to local residents; 2. the use of renewable energy does not contribute to reducing greenhouse gas emissions; 3. the climate conspiracy is aimed at destroying countries based on the coal economy (Digital Poland, 23.02.2024). Referring to a report by Polskie Towarzystwo Elektrociepłowni Zawodowych (PTEZ) (the Polish Association of Combined Heat and Power), it should be noted that in the first half of 2023, 20% of the content on sourcing energy resources, energy prices, nuclear energy, and climate policy on the Polish Internet was disinformation (Executive Magazine, 2023).

According to the European Commission's (EC) definition, disinformation meant information that can be verified as false or misleading, created and disseminated for economic gain or to cause public harm. This phenomenon posed a threat to democratic political processes, the security of citizens of member states, and the environment (European Commission, 26.04.2018). It is noteworthy that for the first time, the scientific community on a large scale identified disinformation as a significant factor affecting democratic processes in the context of the Brexit referendum campaign.

Energy and climate issues were and are among the areas of the economy that arouse the interest of Poles. One example of disinformation was false information about the government's planned increases in electricity and gas prices and ways to avoid them, which contributed to the radicalization of public sentiment and increased distrust in the Polish government. Referring to the theory of agenda-setting, the Polish government has taken steps to make citizens aware of the false information circulating on the Internet and to focus their attention on reliable information in the energy and climate areas. In June 20024, the Ministry of Climate and Environment warned Internet users against clicking on "how to avoid gas/electricity/heat price hikes" links containing false content about prices for energy resources and encouraging applications for energy vouchers. The purpose of these actions was to trick Internet users into providing personal information and to cause panic among recipients of this information. At the same time, the Ministry of Climate and Environment's website announced that as of July 1, 2024, the Law on Temporary Price Restrictions for Electricity, Natural Gas, and System Heat and the Energy Voucher would be in effect. Another wave of false information took place in July 2024 on social media. It concerned the possibility of paying PLN 1,000 in exchange for avoiding changes in energy bills. The Ministry of Climate and Environment warned consumers about such proposals in the form of links to unsafe websites and their consequences, and provided information on the current situation of energy consumers.

In 2024, the issue of RES regulations was another topic of misinformation. The false information that appeared on the Internet concerned the introduction by the amended RES law of an obligation to install photovoltaic panels on newly constructed residential buildings. In fact, the legislative changes were aimed at promoting the use of RES through tax credits and subsidies, without making these solutions mandatory. In response to the false content appearing, the Ministry of Climate and Environment published announcements about false information appearing on the Internet regarding the conditions for installing photovoltaic panels and the actual requirements for their installation.

In December 2024, Climate and Environment Minister – Paulina Henning-Kloska – announced that the ministry under her leadership would focus on tools to combat climate and energy misinformation during the Polish Presidency of the EU Council (Ministerstwo Klimatu i Środowiska, 20.12.2024).

At the same time, the Minister of Foreign Affairs – Radosław Sikorski – also announced that during the Polish Presidency, Poland would seek to establish the European Resilience to Disinformation Council to strengthen resilience and response to cyber threats. The newly established institution was to be financed by EU funds, and its tasks were to protect against disinformation and monitor the information space. In addition, the European Council was also supposed to support member states in building societies' resilience to information manipulation and in implementing fact-based communication strategies.

In turn, the Minister of Digitization – Krzysztof Gawkowski – stressed that his ministry would strive to strengthen cybersecurity and develop artificial intelligence during the Polish Presidency. The ministry's goal was to increase the EU's resilience to cyber threats and develop artificial intelligence as a technology to support combating fake news and the innovation of European economies (Ministerstwo Cyfryzacji, 19.12.2024). A joint declaration by the Ministers of Climate and Environment, Foreign Affairs, and Digitization confirmed that countering disinformation, especially in the area of energy transition and information security, has become one of the priorities of the Polish Presidency. The approach also fits in with the EU's strategy, an important goal of which is to protect democratic values in the face of hybrid threats and information manipulation by external actors such as the Russian Federation.

3 European Union Efforts to Combat Disinformation in the Energy and Climate Sector

Based on an analysis of the European Digital Media Observatory (EDMO) 2022–2024 reports, it should be noted that there has been an increase in disinformation content related to the EU energy crisis, especially after the start of the war in Ukraine. False information appeared in the online space with the aim of causing public anxiety, undermining confidence in EU energy policy, and weakening solidarity among member states. As a result of the analysis of disinformation content, it should be noted that topics of particular interest were:

1. the planned shutdown of electricity supplies to the public due to a "catastrophic" shortage of gas in the EU;
2. false reports about the violent suppression of protests against the gas price hike, which did not actually occur;
3. conspiracy theories about forest fires that were supposedly intentionally started by helicopters (acting on behalf of the EU) to distract attention from the real cause, which is climate change;
4. comments blaming EU sanctions against Russia for rising energy prices, inflation, and a deepening economic crisis;
5. promoting Russian gas as the only stable and cheap source of energy, while undermining the effectiveness and cost-effectiveness of RES.

Typically, each of these misinformation pieces was accompanied by photographs that reassured the message's recipients of its credibility. It is worth noting that in reality, the photographs bore no relation to the content of the message, which misled the recipients. In addition to the EDMO, EUvsDisinfo, operating within the European External Action Service (EEAS), also took action in combating disinformation, including that related to the energy sector (Disinfo, 2022). EUvsDisinfo's analyses provided substantive support to, among others, the EC and the European Parliament (EP) in formulating policies against disinformation and energy transition. In addition, the EUvsDisinfo platform unmasked false information on energy by pointing out that the energy crisis in the EU was not solely the result of the war in Ukraine, but was due to Europe's dependence on Russian energy resources.

In response to the growing phenomenon of disinformation and the resulting threats to democratic processes in the EU, in March 2022 the EP adopted a resolution on foreign interference in all democratic processes in the EU, including disinformation. The document laid out directions for the EC and member states in combating false information on social media. The resolution draws attention to the scale and effects of disinformation activities undertaken, among others, by the Russian Federation to manipulate public opinion, destabilize and internal situation of member states, and weaken the unity of European society. It stressed the need for a systemic EU response to the threats posed by disinformation campaigns and the creation of mechanisms to counter such practices (European Parliament, 2022). According to the EP resolution, EU policy was to be aimed at promoting digital education among the public, which would strengthen its resistance to false information on the Internet and tighten cooperation between member states in combating disinformation. In light of agenda-setting theory, the EU used its power to influence the public and media agenda by issuing regulations and organizing information campaigns to counter the expanding Russian propaganda on energy and climate policy. At the same time, the EU imposed topics in the public debate, such as energy security and transition, in order to counter the influence of false information and strengthen public confidence in reliable sources of information.

It should be noted that EP Resolution of 2022 marked the beginning of work on EU acts to combat disinformation. At the same time, the European External Action Service developed a Strategic Compass relating, among other things, to strengthening the EU's resilience to fake information on the Internet by developing tools to detect and counter disinformation campaigns. In response to growing disinformation, in the context of Russian misinformation regarding the war in Ukraine, energy and climate security, and EU policy, the EC updated the code of conduct on combating disinformation in June 2022. The code, which has been a voluntary agreement, has been signed by 34 entities, including digital platforms (Google, Facebook, Instagram, TikTok, Microsoft), social and research organizations, and digital media companies. Signatories to the code pledged to effectively combat disinformation by limiting the reach of false content, blocking the placement of ads on websites spreading fake news, and working with fact-checkers (Shaping Europe's digital future, 2022).

Another EU action to combat disinformation on the Internet was the adoption of the Digital Services Act (DSA) by the EP and EU Council in November 2022. The document came into force in February 2024 and, unlike the code of conduct on combating disinformation, imposed legal obligations on digital platforms operating in the EU and provided for penalties for non-compliance (European Commission, 16.11.2022). The code focused mainly on combating disinformation, while the DSA addressed the responsibility of online platforms in terms of content moderation, advertising, and algorithm transparency. The legal solutions adopted in the DSA were intended to ensure fair competition, transparency in the operation of digital platforms, and curb the spread of fake news.

As a result of the DSA's implementation, the Code of Conduct on Combating Disinformation was incorporated into its framework, allowing it to become more formal. In January 2023, the European Declaration of Digital Rights and Principles, adopted by the EP Council of the European Union and the EC, was reinforced. The content of the Declaration and the DSA complemented each other in an effort to create a digital environment to protect citizens from disinformation. The Declaration pointed out the need to conduct digital education and develop media competence in an effort to strengthen the public's resilience against false information. In conducting a comparative analysis of the European Declaration of Digital Rights and Principles and the DSA, it should be noted that the Declaration emphasized the protection of Internet users, while the DSA introduced specific legal solutions to combat disinformation. As a part of efforts to combat disinformation in the energy and climate sector, in June 2023 the EP adopted a resolution on foreign interference in all democratic processes in the EU, including disinformation (European Parliament, 2023). The resolution set out directions for countering misinformation that undermines confidence in the EU's energy and climate policy, including energy transition efforts and diversification of energy sources. The resolution, without having the force of law, only recommended that the EC, member states, and EU institutions take measures aimed at deepening cooperation with digital platforms and responding promptly to harmful and untrue content. It also noted the need to expand citizens' knowledge of disinformation techniques through the implementation of educational programs. The EP Resolution of 2023 points to the DSA as an important EU document in combating disinformation in the digital space, demanding its implementation by digital platforms within the union.

Since the start of the war in Ukraine, the EU has supported member states in combating disinformation. It has taken action on the institutional level (using reports, EDMO analyses, EUvsDisinfo), legislative level (creating EU law), and conducted information and education campaigns to combat false Russian information, especially in the energy and climate sector, which undermined credibility of the EU energy market. In 2022–2023, the EC's response to false information in the context of the energy crisis was the #EnergyEU campaign. The goal of the campaign was to ensure energy security through diversification of energy sources, promotion of RES, and independence from Russian gas supplies. At the same time, the EU was running the #ActOnClimate campaign whose task was to encourage citizens to take action to invest in RES and to make the public aware of the rightness of reducing the supply of raw materials from Russia. Also, the Energy Efficiency First campaign promoted measures to increase independence from Russian raw material imports and stressed the importance of energy conservation by citizens, companies, and public administration (Enefirst Plus, 2024). The EU's activities in the framework of information and education campaigns also included cooperation with organizations and institutions involved in combating disinformation in the energy and climate area at the national level. With regard to Poland, the EU institutions cooperated with, among others, the "Demagog" Foundation, the Polish Press Agency, the Ministry of Climate and Environment, the Scientific and Academic Computer Network, the Center for Research on Disinformation, as well as the national media.

Considering the EU's actions toward combating Russian disinformation, it is important to point out the significant importance of the funds that have been provided for this purpose as part of the implementation of various EU programs. Between 2021 and 2024, the EC funded a number of projects under the Creative Europe, Horizon Europe, and Digital Europe programs aimed at combating disinformation. These projects supported fact-checking initiatives, independent media, NGOs, and local EDMO hubs related to the energy and climate sector, among others. For example, EDMO Poland received EU funds to prepare analyses, tools, and training for journalists and experts. The aim of these activities was to increase the public's resilience to disinformation, especially in the area of energy and climate security. Between 2021 and 2023, the EC earmarked more than €20 million for combating disinformation under the Digital Europe program alone.

4 Countering Disinformation as Part of Building Information Resilience in the Energy and Climate Sector in Poland

Since the beginning of the war in Ukraine, Polish governmental and non-governmental institutions have taken measures to protect the public from false information, including in the energy-climate sector. Referring to the agenda-setting theory, it can be noted that the goal of the Polish government and non-governmental institutions was to direct public attention to the issues of energy transition and protection from information manipulation. Their initiatives were aimed at consciously shaping the topics dominating the public debate, thus strengthening the public's information resilience and state security. As a part of these activities, in February 2024, the Ministry of Digitization was working on the draft law on the provision of electronic services and certain other laws (hereinafter: the draft law on the provision of electronic services), creating the possibility of implementing the DSA into the Polish legal order (Ministerstwo Cyfryzacji, 2025). The government's draft law provided for the appointment of a digital services coordinator who was to be responsible for overseeing the prevention of false information, as well as the National Council for Digital Services, whose task was to advise the government in shaping policy on digital services. After a year of work by the government, the draft law on the provision of services by electronic means has reached the Polish Parliament, where it is currently undergoing legislative work. Although the government's bill does not directly address combating disinformation in the energy and climate sector, some of its solutions, such as the establishment of an institution to combat this phenomenon, may reduce the spread of false information in this area.

Another form of the government's efforts to combat disinformation in 2022–2025 was the Ministry of Climate and Environment's activity of promoting facts and data on the energy transition on its website to raise public awareness and counter the spread of false narratives.

In addition, the ministry has been actively supporting a number of NGO projects related to energy transition, environmental protection, and countering misinformation in these areas. One example of the Ministry of Climate and Environment's educational campaign in response to the energy crisis caused by the war in Ukraine was the "We Save Energy" initiative. (Ministerstwo Klimatu i Środowiska, 23.09.2022). The goal of the campaign was to make the public aware that Poland's energy security does not have to depend on Russian energy resources. That is why it was so important for the campaign to promote rational management of heat and electricity in households as well as public institutions and companies, which was expected to contribute to environmental protection and lower energy consumption. The educational campaign "We save energy" included TV spots, radio broadcasts, newspaper articles, and educational activities aimed at young people.

In November 2023, as part of combating disinformation in the energy and climate sector, the Polish Press Agency, together with the Ministry of Climate and Environment, announced the #FakeHunter Challenge: energy and climate. The contest was aimed primarily at Internet users (youth and adults) interested in countering disinformation in the areas of energy transition and climate change. The task of the contestants was to find and report false information on the Internet regarding energy and climate, which was then subjected to verification by a jury. The contest was educational in nature, promoting skills in recognizing manipulation and thinking critically about content published in the online space (Nowa Energia, 6.11.2023). The need to combat climate disinformation was the main topic of the international meeting "Communication in European Terms," organized by the Ministry of Climate and Environment, and was attended by representatives of ministries, offices, and NGOs of EU countries. The Polish side noted that one of the goals of its presidency of the EU Council will be to quickly identify and counteract fake news using artificial intelligence and cooperation with the media. Although issues related to the energy sector were not the main topic of the meeting, they were naturally linked to climate disinformation, which often concerned processes related to moving away from fossil fuels, investments in RES, and energy pricing policies.

Between 2022 and 2025, the fight against disinformation in Poland involved not only government institutions, but also entities outside the public administration. An example of such activities was the initiative of the Polish Electricity Committee using the "We Fight for Polish Independence" slogan. Its goal was to counter false information exploiting the rise in energy prices and to provide knowledge about the Polish government's decisions aimed at protecting Poles from high electricity prices. The campaign's task was to make the public aware of the dangers of disinformation. As highlighted in the campaign materials, "Russian disinformation used high energy prices, caused by Putin's war and the Kremlin's aggressive energy policy, to sow uncertainty and fear. The goal of Russian propaganda is simple – to destabilize Poland and its allies in the European Union and NATO" (Polish Electricity Committee, 2022). The campaign used billboards featuring an image of Russian Federation President Vladimir Putin, a graphic depicting an energy pole that cast a shadow in the shape of a machine gun, and the slogan "High energy prices are also his weapon." According to the campaign's authors, the billboard's message indicated that one of the elements of the war in Ukraine was electricity, and that

Poles were not passive victims of Russia's energy policy. During the "We Fight for Poland's Independence" campaign, attention was drawn to the Polish government's efforts to protect citizens from high energy prices, which included the anti-inflation and solidarity shield and the carbon allowance (Polski Komitet Energii Elektrycznej, 3.09.2022). It was also emphasized that Poland has been implementing infrastructure projects to strengthen the country's energy security for more than a decade.

In an effort to build the Polish public's informational resilience to fake news in the energy sector, the Polish Association of Combined Heat and Power organized the "Energy For Real" campaign in 2023. The initiative was also a response to the low level of knowledge about the energy transition among Poles, and therefore sought to provide the Polish public with knowledge about the energy sector in an accessible way, through appropriately selected communication tools. As part of the "Energy For Real" campaign, the Polish Association of Combined Heat and Power presented a report titled "Mądry Polak po fake'u" ("A wise Pole after a fake"). The research showed that in the infosphere, the daily reach of online publications on energy was about 14.5 million potential contacts. In turn, one in five energy-related messages appearing online was disinformation. Topics related to coal, gas, nuclear power, energy prices, energy company profits, and EU climate and energy policy assumptions were most vulnerable to disinformation efforts (Polskie Towarzystwo Energetyki Cieplnej, 3.04.2024). The complexity of energy issues has fostered the dissemination of untruthful content in the digital space. The strength of fake news was and still is the speed of its dissemination, as well as Internet users' trust in the credibility of the message and limited willingness to verify it. That is why it was so important to organize campaigns that conveyed reliable knowledge about the energy industry to the public in an accessible way. All the more so because a social survey conducted by the Polish Association of Combined Heat and Power Plants in August 2023 showed that Poles support the energy transition, as evidenced by their high support for nuclear power development (56%), RES development (80%) (Polskie Towarzystwo Elektrociepłowni Zawodowych, 1.10.2024).

Given the importance, timeliness, and challenges of the energy transition for the Polish economy and society, in March 2025, the Energy Market Agency and the industry portal CIRE launched an educational and social campaign "Energetyka bez znaków zapytania" ("Energy without question marks"). The authors of the campaign have not yet stated how long it will last, but they indicate that they plan a series of informative and expert activities. Since the start of the war in Ukraine, this has been another initiative focused on providing reliable and understandable information to help citizens better understand the directions of change in the energy sector. The aim of the "Energy without question marks" campaign, as well as the "We fight for Poland's independence," "We save energy" campaign, is to increase public awareness of the challenges posed by the energy transition and to counter misinformation in the energy and climate sector (Centrum Informacji o Rynku Energii, 31.03.2025). Taking into account the initiatives presented above, it should be noted that all of them are part of the process of building the country's energy resilience, combating misinformation in this area, and forming public support for Poland's energy and climate policy.

5 Conclusions

In 2022–2025, disinformation in the energy and climate area was a fairly common phenomenon in the Polish digital space. Disinformation activities carried out by the Russian Federation concerned the negative consequences of the energy transition, high costs for energy, and difficulties with the supply of energy resources. The dissemination of false information in the digital space was aimed at creating public fear, distrust of state institutions, and undermining support for green transition efforts. In response to the growing threat of disinformation, the European Union has taken decisions aimed at protecting the energy and climate sector. The adoption of the Digital Services Act (DSA), which introduced new obligations for online platforms to combat disinformation, as well as numerous information campaigns and public awareness initiatives in the area of information security, was an important factor. The Polish government has been working on the implementation of the DSA regulations and has conducted educational campaigns to warn the public against false content appearing in the digital space. The Ministries of Climate and Environment and Digitization have demonstrated a special role by focusing on increasing citizens' information resilience. Combating disinformation has contributed to reducing citizens' vulnerability to information manipulation. However, despite intensive efforts at both the EU and national levels, the thesis formulated should be considered valid. The effectiveness of the actions taken by the EU and Poland in combating disinformation in the energy and climate sector was limited. The reason for this was the intensive nature of disinformation activities carried out by the Russian Federation, which effectively used social media as a propaganda tool to undermine citizens' trust in the institutions responsible for the energy transition and climate policy.

References

Centrum Informacji o Rynku Energii (2025, March 31). *Rusza kampania edukacyjna "Energetyka bez znaków zapytania."* https://cire-cafe.cire.pl/artykuly/serwis-informacyjny-cire-24/rusza-kampania-edukacyjna-energetyka-bez-znakow-zapytania.

Chomsky, N., & Herman, E. S. (2002). *Manufacturing consent: The political economy of the mass media.* Pantheon Books.

Digital Poland (2024, February 23). *Młodzi Polacy bezradni wobec dezinformacyjnego chaosu.* https://digitalpoland.prowly.com/178556-mlodzi-polacy-bezradni-wobec-dezinformacyjnego-chaosu.

Disinfo (2022). *The EU energy sanctions hit Europe, while Russia is drowning in money.* https://euvsdisinfo.eu/report/the-eu-energy-sanctions-hit-europe-while-russia-is-drowning-in-money/.

Enefirst Plus (2024). *Energy Efficiency First: 2023 status. EU policy update and pilot country analysis.* https://ieecp.org/wp-content/uploads/2024/04/D2-1_ENEFIRSTplus_PolicyReview2024.pdf.

European Commission (2018, April 26). *Tackling online disinformation: A European approach* [Communication]. https://eur-lex.europa.eu/legal-content/EN/TXT/PDF/?uri=CELEX:52018DC0236.

European Commission (2022, November 16). *Digital Services Act: EU's landmark rules for online platforms enter into force.* https://ec.europa.eu/commission/presscorner/detail/en/ip_22_6906.

European Parliament (2022). *European Parliament resolution of 9 March 2022 on foreign interference in all democratic processes in the European Union, including disinformation (2020/2268(INI))* [Resolution]. https://eur-lex.europa.eu/legal-content/EN/TXT/?uri=CELEX:52022IP0064.

European Parliament (2023). *European Parliament resolution of 1 June 2023 on foreign interference in all democratic processes in the European Union, including disinformation (2022/2075(INI))* [Resolution]. https://eur-lex.europa.eu/legal-content/EN/TXT/?uri=CELEX:52023IP0219.

Executive Magazine (2023). *Około 20 proc. treści dotyczących energetyki w polskim Internecie to dezinformacja. Branża postanowiła zareagować.* https://executivemagazine.pl/przemysl-logistyka/okolo-20-proc-tresci-dotyczacych-energetyki-w-polskim-internecie-to-dezinformacja-branza-postanowila-zareagowac/.

IBIMS & IBRIS (2021). *Skąd Polacy czerpią informacje? Raport IBIMS i IBRIS.* https://ibims.pl/skad-polacy-czerpia-informacje-o-polsce-i-swiecie-raport-ibims-i-ibris/.

Ministerstwo Cyfryzacji (2024, December 19). *Cyfrowe priorytety polskiej prezydencji w Radzie Unii Europejskiej.* https://www.gov.pl/web/cyfryzacja/cyfrowe-priorytety-polskiej-prezydencji-w-radzie-unii-europejskiej/.

Ministerstwo Cyfryzacji (2025). *Prace nad ochroną użytkowników internetu – implementacja unijnego Aktu o usługach cyfrowych (DSA)*. https://www.gov.pl/web/cyfryzacja/prace-nad-ochrona-uzytkownikow-internetu–implementacja-unijnego-aktu-o-uslugach-cyfrowych-dsa.

Ministerstwo Klimatu i Środowiska (2022, September 23). *"Oszczędzamy energię" – kampania społeczna Ministerstwa Klimatu i Środowiska*. https://www.gov.pl/web/edukacja-ekologiczna/oszczedzamy-energie-kampania-spoleczna-ministerstwa-klimatu-i-srodowiska.

Ministerstwo Klimatu i Środowiska (2024, December 20). *Priorytety Ministerstwa Klimatu i Środowiska podczas polskiej prezydencji w Radzie UE*. https://www.gov.pl/web/klimat/priorytety-ministerstwa-klimatu-i-srodowiska-podczas-polskiej-prezydencji-w-radzie-ue/.

Nowa Energia (2023, November 16). *Rusza PAP FakeHunter Challenge: energia i klimat – internauci na tropie fake newsów w energetyce*. https://nowa-energia.com.pl/2023/11/16/rusza-pap-fakehunter-challenge-energia-i-klimat-internauci-na-tropie-fake-newsow-w-energetyce/.

Polish Electricity Committee (2022). *PKEE campaign against Russian disinformation in the energy sector launches*. https://pkee.pl/en/aktualnosci/rusza-kampania-pkee-przeciwko-rosyjskiej-dezinformacji-w-energetyce/.

Polski Komitet Energii Elektrycznej (2022, September 3). *Rusza kampania PKEE przeciwko rosyjskiej dezinformacji w energetyce*. https://pkee.pl/aktualnosci/rusza-kampania-pkee-przeciwko-rosyjskiej-dezinformacji-w-energetyce/.

Polskie Towarzystwo Elektrociepłowni Zawodowych (2024, October 1). *Rekomendacje Polskiego Towarzystwa Elektrociepłowni Zawodowych dla Parlamentu Europejskiego i nowej Komisji Europejskiej, z uwzględnieniem Polskiej Prezydencji w Radzie UE*. https://ptec.org.pl/rekomendacje-polskiego-towarzystwa-elektrocieplowni-zawodowych-dla-parlamentu-europejskiego-i-nowej-komisji-europejskiej-z-uwzglednieniem-polskiej-prezydencji-w-radzie-ue/.

Polskie Towarzystwo Energetyki Cieplnej (2024, April 3). *PTEZ zaprezentowało raport "Mądry Polak po fake'u."* https://ptec.org.pl/ptez-zaprezentowalo-raport-madry-polak-po-fakeu/.

Shaping Europe's Digital Future (2022). *Code of Practice on Disinformation* (2022). https://digital-strategy.ec.europa.eu/en/policies/code-practice-disinformation.

Skalamera, M. (2023). The geopolitics of energy after the invasion of Ukraine. *The Washington Quarterly*, *46*(1), 7–24. https://doi.org/10.1080/0163660X.2023.2190632.

Vakulchuk, R., Overland, I., & Scholten, D. (2020). Renewable energy and geopolitics: A review. *Renewable and Sustainable Energy Reviews*, *122*, 109547. https://doi.org/10.1016/j.rser.2019.109547.

CHAPTER 25

EU Hydrogen Dreams Are Central European Nightmare: Lack of Hydrogen and Many Challenges for Its Import

Šárka Waisová

Abstract The chapter analyzes the EU's policy on Critical Raw Materials (CRM) and its efforts to reduce dependencies and vulnerabilities across different stages of the CRM value chain. It focuses on materials essential for the green and digital transition, as well as for defense and space applications, which are characterized by high supply risk and limited substitutes. The study identifies key challenges in EU CRM policy, including insufficient monitoring of supply chains and risk preparedness, underutilization of domestic extraction, processing, and recycling potential, and limited access to finance due to public perceptions of the mining sector and the long-term nature of investment returns. These issues highlight persistent structural gaps that future EU Presidencies and institutions will need to address to strengthen the resilience and sustainability of CRM supply chains.

Keywords Critical Raw Materials – EU – CRM's value chain – strategic agenda

1 Introduction

Renewable hydrogen is a key building block in the *EU's 2040 Climate* goals and energy security policy. The EU Hydrogen Strategy 2020 envisaged that the EU would produce up to 1 million tons of renewable hydrogen in 2024 and set a target of up to 10 million tons of renewable hydrogen in 2030. REPowerEU, which responded to Russia's invasion of Ukraine, proposed 10 million tons to be produced in the EU in 2030 and another 10 million tons to be imported. In February 2024, the European Commission (EC) confirmed that hydrogen will play a key role in decarbonizing hard-to-abate industries and transport, with production of 20 to 35 million tons (Mt) of renewable hydrogen expected by 2040, depending on the scenario chosen. To achieve the hydrogen targets, the EC has adopted more than a dozen legislative and non-legislative acts.

 | HTTPS://DOI.ORG/10.1163/9789004763623-027

Some of them (FuelEU Maritime, ReFuelEU Aviation, or the RED 3) directly oblige Member States to use a certain amount of renewable hydrogen.

The development of hydrogen use in the EU requires that Member States have sufficient volume of hydrogen at affordable prices within a reasonable time. This requirement is linked to both the European Green Deal and security issues, including technological sovereignty. However, there exist differences among Member States in terms of how much hydrogen will be needed, by whom, and the starting points for domestic production. The demand for hydrogen is determined by the nature and intensity of industry, and the possibilities for domestic production by climatic and physical-geographical conditions. EU industrialized countries with limited RES production capacity will have to import hydrogen.

Hydrogen imports were already foreseen in the first EU hydrogen plans. Hydrogen trade and transport are planned both between EU countries and from outside the EU. In the last two years, there have been a dozen techno-economic studies looking at the development of transport technologies and their level of commercialization, and transport costs, as well as a dozen studies looking at the production potential of future export destinations. However, there is a lack of studies addressing the geopolitical aspects of the emerging hydrogen market (exceptions are Pepe et al., 2023; Quitzow & Zabanova, 2024), hydrogen import policies, and, finally, the challenges that EU countries relying on hydrogen imports will have to face.

The aim of this chapter is to fill one of these gaps and to identify and analyze the challenges that hydrogen-importing countries will face around 2030. A sufficient supply of hydrogen is essential to meeting EU and national targets and commitments regarding decarbonization, the energy transition, and energy security. It is already clear that hydrogen importers will include (in order of expected import volumes) Germany, the Netherlands, Poland, Belgium, Italy, and the landlocked Central European countries, mainly the Czech Republic, Slovakia, and Austria (Ausfelder et al., 2024).

To thoroughly explore the challenges of hydrogen import and threats to hydrogen value chain, we have chosen one of the future importers as a case study. The Czech Republic is the country with the comparatively worst conditions for domestic renewable hydrogen production (insufficient year-round renewable capacity), and at the same time, due to its geographical location and the fact that none of its neighbours will produce enough hydrogen to supply, the country will always be dependent on transit countries. We have chosen the 2030 outlook, when the first of the mandatory hydrogen targets set by EU legislation will come into effect. For all importers, it is important that secure and well-functioning hydrogen value chains exist. However, a single state does not have the capability to establish and maintain these chains; the European Commission and the EU Presidency play a crucial role here. The EU has already taken initial steps to secure hydrogen value chains. The Gas Package has been approved, the Commission is supporting the construction of the European Hydrogen Backbone, and a trade agreement has been concluded with Saudi Arabia, a potential supplier of renewable hydrogen. However, these are only initial steps, and more will have to follow.

The chapter is structured in five parts. The first part deals with the EU policy towards a hydrogen economy, including the policy-strategic and legislative framework of the EU hydrogen ecosystem. These provide the main framework for the development of the hydrogen market and the import of hydrogen and its derivatives. The second part explores the current EU hydrogen landscape, in particular the status and prospects for renewable hydrogen production and demand. The third part briefly introduces the issue of ensuring the security of supply chains for energy raw materials. Hydrogen serves as both an energy carrier and an essential raw material for the chemical industry.

The fourth part presents the core and consists of a case study. We present the Czech hydrogen situation and plans, and then focus on the challenges the Czech Republic will have to face to secure its hydrogen supply from abroad. The fifth part concludes by discussing the findings on import challenges.

Hydrogen import policies in the EU present a new research topic, and the number of studies dealing with them is still limited. The most developed topics are technological, e.g. in which form and with which technological solutions renewable hydrogen can be transported and what transport costs can be expected (Kappeler et al., 2023; Galimova et al., 2023). Less attention has so far been paid to the source countries and the conditions under which these countries can supply renewable hydrogen to the EU (Abdelshafy et al., 2024; Ourya et al., 2023). The least attention has been paid to the analysis of hydrogen import policies and the challenges that international hydrogen trade will face. Most of the studies on this topic could be found in Germany, whereas Berlin will be by far the largest hydrogen importer in Europe (Wietschel et al., 2024).

Due to the novelty of the topic and its diverse reflections, this chapter is designed as a qualitative, explorative, evidence-based case study. This approach allows empirical investigation that examines a current hydrogen import situation in its real-world context and opens the door for a comprehensive overview of the context and subject of the study.

2 EU Policy for a Hydrogen Economy

The EU renewable hydrogen ecosystem started to take shape around 2019, with the adoption of the European Green Deal. It currently consists of a network of strategies, policies, legislative and non-legislative acts, and support programs. These include the European Hydrogen Bank (producer support), Important Projects of Common European Interest (demand support), and the Clean Hydrogen Partnership (R&D support)[1]. A specific tool is the promotion of so-called hydrogen valleys, which brings together local hydrogen producers and consumers[2]. The Commission envisages 50 hydrogen valleys in the EU by 2040 (SWD, 2024, p.159).

Looking at the development of the EU hydrogen economy since 2019, we are currently at a turning point. Three major reports assessing the development of the EU hydrogen economy were published in the second half of 2024: the Draghi report, the Special Report of the European Court of Auditors, and the ACER Report. The reports agree that the Commission's initial hydrogen plans were overly optimistic and ambitious and call on the Commission to realistically reassess (moderate) both the hydrogen targets and the policy and regulatory measures. The fact that the development of the hydrogen economy is facing problems was also indicated by developments in several Member States. For example, in September 2024, the Norwegian company Equinor cancelled a project to build a hydrogen pipeline to Germany, arguing that the demand for renewable hydrogen has been growing more slowly than expected. In the same month, Denmark's Energinet announced that it would delay the rollout of a hydrogen pipeline between western Denmark and northern Germany by three years, from 2028 to 2031. In December 2024, Dutch network developer Gasunie announced that the planned national hydrogen pipeline network will be delayed by three years compared to previous plans and will not be completed until 2033. Expert reports and information on project cancellations and postponements suggest that a fundamental debate on the future direction of EU hydrogen policy is urgently needed. The EU aims to reduce its dependence on imports of Russian fossil fuels. Renewable hydrogen has been identified as one of the key tools for diversifying energy supply. In

1 In 2024, the Bank launched its first hydrogen auction. Seven projects were selected to produce a total of 1.58 million tons of renewable hydrogen over 10 years. The winning projects are from Spain, Portugal, Finland, and Norway, countries with abundant RES. https://ec.europa.eu/commission/presscorner/detail/en/ip_24_2333.

2 Hydrogen valleys are defined as geographic areas where several hydrogen applications are combined into an integrated hydrogen ecosystem that consumes a significant amount of hydrogen.

TABLE 25.1 Key EU Legislative Acts supporting the development of the hydrogen economy and climate transition (2020–2024) (Source: Own elaboration.)

Abbreviated title/ EU initiative	Full title of the EU Legal Act
EU Taxonomy	Regulation 2020/852 on the establishment of a framework to facilitate sustainable investment
EU ETS 2	Directive 2023/959 establishing a system for greenhouse gas emission allowance trading within the Union
AFIR	Regulation 2023/1804 on the deployment of alternative fuels infrastructure
CBAM	Regulation 2023/956 establishing a carbon border adjustment mechanism
FuelEU Maritime	Regulation 2023/1805 on the use of renewable and low-carbon fuels in maritime transport
RED 3	Directive 2023/2413 as regards the promotion of energy from renewable sources
ReFuelEU Aviation	Regulation 2023/2405 on ensuring a level playing field for sustainable air transport
CRMA	Regulation 2024/1252 establishing a framework for ensuring a secure and sustainable supply of critical raw materials
NZIA	Regulation 2024/1735 on establishing a framework of measures for strengthening Europe's net-zero technology manufacturing ecosystem
Gas Package	Regulation 2024/1789 on the internal markets for renewable gas, natural gas, and hydrogen Directive 2024/1788 on common rules for the internal markets for renewable gas, natural gas, and hydrogen
DA RFNBO	Delegated Regulation 2023/1184 establishing a Union methodology setting out detailed rules for the production of renewable liquid and gaseous transport fuels of non-biological origin

TABLE 25.1 Key EU Legislative Acts supporting the development of the hydrogen (*cont.*)

Abbreviated title/ EU initiative	Full title of the EU Legal Act
DA GHG	Delegated Regulation 2023/1185 establishing a minimum threshold for greenhouse gas emissions savings of recycled carbon fuels and by specifying a methodology for assessing greenhouse gas emissions savings from renewable liquid and gaseous transport fuels of non-biological origin and from recycled carbon fuels
DA EU ETS	Delegated Regulation 2024/873 regards transitional Union-wide rules for harmonized free allocation of emission allowances
DA LCH	Delegated Regulation 2025/4674 final supplementing Directive (EU) 2024/1788 specifying a methodology for assessing greenhouse gas emissions savings from low-carbon fuels

2023, the European Commission presented the Green Deal Industrial Plan, which aims to strengthen the competitiveness of the European climate-neutral industry. Renewable hydrogen plays an important role in this context as a driver of industrial decarbonization. Among the EU's legislative actions is the "Fit-for-55" package, which sets out to reduce greenhouse gas emissions by 55% by 2030 compared to 1990 levels, with renewable hydrogen being one of the main tools to achieve this goal. Additionally, in 2023–2024, a series of legal acts were adopted to support the development of the hydrogen market and low-emission technologies, including: ReFuelEU Aviation, FuelEU Maritime, the Critical Raw Materials Act (CRMA), the Net-Zero Industry Act (NZIA), and the revision of the EU Emissions Trading System (EU ETS 2). These legislative actions are complemented by delegated acts, such as those to the RED II directive (e.g. concerning the definition of renewable fuels of non-biological origin – RFNBOs – and methodologies for assessing greenhouse gas emissions). Work is also ongoing on the Gas Package, which includes new regulations to decarbonize the gas sector, including the integration of hydrogen into existing transmission networks. The EU's actions on hydrogen therefore include both strategic and political initiatives as well as specific legislative mechanisms and sectoral regulations. This approach is integrated and long-term, addressing the needs of energy transition, energy security, and industrial competitiveness.

3 The EU Hydrogen Landscape

Hydrogen and its derivatives are currently used in the EU mainly as a raw material for the chemical industry (production of fuels and fertilizers). The largest producers and consumers of hydrogen (in order of consumption volume) are Germany, the Netherlands, Poland, Spain, and Italy. Most hydrogen (90%) is produced by steam reforming of natural gas and by partial oxidation of oil residues in Central Europe (the Czech Republic and Poland). In 2023, 11.3 Mt of hydrogen were produced in the EU, of which 39 kt was renewable hydrogen (EHO, 2024). Renewable hydrogen consumption is growing slowly. The greatest progress is being made in the chemical industry, where hydrogen was already being used and is now "only" being replaced by renewable hydrogen.

Within the EU, Germany and the Netherlands belong among the countries with the most active hydrogen policies. Berlin has already updated its national hydrogen strategy twice, was the first country in Europe to introduce a hydrogen import strategy (July 2024), and has included support for the construction of hydrogen projects in developing regions in its international development agenda. Domestically, the German government has subsequently developed a comprehensive system of instruments and measures to support both production and demand (NOW, n.d.). The Netherlands has taken a similar approach. Amsterdam is a signatory to more than three dozen international hydrogen agreements, in particular pre-contractual agreements on hydrogen offtake and cooperation in hydrogen transport and R&D of hydrogen applications. The country is also home to Europe's oldest large-scale hydrogen valley, "HAEVENN" (2019). In both countries, the first kilometres of new hydrogen pipelines are under construction, and dozens of kilometers of pipelines are being rebuilt.

The localization of the European industry suggests that renewable hydrogen will be most in demand in northeastern France, Belgium, the Netherlands, northwestern Germany, and central Poland. Based on national hydrogen strategies, Belgium is expected to require 2–6 TWh of renewable hydrogen in 2030, France 23 TWh, Germany 95–130 TWh, and Poland 13.4 TWh (IEA, 2024). The Netherlands does not provide demand projections for 2030. In addition, almost 80% of the total planned electrolyzer capacity is expected to be installed in the industrial regions of Northwest Europe (ECA, 2024, p. 27). However, calculations indicate that the amount of renewable hydrogen production that the Northwest European countries and Poland will be able to achieve will not be sufficient to meet EU demand.

The only solution is to import renewable hydrogen from non-EU regions where renewable hydrogen can be produced all year round and at prices close to 3€ per kilogram thanks to abundant renewable resources (the UK, Norway, Saudi Arabia, Morocco, Algeria, Canada) (Agora Industry, 2024; IEA, 2023b). In order for renewable hydrogen and its derivatives to be imported into Europe, several key issues need to be addressed. These are addressed in the Czech Republic case study.

4 What Is Necessary to Ensure the Security of Hydrogen Value Chains?

Value Chains for energy resources consist of the interconnected, international network of activities – from design and extraction to refining, distribution, and consumption – that create and deliver energy products, involving dispersed firms, countries, and stages like fuel procurement (e.g. hydrogen), generation, transmission, and final market delivery, shaping global trade, technology transfer, and material infrastructure (IEA, 2025; IRENA, 2024). Global and regional hydrogen value chains are still in their infancy. Currently, we are at a stage where physical transport infrastructure, production capacities, and rules for the functioning of the hydrogen market are being established in both supplier and consumer countries. In Europe, the European Hydrogen Backbone as well as port terminals for liquid ammonia, methanol, and hydrogen in coastal countries such as the Netherlands, Germany, and Italy are being built (IEA, 2025). Based on the Gas Package, the European Network of Network Operators for Hydrogen (ENNO-H) will be the organization for the cooperation of the EU Hydrogen Transmission Network Operators.

Despite all the steps taken, there are uncertainties and threats regarding the existence and functioning of the hydrogen value chain in Europe and non-European regions, such as water demand and supply for hydrogen production, regulatory frameworks, or jobs and skill development, to name but a few (IRENA, 2024). Ensuring the security of hydrogen value chains is more challenging than ensuring the security of existing energy resource value chains, such as those for oil or gas, because new business links, production capacities, transport routes, and consumer bases must be established from ground zero. Thus, hydrogen value chains are exposed to extraordinary pressure and challenges. Key threats include delays in constructing production capacities in future supplier countries, delays in constructing the European hydrogen transport network and associated port terminals, delays in constructing large tankers capable of transporting hydrogen and its derivatives, and the emergence of new armed conflicts along transport routes. Many uncertainties and threats can be mitigated or eliminated through joint EU measures, cooperation between Member States, and a more active approach by the European Commission and Presidency countries. When discussing hydrogen value chains, the Presidencies and the Commission should consider the situation of EU countries with the worst conditions for producing domestic renewable hydrogen, rather than those with favorable climatic conditions and sufficient renewable energy sources (RES) that allow them to produce large volumes of cheap renewable hydrogen all year round. As demonstrated by the second auction of the European Hydrogen Bank (2025), even in countries with sufficient RES, there are not enough interested investors in building production capacities, and even if there are, they are unable to produce renewable hydrogen at a price and in quantities that would be satisfactory for European industry.

Using the case of the Czech Republic, we will analyze the threats and challenges that future importers of renewable hydrogen will have to address.

5 Challenges for EU Hydrogen Importers: Case Study of the Czech Republic

In this subsection, we will first present the Czech hydrogen situation and plans for 2030 and then focus on analyzing the challenges and barriers to hydrogen imports. The Czech Republic presented its first hydrogen strategy in 2021; the strategy was significantly updated in 2024. Between 2020 and 2024, about 120,000 tons of hydrogen were produced and consumed annually in the Czech Republic (Waisová, 2024). The largest producer and consumer (about 85%) is Orlen Unipetrol. The share of renewable hydrogen in 2024 was less than 1%. According to the Czech Hydrogen Strategy (Vodíková strategie ČR, 2024, p. 7), the consumption of renewable hydrogen should reach 20 thousand tons per year around 2030, half of which should be imported. In the Czech Republic, new electrolyzers have been built in the last two years, new hydrogen projects are under construction, and the government has approved several programs to support the construction of electrolyzers and the use of renewable hydrogen in industry and transportation. However, the production capacity of existing and planned projects is still significantly lower than mentioned in the strategy. To meet mandatory EU climate and hydrogen targets, the country will need to purchase and import renewable hydrogen from the international market.

For the analysis of challenges and constraints, we chose the four-dimensional framework of the US National Research Council (2008). NRC identified four dimensions of commodity availability challenges: technological, economic, political, and socio-environmental. The NRC has also identified a geological dimension, i.e. there are mineral deposits. However, this does not apply to renewable hydrogen, and we have excluded this dimension.

5.1 Technological Challenges to Hydrogen Availability

The analysis of technological challenges considers the state of technology, knowledge, and infrastructure for processing and transporting hydrogen. We will focus first on the hydrogen production challenges and then on the transportation challenges. Renewable hydrogen is produced in electrolyzers powered by renewable electricity. Several types of electrolyzers are now available, and large-scale production of renewable hydrogen is already underway in many parts of the world. One of the challenges that the EU will have to address is the availability of electrolyzers. Currently, the largest producer is China, which can produce electrolyzers two to five times cheaper than the EU, the United States, or Canada (Lichner, 2024). The Commission responded to China's dominance in the electrolyzer market in 2024 by adopting a regulation establishing a framework of measures for strengthening Europe's net-zero technology manufacturing ecosystem (NZIA), which aims to increase European manufacturing capacity for net-zero technologies, including hydrogen technologies such as electrolyzers and fuel cells, and by announcing the second hydrogen auction, which restricts the use of Chinese components in EU-funded projects. However, this does not change the fact that China is able to produce large electrolyzers more cheaply. While bans on Chinese electrolyzer purchases may partially support EU manufacturers, for the hydrogen project owners themselves, such a practice will lead to higher project prices and thus higher hydrogen production costs.

The second group of technological challenges is the transport of large volumes of hydrogen over long distances. Several solutions can be used, but not all of them are sufficiently developed to process and transport large volumes of hydrogen (e.g. cracking large volumes of ammonia). In addition, some solutions are costly (hydrogen liquefaction and liquid hydrogen transport). Hydrogen imports and transport costs will depend heavily on technological innovation in the coming years. Without it, it will not be possible to diversify the means of transporting hydrogen and make the whole process sustainable.

Another technological or techno-political challenge is the development of the transportation infrastructure. Given the physical and geographical location of the Czech Republic and the hydrogen situation in neighbouring countries, the country will use hydrogen pipelines or a combination of shipping and hydrogen pipelines to import hydrogen. To support the development of the transport infrastructure, the Commission has initiated the creation of the so-called European Hydrogen Backbone (EHB), a pan-European hydrogen transport network. This will consist partly of upgraded pipelines and partly of new hydrogen pipelines. The national gas network operators are responsible for the construction of the pipelines. However, the construction of the EHB is being delayed in some places. According to the original plans, the northern route of the EHB from northeast Germany to the Czech Republic and the northwest route from the Netherlands via northwest Germany to Central Europe should be completed between 2030 and 2032 and around 2032–2035, respectively. However, Berlin announced in April 2024 in an update to the German Energy Management Act that some parts of the network on its territory may not be completed until 2037 (FNB Gas, 2024).

It is also unclear whether the German hydrogen pipelines will have sufficient transport capacity to satisfy both German and Czech customers. The completion of the southern route of the EHB from Italy via Austria to Central Europe is planned for 2035 and is irrelevant for hydrogen imports to the Czech Republic in 2030. Prague foresees hydrogen imports via Germany in the first phase and has already concluded several agreements with Berlin for the construction of cross-border interconnectors. Ship imports are a bit more complicated. As a landlocked country, the Czech Republic is dependent on port capacity in third countries. Prague will therefore have to negotiate access to port terminals, port storage capacity, and the reservation of relevant capacity on transport routes. Another challenge is the availability of large tankers to transport ammonia, methanol, or liquid hydrogen. There are only a few dozen large tankers, and current estimates suggest that new tankers will not be available before 2030.

5.2 Economic Challenges for the Hydrogen Availability

Economic challenges include the cost of hydrogen production and transportation. First, we introduce the production price problem. For renewable hydrogen to be competitive, its price must fall to around 3€ per kilogram. There is currently no location in the EU where such a production price can be achieved. The lowest production prices are announced by projects in Spain, southern Norway, and Scotland, but even here, the price is above this value

(5 to 7€ per kilogram) (Agora Industry, 2024; Kerle et al., 2023). In Central Europe, the production price of renewable hydrogen in 2024 was 12 to 15€ per kilogram Waisová, 2024). The reason for the high renewable hydrogen prices in the EU is the low production yield of RES, the cost of project construction, and the strict regulations for renewable hydrogen production, which do not allow sufficient year-round utilization of electrolyzers. Non-EU projects achieve lower production prices, but transport costs must be added.

Transportation costs are the second economic challenge. Transporting hydrogen over short distances by pipeline is the cheapest and most efficient way. However, once a certain distance is exceeded (2,000 to 3,000 km), pipeline transport becomes uncompetitive compared to the transport of hydrogen derivatives (Staudt et al., 2024; Ortiz Cebolla et al., 2022). At current prices and available technological solutions, it is economically feasible for the Czech Republic to purchase hydrogen (ammonia, methanol) only from the EU and its immediate surroundings (Scotland, Norway, North Africa, Saudi Arabia). In addition, the country will have to include transit fees in the cost of hydrogen imports. However, none of the EHB members has yet published data or plans on the value of transit fees.

5.3 Political Challenges for the Hydrogen Availability

Political challenges involve a wide range of factors and processes at national, EU, and international levels. At the national level, the biggest challenge is that the Czech Republic has not yet concluded any preliminary offtake agreements or memoranda of cooperation for hydrogen imports. This fact creates two risks: firstly, the available production capacities may be exhausted by other countries, and secondly, the Czech Republic will not have a guaranteed price for hydrogen. The second national challenge is to start the conversion of the gas network. No new hydrogen pipelines will be built in the country; converted pipelines will be used to connect to the EHB. The conversion has not yet started, although the owner of the gas system, NET4GAS, states that the conversion will be completed in 2029. The final national challenge is sufficient demand. For hydrogen imports to make economic sense and be feasible, a certain amount of hydrogen must be supplied domestically. In the Czech Republic, demand for renewable hydrogen is growing slowly, and some potential users have postponed projects due to the high price of hydrogen. So far, government support has been directed at producers, not at stimulating demand. The Czech Republic is thus stuck in a chicken-and-egg trap: demand is not growing because the price of hydrogen is high, and potential producers are postponing projects because demand is uncertain.

At the EU level, the main policy challenge is to create a friendly and functioning framework for the hydrogen market. Although the so-called Gas Package adopted in the summer of 2024 created the framework conditions for the renewable gases market, including the establishment of the European Network of Network Operators for Hydrogen (ENNOH) and the definition of its tasks and responsibilities, as well as the appointment of a supervisory authority for the hydrogen market – ACER, several issues regarding the functioning of the EU hydrogen market remain unclear.

Several challenges for Czech imports also arise from the international level. These are essentially the same challenges that face the international trade in natural gas and oil: geopolitical disruptions such as blockage of sea lines of communication and choke points, attacks by pirates, insurgents or terrorists on ships or hydrogen facilities, the creation of new oligopolies where market power is in the hands of a small number of powerful producers, import dependence, export restrictions, and dominance of one actor in one segment of the hydrogen value chain. Here the country will have to rely on the ability of the EU and other international organizations (e.g. WTO) to organize and regulate the international hydrogen market.

5.4 Socio-environmental Challenges for the Hydrogen Availability

Socio-environmental challenges include ecological attributes and the community acceptance of hydrogen project development. The socio-economic challenges of the hydrogen market are among the least explored topics. It is only in the last two years that the first studies on the so-called environmental footprint of the hydrogen value chain have been published (Arrigoni et al., 2024), and the first debates are emerging, e.g. on the hydrogen projects' impacts on water resources. In many areas with abundant renewable resources, there is a shortage of drinking water (North Africa, Namibia, Chile), and where desalinated seawater is planned to be used, it is not clear what impact the discharge of brine concentrate will have on local ecosystems. The issue of social acceptance of hydrogen projects is hardly explored. However, it can be expected that hydrogen projects will face the same challenges as the construction of other facilities (Ausfelder et al., 2024, p. 53). For the Czech Republic and other importers, this necessarily implies a well-considered selection of future suppliers, especially in terms of project location and level of project completion.

6 Conclusion

This chapter has reviewed the EU hydrogen policy, the situation in the emerging EU hydrogen market, and the security of the hydrogen value chain. EU legislation calls for the progressive achievement of full decarbonization and sets various mandatory hydrogen targets for Member States, which require having sufficient hydrogen available. Countries in the Central European region have limited capacity to produce sufficient renewable hydrogen throughout the year and will not be able to meet decarbonization or hydrogen targets without imports. The main objective of this chapter was to identify and explore the challenges that importers in Central Europe will need to address. The analysis identified more than a dozen challenges at all stages of the hydrogen value chain and showed that the development of hydrogen imports is a complex terrain. The large number of challenges and barriers puts Central European importers in a much more complicated situation than countries in Northwest and Southern Europe. Central Europe is dependent on the cooperation of/with other countries to build up production capacity and transport infrastructure, and in the period up to 2035 this will mainly involve cooperation with Germany and the Netherlands. The Czech Republic (but also Austria and Slovakia) will not be able to meet the EU hydrogen targets from domestic resources, in which case it will become a nightmare. Strengthening the security of hydrogen value chains and ensuring that all Member States have similar starting conditions for building a hydrogen ecosystem will be a major task for the current European Commission and future EU Presidencies.

References

Abdelshafy, A., Lambert, M., & Walther, G. (2024). *MENA region as a potential hydrogen supplier for the European market: Analysing a prospective route between Kingdom of Saudi Arabia and Germany.* Oxford Institute for Energy Studies.

Agora Industry (2024): *EU map of hydrogen production costs.* Model version 1.0, Berlin, 11.06.2024. www.agora-industry.org/data-tools/agoras-eu-map-of-hydrogen-production-costs#downloads.

Arrigoni, A., Dolci, F., Ortiz Cebolla, R., Weidner, E., D'agostini, T., Eynard, U., Santucci, V. & Mathieux, F. (2024). *Environmental life cycle assessment (LCA) comparison of hydrogen delivery options within Europe.* Publications Office of the European Union, Luxembourg. doi:10.2760/5459, JRC137953.

Ausfelder, F., Ragwitz, M., Förster, M.Y., & Nolden, Ch. (Eds.) (2024). *European Hydrogen Infrastructure Planning. Insights from the TransHyDE Project System Analysis.*

https://www.wasserstoff-leitprojekte.de/lw_resource/datapool/systemfiles/elements/files/148FFEF003673B67E0637E695E8625E5/live/document/20240321_European_Hydrogen_Infrastructure_Planning.pdf.

EC (2022). *State Aid: Commission approves up to €5.4 billion of public support by fifteen Member States for an Important Project of Common European Interest in the hydrogen technology value chain.* ec.europa.eu/commission/presscorner/detail/en/ip_22_4544.

ECA (2024). *The EU's industrial policy on renewable hydrogen legal framework has been mostly adopted – time for a reality check.* Special report 11/2024.

EHO (2024). *The European Hydrogen market landscape.* observatory.clean-hydrogen.europa.eu/hydrogen-landscape/end-use/hydrogen-demand.

Galimova, T., Fasihi, M., D., & Breyer, Ch. (2023). Impact of international transportation chains on cost of green e-hydrogen: Global cost of hydrogen and consequences for Germany and Finland. *Applied Energy,* 347, 1–19. https://doi.org/10.1016/j.apenergy.2023.121369.

IEA (2024). *Northwest European Hydrogen Monitor 2024.* https://www.iea.org/reports/northwest-european-hydrogen-monitor-2024.

IEA (2023a). *Global Hydrogen Review 2023.* https://www.iea.org/reports/global-hydrogen-review-2023.

IEA (2023b). *Levelised Cost of Hydrogen Map.* www.iea.org/data-and-statistics/data-tools/levelised-cost-of-hydrogen-maps.

IEA (2024a). *Global Hydrogen Review 2024.* www.iea.org/reports/global-hydrogen-review-2024.

IEA (2025). *Global Hydrogen Review 2025.* www.iea.org/reports/global-hydrogen-review-2025.

IRENA (2024). *Shaping Sustainable International Hydrogen Value Chains.* International Renewable Energy Agency, Abu Dhabi.

Kapeller, J., Reuter, S. & Ortmann, P. (2023). *Zusatzbetrachtungen zur Studie Importmoeglichteiten fuer erneubaren Wasserstoff. Endbericht.* Austrian Institute of Technology. www.bmk.gv.at/themen/energie/publikationen/importmoeglichkeiten.html.

Kerle, F., Herborn, M. & Prickett S. (2023). Cost reduction pathways of green hydrogen production in Scotland – total costs and international comparisons. *Climate Change.* http://dx.doi.org/10.7488/era/3841.

Lichner, C. (2024). Electrolyzer prices – what to expect. *PV Magazine.* www.pv-magazine.com/2024/03/21/electrolyzer-prices-what-to-expect/.

NOW (n.d.). www.now-gmbh.de/ueber-uns/.

NRC (2008). *Minerals, Critical Minerals, and the U.S. Economy.* Washington, DC: The National Academies Press.

Ortiz Cebolla, R., Dolci, F., & Weidner, E. (2022). *Assessment of Hydrogen Delivery Options. Feasibility of Transport of Green Hydrogen Within Europe.* JRC Technical Report.

Ourya I., Nabil, N., Abderafi, S., Boutammachte, N., & Rachidi, S. (2023). Assessment of green hydrogen production in Morocco, using hybrid renewable sources (PV and wind). *International Journal of Hydrogen Energy*, 96, 37428–37442.

Pepe, J.M., Ansari, D., & Gehrung, R.M. (2023). Die Geopolitik des Wasserstoffs. *SWP-Studie 2023/S 14.* doi:10.18449/2023S14v02.

Quitzow, R., & Zabanova, Y. (Eds.). (2024). *The Geopolitics of Hydrogen. Studies in Energy, Resource and Environmental Economics.* Cham: Springer.

Staudt, Ch., Hofsaess, C., von Lewinski, B. & Moers, F. (2024). *Kurzstudie zur Transportoptionen von Wassertoff.* DVGW-EBI.

SWD(2024) 159: COMMISSION STAFF WORKING DOCUMENT Towards a roadmap for accelerating the deployment of Hydrogen Valleys across Europe: challenges and opportunities. research-and-innovation.ec.europa.eu/document/download/e5e75789-d4d8-42aa-8c99-3f33b2f5b935_en?filename=ec_rtd_swd-2024-159-f1.pdf.

Vodíková strategie ČR (2024). https://mpo.gov.cz/assets/cz/prumysl/strategicke-projekty/2024/7/Vodikova-strategie-CR-aktualizace-2024.pdf.

Waisová, Š. (2024). Legislativní a politicko-strategické rámce EU s vlivem na rozvoj vodíkové mobility (2023/2024). *Czech Hydrogen Technological Platform.* https://www.cistadoprava.cz/files/NAHYC_DP010-V05-O-Legislativni_a_politicko-strategicke_ramce_EU.pdf.

Wietschel, M., Weißenburger, B., Wachsmuth, J. & Müller, V.P. (2024). *Was wissen wir über Importe von grünem Wasserstoff und seinen Derivaten und was lässt sich daraus für eine deutsche Importstrategie ableiten?* Karlsruhe: Fraunhofer ISI.

FNB Gas (2024, October 22). *Wasserstoff-Kernnetz: Ein Meilenstein für die Energiewende.* https://fnb-gas.de/wasserstoffnetz-wasserstoff-kernnetz/.

CHAPTER 26

How the French Presidency Decisively Shaped the Critical Raw Materials Act

Lucas Miailhes and Andrew Glencross

Abstract The EU Critical Raw Materials Act (CRMA), prioritised during the French Presidency of the EU Council in early 2022, aimed to secure supplies of materials essential for clean energy technologies such as batteries and solar panels. The French government promoted the act as part of a broader strategy of European strategic autonomy, emphasising reduced dependency on external suppliers and strengthening EU-based mining. Based on interviews and policy analysis, the chapter shows how France advanced this agenda at EU level, contributing to a more geopolitical and geoeconomic approach to European integration, even if not all objectives were fully achieved.

Keywords critical raw materials – security of supply – EU strategic autonomy – geopolitical Europe – geoeconomics

1 Introduction

When Ursula von der Leyen took office as incoming European Commission President in 2019, she promised to lead a "geopolitical commission" that would leverage the EU's economic heft in order to become more assertive in international affairs (Haroche, 2023). This ambition built on the fact that the EU, thanks to its internal market, has an under-utilized market or regulatory power (Damro, 2012) that could complement its diplomatic strategy, especially in a time when the US-led liberal international order is under strain (Ikenberry, 2018). From a French perspective, this "geoeconomic" gambit corresponded with long-standing ambitions to shape European integration in a way that would integrate EU economic and foreign policy in order to create a distinct European position in the international order. Hence, when France assumed the rotating EU Council presidency in the first half of 2022, it did so

 HTTPS://DOI.ORG/10.1163/9789004763623-028

with the clear intention of promoting an EU-focused industrial policy that would be linked to a global diplomatic strategy.

The key to understanding why this move was considered necessary by French President Emmanuel Macron lies in the term "strategic dependence." This expression was used in the 2022 French presidency's policy programme in the section on defence, representing a deeper and essentially Gaullist tradition of conceptualizing international politics. Strategic dependence is a concept that goes hand-in-hand with another idea central to French defence policy: strategic autonomy. Taken together, these signify that a country such as France should have the means to defend its territory and interests without having to rely on foreign-sourced matériel. Since the late 1990s, France has promoted the development of more autonomous EU-capacities in foreign and defence policy, beginning with the Saint-Malodeclaration in 1998 (Bernard, 2024). The importance of having a common European approach to strategic autonomy was a key message in Macron's 2017 Sorbonne speech on EU integration. In it, he defined strategic autonomy to include key domains such as defense, border control, foreign relations, environmental transition, digital transformation, and economic and monetary power, thereby setting the tone for the future geopolitical commission of von der Leyen (Vicard & Wibaux, 2022).

This contribution does not intend to delve into all the complexities of the interconnected French ambitions in European industrial and defence policy. Rather, the ambition is limited to highlighting how French policy ideas shaped one specific instance where this interconnection is especially evident and powerful: critical raw materials. The evidence that the supply chain in critical raw materials (CRM) – the inputs for a host of industrial goods crucial for clean technology, industrial processes, and infrastructure – currently places the EU in a state of serious dependency is overwhelming. The EU produces only a tiny fraction (3%) of the 30 materials it has classified as critical, with more than half of these sourced from China (Danino-Perraud, 2021). In sharp contrast to the EU mining decline that started in the 1970s, China has been consolidating its leading position in various CRM markets through investments in midstream and downstream sectors, particularly since the early 2000s (Bonnet et al., 2022). It is precisely in this context that the French state began to develop a coalition of industrial and mining actors around the issue of security of supply beginning in the early 2010s (Chailleux, 2022). The timing coincided with the start of geopolitical concerns around sourcing rare earth materials from China, which the EU initially responded to by seeking remedial measures at the World Trade Organization. This move was characteristic of the EU's broader "raw material diplomacy" from 2008 to 2020 that focused on securing diverse supply sources through trade liberalization

(Glencross, 2024). The overall objective was to reduce dependency on any single supplier by expanding trade relationships across multiple countries.

However, the 2023 European Commission's Critical Raw Materials Act (CRMA) marked a sea change in strategy. The new legislation retained the long-standing goal of reducing import dependency on a single country, yet also included ambitious goals for EU-sourced extraction, refining, and recycling capacity. Drawing on interview data with a range of officials and stakeholders, along with documentary analysis of French mining strategy, this chapter shows how the 2022 French presidency of the EU Council provided a crucial platform for incorporating French pro-mining preferences into the CRMA. In particular, the French government's insistence on getting this legislation passed was the result, we argue, of the special attention paid to security of supply in the previous decade by French policy-makers.

2 The French Stance on the Critical Raw Materials Supply Chain

French concerns regarding the security of supply in critical raw materials date back to the early 2010s, triggering the construction of a coalition of domestic industrial and mining actors around the issue (Chailleux, 2022). France, once a significant mining power in Europe, experienced a progressive decline in its mining industry starting in the 1970s due to intense international competition and increasingly strict environmental regulations. The decline of the mining industry in France and Europe was not perceived as a problem because it was associated with a structural shift toward higher-value economic activities (Herrendorf et al., 2014). This perception began to shift in the 2010s, when CRM supply crises alerted French authorities to their industrial sector's dependence on foreign sources.

The first governmental initiative to build a coalition of actors around supply security issues was the creation of the Comité pour les Métaux Stratégiques (Committee for Strategic Metals - COMES) in 2011, after the rare earth crisis between China and Japan. One of COMES's key roles was to develop criticality assessments for various metals, providing a tool to evaluate the criticality of metals in France. Such assessments were tools for businesses to conduct their own vulnerability analyses. Following the creation of COMES, the publication of around ten reports by various French public institutions repeatedly warned the French economy of its dependency on raw materials imports, especially those originating from China (Boudinet & Khater, 2021). Arnaud

Montebourg, as Minister of Industry from 2012 to 2014, advocated for a more interventionist approach, proposing the creation of the Compagnie Nationale des Mines de France (CMF). Although this project did not materialize, it set the stage for future initiatives. Emmanuel Macron, first as Minister of Industry (2014–2016) and then as President (2017–2022), continued to refine France's mining strategy. Macron's approach focused on creating stable conditions for mining activities, including reforming the mining code and encouraging best practices (Chailleux, 2022).

In 2018, the creation of a Comité Stratégique de Filière (CSF – Strategic Sector Committees) dedicated to the mining industry enabled closer public-private cooperation in the mining sector. CSFs are platforms to enhance coordination and competitiveness within specific industrial sectors. They operate under the authority of the French Prime Minister and aim to reinforce mutual commitments between the state and private actors. It is within the CSF dedicated to mining and metallurgy activities that concerns about the security of supply for French manufacturers appeared most strongly. In 2019, the Conseil Général de l'Économie conducted a study on the security of supply for the French industry by surveying members of the CSF. The survey revealed that only 18% of industrial companies had complete visibility of their supply chains, while almost half relied heavily on a single supplier (Homobono & Vignolles, 2019). Three years later, the same observation remains unchanged in the Varin report submitted to the French government in January 2022. In close collaboration with members of the CSF, this report represented the French domestic coalition of actors' concerns and recommendations for securing raw materials. It recommended various measures to improve the resilience of industrial value chains, notably through domestic mining.

In an EU context, the CSF for mining and metallurgy enabled a strong coordination effort ahead of the CRMA negotiations, allowing French governmental actors to be particularly active in the pre-formulation stage of the CRMA. That phase before drafting the Commission's legislative proposal is the most important one for influencing an EU law because it lays the technical groundwork for the legislation (Kluver, 2013). In a centralized political system such as France, the CSF was not just a conduit for policy-relevant information; it was also a key channel for private actors' EU lobbying activities because of limited alternatives for accessing Brussels (Saurugger, 2009). The Varin report further recommended the creation of the DIAMMS (délegation interministérielle aux approvisionnements en minerais et métaux stratégiques – Interministerial Delegation for the Supplies of Strategic Minerals and Metals) to coordinate the country's policy and international negotiations on the supply of strategic minerals and metals. It was designated as the responsible directorate for forming a common position by gathering the demands of various actors from mining and metals companies. Through regular meetings with the CSF and ministries involved in mining policy, DIAMMS was tasked with ensuring a unified French stance. Ministries that negotiated at the EU Council previously aggregated the demands of a variety of actors through various cooperation mechanisms. As noted by an interview with a DIAMMS representative, "there was almost no disagreements across Ministries on our position which allowed us to be very active in the discussions" (Interview 28/8/2024). That coordination effort was essential in ensuring the French perspective was well-represented in the CRMA discussions.

The unified position of French actors involved in the CRMA closely aligned with the positions put forward by Euromines and Eurometaux, the two main industry associations representing the sectors of metals and mining in Europe. Both shared the same views on the need for funding, high ESG standards, faster permitting, and the importance of information on European mining deposits. The alignment of French actors' demands with Euromines and Eurometaux can be explained by the fact that both the French State and mining and metals lobbies have been promoting the idea of a common European industrial policy. The French had long been advocating for a common EU approach to industrial policy, arguing this would reinforce the EU's position in the geopolitical competition. This was reflected in the 2019 French-German common Manifesto for a European industrial policy fit for the twenty-first century, calling for "a more ambitious European industrial strategy with clear objectives for 2030."

3 The French Approach to Lobbying the EU during the CRMA Negotiations

3.1 In the Run-Up and during the Presidency

The COVID-19 pandemic and the war in Ukraine brought the issue of CRM supply security to the forefront of the European Union's agenda. These crises exposed the EU's vulnerabilities and over-reliance on external sources for essential goods. Broader disruptions in supply chains catalyzed a shift in EU policy toward securing a sustainable, resilient, and diversified supply of CRM. Russia's invasion of Ukraine coincided with the French EU Presidency, allowing France an opportunity to use strategic autonomy as an argument to support policies aimed at ensuring future access to critical raw materials. The French domestic coalition's concerns over CRM supply security had already been incorporated in official discussions with CSF mining and metallurgy and found their way into the French authorities' "Made in Europe" strategy, issued at the outset of the French EU presidency in January 2022. So it was no surprise that when EU leaders issued the Versailles Declaration as a response to Russia's full-scale invasion of Ukraine, this document emphasized that the time had come to "reduce our strategic dependence" on external CRM sources.

Hence, France took the opportunity of the EU presidency to promote the recommendations of the Varin report. Recommendations from the report were also promoted at the European level by Eurometaux, the professional organization of the European metals industry. When the report was issued, Eurometaux praised the French government for writing the report, stating:

> We applaud France for introducing its own raw materials strategy earlier this month, and endorse the recommendations given by Philippe Varin in his report to your government. It's imperative that your national urgency is now transferred to the EU level, building on the progress made with the European Raw Materials Alliance (Eurometaux, 2022).

That promotion was crucial in advocating for the need to create the CRMA. As noted during an interview with a French representative of the metals industry, "France was perceived as a precursor to the Act because of the promotion of this report, having been proactive in addressing the issue of critical raw materials" (Interview 21/2/2024). Following the report's publication and promotion, Varin himself gained an important aura in the European CRM security of supply enterprise and became a media advocate to promote the need for European extraction.

The French state used a variety of means to promote an EU policy position in line with domestic mining actors' preferences regarding a putative CRM law. It firstly heavily engaged its commissioner, Thierry Breton, who held the crucial internal market portfolio, in promoting the domestic coalition's positions. Although EU commissioners do not represent their home member states but the interests of the European Union as a whole, the literature on EU lobbying has consistently shown that states tend to appoint commissioners who have congruent positions to theirs (Wonka, 2007). This seems particularly true for Breton, who was a strong advocate for the Commission to develop a new focus on sovereignty and the protection of Europe's industrial base, diverging from its traditionally open and liberal economic approaches. Prior to the CRMA negotiations, Breton was also one of the architects responsible for the creation of the European Battery Alliance and the European Raw Materials Alliance in close cooperation with European industrial lobbies.

Throughout his tenure, the French Commissioner for the Internal Market played a pivotal role in shaping the discourse on CRM security of supply, framing it as a problem rooted in insufficient mining capacity (Tansey, 2021). Breton repeatedly advocated for the inclusion of mining in the green taxonomy in line with the recommendations of the Varin report, arguing that such inclusion was necessary to ensure funding for mining companies, a position echoed by French industrial actors during the CRMA consultation process. By prioritizing certain perspectives and excluding others, Breton influenced which issues were understood as priorities, thereby shaping the policy instruments developed and the problems that gained attention and reached the decision-making agenda.

In his blog posts and media appearances, Breton frequently referenced the risk of losing competitive advantage to China and the potential economic and strategic consequences of failing to secure CRM supply. In September 2022, he warned that lithium and rare earths would soon be more important than oil and gas, and stressed the need to avoid dependency similar to that on oil and gas (European Commission, 2022). His strategic framing positioned mining as a crucial component of Europe's future, aligning with the interests of the French mining coalition. Breton also consistently cited the study of Katholieke Universiteit Leuven (KU Leuven) to support his position on the necessity of mining on European soil. The 2022 study conducted by KU Leuven, titled "Metals for Clean Energy: Pathways to solving Europe's raw materials challenge" (Gregoir & van Ackler, 2022), was commissioned by the mining interest group Eurometaux. The KU Leuven report emphasized the critical need for metals to achieve Europe's 2050 climate-neutrality goals. It highlighted the theoretical potential for new European mines to meet between 5% and 55% of Europe's raw material needs by 2030.

Interviews with the European Commission's DG GROW.I.1 revealed Thierry Breton particularly pushed for a quick legislative process (Interview 6/3/2024). The French Commissioner wanted the legislation passed before the end of the Commission's term of office in 2024. This prompted the various European officials to act quickly to negotiate the regulation. Breton was successful in that regard since it took 12 months to adopt the legislation, while the average duration of the legislative process usually takes about 2 years.

3.2 After the French Presidency

As much as France was proactive in the build-up to and during the EU Council Presidency, it also sought to provide leadership and coordination on the CRMA afterwards. French actors were particularly keen, along with Euromines and Eurometaux, to impose the need for substantial funding to support mining projects, given the capital-intensive nature of these activities. One of the key events that showcased France's influence efforts on this topic was the workshop on financial solutions for the mining industry that took place on June 9, 2023 in Brussels. Organized by the French representation, the event's agenda focused on how to develop financing to secure critical metals. The workshop exposed the complexities mining companies had in securing EU funding for their extraction projects.

The discussions on mining funding needs were a strategic move to advance the idea of an EU sovereign fund, an idea that France had long advocated for in line with its promotion of a common European industrial policy. That matter was on the agenda of the French authorities' "Made in Europe" strategy in 2023, which was prompted by the US Inflation Reduction Act. The idea of a European Sovereignty Fund was framed as a strategy to avoid further budgetary fragmentation among member states. Presented by Commission President Ursula von der Leyen in her State of the Union address in September 2022, it gained momentum under French impetus. Thierry Breton suggested a target of EUR 350 billion that would be partly funded by common borrowing, which was opposed vehemently by Germany. Discussions rapidly reached a gridlock against a backdrop of the autonomy-austerity conflict that has hampered EU geoeconomic policy-making for numerous years (Abels, 2025). The inability of France to convince Germany, a defender of EU-level austerity, about the value of such a sovereign fund largely explains the failure of this project.

In response to the failure to consider funding at the EU level, France opted for a more national approach to CRM supply security funding. Enacting domestic measures and advocating for them to be replicated in the European arena was a key aspect of the French advocacy strategy to influence the EU throughout the CRMA negotiations that followed its Presidency. France established a national fund for critical raw materials in May 2023 to respond to this key industry concern. The French government backed this initiative by committing EUR 500 million through the France 2030 plan. It committed to match private investments, aiming to raise a target of EUR 2 billion. Following the launch of the fund and the incentive schemes, France deepened its cooperation with Euromines. The French fund for critical minerals was discussed with other European officials during round tables organized by Euromines, further reinforcing France's leadership position on that issue. B.Gallezot, the head of DIAMMS, was invited several times to present the French fund at various Euromines' conferences. France also implemented a national incentive scheme for the mining industry during the CRMA negotiations. Officially opened for applications on September 27, 2023, the Green Industry Tax Credit

(*Crédit d'Impôt pour l'Industrie Verte*) focuses on key sectors crucial for the energy transition, including the extraction of raw materials. It provides for a tax credit to mining companies ranging from 20% to 45% of eligible expenses, as long as the company is able to demonstrate that at least half of its revenue will be generated by companies involved in downstream production activities.

The French state was also particularly efficient at finding allies in other member states to put forward its position in the EU arena. In that regard, the coalition between France, Germany, and Italy played a pivotal role in shaping the CRMA, leveraging their combined influence to address common strategic concerns. Although cooperation failed on issues such as the EU Sovereign Fund, France, Italy, and Germany coordinated their proposals and positions during the Council's meetings, advocating for specific measures. They pushed for the expansion of the lists of strategic and critical raw materials to include materials like aluminum, as well as the establishment of specific extraction, processing, and recycling targets for each strategic raw material.

4 After the CRMA: Enhancing EU Strategic Autonomy?

Although the EU, in large part thanks to French impetus, embarked on a new geopolitical direction with the adoption of the CRMA, it is important to evaluate the EU's real capacity to meet the objectives of this landmark piece of legislation. Here, the failure of the French government to succeed in building a coalition that would break the austerity deadlock is significant. In particular, the inability to include a financial component to the CRMA stands in marked contrast to recent US attempts to address similar security of supply concerns, namely the Biden administration's Inflation Reduction Act (IRA). The 2022 IRA provided hundreds of billions of dollars in subsidies and tax credits to make the U.S. a more appealing destination for investments in clean technologies. In contrast, the EU's CRMA shifts the responsibility for financing CRM initiatives to member states and the private sector, lacking a unified funding framework that could offer the same predictability and stability found in the USA. The CRMA may thus fall short of its ambitions due to its insufficient financial power compared to programs like the IRA. Without greater public financial support, as the French domestic coalition argued in the lead-up to the CRMA, such as equity and investment guarantees, companies might remain hesitant to commit to the high-risk investments required for CRM projects.

In addition, without an equivalent to the European Sovereignty Fund, there is a risk of creating a fragmented European approach to security of supply.

Given the current budgetary rules within the EU, reliance on a state aid-based approach to funding strategic projects could exacerbate investment disparities across the member states (De Ville, 2023; Findeisen & Wernert, 2023). States with stronger fiscal positions are better positioned to fund CRM mining projects, potentially deepening the uneven development landscape across the EU, where only a handful of countries can advance their CRM strategies effectively, leaving others behind and weakening the collective effort to secure a resilient and self-sufficient supply chain for critical raw materials.

The fragmentation of budgetary efforts in the European Union's strategy to meet the CRMA targets is also indicative of differing views among member states on the nature of "open strategic autonomy." The implementation and acceptance of this concept vary significantly across the EU, reflecting deep-rooted economic and political divergences (Vicard & Wibaux 2022), which a common funding approach could have helped counteract. While countries such as France advocate for a strong and somewhat protectionist approach to strategic autonomy, other member states hold contrasting views. Notably, some Eastern European countries, such as Hungary, with substantial Chinese investments, express concerns over the potential repercussions of the EU's strategic shift in their relationships with China. These countries are apprehensive about the strategies to reduce dependence on Chinese imports of CRMs. This fragmentation reflects a broader conflict within the EU about how to balance national interests with collective European goals in areas not limited to critical raw materials. The presence of such disagreements highlights the challenges the EU faces in harmonizing its internal policies with the broader goals of strategic autonomy.

Another fundamental challenge posed by the CRMA is the ability to meet its ambitious and explicit targets for ensuring CRM security of supply. The CRMA sets ambitious targets for domestic extraction (10%), processing (40%), and recycling (25%) of CRMs by 2030. However, there are legitimate concerns as to whether the EU is even able to reach those targets, especially the 10% extraction target. Despite France's backing for an extractivist approach, achieving this particular target appears unattainable because of insufficient metal reserves and a lack of knowledge covering the presence of 13 critical metals identified by the Commission in 2023 (Capliez et al., 2024). Decades of limited research and exploration activity have left the EU with inadequate data on its mineral reserves, significantly hindering the ability to develop and implement viable mining projects. Similarly, the lengthy process from initial exploration to the start of commercial production can take up to 17 years for certain metals. Even though the CRMA will accelerate permitting procedures to expedite the opening of new mines, this effort may fall short. The current primary delays in permitting processes stem not from bureaucratic overreach but from financial constraints and staffing shortages in regulatory bodies, alongside the industry's underinvestment in project preparation stages, which often leads to legal challenges that further burden these authorities (European Environmental Bureau, 2023; Findeisen & Wernert, 2023). Moreover, mining activities suffer from low social acceptance in Europe, which jeopardizes the economic feasibility of CRM projects.

There are also significant challenges associated with achieving the EU's goal of processing at least 40% of its CRM consumption. Eurometaux estimates suggest that by 2030, the EU is likely to meet its demand for processed metals from raw materials only in the cases of nickel and cobalt. Refining and processing are highly energy-intensive and polluting activities, and their relocation to Europe requires both acceptance and substantial logistical planning (Capliez et al, 2024). High energy costs and environmental and labor standards increase operational costs for processing facilities (Righetti & Rizos, 2024). To compete effectively with countries such as the United States and China, which have access to unconventional gas reserves and lower electricity costs, Europe must secure highly competitive energy prices.

Lastly, the recycling target is unlikely to be met for key battery metals (Transport & Environment, 2024). This shortfall is attributed to several factors, including the nascent stage of many recycling projects and the technological and logistical challenges associated with collecting and processing end-of-life batteries efficiently. Recycling is an industry that must take into account the expensive infrastructure necessary for collection, sorting, pretreatment, as well as ongoing technological adaptations to accommodate the evolving battery chemistries used in electric vehicles. The CRMA's emphasis on mining activities as the primary strategy for securing raw material supply means recycling projects risk not being prioritized to the same extent. Mining projects represent 45% of the applications for "Strategic Project" status under the CRMA, while recycling projects account for only 16%, indicating a skewed focus that could undermine the recycling sector's development. The mining sector benefits not only from regulatory facilitation but also from larger scale operations that can produce materials at a lower cost. Recyclers find themselves at a significant competitive disadvantage even though they produce the same output as miners. The CRMA's lack of specific incentives for recycling, such as subsidies or tax breaks to offset these higher costs, exacerbates the issue. As a result of differing cost structures and output price imbalances, extracting lithium from used batteries can be up to five times more expensive than producing it from raw materials (Goyal et al., 2023). Without significant adjustments to this imbalance to include substantial support for recycling, the

EU risks failing to meet its 25% recycling target, continuing its reliance on primary extraction.

5 Conclusions

This chapter argued that the French government used the opportunity provided by the rotating EU Council Presidency to upload a specific policy preference into the design of a major piece of EU legislation. Yet this strategy did not come out of nowhere: it was the product of almost a decade of domestic public-private coordination based on recognizing the importance of critical raw materials and the important contribution mining could make in this area. Building a winning coalition around the CRMA meant reprising arguments about strategic autonomy, which had in the meantime become all the more salient in the aftermath of the COVID pandemic and the full-scale invasion of Ukraine. The French government also continued to provide leadership over the CRMA negotiations after the end of the Presidency, although this did not mean they were able to achieve all their objectives, as evidenced by the stillborn plan for a European sovereignty fund to back new CRMA-related projects.

However, the effectiveness of the major piece of legislation desired by the French Presidency remains open to question. Without a coherent and unified approach, the EU will struggle to effectively utilize its combined economic weight to secure a resilient supply chain for CRMs, potentially undermining its capacity to compete globally while ensuring its independence in key strategic sectors. By adopting measures at the EU level, the Union can present a collective front, leveraging its economic and regulatory influence to negotiate more favorable terms with China while protecting its industries. Such tools offer a structured and unified approach that could alleviate the fears of individual member states about provoking economic retaliation, a worry that fragments the EU's efforts to pursue strategic autonomy efforts (Vicard & Wibaux, 2021). Furthermore, in addition to the fragmentation it introduces among member states, the CRMA might contribute to an increasing fragmentation of the CRM international landscape. Analyses suggest that the current trajectory in international CRM strategies is leading to a more contentious global landscape (Evenett & Fritz, 2023; Zhou & Månberger, 2024). The international scramble for raw materials is marked by a shift from cooperative global governance of resource management to a more adversarial approach centered on national gains and strategic advantages (Zhou & Månberger, 2024). In the past years following the COVID-19 crisis, the emergence of different countries' policies toward CRMs is informed by a mix of economic, security, and strategic considerations, often resulting in contradictory and competing national agendas. This has led to a scramble for materials, with countries employing a range of policy tools that, while aimed at securing national interests, often contribute to market fragmentation and reduce global cooperation on mineral security (Evenett & Fritz, 2023). These and similar geopolitical challenges will be high on the agenda of future EU Council Presidencies.

References

Abels, J. (2025) 'Making "strategic autonomy" rhyme with "fiscal austerity?" Unresolved conflicts of (geo)economic ideas in EU infrastructure policy', Competition & Change, 29(2), pp. 204–224. https://doi.org/10.1177/10245294241277255.

Bernard, E. (2024). Les Spécificités de l'autonomie stratégique européenne en matière de défense. In A. Hamonic (Ed.), *Quelle spécificité de la défense en droit de l'Union européenne ?* Presses universitaires de Rennes.

Bonnet, T., Grekou, C., Hache, E., & Mignon, V. (2022). Métaux stratégiques : La clairvoyance chinoise. *La Lettre du CEPII,* (428). https://www.cepii.fr/PDF_PUB/lettre/2022/let428.pdf.

Boudinet, L., & Khater, N. (2021). *Comment sécuriser nos approvisionnements stratégiques ?* (Les Docs de la Fabrique 16). Paris: Presses des Mines–Transvalor La Fabrique de l'Industrie.

Capliez, R., Hache, E., & Mignon, V. (2024). Batteries lithium-ion : Cartographie dynamique de la chaîne de valeur et perspectives. *La Lettre du CEPII,* (48). https://www.cepii.fr/PDF_PUB/pb/2024/pb2024-48_FR.pdf.

Chailleux, S. (2022). La Trajectoire de la politique minière française depuis 1990: Survivance et marginalisation du régime de justification des acteurs miniers. In S. Chailleux, S. Le Berre, & Y. Gunzburger (Eds.), *Ressources minérales et transiti ons – Trajectoires politiques du sous-sol français au 21ème siècle* (Vol. 38, pp. 151–179). Peter Lang Verlag. https://hal.science/hal-03857622.

Damro, C. (2012). Market power Europe. *Journal of European Public Policy, 19*(5), 682–699. https://doi.org/10.1080/13501763.2011.640800.

Danino-Perraud, R. (2021). *Géoéconomie des chaînes de valeur: Les matières premières minérales de la filière batterie.* https://www.ifri.org/sites/default/files/migrated_files/documents/atoms/files/danino-perraud_geoeconomie_chaines_de_valeur_2021.pdf.

De Ville, F. (2023, February 7). The return of industrial policy in the European Union. *Green European Journal.* https://www.greeneuropeanjournal.eu/the-return-of-industrial-policy-in-the-european-union/.

Donnen, V. (2022). Vers une ère métallisée: Renforcer la résilience des industries par un mécanisme de stockage stratégique de métaux rares. *Institut français des relations internationales (Ifri)*. https://www.ifri.org/fr/notes/vers-une-ere-metallisee-renforcer-la-resilience-des-industries-par-un-mecanisme-de-stockage.

Eurometaux (2022). *Letter in support of France's actions on EU raw materials resilience.* Euromines & Eurometaux. https://www.euromines.org/files/2022_01_26_euromines_and_eurometaux_letter_in_support_of_france_actions_on_eu_raw_materials_resilience.pdf.

European Commission (2022). *Critical Raw Materials Act.* https://ec.europa.eu/commission/presscorner/detail/en/STATEMENT_22_5523.

European Environmental Bureau (2023, July 10). *A turning point: The Critical Raw Material Act's needs for a social and just green transition* [Position paper]. https://eeb.org/wp-content/uploads/2023/07/CRMA-Position-Paper-Final-4.pdf.

Evenett, S., & Fritz, J. (2023, July 5). *The scramble for critical raw materials: Time to take stock?* (31st Global Trade Alert Report). https://www.globaltradealert.org/reports/gta-31-report.

Findeisen, F., & Wernert, Y. (2023). Meeting the costs of resilience: The EU's critical raw materials strategy must go the extra kilometer. *Jacques Delors Centre*. https://www.delorscentre.eu/en/publications/eu-critical-raw-materials.

Glencross, A. (2024). The geopolitics of supply chains: EU efforts to ensure security of supply. *Global Policy*. https://doi.org/10.1111/1758-5899.13388.

Goyal, M., Kulwant, S., & Nitu, B. (2023). Circular economy conceptualization for lithium-ion batteries: Material procurement and disposal process. *Chemical Engineering Science, 272*, 119080. https://doi.org/10.1016/j.ces.2023.119080.

Gregoir, L., & van Acker, K. (2022). *Metals for clean energy: Pathways to solving Europe's raw materials challenge.* KU Leuven. https://eurometaux.eu/media/rqocjybv/metals-for-clean-energy-final.pdf.

Hache, E., Grekou, C., & Mignon, V. (2024). Batteries lithium-ion: L'Europe peut-elle s'extraire de la dépendance chinoise ? *The Conversation France*. https://theconversation.com/batteries-lithium-ion-leurope-peut-elle-sextraire-de-la-dependance-chinoise-238335.

Haroche, P. (2023). A "geopolitical commission": Supranationalism meets global power competition. *JCMS: Journal of Common Market Studies, 61*(4), 970–987. https://doi.org/10.1111/jcms.13396.

Herrendorf, B., Rogerson, R., & Valentinyi, Á. (2014). Growth and structural transformation. (*Uzupełnij źródło – czy to rozdział, raport czy artykuł?*).

Homobono, N., & Vignolles, D. (2019). *Analyse de la vulnérabilité d'approvisionnement en matières premières des entreprises françaises.* Conseil Général de l'Économie.

Ikenberry, G. J. (2018). The end of liberal international order? *International Affairs, 94*(1), 7–23. https://doi.org/10.1093/ia/iix241.

Klüver, H. (2013). *Lobbying in the European Union: Interest groups, lobbying coalitions, and policy change.* Oxford University Press.

Ministère de l'Économie et des Finances (2019). *A Franco-German manifesto for a European industrial policy fit for the 21st century.* https://presse.economie.gouv.fr/wp-content/uploads/2020/11/fd32d63828617cc973af75261e66209d.pdf.

Mononen, T., Kivinen, S., Kotilainen, J.-M., & Leino, J. (2022). *Social and environmental impacts of mining activities in the EU.* Publication for the Committee on Petitions, Policy Department for Citizens' Rights and Constitutional Affairs, Directorate-General for Internal Policies, European Parliament.

Righetti, E., & Rizos, V. (2024). *Reducing supply risks for critical raw materials: Evidence and policy options* (CEPS In-Depth Analysis No. 2024–01). Centre for European Policy Studies. https://cdn.ceps.eu/wp-content/uploads/2024/01/CEPS-InDepthAnalysis-2024-01_Reducing-supply-risks-for-critical-raw-materials.pdf.

Saurugger, S. (2009). COREPER and national governments. HAL Working Paper. https://shs.hal.science/halshs-00474687

Tansey (2021). *The EU's industry alliances: The new corporate capture that threatens democracy and the environment.* Policy Commons. https://policycommons.net/artifacts/4398231/mining-the-depths-of-influence/5194850/.

Tansey (2023). *Mining the depth of influence: How industry is forging the EU Critical Raw Materials Act.* Policy Commons. https://policycommons.net/artifacts/4398231/mining-the-depths-of-influence/5194850/.

Taylor, K. (2022, April 27). Sans matières premières, il n'y aura pas de transition écologique et numérique. *Euractiv France.* https://www.euractiv.fr/section/energie-climat/news/sans-matieres-premieres-il-ny-aura-pas-de-transition-ecologique-et-numerique/.

Transport & Environment (2024, September). *What strategic projects to select: T&E recommendations on the implementation of the EU Critical Raw Materials Act.* https://www.transportenvironment.org/uploads/files/2024_09_CRMA_implementation_briefing.pdf.

Vicard, V., & Wibaux, P. (2022). L'économie mondiale: De quoi l'autonomie stratégique ouverte est-elle le nom ? *CEPII.* https://www.cepii.fr/CEPII/fr/publications/em/abstract.asp?NoDoc=13464.

Wonka, A. (2007). Technocratic and independent? The appointment of European Commissioners and its policy implications. *Journal of European Public Policy, 14*(2), 169–189. https://doi.org/10.1080/13501760601122241.

Zhou, J., & Månberger, A. (2024, October). *Critical minerals and great power competition: An overview* (SIPRI Research Report). https://www.sipri.org/sites/default/files/2024-10/critical_minerals.pdf.

CHAPTER 27

Advancing a Water-Smart and Resilient Europe: A Critical Analysis of the EU Water Governance

Samara López-Ruiz

Abstract This chapter examines the evolution of EU water resilience policy, focusing on key frameworks such as the Water Framework Directive, the Circular Economy Action Plan, and Horizon Europe. It assesses their effectiveness in addressing water scarcity, pollution, and governance disparities across Member States, highlighting persistent challenges including weak enforcement, cybersecurity risks in digital water infrastructure, and transboundary management issues. The analysis also considers the role of innovation, particularly smart water technologies and digital twins, in improving adaptive water governance. It concludes that despite significant progress, uneven implementation and structural limitations continue to hinder EU water resilience, and calls for stronger policy integration and enhanced climate adaptation strategies.

Keywords Water governance – EU policy – Water Framework Directive – climate adaptation – digitalization – transboundary water management

1 Introduction

Water resilience has emerged as a cornerstone of the European Union's (EU) governance agenda, reflecting the critical importance of water as both a resource and a system upon which environmental, economic, and societal stability depend. The EU Blue Deal call, launched by the European Economic and Social Committee in 2023, and the Polish Programme for the Presidency of the Council of the EU (Jan–Jun 25) exemplify its growing prioritization within EU policy. Resilience, in this context, refers to the ability of water systems to adapt to and recover from shocks while maintaining essential functions. These shocks range from climate change-induced water scarcity and extreme weather events to growing urban demands and pollution pressures. Ensuring resilient water systems is not only an environmental necessity

 | HTTPS://DOI.ORG/10.1163/9789004763623-029

but also a fundamental element of sustainable development and regional security (Rockström et al., 2014).

In recognition of these challenges, the EU has implemented ambitious policy frameworks to manage its water resources, including the Water Framework Directive (WFD), the Circular Economy Action Plan (CEAP), and Horizon Europe. These frameworks reflect the EU's commitment to achieving "good ecological status" for water bodies, promoting resource efficiency through circularity, and fostering innovation to address emerging challenges. However, increasingly complex and interconnected pressures test the effectiveness of these policies. Climate change has intensified water scarcity in vulnerable regions such as Southern Europe, while pollution from agriculture and urban runoff continues to threaten ecosystems. At the same time, aging infrastructure and uneven enforcement across member states hinder the full realization of these frameworks' goals (OECD, 2016; Water Europe, 2024). Emerging technologies, including predictive analytics, smart water systems, and digital twins, offer new opportunities to enhance water resilience. These innovations enable real-time monitoring, adaptive management, and a more efficient use of resources. However, they also introduce new risks, particularly in the realm of cybersecurity, where critical water infrastructure has become increasingly vulnerable to attacks (Markopoulou & Papakonstantinou, 2021). Furthermore, water governance intersects with broader security concerns, such as transboundary water disputes, and the protection of critical infrastructure. These challenges underscore the need for a more integrated and forward-looking approach to EU water policy.

This chapter critically examines the evolution and effectiveness of EU water governance, including the strategic role of the Polish Presidency of the Council of the EU in shaping water priorities, focusing on how key frameworks address these contemporary challenges. It explores the successes and limitations of the WFD, CEAP, and Horizon Europe in fostering water resilience, particularly in the face of climate change, regional disparities, and digital transformation. The chapter also investigates how water policies intersect with security concerns. By critically analyzing the intersection of water resilience, digital innovation, and security, this chapter seeks to contribute to ongoing efforts to position water governance as a central pillar of sustainable and secure EU integration. In doing so, it underscores the urgency of advancing a water-smart Europe capable of addressing the challenges of the twenty-first century.

This chapter employs a critical review approach, synthesizing insights from policy documents, case studies, and secondary literature. The case studies are used comparatively to illustrate how EU water governance frameworks are mediated by differing institutional capacities, hydrological pressures, and governance traditions across member states. The rotating Presidency of the Council of the EU is treated analytically as an agenda-setting and coordinating actor, shaping policy priorities and cross-sectoral integration rather than as a legislative decision-maker. By analyzing these sources, the chapter identifies key gaps in EU water governance and offers actionable recommendations to strengthen policy coherence, foster innovation, and address regional disparities.

2 Evolution of EU Water Governance

Water governance in the EU has undergone significant transformation over the past few decades, evolving from fragmented national policies to an integrated, supranational framework aimed at achieving sustainable water management. At the heart of this transformation is the recognition that water resources are essential not only for environmental sustainability but also for economic development, public health, and geopolitical stability. The EU has sought to harmonize water policies across member states through legally binding directives, strategic action plans, and research-driven innovation programs. However, the implementation and effectiveness of these frameworks have been uneven, revealing critical gaps in enforcement, regional equity, and policy integration.

2.1 From National Policies to Supranational Water Governance: The Shift toward Ecosystem-Based Management

The transition to a unified EU water governance framework was driven by the need to address transboundary water challenges, growing pollution levels, and increasing water scarcity across member states. Prior to the 1990s, water policies were largely developed at the national level, with varying degrees of environmental protection and resource management. The introduction of the WFD in 2000 (Directive 2000/60/EC) marked a paradigm shift toward an integrated, ecosystem-based approach, requiring member states to manage water resources at the river basin level rather than through political or administrative boundaries.

The WFD, alongside other related regulations, set ambitious targets, including achieving "good ecological status" for all surface and groundwater bodies by 2027, reducing pollution, and ensuring sustainable water use through economic instruments such as water pricing (WFD, Art. 9). While the directive provided a robust legal framework, its implementation has been uneven. By

2024, only 37% of surface water bodies in the EU had reached good ecological status, reflecting persistent challenges in enforcement and compliance (EEA, 2024). Southern and Eastern European countries, such as Spain and Romania, continue to face difficulties in meeting WFD and UWWTD targets due to limited administrative capacity, high agricultural water demand, and ongoing pollution from industrial and agricultural sources (Kampa et al., 2020; EEA, 2024). Transboundary water governance has further tested the WFD's effectiveness. The directive mandates that shared river basins – such as the Danube, Rhine, and Elbe – be managed cooperatively through River Basin Management Plans (RBMPs). While this has fostered collaboration among member states, disparities in technical expertise and financial resources have led to inconsistent implementation. Upstream countries, such as Austria, have made substantial progress in reducing pollution, while downstream nations, including Romania and Serbia, continue to struggle with inadequate wastewater treatment and agricultural runoff (World Bank, 2024).

2.2 The Circular Economy and Water Reuse: Expanding the Scope of Water Governance

Beyond the WFD, the EU has sought to align water governance with broader sustainability goals through the CEAP. The CEAP, introduced in 2020, represents a shift from traditional resource management toward water reuse, efficiency, and waste reduction. Regulation (EU) 2020/741 specifically promotes the reuse of treated wastewater for agricultural irrigation, addressing water stress in arid regions while reducing reliance on freshwater sources.

However, the adoption of water reuse practices has been slow. As of 2021, less than 2.4% of treated wastewater in Europe was reused, despite its potential to alleviate water scarcity in highly stressed regions such as Spain, Italy, and Greece (WISE-Freshwater, 2024). The slow uptake is attributed to regulatory fragmentation, financial constraints, and public perception barriers (López-Ruiz & González-Gómez, 2023). While some member states, such as Cyprus and Malta, have successfully integrated wastewater recycling into their national water strategies, others lack the necessary infrastructure and policy incentives.

The CEAP's focus on water reuse has also been criticized for its narrow scope. While it prioritizes agricultural irrigation, it overlooks other sectors where treated wastewater could be valuable, such as industrial cooling, urban landscaping, and groundwater recharge. However, there are 8 countries in the EU, such as Poland or Finland, that do not allow water reuse for agricultural irrigation (WISE-Freshwater, 2024). Moreover, countries like Germany and the

Netherlands have made strides in integrating water reuse into industrial processes, but a lack of uniform guidelines across the EU has hindered broader adoption. Addressing these challenges requires stronger policy alignment between the CEAP and WFD, along with financial incentives to accelerate infrastructure investment in water reuse technologies.

2.3 Innovation and Digitalization in Water Governance

Technological advancements are reshaping water governance, with digital tools playing an increasingly important role in optimizing water resource management. Strategic investments in water-related digital innovations have proven critical for enhancing efficiency and sustainability across member states (Water Europe, 2024). Horizon Europe, the EU's flagship research and innovation program, has prioritized digital innovation to tackle emerging water challenges, particularly in areas such as pollution monitoring, predictive analytics, and smart water distribution systems.

One of the most promising developments is the use of digital twins, virtual replicas of water infrastructure that allow real-time simulation and risk assessment. In Rotterdam, a digital twin was deployed to optimize flood barrier operations, reducing flood risks and saving thousands of euros annually in operational costs (World Economic Forum, 2024). Similarly, smart water management initiatives in Copenhagen, including the use of advanced metering systems, have significantly contributed to reducing water losses (Pederson et al., 2013).

Despite overall successes, the integration of digital tools into EU water governance remains uneven. A 2024 study by the European Parliamentary Research Service (2024) highlights ongoing challenges in bridging the research and innovation gap between leading and lagging member states, which may affect the equitable distribution of Horizon Europe funding for water-related digital innovations. Moreover, digitalization introduces new risks, particularly in the realm of cybersecurity. Recent ransomware attacks on water utilities in Ireland and Germany highlight the vulnerabilities of digitalized water infrastructure, emphasizing the need for stronger cybersecurity measures within EU water frameworks (Markopoulou & Papakonstantinou, 2021).

The lack of explicit cybersecurity provisions in the WFD and CEAP is a major gap in EU water governance. As digital tools become more widespread, ensuring cyber resilience in water infrastructure must be a priority, as stated by the Blue Deal call. However, the Presidency's emphasis on critical infrastructure protection, including the application of the NIS2 and CER Directives to the water sector, marked a significant step toward integrating cybersecurity into water governance. This political focus exposed a key gap in existing frameworks such as the WFD and CEAP, which remain largely silent on digital security despite increasing reliance on digital tools.

2.4 The Need for a More Integrated Approach

Despite the breadth of existing EU frameworks addressing water governance, significant gaps in enforcement, policy coherence, and resource allocation persist, limiting their overall effectiveness. While the WFD, CEAP, Horizon Europe, and additional regulations each contribute to different aspects of water resilience – regulation, circularity, and innovation, respectively – their impact remains fragmented without stronger integration and cross-framework coordination.

During the 2025 Polish Presidency, these implementation gaps were explicitly acknowledged in Council discussions and high-level meetings of EU Water Directors. By organizing a Strategic Dialogue on Water, the Presidency placed renewed political emphasis on the persistent failure of many Member States to achieve good ecological status and on the need to modernize water quality standards in line with scientific evidence. Poland also prioritized negotiations on updating priority substances under the WFD, facilitating inter-institutional progress that ultimately led to the inclusion of emerging pollutants such as PFAS and pharmaceuticals.

The Presidency's approach underscored that ecosystem-based governance cannot be decoupled from institutional capacity and security considerations. By framing weak enforcement and pollution as risks to societal stability and resilience, the Polish Presidency strengthened the political legitimacy of further regulatory tightening and coordination. In this regard, the 2027 review of the WFD presents a key opportunity to align these initiatives, ensuring that regulatory objectives under the WFD are reinforced by the financial mechanisms of the CEAP and the technological advancements driven by Horizon Europe. Strengthening enforcement mechanisms, improving financial accessibility for less-resourced member states, and embedding climate adaptation and cybersecurity considerations into digital water governance will be essential for translating policy ambition into tangible resilience.

3 Strengthening EU Water Governance: Policy Gaps, Security Challenges, and Transboundary Cooperation

3.1 Persistent Policy and Implementation Gaps in EU Water Governance

Despite the European Union's sustained efforts to develop a comprehensive framework for water governance, policy effectiveness continues to be limited by uneven implementation across Member States. These shortcomings arise from a combination of regional disparities in administrative and technical capacity, the insufficient integration of climate adaptation into water policy, and emerging vulnerabilities associated with the digitalization of water systems. Although key legislative and strategic instruments like WFD, CEAP, and Horizon Europe have strengthened specific dimensions of water resilience, their practical impact remains inconsistent, reflecting the structural constraints of EU multi-level governance and divergent national priorities.

Table 27.1 maps the principal contemporary challenges confronting EU water governance and the legislative and policy instruments through which they are addressed. The overview highlights the breadth of regulatory responses currently in place, spanning water quality protection, infrastructure renewal, climate adaptation, digital security, and transboundary management. Importantly, these policy gaps do not stem solely from weak enforcement or inadequate funding; rather, they are rooted in the structural limitations of EU multi-level governance, the competing interests of member states, and the tension between market-based and regulatory approaches to environmental governance. Against this backdrop, this chapter focuses on three core challenges that persist and have been highlighted in the Blue Deal Call:

1. Weak enforcement and regional disparities in policy implementation.
2. Security vulnerabilities in digitalized and aging water infrastructure.
3. Gaps in transboundary water management and external water diplomacy.

To illustrate how EU water governance outcomes vary under a shared regulatory framework, this chapter draws primarily on three illustrative national cases: Spain, Romania, and the Netherlands. These cases are selected to reflect contrasting combinations of hydrological pressure, institutional capacity, and governance approaches. Spain represents a water-stressed member state that has pursued technologically intensive responses to scarcity; Romania illustrates persistent implementation challenges linked to administrative and infrastructure constraints; and the Netherlands exemplifies a proactive, prevention-oriented model grounded in long-standing flood risk governance and

TABLE 27.1 Contemporary challenges in EU water governance and corresponding legislative frameworks (Source: Own elaboration).

Contemporary challenge	Relevant EU legislation/regulation
Climate-induced water scarcity	WFD; European Green Deal, EU Climate Adaptation Strategy
Pollution and water quality decline	WFD; Urban Wastewater Treatment Directive (2024/3019); Zero Pollution Action Plan; Environmental Quality Standards Directive (2008/105/EC)
Regional disparities in policy implementation	WFD; European Regional Development Fund
Aging water infrastructure	Drinking Water Directive (2020/2184); Urban Wastewater Treatment Directive (2024/3019)
Transboundary water Management conflicts	WFD; EU Strategy for International River Basin Management
Insufficient integration of climate adaptation	EU Adaptation Strategy (2021); European Green Deal
Digitalization and cybersecurity risks in water systems	NIS2 Directive (EU 2022/2555); Horizon Europe Cybersecurity Programs
Low adoption of circular water use and reuse	CEAP; Regulation 2020/741 on Minimum Requirements for Water Reuse
Groundwater pollution and deterioration	Groundwater Directive (2006/118/EC)
Chemical pollution in surface waters	Environmental Quality Standards Directive (2008/105/EC) as amended by Directive 2013/39/EU (EQS)
Marine pollution and ecosystem protection	Marine Strategy Framework Directive (2008/56/EC)
Flood risk management	Floods Directive (2007/60/EC)

nature-based solutions. Together, they enable a comparative assessment of how EU water policies are mediated by national capacities and strategic priorities. A central source of these challenges lies in the design of the WFD itself. Its decentralized implementation model, whereby Member States retain primary responsibility for achieving objectives through RBMPs, has enabled flexibility and contextual adaptation but has also resulted in weak enforcement and inconsistent compliance (Kaika, 2003; Moss, 2012). Unlike regulatory frameworks such as the EU Emissions Trading System, which impose clear financial penalties for non-compliance, the WFD relies largely on soft governance mechanisms and cooperative compliance. Consequently, progress toward WFD objectives varies significantly across the Union, with outcomes shaped as much by institutional capacity and resource availability as by formal legal obligations.

Comparative evidence illustrates these disparities. Spain, despite experiencing acute water stress, has developed a relatively structured water governance system that incorporates WFD principles into national hydrological planning. Through economic instruments such as water pricing under Article 9 of the WFD, Spain has improved efficiency in several river basins (Moss, 2012). Nonetheless, persistent challenges remain, particularly in ecologically sensitive areas such as Doñana, where groundwater over-extraction for agriculture continues despite clear regulatory constraints. In contrast, Romania has faced greater difficulty in meeting WFD requirements, largely due to insufficient investment in wastewater treatment, agricultural runoff management, and weak coordination between national and regional authorities. Although substantial EU funding has been allocated to infrastructure upgrades, limited institutional capacity has slowed effective implementation (ICDR, 2021; World Bank, 2024).

These cases highlight that compliance with EU water legislation cannot be reduced to questions of political will alone. Rather, it depends on a combination of governance capacity, technical expertise, financial resources, and the ability to integrate EU objectives into domestic administrative systems. Addressing these structural shortcomings requires moving beyond a uniform compliance model toward a more differentiated approach, combining stronger enforcement mechanisms with targeted financial and technical assistance. Such an approach would enable Member States with weaker institutional capacity to meet regulatory requirements more effectively, while enhancing overall coherence and resilience within EU water governance.

3.2 Integrating Climate Adaptation into Water Governance

Although climate change is disrupting hydrological cycles across Europe, adaptation remains secondary in EU water policy. The European Green Deal and EU Adaptation Strategy acknowledge the need for climate-resilient water management, yet these measures lack direct integration within the WFD and CEAP, leading to gaps in preparedness and response capacity.

The consequences of this policy gap are evident in the divergent adaptation pathways adopted across Member States. In the Netherlands, the Room for the River programme illustrates a proactive, nature-based approach to flood risk management, in which spatial planning has been reoriented to allow rivers to expand naturally and absorb excess water (Zevenbergen et al., 2016). Rather than prioritizing structural flood defenses alone, Dutch authorities have integrated ecosystem-based solutions into both urban and rural landscapes. By contrast, several Southern European countries, notably Spain, have relied heavily on desalination and inter-basin water transfers to address water scarcity. While these measures offer short-term relief, they raise significant concerns regarding energy intensity, ecological degradation, and long-term sustainability (López-Ruiz et al., 2021). An over-reliance on capital-intensive technological solutions, without complementary investments in nature-based adaptation, risks exacerbating regional disparities in climate resilience (Water Europe, 2024).

Addressing these inconsistencies requires structural reform at the EU level. A key priority would be to mandate the explicit integration of climate adaptation measures within all RBMPs, ensuring that water governance extends beyond compliance with WFD objectives to encompass long-term resilience planning. Recent political initiatives suggest growing recognition of this need. For example, the Blue Deal Congress 2025, held in Warsaw under the Presidency's patronage, focused on water security amid economic and climate pressures, financing needs for water infrastructure, and efficiency-enhancing technologies. The Presidency's involvement in this forum signals increasing high-level political support for elevating water adaptation within the EU policy agenda; however, translating such momentum into binding regulatory reform remains a critical challenge.

3.3 Security Dimensions of Water Governance: Digitalization and Infrastructure Resilience

The rapid expansion of digital water management systems – including smart meters, predictive analytics, and digital twins – has significantly enhanced efficiency, real-time monitoring, and data-driven decision-making in EU water governance. At the same time, digitalization has introduced new security vulnerabilities, exposing critical water infrastructure to cyber threats with potentially systemic consequences (Grievson et al., 2022). Although the NIS2 Directive (EU 2022/2555) formally recognizes water utilities as critical infrastructure, cybersecurity requirements remain weakly integrated into core EU water governance frameworks such as the WFD and CEAP. As a result, most cybersecurity measures in the water sector remain voluntary, leaving implementation largely at the discretion of Member States and water operators. Unlike sectors such as energy or finance, where cybersecurity obligations are more clearly defined, water systems remain disproportionately exposed to digital risks (Markopoulou & Papakonstantinou, 2021). According to the European Court of Auditors (2024), disparities in cybersecurity capacity and investment across Member States further undermine the EU's collective resilience to cyber threats.

These risks are no longer merely hypothetical. The growing frequency and sophistication of cyberattacks targeting water infrastructure highlight the urgency of regulatory reform. In 2023 alone, Spain recorded 107,777 cyber incidents, representing a 94 per cent increase compared to the previous year (Idrica, 2024). Internationally, attacks on Israel's water supply in 2020 and Florida's water treatment facilities in 2021, alongside incidents in Europe linked to Russian cyber groups, demonstrate that water utilities are increasingly viewed as strategic targets. Such cases underscore the vulnerability of digitalized water systems in the absence of robust, mandatory cybersecurity standards.

Security challenges, however, extend beyond the digital domain. Physical threats to water infrastructure are intensifying due to climate change, extreme weather events, and aging assets. While the WFD acknowledges the need for resilience to natural hazards, implementation has been uneven, particularly in regions characterized by outdated infrastructure and limited technical capacity (OECD, 2016). The 2021 floods in Central Europe, which caused damages exceeding €30 billion, exposed structural weaknesses in drainage and flood management systems, reinforcing the need for climate-resilient water infrastructure (Birkmann et al., 2023). Taken together, cyber and physical risks represent interconnected dimensions of infrastructure security that current EU water governance frameworks address only partially.

Addressing these challenges requires a more comprehensive EU security strategy that integrates digital and physical resilience within water governance. First, cybersecurity risk assessments should be mandated within RBMPs, ensuring that digital risks are systematically evaluated alongside hydrological and environmental pressures. Mandatory, periodic cyber-resilience audits for water utilities would further enhance early detection of vulnerabilities and reduce the likelihood of operational disruptions. Second, expanded public–private partnerships could facilitate collaboration between water operators, technology providers, and cybersecurity specialists, embedding security-by-design principles into digital water systems and fostering innovation in risk management (Markopoulou & Papakonstantinou, 2021). Finally, increased EU investment in climate-resilient water infrastructure is essential, particularly in flood-prone regions where cyber and physical vulnerabilities may compound one another.

In this context, initiatives under the Polish Presidency marked an important step toward reframing water governance as a security issue. Discussions at the Warsaw conference in June 2025 emphasized that protecting water infrastructure requires robust regulatory frameworks and coordinated tools to address emerging security threats, including cyber risks and hybrid disruptions. The conference highlighted the need to integrate civil protection, cybersecurity, and emergency planning more closely into water policy, signaling a shift toward a more holistic EU security governance approach. This emerging framework links water governance with crisis management, critical infrastructure protection, and defense-related capabilities, including the use of satellite data for monitoring droughts and infrastructure threats. Strategically, Poland's agenda helped consolidate the view that water security constitutes a core component of EU strategic resilience – essential for safeguarding populations, economic activity, and critical installations during crises. By the end of 2025, the concept of "water resilience" had entered the EU security lexicon, aligning environmental governance with broader objectives of strategic autonomy and systemic resilience.

3.4 Strengthening Transboundary Water Management and EU Water Diplomacy

3.4.1 Internal Transboundary Water Challenges

The EU is home to some of the world's most complex transboundary river basins, which require coordinated governance among multiple member states due to the high level of decentralization of transboundary water governance in the EU (Baranyai, 2020). In this regard, the WFD established a legal foundation for transboundary cooperation, requiring member states to develop RBMPs for shared water bodies. For example, the Danube River Protection Convention and the International Commission for the Protection of the Rhine exemplify successful governance models where multilateral cooperation has led to significant improvements in water quality and pollution control (ICPDR, 2021).

However, despite these achievements, the effectiveness of transboundary governance is undermined by enforcement challenges and capacity disparities. In the Danube Basin, for example, Austria and Germany have significantly reduced industrial discharges, while downstream countries like Romania and Bulgaria continue to struggle with untreated wastewater and agricultural runoff. Meanwhile, the Rhine Basin, while benefiting from strong institutional coordination, faces ongoing challenges in microplastic pollution (Shrank et al, 2022), which require more stringent regulatory alignment among member states.

Although climate change is increasingly recognized in transboundary water policy discussions, its practical integration remains inconsistent and secondary to pollution control, leaving significant gaps in preparedness. A Blue Deal that positions water resilience as a central pillar is essential to ensuring long-term sustainability. For example, rising temperatures are altering hydrological cycles in the Elbe and Meuse Basins, exacerbating water scarcity and competition between urban and agricultural users. However, most transboundary agreements continue to prioritize pollution control, overlooking long-term adaptation challenges posed by climate variability.

To enhance transboundary water resilience, the EU must move beyond a pollution-control-centric approach and systematically embed climate adaptation into RBMPs. Expanding financial and technical support for downstream countries facing persistent water quality challenges would help reduce capacity disparities, which remain a key barrier to effective adaptation. Furthermore, strengthening monitoring and data-sharing mechanisms is essential for coordinated climate responses across transboundary basins. However, persistent challenges – including data standardization issues, unequal access to digital technologies, and capacity gaps – must be addressed to ensure that digital innovations contribute to more equitable and effective transboundary water governance. Addressing these gaps collectively highlights the interdependence between policy implementation, technical capacity, and governance efficiency in achieving long-term water resilience.

3.4.2 The EU'S Limited Role in Global Water Diplomacy

Water governance is not just an internal EU issue; it should play a central role in EU foreign policy and security strategy. Environmental changes, including water scarcity, have been identified as contributing factors to migration patterns from neighboring regions, particularly in North Africa, the Middle East, and Central Asia. The Sahel region, for example, faces severe desertification and groundwater depletion, forcing communities to migrate toward the EU. Reports suggest that environmental degradation, including water-related issues, plays a role in influencing migration decisions to Europe (IOM, 2024).

Despite these risks, the EU's migration policies have largely focused on border security and asylum management, with insufficient attention to addressing the root causes of displacement. While the European Green Deal and Global Gateway Initiative support climate adaptation projects in Africa, investment in water infrastructure remains underfunded and fragmented. Morocco's Noor Ouarzazate solar plant, co-financed by the EU, demonstrates how renewable energy and desalination can be integrated to promote water security; yet such initiatives remain limited in scale.

The Nile Basin has been a persistent source of diplomatic tension between Egypt, Sudan, and Ethiopia, primarily due to the construction of the Grand Ethiopian Renaissance Dam, which raises concerns over downstream water availability. Similarly, disputes over the Euphrates-Tigris system, involving Turkey, Syria, and Iraq, continue to exacerbate regional tensions, driven by competing demands for water resources and unilateral water development projects. While global powers such as the United States and China have engaged in hydro-diplomacy through strategic investments and direct mediation, the EU role has been more nuanced. Although often perceived as reactive, the EU has increasingly positioned water diplomacy as a strategic tool within its foreign policy, leveraging soft power to promote cooperative water governance frameworks rather than engaging in direct conflict mediation (Álvarez-Arcá, 2024). Considering this state of the art, the EU is encouraged to enhance its global role in water governance. In doing so, it should (1) Integrate water resilience into migration and foreign policy strategies, focusing on water-scarce regions that drive displacement; (2) Expand investment in transboundary water infrastructure, particularly through joint desalination and irrigation projects in Africa and the Middle East; (3) Strengthen EU diplomatic engagement in water disputes, using its convening power to mediate transboundary conflicts.

4 The 2025 Polish Council Presidency as an Agenda-Setter in EU Water Governance

The 2025 Polish Presidency of the Council of the European Union played a distinctive agenda-setting role in EU water governance by reframing water policy as a strategic issue linked to security, resilience, and crisis preparedness

(Council of the EU, 2025a). While the rotating Presidency lacks formal legislative initiative, it exercises significant influence through agenda control, political sequencing, and discursive framing (Vaznonytė, 2020; Wurzel et al., 2020). In this regard, the Polish Presidency exemplifies how presidencies can shape the policy salience of water governance without directly altering regulatory instruments.

A central contribution of the Polish Presidency was its explicit alignment of water governance with the broader EU security agenda, encapsulated in the Presidency's motto "Security, Europe!" Water resilience was consistently presented as a prerequisite for climate adaptation, societal stability, and the protection of critical infrastructure. This framing resonated strongly with the European Commission's launch of the European Water Resilience Strategy in June 2025, which the Polish Presidency immediately placed on the Environment Council agenda, facilitating early political endorsement and coordinated discussion among Member States (EEB, 2025a). By doing so, the Presidency accelerated the integration of water into the EU's strategic resilience discourse, moving it beyond a narrowly environmental domain.

Across policy subfields, the Presidency acted as a connector rather than an initiator. In relation to the Water Framework Directive, Poland prioritized political discussions on persistent implementation gaps and supported negotiations on updating priority substances, including emerging pollutants such as PFAS and pharmaceuticals. Although final agreements were concluded under the subsequent Presidency, Poland's role in maintaining momentum illustrates how presidencies influence legislative trajectories through continuity and coordination (EEB, 2025a; EEB, 2025b).

Similarly, in the context of the Circular Economy Action Plan, the Polish Presidency advanced files with indirect but significant implications for water quality, such as measures on microplastic pollution and industrial waste streams. These initiatives were framed as contributing to water resilience by reducing pollution at source, reinforcing the growing policy consensus that water governance must be embedded within broader resource and industrial strategies (EEB, 2025a; Water Europe, 2024).

Perhaps most notably, the Presidency strengthened the security dimension of water governance by explicitly linking water infrastructure to EU critical infrastructure and cybersecurity frameworks, notably NIS2 and the Critical Entities Resilience Directive. High-level discussions and Presidency-backed events underscored the vulnerability of digitalized water systems to cyber and hybrid threats, contributing to the securitization of water policy at EU level (Markopoulou & Papakonstantinou, 2021; EEB, 2025a; Council of the EU, 2025b).

Overall, the impact of the 2025 Polish Presidency lies less in substantive regulatory change than in its successful reframing of water as a strategic, cross-sectoral policy issue. By embedding water governance within the EU's resilience and security agenda, the Presidency reshaped political priorities in ways likely to influence future reforms, funding decisions, and institutional coordination beyond its formal term.

5 Conclusion

Water governance in the EU has evolved into a highly sophisticated, yet increasingly strained policy domain. The 2025 Polish Council Presidency played a pivotal role in highlighting both the strengths and the structural limitations of the existing framework, particularly by elevating water governance within broader debates on resilience and security.

Although the current legislative framework provides a solid foundation, its implementation remains highly uneven across the Union. While some Member States have benefited from advanced governance models, digital innovation, and market-based instruments, others continue to face persistent challenges linked to limited institutional capacity, aging infrastructure, and financial constraints. These disparities indicate that persistent non-compliance cannot be understood solely as a legal failure, but rather as a structural governance challenge that requires targeted EU-level responses.

At the same time, climate change has introduced new and intensifying pressures on European water systems that existing policies are only partially equipped to address. The limited integration of climate adaptation into core water governance instruments weakens the EU's ability to anticipate and manage droughts, floods, and long-term hydrological change. Strengthening the systematic incorporation of adaptation measures into EU water policy is therefore essential to ensure that resilience becomes a central, rather than ancillary, objective of future reforms.

By reframing water governance through the lenses of resilience and security, the Polish Presidency contributed to a more integrated and forward-looking policy agenda that recognizes water as a strategic asset for the Union. Despite enduring challenges in enforcement, regional disparities, and climate adaptation, the Presidency's emphasis on agenda-setting, coordination, and securitization has left a lasting imprint on EU water policy debates. Future reforms are likely to build on this legacy, further embedding water governance within the EU's broader objectives of resilience, security, and strategic autonomy.

References

Álvarez-Arcá, I. (2024). EU water diplomacy: A tool to strengthen the EU's soft power on environmental governance. *Global Policy*, 15(S8), 7–14. https://doi.org/10.1111/1758-5899.13443.

Baranyai, G. (2020). *European Water Law and Hydropolitics: An inquiry into the resilience of transboundary water governance in the European Union*. Netherlands: Springer.

Birkmann, J., Schüttrumpf, H., Handmer, J., Thieken, A., Kuhlicke, C., Truedinger, A., & Kirschbauer, L. (2023). Strengthening resilience in reconstruction after extreme events – Insights from flood affected communities in Germany. *International Journal of Disaster Risk Reduction*, 96, 103965. https://doi.org/10.1016/j.ijdrr.2023.103965.

Council of the European Union (2025a). *Programme of the Polish Presidency of the Council of the European Union (January–June 2025)*. https://www.parlament.gv.at/dokument/XXVIII/III/145/imfname_1676568.pdf#:~:text=48%20Directives%20on%20priority%20substances,the%20communication%20on%20water%20resilience.

Council of the European Union (2025b). *Council pushes for a holistic water strategy to safeguard water resources and boost resilience*. https://www.consilium.europa.eu/en/press/press-releases/2025/10/21/council-pushes-for-a-holistic-water-strategy-to-safeguard-water-resources-and-boost-resilience/#:~:text=The%20Council%20also%20reiterates%20the,remote%20areas%20and%20on%20islands.

EEB (2025a). *The EEB's Assessment of the environmental performance of the Polish Presidency of the EU*. https://eeb.org/wp-content/uploads/2025/07/EEB_presidency-papers-1-merged-4.pdf.

EEB (2025b). *The EEB's Assessment of the environmental performance of the Danish Presidency of the EU*. https://eeb.org/wp-content/uploads/2025/12/EEB_presidency-papers-1-combined.pdf.

European Court of Auditors (2024). *Challenges to effective EU cybersecurity policy*. https://www.eca.europa.eu/lists/ecadocuments/brp_cybersecurity/brp_cybersecurity_en.pdf.

Grievson, O., Holloway, T., & Johnson, B. (2022). *A strategic digital transformation for the water industry* (p. 117). IWA Publishing.

ICDR (2021). *Danube flood risk management plan – Update 2021*.

Idrica (2024). *Cybersecurity in European water utilities: A growing challenge*. https://www.idrica.com/blog/cybersecurity-european-water-utilities/.

International Organization for Migration [IOM] (2024). *Impacts of environmental changes on mixed migration to Europe: Insights from flow monitoring surveys in Italy, Malta, and Spain*. IOM.

Kaika, M. (2003). The Water Framework Directive: A new directive for a changing social, political and economic European framework. *European Planning Studies*, 11(3), 299–316. https://doi.org/10.1080/09654310303640.

Kampa, E., Völker, J., Stein, U., & Mohaupt, V. (2020). *European significant water management issues*. European Environment Agency (EEA).

López-Ruiz, S., & González-Gómez, F. (2023). Regenerate and Reuse Water in Spain: Facts and Politics. In *Water management and circular economy* (pp. 177–196). Elsevier.

López-Ruiz, S., Moya-Fernández, P. J., García-Rubio, M. A., & González-Gómez, F. (2021). Acceptance of direct potable water reuse for domestic purposes: Evidence from southern Spain. *International Journal of Water Resources Development*, 37(5), 772–792. https://doi.org/10.1080/07900627.2020.1799762.

Markopoulou, D., & Papakonstantinou, V. (2021). Digitalisation of water services and the water sector cyber threat landscape: Is the EU Regulatory Framework adequate?. *Journal of Water Law*, 27(4).

Moss, T. (2012). Spatial fit, from panacea to practice: Implementing the EU Water Framework Directive. *Ecology and Society*, 17(3).

OECD (2016). *Water governance in cities*. OECD Publishing. 10.1787/9789264251090-en.

Pedersen, J. Baadsgaard (Ed.) & Klee, P. (Ed.in C.), 2013. *Meeting an increasing demand for water by reducing urban water loss – Reducing Non-Revenue Water in water distribution. The Rethink Water network and Danish Water Forum White Papers, Copenhagen.* https://copenhagenwater.wordpress.com/wp-content/uploads/2013/11/white-paper-reducing-urban-water-loss.pdf.

Rockström, J., Falkenmark, M., Allan, T., Folke, C., Gordon, L., Jägerskog, A., ... & Varis, O. (2014). The unfolding water drama in the Anthropocene: Towards a resilience-based perspective on water for global sustainability. *Ecohydrology*, 7(5), 1249–1261. https://doi.org/10.1002/eco.1562.

Schrank, I., Löder, M. G., Imhof, H. K., Moses, S. R., Heß, M., Schwaiger, J., & Laforsch, C. (2022). Riverine microplastic contamination in southwest Germany: A large-scale survey. *Frontiers in Earth Science*, 10, 794250. https://doi.org/10.3389/feart.2022.794250.

Vaznonytė, A. (2020). The rotating Presidency of the Council of the EU – Still an agenda-setter?. *European Union Politics*, 21(3), 497–518.

Water Europe (2024). *Socio-economic study on the value of the EU investing in Water*. Water Europe.

World Bank (2024). *Danube Region. Regional Water Security Assessment*. World Bank.

World Economic Forum (2024). How digital twins are transforming the world of water management. https://www.weforum.org/stories/2024/11/why-digital-twins-might-transform-the-world-of-water-management/.

Wurzel, R. K., Liefferink, D., & Di Lullo, M. (2020). The European Council, the Council and the Member States: Changing Environmental Leadership Dynamics in the European Union. In *The future of European Union environmental politics and policy* (pp. 62–84). Routledge.

Zevenbergen, C., Rijke, J., van Herk, S., Chelleri, L., & Bloemen, P. J. T. M. (2016). Towards an adaptive, flood risk management strategy in the Netherlands: An overview of recent history. *International Journal of Water*, 3, 140.

CHAPTER 28

Between Innovation and Regulation: The EU's AI Governance in the Global Technological Rivalry

Danni Zhang, Xuechen Chen and Xinchuchu Gao

Abstract As artificial intelligence (AI) rapidly reshapes global economic, social, and political landscapes, the European Union (EU) has emerged as the first major political entity to develop a comprehensive regulatory framework for AI–the AI Act. Building on its established leadership in data protection through the Data Protection Directive and the GDPR, the EU seeks to extend its regulatory influence into AI governance. While the EU aims to position itself as a global standard-setter in ethical AI regulation, its effectiveness is challenged by internal policy fragmentation and external competition from the United States and China. This chapter applies role theory to analyse the EU's role conception, enactment, and contestation in AI governance, examining how internal cohesion shapes its international regulatory authority and influence.

Keywords EU AI governance – global AI regulation – AI governance – role theory

1 Introduction

Artificial intelligence (AI) has emerged as a critical issue in global governance, raising significant concerns related to ethics, security, and economic competitiveness. While it has been argued that the United States (US) promotes a market-driven model and China advances a state-controlled approach to AI (Bradford, 2023), the European Union (EU) seeks to differentiate itself by advocating a human-centric AI governance model (European Commission, 2020). The EU's Artificial Intelligence Act (AI Act), introduced as the world's first comprehensive regulatory framework for AI (European Parliament, 2023), embodies this approach. This research demonstrates that the EU's ambition in AI governance is founded upon three key aspects: First, it emphasizes normative leadership, aiming to establish ethical and transparent AI governance aligned with its existing approach to data governance. Second, it aims to

 HTTPS://DOI.ORG/10.1163/9789004763623-030

extend extraterritorial regulatory influence by promoting EU standards globally through market mechanisms that are termed as the 'Brussels Effect' (Bradford, 2020), along with proactive engagement with multilateral fora at the global level. Third, it increasingly prioritizes strategic autonomy in the AI domain, seeking to reduce dependence on external technological providers.

Despite the EU's ambitions in this emerging domain, it faces two critical challenges, both internally and globally. First, internal policy fragmentation – EU member states exhibit varied approaches regarding the stringency of AI regulations, resulting in inconsistent implementation across the Union. This regulatory divergence complicates efforts to harmonise AI governance within the EU. Second, intensifying global AI competition presents additional hurdles; the international AI landscape remains largely dominated by US firms such as OpenAI, while the rise of Chinese AI companies like DeepSeek has intensified debates on technological leadership in AI development. To date, the EU has not produced globally competitive AI enterprises, raising concerns that its regulatory framework could potentially hinder innovation and place European AI development at a structural disadvantage in the global market (Bermann, 2022; Calderaro & Blumfelde, 2022; Mügge, 2024).

This research employs role theory (Harnisch, 2012; Klose, 2018) to analyze the EU's role-making process in AI governance. Utilizing this theoretical framework, the study investigates how the EU has (re)conceptualized its international role in AI governance, how it has operationalized this role through the external dissemination of its regulatory frameworks and diplomatic engagement, and the implications of this role-making process for internal cohesion and global influence. Methodologically, the study applies a content analysis approach, examining various primary data sources, including official EU documents and policy papers, alongside secondary sources such as policy reports and scholarly articles. This chapter is structured as follows. The first section introduces the role-theoretic approach and outlines the analytical framework. The second section applies this framework to examine the EU's role in global AI governance and assesses how its role-making process in AI governance influences its pursuit of political and economic interests in the evolving AI landscape. The final section provides the key findings of the study and discusses their policy implications.

2 Theoretical Framework

Over the past few decades, a substantial body of scholarship has investigated the European Union's (EU) role in global affairs (Tonra & Christiansen, 2004).

Within this literature, a subset of scholars has adopted the concept of actorness to explore the EU's significance in various international contexts (Carbone, 2013). According to one of the earliest proponents, Sjöstedt (1977, p. 16), actorness denotes an actor's 'capacity to behave actively and deliberately in relation to other actors in the international system.' Since the late 1990s, increasing scholarly attention has been directed toward refining the concept of the EU's international actorness, generating multiple waves of theoretical and empirical inquiry (Klose, 2018). Nevertheless, as Klose (2018) observes, existing research has tended to overlook the relative importance of and interactions among different determinants of the EU's actorness, as well as the social processes underpinning the Union's emergence as a key international actor.

To address these gaps, Klose (2018) introduces an interactionist role theory approach, reframing the EU's international actorness as a process of role-making. Within this framework, the EU's actorness is conceptualized as its 'capacity to imagine and realise roles for its "self" in (specific contexts) of international affairs' (ibid., p. 1146), a capacity that emerges through the interplay of internal and external role expectations, creative actions, and the material and social resources at the Union's disposal (ibid., p. 1146). A central tenet of interactionist role theory is the notion that political actors project themselves in international society via two interlinked forms of agency – 'me' and 'I.' The 'me' signifies an actor's ability to understand its own identity by adopting others' perspectives (role-taking), whereas the 'I' represents its capacity for generating creative impulses in response to these external perceptions (Klose, 2018). The dialogue between these two dimensions enables a political actor to realise its own identity within a particular international context and to reflect on the roles it plays in the broader international community. This iterative process ultimately fosters learning, thereby allowing the actor to refine its self-positioning in world politics (Harnisch, 2012).

Within Klose's interactionist role theory framework, several elements shape the EU's international actorness. First, internal role expectations – established by member states and various EU institutions – determine the kinds of roles the EU aspires to occupy in a specific situation (Klose, 2018, p. 1148). Second, cohesion, or the level of internal agreement among EU actors, not only affects the Union's capacity to mobilize resources but also influences other international actors' expectations regarding the EU's role (ibid.). Additionally, social and material resources – such as economic and military capabilities and expertise – alongside the ability to exercise creative agency inform the EU's actorness by delimiting what the Union can feasibly do and how effectively it can harness available resources to actualize its envisioned roles (ibid.).

Klose (2018) outlines three steps to guide empirical application of interactionist role theory. The first step involves examining how the EU (re)imagines its international role in the face of 'problematic situations,' or moments of uncertainty that undermine established role-playing routines (Klose, 2018, p. 1149; Harnisch, 2012). These disruptions may stem from internal contestations, shifts in the expectations of external actors, or transformations in relative social and material resources (Harnisch, 2012, pp. 50–51; Klose, 2018). During such junctures, the EU's constituent units reassess potential responses, driven by internal impulses for particular role expectations and external efforts by other actors to assign roles to the EU (Klose, 2018, p. 1150).

The second step focuses on the EU's endeavours to implement these roles in social interaction. This involves scrutinizing how the EU strategically employs social and material resources to position significant others in either complementary or corresponding roles, as well as assessing how other actors react to the EU's role-making (Klose, 2018, p. 1150). Subsequently, the third step of the theoretical framework evaluates the broader ramifications of this role-making process, determining whether the EU has successfully convinced relevant actors to accept the roles it envisions for itself and whether it effectively asserts itself as an international actor in a specific context. It also investigates whether these efforts alter the social structure within which the EU operates (ibid., p. 1150–1151).

3 An Analysis of the EU's AI Governance

By adopting the framework outlined above, this section seeks to investigate the EU's desired role in the field of global AI governance, as well as the extent to which the EU has managed to realize and solidify this role in practice. The first part of this section analyses the EU's desired role in the field of AI within the EU as well as other roles that it can achieve through its various activities around the globe. The second part explores how the EU has used its resources to realize its desired role. The final part discusses the extent to which the EU has succeeded in achieving and/or solidifying its desired roles.

3.1 The EU's Desired Role in the Field of AI

A close examination of the EU's key strategic documents on AI published since 2018 reveals three major dimensions underpinning the EU's intended role in the field of AI. First, the EU's articulation of strategic autonomy and digital sovereignty in the field of AI emerged from a sober assessment of its relative position within the global technological landscape. Compared to the US and China, the EU has long been perceived as lagging behind in core AI capabilities, large-scale data ecosystems, and platform-based innovation (European Commission, 2018a; 2018b; Mügge, 2024). In the first two Common strategies – namely the Coordinated Plan on Artificial Intelligence (European Commission, 2018a) and Artificial Intelligence for Europe (European Commission, 2018b), the Commission acknowledged and highlighted that 'Europe is currently behind in private investments in AI' and 'risks losing out the opportunities offered by AI' (European Commission, 2018a). The second Commission strategy on AI explicitly featured the EU's position as a laggard in AI vis-à-vis other major economies, particularly in the context of the US-China rivalry, stating that Europe maintains an inferior position in private investments in the development of AI, with a total of EUR 2.4–3.2 billion in 2016, in comparison with EUR 6.5–9.7 billion in Asia and EUR 12.1–18.6 billion in North America (European Commission, 2018b). Moreover, the EU has found itself in a disadvantaged position in relation to the US and China, given that American and Chinese large corporations have been significantly investing in AI while exploiting large amounts of data (European Commission, 2018b). By realistically assessing its own position in an increasingly competitive and politically contested technological landscape, the EU has gradually articulated strategic autonomy and digital sovereignty not merely as political objectives, but as constitutive elements of its desired role in global AI governance. While early AI strategies emphasized coordination and competitiveness, subsequent policy discourse increasingly framed autonomy as a prerequisite for the EU's capacity to act under conditions of uncertainty, crisis, and technological dependence. This shift became particularly salient after 2020, when the COVID-19 pandemic exposed structural vulnerabilities in digital supply chains and data infrastructures, reinforcing concerns about Europe's reliance on external technological providers (Council of the European Union, n.d.). These developments were accompanied by political reinforcement at the Council level. During Germany's Presidency of the Council of the EU in 2020, 'digital sovereignty' was explicitly foregrounded as a strategic response to crisis, with artificial intelligence presented as a key domain through which Europe could strengthen its resilience, safeguard its values, and retain control over critical digital capacities (Council of the European Union, n.d.). In this context, 'strategic autonomy' and 'digital sovereignty' came to function as stabilising reference points for the EU's global ambition in AI governance, linking economic competitiveness, technological capacity, and political resilience within a coherent role narrative (e.g. Bermann, 2022; Mügge, 2024).

The second dimension of the EU's self-imagined role in the sphere of AI is concerned with its growing ambition to act as an 'ethical power'– a self-perception as a benign international actor that is capable of pursuing ethical goals at the global stage (Aggestam, 2008). A scrutiny of the key policy documents shows that the EU's desired role in the domain of AI has been consistently articulated in a discourse of universal values and ethics that defines the EU as a distinctive 'power for good' in global AI governance. The first 2018 Commission strategy ambitiously claims that 'the EU can lead the way in developing and using AI for good and for all, building on its values and its strengths' and that the EU must ensure that any new AI technologies should be developed and practices within an appropriate framework compliant with the EU's core values, fundamental rights, as well as ethical principles such as democracy, human rights, sustainability, rule of law, and transparency (European Commission, 2018a). The second 2018 Commission strategy further added that the EU's AI approach should aim at promoting a 'human-centric approach' based on ethics-by-design principles, placing people at the centre of the development of AI (European Commission, 2018b). Furthermore, the EU has sought to consolidate its role as an ethical power in AI governance through the Commission's 2019 Joint Communication, which articulated the ambition to promote human-centric and trustworthy AI (European Commission, 2019a). In this early phase, ethical considerations were primarily framed as guiding principles shaping the conditions under which AI technologies should be developed and deployed, rather than as binding regulatory requirements. This ethical orientation was subsequently reinforced and given greater political visibility after 2020. The White Paper on Artificial Intelligence: A European Approach to Excellence and Trust (hereinafter referred to as the 'White Paper') explicitly linked ethical principles to policy objectives such as trust, legal certainty, and societal acceptance, thereby reframing ethical compliance as a perquisite for sustainable innovation (European Commission, 2020). Ethical principles were also progressively embedded within EU-level political agendas. During Germany's Presidency in 2020, Council conclusions anchored AI governance within the framework of the EU Charter of Fundamental Rights, elevating ethical considerations from abstract principles to politically salient reference points (Council of the European Union, n.d.). This emphasis on ethics was subsequently reinforced during France's Presidency of the Council of the EU in 2022, which highlighted the promotion of reliable and human-centred AI and signalled continued political commitment to ethics as a core element of the EU's AI governance agenda (Council of the European Union, n.d.). These developments indicate that ethical power has evolved from a primarily declaratory commitment into a stabilized dimension of the EU's desired role in AI governance, closely intertwined with the EU's broader regulatory ambitions.

The third aspect of the EU's aspired role lies in its broader ambition to act as a 'regulatory power' with the capability to exert regulatory influence unilaterally through the EU's market mechanisms outside of its territorial borders (Bradford, 2020; Kennis & Liu, 2024) so as to reshape the global standards and regulatory frameworks on AI governance. By doing so, the EU aims to overcome its previously mentioned disadvantaged position in the global AI race, with the ultimate objectives of 'exercising global leadership' and promoting 'regulatory convergence' with like-minded actors at the international level, based on its regulatory framework and values (European Commission, 2020). While the EU exhibits uncertainty regarding its investments in AI development, it demonstrates considerable confidence in its regulatory framework in digital governance. The Commission explicitly stated that the EU is equipped with 'a strong and balanced regulatory framework to build on, which can set the global standard for a sustainable approach' to AI, referring to the EU's successful experience in exercising its regulatory power through the General Data Protection Regulation (European Commission, 2018b).

3.2 EU Attempts to Realize Its Desired Role in the Field of AI

After analyzing the EU's aspirational role in AI governance, this section examines how the EU leverages its resources to translate these ambitions into concrete actions. To achieve strategic autonomy in AI technology, the EU has increased its investments in technological development. Due to the EU's lack of major technology firms, it faces challenges in establishing a competitive edge in the global race for technological leadership (Calderaro & Blumfelde, 2022). According to a report by the European Union Agency for Cybersecurity (European Parliament, 2017a), the EU exhibits a high degree of dependence on third-party technologies. This report further emphasized that to achieve digital sovereignty and safeguard EU security, the EU must rely on strict regulatory frameworks and the capacity to develop core technologies (European Parliament, 2017a). In response, to advance AI technological development and address the shortfall in private investment, the European Commission adopted a Coordinated Plan on Artificial Intelligence in 2018, committing to allocate €2 billion annually to support entrepreneurs and startups in developing AI-based applications (European Commission, 2018a). Moreover, at the 2025 AI Action Summit held in Paris, the European Commission (2025) launched the InvestAI initiative, aiming to mobilize €200 billion in AI investments to make Europe an 'AI continent'. These examples illustrate the EU's ambition to achieve strategic autonomy in AI technology.

Within the EU, efforts to realize ethical power and regulatory power have been closely intertwined. Ethical principles such as trustworthiness, human oversight, and respect for fundamental rights have not operated in isolation, but have increasingly informed regulatory design and policy coordination. More specifically, the EU embedded its core values – such as privacy, legality, transparency, and ethics – into AI-related guidelines and policies and to promote AI governance standards' setting in transnational cooperation, thereby fulfilling its role as a norm shaper in shaping global AI governance. First, the EU has established core values for AI regulations within its internal governance framework, laying the foundation for its human-centric approach to AI governance. Tracing back to the European Parliament resolution of 16 February 2017 with recommendations to the Commission on Civil Law Rules on Robotics, the resolution emphasizes the importance of drafting guidelines addressing the ethical dimensions of AI and its human rights implications (European Parliament, 2017b). It highlighted the need for AI governance to align with the principles and values 'enshrined in Article 2 of the Treaty on the European Union and the Charter of Fundamental Rights of the European Union' (European Parliament, 2017b, p. 8). Furthermore, the European Commission appointed a High-Level Expert Group on Artificial Intelligence (AI HLEG) to provide strategic guidance on its AI governance framework (European Commission, 2019b). In its Ethics Guidelines for Trustworthy AI (hereinafter referred to as the 'Ethics Guidelines'), the group outlined seven key requirements that AI systems must meet to be deemed trustworthy, underscored the necessity of upholding fundamental rights and freedoms throughout the development and deployment of AI technologies (European Commission, 2019b). The practical implementation of these ethical principles was later acknowledged in the White Paper, further reinforcing the EU's commitment to responsible AI governance. In the White Paper, the EU further reaffirmed its objective of 'becoming a global leader in fostering the development of trustworthy and ethical AI' (European Commission, 2020, p. 8). These ethical commitments were subsequently reinforced through political coordination within the Council of the EU, where successive rotating presidencies helped sustain ethical considerations within the implementation and negotiation of AI-related policy initiatives. By foregrounding issues such as fundamental rights, trustworthiness, and human-centred AI alongside regulatory files, Council presidencies contributed to consolidating ethics as an operational rather than merely aspirational component of the EU's AI governance framework (Nikolinakos, 2023; Christou et al., 2025).

Second, the EU promotes its established human-centric approach through bilateral and multilateral cooperation as well as engagement with international organizations, reinforcing its commitment to ethical and trustworthy AI governance on the global stage. The EU initially engaged in cooperation on AI standards with like-minded countries and intergovernmental organizations that share its values and regulatory principles. For instance, the Norwegian government adopted the declaration The Nordic-Baltic Region: A Digital Frontrunner, endorsing the Cooperation on Artificial Intelligence declaration signed by 24 EU member states and Norway (Nordic Council of Ministers, 2018). This cooperation encompassed multiple AI-related areas, including the development of an AI regulatory framework, with a particular emphasis on 'developing ethical and transparent guidelines, standards, principles and values to guide when and how AI applications should be used' (Nordic Council of Ministers, 2018). Another example of EU cooperation with a like-minded country can be observed in the EU-US collaboration on AI risk through the Trade and Technology Council (TTC). At the first ministerial meeting of the US-EU TTC in Pittsburgh in 2021, the EU and the US agreed to launch three initiatives under the framework of Trustworthy AI (European Commission, 2022). Among them, the TTC Joint Roadmap for Trustworthy AI and Risk Management includes the establishment of a common repository of metrics for assessing AI trustworthiness and risk management approaches (European Commission, 2022). The EU's efforts can also be observed within intergovernmental organization. Ingersleben-Seip (2023) pointed out that, building on the EU's efforts within the Organization for Economic Cooperation and Development (OECD), the OECD adopted the Recommendation of the Council on Artificial Intelligence in 2019, in which the 'Principles for Responsible Stewardship of Trustworthy AI' align closely with the EU's Ethics Guidelines. These examples illustrate the EU's efforts to shape AI governance norms by developing regulatory frameworks and promoting its approach through cooperation with other countries and intergovernmental organizations.

Furthermore, it has also reinforced its extraterritorial regulatory power in AI governance by implementing AI governance frameworks and regulations, thereby strengthening its global influence in shaping AI standards. First, the EU extends its extraterritorial regulatory power from data governance under the GDPR to influence AI regulation. As the development and operation of AI are inherently dependent on access to and processing of (personal) data, the application and interpretation of the GDPR have significantly shaped regulatory conditions in digital technologies and data protection. For instance, within the GDPR, the transparency requirements for personal data processing (see Articles 12–14), the right to data portability (see Article 20), and the right not to be subject to automated individual decision-making (see Article 22) all reflect the EU's regulatory approach to AI governance (GDPR, 2016). As Bradford (2020) argues, the EU's single market strategy and its relatively stringent regulatory standards resulted in EU legislation directly or indirectly influencing the legal frameworks of third countries. In other words, any operators or intermediaries developing or using AI-driven data processing, whether operating within or outside the EU, are subject to the GDPR if their activities involve monitoring EU's citizens or offering goods and services to individuals located in the EU.

Besides, to further extend its extraterritorial regulatory power in AI governance, the European Commission proposed the AI Act in April 2021. At the hearing for 'A Europe Fit for the Digital Age,' Margrethe Vestager (2019) emphasized: 'Some say that the Chinese have all the data and the Americans have all the money. But when I see what we have going for us in Europe, it's what we have purpose.' The EU reaffirmed its objective of establishing itself as a global leader in AI domain by developing a robust regulatory framework rooted in its core values and human-centric approach, while promoting AI innovation within the EU and fostering international cooperation. Taken together, these internal and external efforts illustrate that the EU's ethical and regulatory roles in AI governance have been co-constructed through institutional interaction and political coordination, rather than pursued as separate or sequential strategies.

4 The Extent to Which the EU Has Succeeded in Realising Its Desired Role in the Field of AI

According to the third step of this analysis, this section assesses whether, and to what extent, the EU has succeeded in realising its desired role in governance. Specifically, the EU's ambition to achieve strategic autonomy in AI technology has been subject to scrutiny, given its technological lag in AI development and the constraints imposed by its stringent data and regulatory frameworks. However, the EU's role as an ethical power has gained broad recognition. Additionally, divergences among EU member states regarding AI regulatory approaches may constrain the EU's capability to practice extraterritorial regulatory influence beyond its borders. As mentioned above, since AI development is inherently dependent on data, overly stringent data and AI regulations – while shielding the EU's technology sector from huge foreign high-tech companies' monopolisation and protecting European enterprises – may also hinder AI technological advancement within the EU itself. More specifically, excessively strict regulations have compelled EU-based AI companies to relocate to other countries, thereby hindering the development of AI technologies within the EU. For instance, Bird, a Netherlands-based cloud communication software company, announced its decision to relocate most of its operations outside Europe, because of the EU's inability to provide an environment for AI-driven innovation as well as the challenges in recruiting skilled technology workers within the region (Reuters, 2025a). Another example is Mistral, a Paris-based leading AI startup, which established an office in California to attract top AI talent and leverage the market-driven AI regulatory approach advocated by the US to foster its technological development (Financial Times, 2024).

Furthermore, despite significant EU investments in AI technologies, these efforts pale in comparison to the scale of funding allocated to AI development in China and the US. In terms of private investment in AI companies, between 2013 and 2021 the US led with $52 billion, followed by China with $17, while the EU lagged behind with only $6 billion (Calderaro & Blumfelde, 2022). To cover the shortage of private investment, as of 2019, the EU has invested nearly €9 billion in the area of AI, yet half of this funding was allocated to infrastructure and skills development (European Commission, 2018c). Despite the EU's investment efforts, it still lags significantly behind the US and China in AI technological development. For instance, in the field of open-source large language models, the US-based Meta released Llama in 2023, followed by China's DeepSeek, which launched DeepSeek-R1 in 2025. In contrast, the EU only announced the launch of OpenEuroLLM in February 2025, highlighting its delayed entry into the AI race.

Despite the EU's relative lag in AI technology, its role as an ethical power has been widely recognized in global AI governance. As an emerging technology, AI has a broad range of potential applications, encompassing critical infrastructure, public life, and the military sector. Given that standard-setting is a key aspect of emerging technology governance, it enables international actors such as governments, intergovernmental organizations, and private sector actors to advance their political agendas and commercial interests (Mattli, 2001). Consequently, these international players are competing to develop their own regulatory standards in order to shape the AI technology and its governance (Cheng & Zeng, 2023). Due to the lack of major European high-tech companies, the EU is unable to exert significant influence in the competition over technical standards for AI – a domain primarily dominated by technological powers such as the US and China (Ingersleben-Seip, 2023). Instead, the EU extends its European values and human rights-based data governance model into the area of AI governance, securing a position in the global discourse on ethical AI standards (Council of the European Union, 2022).

Moreover, the Biden administration's policies and strategies have reinforced the EU's role in establishing itself as an ethical power in global AI governance. For instance, during a Group of Seven (G7) meeting, Canada and France announced plans to establish the International Panel on Artificial Intelligence (IPAI) to enhance international cooperation and coordination in AI policy-making, aiming to 'promote and protect a human-centric and ethical approach to AI, grounded in human rights' (Government of Canada, 2019). While six G7 members expressed support, the Trump administration rejected the proposal, arguing that 'international rules for AI would be premature and hamper innovation' (Politico, 2020). It was not until 2020 that the Biden administration, in response to China's New Generation Artificial Intelligence Development Plan, which aimed to establish China as a leading global AI innovation hub by 2030, reconsidered the proposal put forward by Canada and France (ibid.). To prevent China from exporting AI technology and shaping global AI standards, the US signed and endorsed the Global Partnership on AI (GPAI) initiative, asserting that 'it was high time for the West to join forces and beat China at writing the global rules for artificial intelligence' (ibid.).

Additionally, the successful convening of the 2025 Paris AI Action Summit reflects the EU's ethical power in AI governance. This summit underscored the EU's approach to artificial intelligence, advocating for an 'open, multi-stakeholder, and inclusive approach' while ensuring that AI development and application are grounded in a 'human rights-based, human-centric, ethical, safe, secure, and trustworthy' framework (Élysée, 2025). Furthermore, the 2025 AI Summit introduced the Statement on Inclusive and Sustainable Artificial Intelligence for People and the Planet (hereinafter referred to as the 'Statement'), which emphasizes key principles such as AI accessibility, transparency, ethical development, and international cooperation (ibid.). However, the United Kingdom (UK) and the US declined to sign the Statement. The Trump administration justified its decision by prioritizing 'pro-growth AI policies', arguing that excessive AI regulation would 'kill a transformative industry just as it's taking off' (BBC, 2025). Although the UK had traditionally aligned with the EU's AI regulatory approach, it ultimately chose not to sign the Statement to preserve opportunities for collaboration with the US (ibid.). Despite the US and UK declining to sign, around sixty countries signed the Statement, including China (Élysée, 2025). The Chinese government expressed its willingness to work with France in calling on the international community to jointly uphold 'the principle of developing AI for good, to deepen innovative cooperation, strengthen inclusiveness and benefits, and improve global governance' (State Council of the People's Republic of China, 2025). This evidence demonstrates that, despite facing certain challenges, the EU continues to assert its role as an ethical power in AI governance.

Finally, although the EU seeks to assert itself as a regulatory power on a global stage and has successfully established extraterritorial regulatory influence through its data regulation framework, empirical evidence suggests that the EU has not demonstrated the same level of regulatory power in AI governance. There are two primary reasons for this limitation. First, as mentioned earlier, overly stringent data and AI regulations within the EU have led both European and foreign technology companies to relocate their operations to jurisdictions with more innovation-friendly regulatory environments to advance their AI technologies. In addition to the previously mentioned relocation of European companies to the US to advance their technological innovation, another example further illustrates the limitation of the EU's regulatory power. In January 2025, the Italian Data Protection Authority (Garante) initiated an inquiry into DeepSeek to assess its collection and processing of Italian users' personal data, including the scope of data collected, its intended use, and whether it was stored in China. The investigation stemmed from concerns over DeepSeek's non-compliance with EU data and AI regulatory frameworks. However, DeepSeek asserted that it does not operate in European countries and is therefore not subject to EU regulations (Reuters, 2025b).

Second, the divergence among EU member states regarding AI regulatory approaches may undermine the EU's capacity to exercise extraterritorial regulatory power in AI governance. For instance, while some EU member states, such as Germany, support the EU's AI regulatory framework, particularly the distinction between high-risk and low-risk AI applications, as it aligns with their specific trade interests, other member states, such as Ireland, argued that excessive regulation in emerging markets could stifle development (Justo-Hanani, 2022). As former European Commission President Jean-Claude Junker (2018) once stated, 'Because [we have] the largest single market in the world, we can set standards for big data, AI we are able to uphold Europeans' values, rights, and identities,' but he emphasized that this ability relies on the EU's internal 'unity.' This internal fragmentation within the EU may undermine its ability to exercise extraterritorial regulatory power in AI governance.

5 Conclusion

To examine how the evolution of the EU's role in AI governance shapes its broader economic and political influence in global AI governance, this study employs role theory to analyze both the EU's desired roles in the international AI domain and the extent to which the EU has sought to realize or consolidate its roles in practice. This analysis has shown that the EU's role in AI governance cannot be understood as static or uniform. Rather, the EU's desired role has evolved through the interaction of strategic, ethical, and regulatory dimensions, shaped by shifting technological conditions and internal political coordination. The first dimension is to achieve 'strategic autonomy' in AI technology development. Second, the EU seeks to assert its 'ethical power' in global AI governance by promoting the development of 'trustworthy and ethical AI.' Finally, the EU aims to extend its regulatory power from data governance to AI governance. To realize its ambitions, the EU has significantly increased investments in AI innovation and development, actively formulated AI regulatory frameworks embedded with European values, and engaged in bilateral and multilateral negotiations and cooperation with like-minded partners. In other words, the EU has relied not only on Commission-led initiatives but also on political coordination within the Council, where successive rotating presidencies have contributed through agenda-setting and procedural steering. Together, these internal efforts have provided the institutional foundation of the EU's external engagement in AI governance. Building on this basis, the EU has actively participated in multilateral forums to promote its human-centric and ethics-based approach and to shape merging global standards for AI. Despite the EU's significant lag in AI technology compared to technological powers such as the US and China, and the growing criticism of its overly stringent data and AI regulatory framework – which has been accused of stifling innovation, hindering technological development, and promoting some European tech companies to relocate to more innovation-friendly regulatory environments – the EU's role as an ethical power in AI governance has been widely recognized by other countries.

Overall, in pursuing its desired role in AI governance, the EU's ethical power has gained widespread recognition. While the EU's AI regulatory approach aspires to balance strict regulation with fostering technological innovation and development, there is no clear evidence that its regulatory framework effectively supports its AI innovation and development. The EU still need to refine this balance to ensure that it does not fall behind in the rapidly evolving global technological race, thereby risking its opportunity to emerge as a global leader in AI governance.

References

Aggestam, L. (2008). Introduction: ethical power Europe? *International Affairs (London)*, 84(1), 1–11. https://doi.org/10.1111/j.1468-2346.2008.00685.x.

BBC (2025). *UK and US refuse to sign international AI declaration.* BBC News. https://www.bbc.co.uk/news/articles/c8ednon58gwo.

Bermann, S. (2022). *European Strategic Autonomy and the US-China Rivalry: Can the EU" Prefer Not to Choose"?.* Istituto Affari Internazionali (IAI). https://institutdelors.eu/wp-content/uploads/dlm_uploads/2021/09/iaip2132.pdf.

Bradford, A. (2020). *The Brussels Effect: How the European Union Rules the World.* Oxford University Press. https://doi.org/10.1093/oso/9780190088583.001.0001.

Bradford, A. (2023). *Digital Empires: The Global Battle to Regulate Technology.* Oxford University Press.

Calderaro, A., & Blumfelde, S. (2022). Artificial intelligence and EU security: the false promise of digital sovereignty. *European Security (London, England)*, 31(3), 415–434. https://doi.org/10.1080/09662839.2022.2101885.

Carbone, M. (2013). Between EU actorness and aid effectiveness: The logics of EU aid to Sub-Saharan Africa. *International Relations*, 27(3), 341–355.

Cheng, J., & Zeng, J. (2023). Shaping AI's Future? China in Global AI Governance. *The Journal of Contemporary China*, 32(143), 794–810. https://doi.org/10.1080/10670564.2022.2107391.

Christou, G., Meyer, T., & Fanni, R. (2025). The European Union: Assessing global leadership through actorness in artificial intelligence. *Journal of European Integration*, 47(3), 383–401. https://doi.org/10.1080/07036337.2024.2377200.

Council of the European Union (n.d.). Timeline of the presidencies of the Council of the EU. https://www.consilium.europa.eu/en/council-eu/presidency-council-eu/timeline-presidencies-of-the-council-of-the-eu/.

Council of the European Union (2022). *EU digital diplomacy: Council agrees a more concerted European approach to the challenges posed by new digital technologies.* Council of the European Union. https://www.consilium.europa.eu/en/press/press-releases/2022/07/18/eu-digital-diplomacy-council-agrees-a-more-concerted-european-approach-to-the-challenges-posed-by-new-digital-technologies/.

Élysée (2025). *Statement on inclusive and sustainable artificial intelligence for people and the planet.* Élysée. https://www.elysee.fr/en/emmanuel-macron/2025/02/11/statement-on-inclusive-and-sustainable-artificial-intelligence-for-people-and-the-planet.

European Commission (2018a). *Communication from the Commission to the European Parliament, the European Council, the Council, the European Economic and Social Committee and the Committee of the Regions: Coordinated plan on artificial intelli-*

gence. EUR-Lex. https://eur-lex.europa.eu/legal-content/EN/TXT/?uri=CELEX%3A52018DC0795.

European Commission (2018b). *Communication from the Commission to the European Parliament, the European Council, the Council, the European Economic and Social Committee and the Committee of the Regions: Artificial intelligence for Europe.* EUR-Lex. https://eur-lex.europa.eu/legal-content/EN/TXT/?uri=celex:52018DC0237.

European Commission (2018c). *EU budget: Commission proposes €9.2 billion investment in first ever digital programme.* European Commission. https://ec.europa.eu/commission/presscorner/detail/en/ip_18_4043.

European Commission (2019a). *Communication from the Commission to the European Parliament, the Council, the European Economic and Social Committee and the Committee of the Regions: Building trust in human-centric artificial intelligence.* EUR-Lex. https://eur-lex.europa.eu/legal-content/EN/TXT/PDF/?uri=CELEX:52019DC0168.

European Commission (2019b). *Ethics guidelines for trustworthy AI.* European Commission. https://digital-strategy.ec.europa.eu/en/library/ethics-guidelines-trustworthy-ai.

European Commission (2020). *White paper on artificial intelligence – A European approach to excellence and trust.* European Commission. https://digital-strategy.ec.europa.eu/en/consultations/white-paper-artificial-intelligence-european-approach-excellence-and-trust.

European Commission (2022). *TTC joint roadmap for trustworthy AI and risk management.* European Commission. https://digital-strategy.ec.europa.eu/en/library/ttc-joint-roadmap-trustworthy-ai-and-risk-management.

European Commission (2025). *EU launches InvestAI initiative to mobilise €200 billion of investment in artificial intelligence.* European Commission. https://ec.europa.eu/commission/presscorner/detail/en/ip_25_467.

European Parliament (2017a). *Cybersecurity in the EU Common Security and Defence Policy (CSDP).* European Parliamentary Research Service (EPRS). https://www.europarl.europa.eu/RegData/etudes/STUD/2017/603175/EPRS_STU(2017)603175_EN.pdf.

European Parliament (2017b). *European Parliament resolution of 16 February 2017 with recommendations to the Commission on civil law rules on robotics.* European Parliament. https://www.europarl.europa.eu/doceo/document/TA-8-2017-0051_EN.html.

European Parliament (2023). *EU AI Act: First regulation on artificial intelligence.* European Parliament. https://www.europarl.europa.eu/topics/en/article/20230601STO93804/eu-ai-act-first-regulation-on-artificial-intelligence.

Financial Times (2024). *Europe's Mistral expands in Silicon Valley in hunt for AI staff.* Financial Times. https://www.ft.com/content/71e8248d-b473-47fa-979d-172bfa75df3e.

General Data Protection Regulation (GDPR) (2016). *General Data Protection Regulation (GDPR)*. GDPR Info. https://gdpr-info.eu/.

Government of Canada (2019). *Canada and France work with international community to support responsible use of artificial intelligence.* Government of Canada. https://www.canada.ca/en/innovation-science-economic-development/news/2019/05/canada-and-france-work-with-international-community-to-support-responsible-use-of-artificial-intelligence.html.

Harnisch, S. (2012). Conceptualizing in the Minefield: Role Theory and Foreign Policy Learning. *Foreign Policy Analysis*, 8(1), 47–69. https://doi.org/10.1111/j.1743-8594.2011.00155.x.

Ingersleben-Seip, N. (2023). Competition and cooperation in artificial intelligence standard setting: Explaining emergent patterns. *The Review of Policy Research*, 40(5), 781–810. https://doi.org/10.1111/ropr.12538.

Juncker, J.-C. (2018). *President Jean-Claude Juncker's State of the Union address 2018.* European Commission. https://ec.europa.eu/commission/presscorner/detail/en/SPEECH_18_5808.

Justo-Hanani, R. (2022). The politics of Artificial Intelligence regulation and governance reform in the European Union. *Policy Sciences*, 55(1), 137–159. https://doi.org/10.1007/s11077-022-09452-8.

Kennis, A., & Liu, X. (2024). The European Union's regulatory power: Refining and illustrating the concept With the case of the transfer of EU Geographical Indication Rules to Japan. *Journal of Common Market Studies*, 62(6), 1578–1593. https://doi.org/10.1111/jcms.13579.

Klose, S. (2018). Theorizing the EU's actorness: Towards an interactionist role theory framework. *Journal of Common Market Studies*, 56(5), 1144–1160. https://doi.org/10.1111/jcms.12725.

Margrethe Vestager (2019). *Hearing of Margrethe Vestager, Executive Vice-President-designate of the European Commission (Europe fit for the digital age)*. European Parliament. https://www.europarl.europa.eu/resources/library/media/20191009RES63801/20191009RES63801.pdf.

Mattli, W. (2001). The politics and economics of international institutional standards setting: an introduction. *Journal of European Public Policy*, 8(3), 328–344. https://doi.org/10.1080/13501760110056004.

Mügge, D. (2024). EU AI sovereignty: For whom, to what end, and to whose benefit?. *Journal of European Public Policy*, 31(8), 2200–2225.

Nikolinakos, Nicos. Th. (2023). A European Approach to Excellence and Trust: The 2020 White Paper on Artificial Intelligence. In N. T. Nikolinakos (Ed.), *EU Policy and Legal Framework for Artificial Intelligence, Robotics and Related Technologies: The AI Act* (*Law, Governance and Technology Series*, Vol. 53, pp. xx–xx). Springer. https://doi.org/10.1007/978-3-031-27953-9_5.s.

Nordic Council of Ministers (2018). *AI in the Nordic-Baltic region*. Nordic Council of Ministers. https://www.norden.org/en/declaration/ai-nordic-baltic-region.

Politico (2025). *Wary of China, the West closes ranks to set rules for artificial intelligence*. Politico (2020). https://www.politico.eu/article/artificial-intelligence-wary-of-china-the-west-closes-ranks-to-set-rules/.

Reuters (2025a). *Dutch software firm Bird to leave Europe due to onerous regulations in AI era, says CEO*. Reuters. https://www.reuters.com/technology/dutch-software-firm-bird-leave-europe-due-onerous-regulations-ai-era-says-ceo-2025-02-24/.

Reuters (2025b). *Italy's regulator blocks Chinese AI app DeepSeek on data protection*. Reuters. https://www.reuters.com/technology/artificial-intelligence/italys-privacywatchdog-blocks-chinese-ai-app-deepseek-2025-01-30/.

Sjöstedt, G. (1977). *The external role of the European Community* (Vol. 7). Gower Publishing Company, Limited.

State Council of the People's Republic of China (2025). *China willing to promote AI development with other countries: Chinese Vice Premier*. The State Council of the People's Republic of China. https://english.www.gov.cn/news/202502/12/content_WS67abd3cbc6d0868f4e8ef92e.html.

Tonra, B., & Christiansen, T. (Eds.). (2004). *Rethinking European Union Foreign Policy*. Manchester University Press.

Conclusion and Recommendations: From Politics, Economy, and Society to the Digitization of the European Union

Marek Górka and Małgorzata Kamola-Cieślik

This book presents an in-depth analysis of political, economic, social, and digitalization-related issues affecting security and technological sovereignty. It offers a unique research perspective by shedding light on the policies of EU Member States such as Spain, France, Germany, Ireland, Poland, the Czech Republic, Croatia, Italy, Greece, Romania, Lithuania, Latvia, and Estonia. The analysis also covers countries at various stages of prospective EU integration – such as Serbia, Bosnia and Herzegovina, Ukraine, Moldova, Georgia, and Turkey – as well as Norway, which maintains a broad partnership with the Union. By combining a wide national perspective with reflection on the EU presidency, security, and digitalization, the book provides a modern reference point for strategic, political, and economic analyses. The monograph addresses key problem areas facing today's European Union (EU), including migration, energy, strategic raw materials, and environmental issues. Another important factor shaping European policy is the development of new technologies, including the emergence of artificial intelligence, which is increasingly influencing the public sphere. This publication responds to these changes by aiming to present the current situation and formulate conclusions for the future – viewed both as opportunities and challenges for the EU.

In the face of an unstable international situation, the EU is now confronted with the need to redefine its role on the global stage. Divergent positions among Member States on issues such as energy, migration, and relations with external partners complicate the development of a unified and coherent policy. One of the key challenges is the future of transatlantic relations. Atlantic policy, traditionally focused on building and strengthening ties between the EU and the United States as liberal democracies, was particularly evident in response to threats from Russia, China, and other authoritarian regimes. However, after 2025, the administration of President Donald Trump has been less engaged in European security. This situation, in turn, forces EU decision-makers to pursue operational autonomy – that is, to build greater economic, technological, and military self-sufficiency independent of the United States.

A turning point in perceptions of security in Europe was the outbreak of the war in Ukraine, the consequences of which made many political leaders

 HTTPS://DOI.ORG/10.1163/9789004763623-031

acutely aware of how dependent European countries are on imported energy resources, particularly from Russia.

Although Europe possesses certain strengths in the digital economy – such as industrial applications of artificial intelligence, 5G technologies, and the digitalization of the public sector – it still largely depends on US technologies and Chinese hardware and infrastructure solutions. This challenge is often overshadowed by the emphasis on military and energy security, raising the question of whether the development of digital capabilities will receive sufficient support. In this context, reducing technological dependence and building resilient, diversified supply chains – including critical raw materials, technological components, and access to a skilled workforce – are becoming key elements of future strategies and programs of the Council of the EU presidencies.

Another equally important situation that highlights the complexity of security policy is a state's dependence on foreign suppliers. The COVID-19 pandemic revealed how crucial independence from external suppliers of basic goods – such as medicines and medical equipment – is for national security. It also underscored the importance of proactive measures and the ability to anticipate threats. The EU did not respond quickly enough, despite warning signals from some Member States – particularly those with post-communist experience. This difficult situation, both socially and in terms of public health, became another reason for the EU to reconsider its identity and previously defined objectives. Fundamental, existential questions began to emerge: Is the European community ready to aggressively defend its interests? And if so, is it also prepared to bear the costs of such an assertive policy? Is Europe therefore ready to adopt a tough trade policy toward the United States or China if necessary? Such questions usually arise in the context of a distant future. Nevertheless, today's political reality increasingly replaces conciliatory, compromise-based approaches with authoritative, coercive, and decisive stances – which in many cases have proven to be effective strategies.

It is worth noting that threats perceived as distant can have very real and immediate consequences. A clear example of this is the war in Ukraine and the aggressive policy of the Russian Federation, which has also affected the Iberian Peninsula – through rising energy prices, the need to diversify energy sources, and increased engagement in security activities within NATO and the EU. In recent years, Europe has faced numerous crises requiring new solutions and improved cooperation among states. The emergence of a new multipolar world order, the crisis of liberal democracy (which constitutes a fundamental value of the EU), as well as Russia's revisionist policy and incitement to war, are among the main challenges forcing the EU to update its approach to European integration.

A key political challenge is ensuring the involvement of all Member States in security matters – including those in Western and Southern Europe, such as Spain, Portugal, Italy, and Greece – which, partly due to geographical distance, tend to perceive Russia's policy differently. In the case of the Iberian Peninsula in particular, internal security related to migration flows from Africa and the Middle East remains a tangible problem. This demonstrates that a crucial, ongoing task for political decision-makers is to seek effective ways to preserve European unity and to foster genuine solidarity in the areas of security and defense.

The difficult and complex nature of the current situation stems from the fact that territorial threats have re-emerged in the form of the war in Ukraine. At the same time, non-conventional threats have also appeared – threats for which decision-makers are still seeking effective responses. It has become clear that traditional, conventional tools used in times of war or peace are insufficient when it comes to the ambiguous grey zone often referred to as hybrid warfare, and more recently, cognitive warfare. In the context of contemporary geopolitical processes, Ukraine's integration into the EU is becoming essential – both for Ukraine itself and for the EU. The consequences of political instability are now also felt in areas that until recently were not considered critical from the perspective of the democratic order. The annulled elections in Romania in December 2024 provide a clear example of how today's reality is increasingly shaped beyond the control of democratic media.

It is also important to note that the conduct of European policy is complicated by the specific structure of the EU and its complex legal and political conditions. Given that the EU is neither a federation nor a confederation, but rather a political entity based on the consent of Member States to act jointly in selected areas, it must be remembered that in some fields – such as defense – Member States retain their own national prerogatives.

When discussing the crisis of democracy, it is impossible not to notice that this phenomenon is particularly evident in the decline of public trust in institutions, the growing sense of powerlessness among citizens, and the belief that political parties no longer represent the interests of specific social groups. This is reflected not only in falling voter turnout and widespread protests, but also in increasing support for radical and anti-EU movements. This trend helps explain the growing popularity of politicians with extreme views, such as Giorgia Meloni in Italy, Marine Le Pen in France, Geert Wilders in the Netherlands, as well as leaders from Central Europe, such as Viktor Orbán in Hungary, Robert Fico in Slovakia, and Jarosław Kaczyński in Poland. These developments are not detached from the broader context of global transformation, including changes in US policy. In the face of mounting geopolitical, economic, and technological tensions, political decision-makers must recognize that the coming years will be marked by strategic rivalry between the United States and China – particularly in the sphere of technology-driven economies.

In a rapidly changing world and amid transformations of political alliances, the very concept of security is evolving. As a result, there is a growing need to redefine what security means today. This stems primarily from a lack of stability and predictability – both in the European and global contexts – where nearly any phenomenon can be perceived as a potential threat. In this context, the Presidency of the Council of the EU may become a platform for building compromise and seeking solutions to challenges such as the so-called cost-of-living crisis in the EU, which affects everyone regardless of social class, influences the labor market, and limits opportunities for development. Does the European community have the potential to pursue proactive policies, or is it condemned to reactive responses? How can the interests of different actors be reconciled – should policy remain national and self-interested, or rather community-oriented and based on solidarity? How can the opinions and positions of individual regions be heard and incorporated into a policy-making process shaped at the center – not only to preserve cohesion, but also to legitimize EU decisions? How can bottom-up voices and top-down mechanisms be balanced so that no side or participant in the European debate feels excluded? What can be done when member states are no longer dependent – on the one hand – on gas imports, but on the other hand experience economic and technological pressure from global actors, while simultaneously striving to pursue their own development ambitions? Moreover, many Europeans believe that the EU does not function efficiently and that regulations sometimes outweigh how citizens actually feel. The question is how not to betray this trust – especially given the EU's multi-layered, long-term, and often arduous decision-making process.

It is also difficult to clearly define the significance of chairing the Council of the EU. On the one hand, since the Treaty of Lisbon, the presidency can be seen as having largely lost its importance, now serving mainly a symbolic role – especially in light of the introduction of the permanent position of President of the European Council, elected for a two-and-a-half-year term with the possibility of renewal. On the other hand, however, the presidency still plays an important role in the day-to-day functioning of the Council of the EU by ensuring the smooth conduct of debates on draft legislative acts. This provides real opportunities to build compromises within the Council and to conduct negotiations with the European Parliament.

The EU Presidency does not operate in a political vacuum – its course largely reflects the current political context and international atmosphere, which further complicate external challenges. Of particular importance for future EU presidencies are transatlantic relations, which may lead to a revision of existing rules governing European policy. A change in the US president, the return of Donald Trump to power, and his stance on NATO and Ukraine raise uncertainty about the durability of American security guarantees. President Trump is demanding that Europe purchase more American military equipment and bear a greater share of the costs of supporting Ukraine.

Another important area of tension concerns trade issues. US announcements of tariff increases could seriously worsen economic relations and trigger a protectionist response from the EU. As a result, the EU may be forced to redefine its trade and geopolitical partnerships. In this context, understanding security in a broad sense becomes particularly important. It turns out that security cannot be separated from economic and technological policy. Investments in advanced technologies strengthen state resilience, support innovation, and create jobs. Security is becoming a multidimensional concept – encompassing access to raw materials, data, digital infrastructure, and media systems.

One of the major challenges the EU has been facing for some time – and will continue to face in the coming years – is migration. On the one hand, uncontrolled migration flows raise security concerns; on the other hand, the Schengen Area, which enables the free movement of people, is one of the EU's most important values. How can these two conflicting demands be reconciled? It is expected that each member state holding the presidency will moderate discussions on this issue and strive for compromises that do not undermine civil liberties while at the same time addressing security needs. This problem becomes particularly acute in light of increasing hostile migratory flows across the border with Belarus – a serious issue for Poland, Lithuania, and Latvia. An important element of EU security policy should also be active cooperation with countries experiencing instability, such as those in the Middle East and Africa. Launching assistance and stabilization programs in these regions could, in the near future, help reduce migratory pressure.

Beyond migration, one of the key challenges for the EU in the coming years remains the issue of the energy raw materials market. Since the outbreak of the war in Ukraine, member states have significantly reduced their dependence on Russian gas, oil, and coal. However, growing energy demand driven by economic development means that full independence from Russian resources will be a long-term and costly process. This entails both the need to import more expensive resources from other sources and investments in renewable energy and nuclear power. Moreover, the EU must confront a growing wave of disinformation accompanying, among other things, the energy transition. After losing the European energy market, Russia has been conducting disinformation campaigns aimed at undermining EU climate policy and the cohesion of member states – both at the political and societal levels.

The international context also had a significant impact on Poland's presidency in 2025. Poland assumed the presidency during the war in Ukraine, amid US–Russian negotiations and considerable uncertainty in relations with the United States. This raised the question of whether Poland would be able to propose a new strategic narrative within the EU. Poland's presidency was an attempt to introduce a culture of strategic thinking into the European debate. "Europe's security" became the main slogan of the presidency – both in military terms and in economic, technological, and energy dimensions. The need to simplify and deregulate administrative regulations that limit the implementation of innovative projects was also emphasized. One of the key initiatives was the proposal of Rearm Europe – strengthening the defense capabilities of EU countries and establishing the SAFE fund (Support for Ammunition and Armaments for Europe). This fund, worth tens of billions of euros, is intended to support joint weapons production and the development of defense infrastructure. Poland's presidency also supported the development of cooperation with external partners – for example, a security agreement was signed with the United Kingdom. In the field of trade policy, a common position toward US actions was achieved, supported by all member states.

It is worth noting that 2025 also marks the 75th anniversary of the Schuman Declaration, which further strengthened the symbolic significance of Poland's presidency. Compared with Poland's first presidency in 2011, a clear shift in emphasis is visible – from a focus on cohesion and solidarity to security as the overriding priority. In 2011, one of the main objectives was to increase development assistance for new member states, whereas in 2025 the focus was on building a common European defense policy.

The outcome of this collective monograph is the demonstration that the contemporary sovereignty of EU Member States is multidimensional and complex, extending far beyond the traditional understanding of the concept. The authors of the individual chapters show how technology intersects with politics, integrating diverse dimensions of Member State security (military, social, cyber, energy, critical infrastructure, and informational). The research conducted clearly indicates that EU security policy today involves development in both the public and digital spheres. The stable development of the

Union requires parallel progress in digitalization and new technologies. Digitalization is becoming one of the main pillars of EU policy, and its significance cannot be overlooked in the context of contemporary political, economic, and social challenges. The EU is actively incorporating initiatives and proposals from Member States in this area into its agenda, which demonstrates a growing awareness of the importance of technology for the security of the community.

Despite differences in levels of integration and the populist rhetoric of some political groupings, cybersecurity remains the common denominator of Member State policies. The search for shared technological solutions unites Member States and constitutes an important element of the Europeanization process – not only within the Union but also in its relations with future candidates such as Turkey and Bosnia and Herzegovina.

In the context of strategic rivalry between the United States and China, which is particularly intense in the technological sphere, the EU, in order to become an independent and significant actor on the international stage, must develop its own solutions to problems related to digital security and the development of new technologies. This applies both to their use in the political and social spheres and in the economic domain. These issues are systematically addressed on the EU agenda, among others, by the Council of the EU, which plays a key role in coordinating the actions of Member States.

The Polish Presidency of the Council of the EU (January–June 2025) made the recognition of digital challenges and the search for solutions at both the EU and regional levels its starting point. It should be noted that the development of the EU in this area is impossible without the active involvement of Member States. It is the Member States that propose concrete solutions that can serve the entire community, contributing their experiences and perspectives to the European debate. The dynamic geopolitical situation in Europe and globally means that traditional notions of sovereignty and digital security are evolving, and their definitions require constant expansion and updating. Dictionary definitions of these terms no longer fully capture the complexity of contemporary phenomena. This monograph contributes to enriching the conceptual framework of political science and security studies by offering an in-depth analysis of these key categories in the European context.

The contribution of this publication to the development of research on security and European policy is multifaceted. First, the monograph integrates the perspectives of scholars from many Member States, making it possible to capture the diversity of experiences and approaches to security and technology issues across different parts of Europe. Second, the publication combines analyses of political, economic, and technological dimensions of security, avoiding the fragmented approach characteristic of many earlier studies. Third, the monograph provides an up-to-date analysis that takes into account recent events, including the Polish Presidency of the Council of the EU in 2025 and the changing transatlantic relations following Donald Trump's election victory. The research presented in this volume shows that the future of the EU largely depends on its ability to build technological autonomy while maintaining the coherence of Member State actions. Digital and technological challenges are no longer a matter of the future but an urgent reality requiring coordinated action at the European level. This monograph constitutes an important voice in the debate on the directions of development of European security policy in the era of digital transformation.

Index

HTTPS://DOI.ORG/10.1163/9789004763623-032

www.ingramcontent.com/pod-product-compliance
Lightning Source LLC
LaVergne TN
LVHW082058151026
850117LV00003B/19